T0199812

Chemical, Biological, and Functional Aspects of Food Lipids

Second Edition

Chemical and Functional Properties of Food Components Series

SERIES EDITOR

Zdzisław E. Sikorski

Chemical, Biological, and Functional Aspects of Food Lipids, Second Edition
Edited by Zdzisław E. Sikorski and Anna Kołakowska

Chemical and Biological Properties of Food Allergens
Edited by Lucjan Jędrychowski and Harry J. Wichers

Food Colorants: Chemical and Functional Properties
Edited by Carmen Socaciu

Mineral Components in Foods
Edited by Piotr Szefer and Jerome O. Nriagu

Chemical and Functional Properties of Food Components, Third Edition
Edited by Zdzisław E. Sikorski

Carcinogenic and Anticarcinogenic Food Components
Edited by Wanda Baer-Dubowska, Agnieszka Bartoszek and Danuta Malejka-Giganti

Methods of Analysis of Food Components and Additives
Edited by Semih Ötleş

Toxins in Food
Edited by Waldemar M. Dąbrowski and Zdzisław E. Sikorski

Chemical and Functional Properties of Food Saccharides
Edited by Piotr Tomasik

Chemical and Functional Properties of Food Proteins
Edited by Zdzisław E. Sikorski

Chemical, Biological, and Functional Aspects of Food Lipids

Second Edition

EDITED BY

Zdzisław E. Sikorski

Gdańsk University of Technology
Gdańsk, Poland

Anna Kołakowska

West Pomeranian University of Technology
Szczecin, Poland

CRC Press
Taylor & Francis Group
Boca Raton London New York

CRC Press is an imprint of the
Taylor & Francis Group, an **informa** business

CRC Press
Taylor & Francis Group
6000 Broken Sound Parkway NW, Suite 300
Boca Raton, FL 33487-2742

First issued in paperback 2019

© 2011 by Taylor and Francis Group, LLC
CRC Press is an imprint of Taylor & Francis Group, an Informa business

No claim to original U.S. Government works

ISBN-13: 978-1-4398-0237-3 (hbk)
ISBN-13: 978-0-367-38344-2 (pbk)

This book contains information obtained from authentic and highly regarded sources. Reasonable efforts have been made to publish reliable data and information, but the author and publisher cannot assume responsibility for the validity of all materials or the consequences of their use. The authors and publishers have attempted to trace the copyright holders of all material reproduced in this publication and apologize to copyright holders if permission to publish in this form has not been obtained. If any copyright material has not been acknowledged please write and let us know so we may rectify in any future reprint.

Except as permitted under U.S. Copyright Law, no part of this book may be reprinted, reproduced, transmitted, or utilized in any form by any electronic, mechanical, or other means, now known or hereafter invented, including photocopying, microfilming, and recording, or in any information storage or retrieval system, without written permission from the publishers.

For permission to photocopy or use material electronically from this work, please access www.copyright.com (http://www.copyright.com/) or contact the Copyright Clearance Center, Inc. (CCC), 222 Rosewood Drive, Danvers, MA 01923, 978-750-8400. CCC is a not-for-profit organization that provides licenses and registration for a variety of users. For organizations that have been granted a photocopy license by the CCC, a separate system of payment has been arranged.

Trademark Notice: Product or corporate names may be trademarks or registered trademarks, and are used only for identification and explanation without intent to infringe.

Visit the Taylor & Francis Web site at
http://www.taylorandfrancis.com

and the CRC Press Web site at
http://www.crcpress.com

Contents

Preface

Lipids are ubiquitous in living organisms and are thus present in most foods, albeit in various concentrations, and differing in composition and properties. They may form large pools almost free of other material, may be intricately bound to other components, or may be dispersed as droplets in the food matrix. Furthermore, their concentration, distribution, and properties may be purposely changed in husbandry and in food-processing operations. Thanks to their diversified chemical composition and corresponding physical and biochemical properties, they perform an array of biological functions in plant and animal organisms, play a crucial role in creating the sensory characteristics of many foods, and contribute to nutritional value and health effects. The interactions of lipids with other food components are the most important for the quality of numerous raw materials and products. Chemical, biochemical, and technological research carried out in universities and in the industry is aimed at learning more about the different roles of lipids in foods.

This book has been devised as an extension of the lipid section of any standard, academic text on food chemistry. It contains a concise, yet well-documented, presentation of the current state of knowledge on the occurrence, chemical and biochemical reactivity, biological role, and functional properties of lipids in food systems. Chapters 1 through 3 provide a general coverage of lipid chemistry, and may serve as an introduction to the more specialized topics dealt with in the book. Chapter 1 familiarizes the reader with the standard chemical nomenclature of a large variety of lipid species, Chapter 2 presents the most important properties of lipids, while Chapter 3 describes the progress in the methods of physical, chemical, and biochemical analyses. Chapter 4 provides a broad outline of the effect of fats on food quality, while Chapter 5 looks into the role of lipids in various food structures. Chapters 6 through 8 provide updated information on phospholipids, sterols, and fat-soluble vitamins. Lipid oxidation and antioxidants, of crucial significance for the sensory and nutritive aspects of food quality, have been dealt with expertly in Chapters 9 and 10 by a biochemist and a food scientist. Chapters 11 through 13 present various aspects of the biological role of dietary lipids in humans as seen by scientists as well as practicing medical specialists. These chapters add a new dimension to the book and justify the modified title of the second edition. The contents of lipids in plants, fish, milk, meat, and eggs, with their role in different foods are described in Chapters 14 through 18. Modified lipids and fat mimetics, as well as those of special biological and physicochemical activity have been treated in Chapters 19 and 20. Frying fats, lipid–proteins and lipid–saccharides interactions, and lipid contaminants, all of high importance for food quality, are covered in Chapters 21 through 23.

The chapters have the character of monographs written by authorities from universities and other research institutions. They are based on the research/teaching experience of the contributors, as well as on a critical evaluation of the current literature in the field. All chapters in the book are new, prepared intentionally for this edition. The book is addressed to food scientists in industry and academia, to graduate students of food science, to nutritionists, and to all those who are interested in the role and attributes of lipids as food components.

We have had the chance as editors to cooperate with renowned scientists from Australia, the Czech Republic, Germany, Greece, Poland, Slovakia, Turkey, and the United States. The contributors accepted our conception of the book and the editorial suggestions—this is highly appreciated. Special thanks are due to several authors who prepared their chapters well ahead of the deadline.

We dedicate this volume to the researchers whose investigations have contributed to disclosing the effects of lipids on food quality and on human well-being.

Zdzisław E. Sikorski
Anna Kołakowska

Editors

Zdzisław E. Sikorski received his BS, MS, PhD, and DSc in food technology/ food chemistry from the Gdańsk University of Technology (GUT), Poland, and Dr *honoris causa* from the University of Agriculture in Szczecin, Poland. He gained practical experience in breweries, food-freezing plants, and fish canneries in Poland and Germany, and on a Polish deep sea fishing trawler. He was organizer, professor, and head of the Department of Food Chemistry and Technology at GUT; served two terms as dean of the Faculty of Chemistry at GUT; was chairman of the Committee of Food Technology and Chemistry of the Polish Academy of Sciences, Warsaw; and chaired the Scientific Council of the Sea Fisheries Institute in Gdynia. He also worked several years as a postdoctoral researcher and professor at Ohio State University, Columbus; Commonwealth Scientific and Industrial Research Organization, Hobart, Australia; Department of Scientific and Industrial Research, Auckland, New Zealand; and National Taiwan Ocean University, Keelung, Taiwan. He has published nearly 200 journal papers, 17 books, and 12 chapters on marine food technology and food chemistry, and holds numerous patents regarding fish and krill processing. Several of his books have also appeared in multiple editions. He is a member of the editorial boards of the *Journal of Food Biochemistry, Polish Journal of Food and Nutrition Sciences*, and of *Food Science Technology Quality*. His research deals mainly with food preservation, functional properties of food proteins, and interactions of food components. In 2003, he was elected a fellow of the International Academy of Food Science and Technology.

Anna Kołakowska received her BS and MS from the Faculty of Food Technology, University of Agriculture in Olsztyn, Poland, and her PhD and DSc from the University of Agriculture in Szczecin, Poland. She is the first woman to have worked two winters in a research station in Antarctica. Her investigations in Antarctica were concerned with the Antarctic krill. She is a professor at the Faculty of Food Science and Fisheries, Department of Food Quality at the West Pomeranian University of Technology, Szczecin, Poland. She has been involved in several large research projects realized in cooperation with various universities and institutions. Her teaching focuses mainly on food quality, while her research is concerned primarily with marine lipids and their changes due to biological factors, storage, and processing. She has published nearly 150 journal papers and 13 chapters in 7 books, dealing mainly with marine food science, polyenoic fatty acids, and food quality. She also holds five patents.

Contributors

Marek Adamczak
Department of Food Biotechnology
University of Warmia and Mazury in
  Olsztyn
Olsztyn, Poland

Grzegorz Bartosz
Department of Molecular Biophysics
University of Łódź
Łódź, Poland

and

Department of Biochemistry and Cell
  Biology
University of Rzeszów
Rzeszów, Poland

Włodzimierz Bednarski
Department of Food Biotechnology
University of Warmia and Mazury in
  Olsztyn
Olsztyn, Poland

Grzegorz Bienkiewicz
Department of Food Quality
West Pomeranian University of
  Technology
Szczecin, Poland

Wioletta Błaszczak
Institute of Animal Reproduction and
  Food Research
Polish Academy of Sciences
Olsztyn, Poland

Dimitrios Boskou
Laboratory of Food Chemistry and
  Technology
Department of Chemistry
Aristotle University of Thessaloniki
Thessaloniki, Greece

Astrid M. Drotleff
Institute of Food Toxicology and
  Analytical Chemistry
Center for Food Science
University of Veterinary Medicine
  Hannover
Hannover, Germany

Józef Fornal
Institute of Animal Reproduction and
  Food Research
Polish Academy of Sciences
Olsztyn, Poland

Zdzisław Florian Forycki
Cardiology Department
Neukölln Krankenhaus
Vivantes Network for Health
Berlin, Germany

Aytaç Saygin Gumuskesen
Food Engineering Department
Ege University
Izmir, Turkey

Timothy B. Jordan
Department of Environment, Parks,
  Heritage and the Arts
Analytical Services Tasmania
Hobart, Tasmania, Australia

Yildiz Karaibrahimoglu
Consultant, Food Scientist
Oakland, California

Neil Kerr
Department of Environment, Parks,
  Heritage and the Arts
Analytical Services Tasmania
Hobart, Tasmania, Australia

Anna Kołakowska
Department of Food Quality
Department of Food Science
West Pomeranian University of
    Technology
Szczecin, Poland

Paweł Lipowski
Department of Opthalmology
Medical University of Gdańsk
Gdańsk, Poland

David S. Nichols
Department of Environment, Parks,
    Heritage and the Arts
Analytical Services Tasmania
Hobart, Tasmania, Australia

Małgorzata Nogala-Kałucka
The Department of Biochemistry and
    Food Analysis
Poznań University of Life Sciences
Poznań, Poland

Semih Otles
Food Engineering Department
Ege University
Izmir, Turkey

Jan Pokorný
Department of Food Chemistry and
    Analysis
Prague Institute of Chemical
    Technology
Prague, Czech Republic

Magdalena Rudzińska
Institute of Food Technology of Plant
    Origin
University of Life Sciences
Poznań, Poland

Štefan Schmidt
Department of Food Science and
    Technology
Slovak Technical University
Bratislava, Slovakia

Grażyna Sikorska-Wiśniewska
Department of Pediatrics, Pediatric
    Gastroenterology, Hepatology and
    Nutrition
Medical University of Gdańsk
Gdańsk, Poland

Zdzisław E. Sikorski
Department of Food Chemistry,
    Technology, and Biotechnology
Gdańsk University of Technology
Gdańsk, Poland

Izabela Sinkiewicz
Department of Catering Technology
    and Food Hygiene
Warsaw University of Life Sciences
Warsaw, Poland

Waldemar Ternes
Department of Analytical Chemistry
Institute of Food Toxicology and
    Analytical Chemistry
Center for Food Science
University of Veterinary Medicine
    Hannover
Hannover, Germany

Michael H. Tunick
Dairy Processing & Products Research
    Unit
Eastern Regional Research Center
Agricultural Research Service
U.S. Department of Agriculture
Wyndmoor, Pennsylvania

Erwin Wąsowicz
Institute of Food Technology of Plant
    Origin
University of Life Sciences
Poznań, Poland

Fahri Yemiscioglu
Food Engineering Department
Ege University
Izmir, Turkey

1 The Nomenclature and Structure of Lipids

David S. Nichols, Timothy B. Jordan, and Neil Kerr

CONTENTS

1.1 INTRODUCTION

Lipids, or those classes of organic compounds that may be, have been, or are referred to as fats, are an essential component of many foods. Their use as ingredients in foods, cosmetics, and medicines dates to prehistoric times. It is perhaps not surprising, therefore, that the knowledge of the chemical nature of some lipids predates that of other food components such as saccharides or proteins.

The term "lipids" is defined as those organic compounds that are insoluble in water, soluble in organic solvents (e.g., chloroform, ether), contain hydrocarbon groups as primary parts of the molecule, and are present in or derived from living organisms. Compound classes covered in this definition include fatty acids (FA), acylglycerols, FA esters (e.g., waxes), and isoprenoid hydrocarbons. Other compounds also included are often considered as belonging to different classes such as carotenoids, sterols, and the vitamins A, D, E, and K.

Lipids tend to be classified as "simple" or "complex," referring to the size or structural detail of the molecule. Simple lipids include FA, hydrocarbons, and alcohols are relatively "neutral" in terms of charge. Complex lipids, such as phospholipids and glycolipids, are relatively more charged and are also referred to as "polar."

1.2   FATTY ACIDS

1.2.1   GENERAL FEATURES

There are many features of FA that can be described in general terms. First, FA are characterized by a long chain of carbon atoms (generally 12–22) ending in a carboxyl functionality (Figure 1.1). Substitutions or variations in chain structure may occur, yielding a diversity of FA types (Figure 1.1 and Table 1.1). These include branched-chain FA, unsaturated FA, and those possessing oxygenated groups. While over 500 individual FA have been described from plant and microbial sources, only a few are significant quantitatively. For example, approximately 95% of the FA from plant leaves or oils consist of only seven components: lauric, myristic, palmitic, stearic, oleic, linoleic, and α-linolenic acids (Tables 1.1 and 1.3).

The nomenclature of FA suffers from the long history of their study and description. Most FA were originally described under "trivial" names prior to the adoption of the international molecular nomenclature rules in 1892 (Table 1.1). Even after the adoption of the current International Union of Pure and Applied Chemistry (IUPAC) system for nomenclature (IUPAC-IUB 1977), the habit of assigning trivial names to FA continues. The basis of the systematic nomenclature system is an extension of that accepted for hydrocarbon (alkane/alkene) naming. Hence, the descriptive name is based on the number of carbon atoms contained in the molecule, with the suffix "-e" replaced with "-oic acid." However, in most instances FA are referred to by their formula notation. As is the case with systematic and trivial names, there exists an IUPAC accepted formula notation nomenclature and several earlier versions (Table 1.2). Each system is interchangeable and it is important to recognize each as earlier literature may have used either system.

The three nomenclature notation systems commonly used for FA are described in Figure 1.2 and Table 1.2. Each has certain advantages in the description of structural features. However, the current IUPAC system of (n-) nomenclature for unsaturated FA should be used in preference to the ω or Δ notations. The systems of formula notation are based on a common core. Each firstly denotes the number of carbon atoms and the number of double bonds in the molecule, with the numerics separated by a colon. A term then follows that describes the location of any

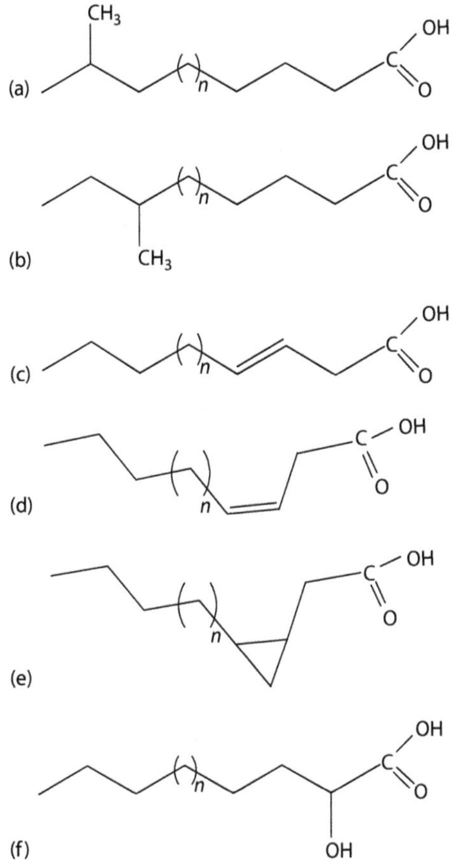

FIGURE 1.1 Generic structures of various FA types: (a) *iso*-branched, (b) *anteiso*-branched, (c) *trans*-monounsaturated, (d) *cis*-monounsaturated, (e) cyclopropane, and (f) hydroxylated.

double bonds in the carbon chain. It is the method of this description that differs between the notation systems. The first describes double bond location from the methyl (ω or n) end of the carbon chain. Where more than one double bond is present, only the first from the (n) position is noted, as subsequent double bonds are assumed to be present in a methylene interrupted series (see Section 1.2.4). The second notation system describes double bond location from the carboxyl (Δ) carbon. Here, the location of each double bond is listed in order. Each system is the same in using a series of general prefix and suffix terms to describe additional structural details. The prefixes "i" and "a" refer to *iso-* and *anteiso-*branching, respectively (see Section 1.2.5). The suffixes "c" and "t" denote *cis* and *trans* double bond geometry, respectively (see Figure 1.1). Specialized terms specific for unusual FA structures or functionalities will be discussed under the relevant sections below.

TABLE 1.1
The Formula, Nomenclature, and Names of Representative FA That May Be Present in Foods

Formula	Notation[a]		Acid Name		Discovered
	Omega	Delta	Systematic	Trivial	
Saturated					
$C_{12}H_{24}O_2$	12:0	12:0	Dodecanoic	Laurie	1842
$C_{14}H_{28}O_2$	14:0	14:0	Tetradecanoic	Myristic	1841
$C_{15}H_{30}O_2$	15:0	15:0	Pentadecanoic	—	1926
$C_{16}H_{32}O_2$	16:0	16:0	Hexadecanoic	Palmitic	1816
$C_{18}H_{34}O_2$	18:0	18:0	Octadecanoic	Stearic	1816
Branched-chain					
$C_{15}H_{29}O_2$	i15:0	i15:0	13-Methyltetradecanoic	—	1960
$C_{15}H_{29}O_2$	a15:0	a15:0	12-Methyltetradecanoic	Sarcinic	1960
$C_{19}H_{38}O_2$	10-Mel 8:0	10-Mel 8:0	10-Methyloctadecanoic	Tuberculostearic	1929
Monounsaturated					
$C_{16}H_{30}O_2$	16:1 ω7c	16:1 Δ9c	*cis*-9-Hexadecenoic	Palmitoleic	1854
$C_{18}H_{32}O_2$	18:1 ω7c	18:1Δ9c	*cis*-9-Octadecenoic	Oleic	1815
$C_{18}H_{32}O_2$	18:1 ω9c	18:1 Δ11c	*cis*-11-Octadecenoic	Vaccinic	1844
Polyunsaturated					
$C_{18}H_{30}O_2$	18:2 ω6	18:2 Δ9,12c	*cis*-9,12-Octadecadienoic	α-Linoleic	1844
$C_{18}H_{30}O_2$	18:2 ω3	18:2 Δ12,15c	*cis*-2,15-Octadecadienoic	γ-Linoleic	1844
$C_{18}H_{28}O_2$	18:3 ω3	18:3 Δ9,12,15c	*cis*-9,12,15-Octadecatrienoic	α-Linolenic	1887
$C_{18}H_{28}O_2$	18:3 ω5	18:3 Δ6,9,12c	*cis*-6,9,12-Octadecatrienoic	γ-Linolenic	1887
$C_{20}H_{32}O_2$	20:4 ω6	20:4 Δ5,8,11,14c	*cis*-5,8,11,14-Eicosatetraenoic	Arachidonic	1909
$C_{20}H_{30}O_2$	20:5 ω3	20:5 Δ5,8,11,14,17c	*cis*-5,8,11,14,17-Eicosapentaenoic	Eicosapentaenoic	1948
Oxygenated					
$C_{16}H_{32}O_3$	16-OH16:0	16-OH16:0	16-Hydroxyhexadecanoic	Juniperic	1909
$C_{18}H_{36}O_6$	9,10-OH18:0	9,10-OH18:0	9,10-Hydroxyoctadeeanoic	Dihydroxystearic	1925
$C_{18}H_{28}O_3$	4-O-18:3 ω5	4-O-18:3 Δ18,3c	4-keto-9,11,13-Octadecatrienoic	Licanic	1931

Source: Data compiled from Deuel, H.J. Jr., *The Lipids, Their Chemistry and Biochemistry, Volume I: Chemistry,* Interscience Publishers Inc., New York, 1951.

[a] The two notation forms displayed are no longer recommended for usage. They are used here to highlight earlier descriptions of FA that may be encountered elsewhere. The currently recommended IUPAC notation system for FA (the (n-) notation) is described in Table 1.2.

TABLE 1.2

Schematic Formulas for the Two Main Types of FA Systematic (n) and Nonsystematic (ω) Structural Notations (Refer to Figure 1.2 as an Example)

y X D–G:H ω Z b

y X D–G:H (n–) Z b

where

y = i (*iso-*) or a (*anteiso-*) branching (if present)

X = carbon number of additional functional group (if present) from the carboxyl carbon

D = abbreviation of additional functional group (if present)

G = total number of carbon atoms

H = total number of double bonds

Z = carbon number of first double bond from the methyl position (if present)

b = c (*cis-*) or t (*trans-*) double bond geometry (if present)

1.2.2 SATURATED FATTY ACIDS

Saturated FA are those containing only single bonds between carbon atoms and, hence, the molecule is "saturated" or contains the maximum possible number of hydrogen atoms per carbon. In practice, the term also tends to be limited to those FA that do not possess any other structural feature or functionality. The formula for such molecules therefore follows the series $C_nH_{2n}O_2$ (Table 1.1) and consists of a linear chain of carbon atoms (Figure 1.1). In the case of most food lipids, saturated FA possess an even carbon number. Odd-numbered FA are produced almost exclusively by certain bacteria, and can be considered as markers of bacterial growth.

1.2.3 MONOENE FATTY ACIDS

Monoene or monounsaturated FA are defined as components containing one double bond between adjacent carbon atoms. They therefore follow the generalized formula of $C_nH_{(2n-2)}O_2$. The presence of even this simple functionality introduces two important variables into monoene structure and nomenclature. First, the double bond may be present in a number of possible locations along the carbon chain, giving rise to different monoene isomers (e.g., 16:1(n-9) and 16:1(n-11); Table 1.1). In practice, Δ9 and Δ11 isomers are the most common, although other variations do occur. Second, the double bond may be in either the *cis* or the *trans* geometry (Figure 1.1). Hence, the addition of the formula notation suffix "c" or "t" is required to define this important difference (Table 1.2). In nature, the vast majority of all double bonds in FA are of the *cis* geometry. However, *trans* components do occur, particularly where food lipids have been exposed to heating (e.g., vegetable oils used for cooking).

1.2.4 POLYENE FATTY ACIDS

Polyene FA (PEFA) or polyunsaturated FA (PUFA) are classed as those containing more than one double bond in the carbon chain. As such, the class does not possess

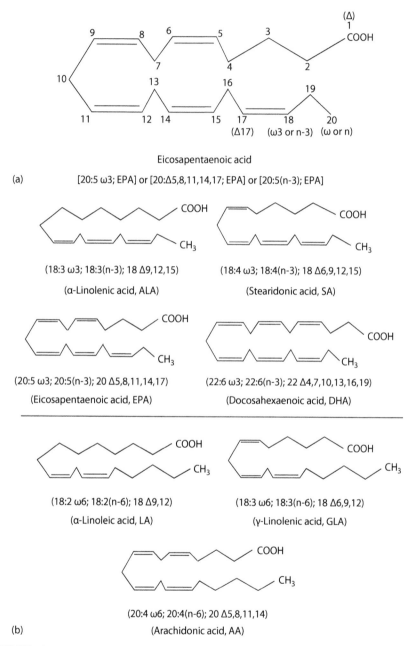

FIGURE 1.2 (a) The relationship between the three nomenclature notations for FA. Eicosapentaenoic acid is shown as an example. (b) Representative structures of the common PEFA of the n-3 and n-6 families.

a generic formula. While *trans* bonds are possible, PUFA from all natural sources contain *cis* double bonds. In addition, the majority of natural PUFAs contain double bonds in a methylene interrupted pattern (Figure 1.2), yielding a generic isomeric distribution of $\Delta x, (x+3), (x+6), (x+9), (x+12)$. Hence, the maximum number of double bonds possible in a PUFA is theoretically limited by the chain length and the position of the first double bond from the carboxyl carbon. In practice, natural PUFA are limited to the last double bond involving the (n-3) position and the first involving either the $\Delta 4$, $\Delta 5$, or $\Delta 6$ position for C_{18}, C_{20}, and C_{22}, respectively (Figure 1.2). Hence, the maximum number of double bonds in naturally occurring PUFAs is four for C_{18}, five for C_{20}, and six for C_{22}.

1.2.5 BRANCHED-CHAIN ACIDS

Branched-chain FA do not commonly occur in food products, but may be widely distributed in the environment and are therefore of relevance. Branched-chain FA can be considered to fall within two categories. The first category occurs where a methyl group replaces one or more of the hydrogen atoms of the carbon chain. The majority are produced by microorganisms and contain a single methyl branch on the second-last (*iso*-branched) or third-last (*anteiso*-branched) carbon of the chain, which is usually odd numbered, i.e., C_{13}, C_{15}, or C_{17} (Figure 1.1). Microorganisms of the genus *Mycobacterium* also produce a range of mid-chain methyl branched FA, such as 10-Me18:0 (tuberculostearic acid, Table 1.1) and a range of polymethyl-branched FA (e.g., 3,13,19-trimethyltricosanoic acid) collectively known as mycolic acids (Harwood and Russell 1984).

The second category of branched-chain FA are the isoprenoid acids, derived from isoprene units (see Section 1.5). They are widely distributed in both terrestrial and marine organisms. Examples include 2,6,10,14-tetramethylpentadecanoic acid (pristanic acid) and 3,7,11,15-tetramethylhexadecanoic acid (phytanic acid).

1.2.6 OXYGENATED FATTY ACIDS

FA containing oxygen functionality exist in a number of forms, most commonly as a hydroxyl or keto group. Both types mirror the general structural theme of branched-chain FA, where an oxygenated functionality replaces one or more hydrogen atoms on one carbon atom of the chain. They occur as natural products from many organisms but may also be produced from the partial oxidation of unsaturated FA during cooking or heating.

Hydroxy FA occur in both saturated and unsaturated forms, with various positional isomers. However, there are three major types. First, the 2-hydroxy (2-OH) or α-hydroxy FA (Figure 1.1). These saturated components occur as a series of even-numbered acids from C_{10} to C_{26} and are components of animal tissues (within cerebrosides) and certain plants. Second, 3-hydroxy (3-OH) or β-hydroxy FA exist as a series of even-numbered acids from C_{10} to C_{18}. They are ubiquitous components of many bacteria and yeasts where they exist as ester-linked residues of extracellular lipids. Third, the hydroxyl group may be present at the penultimate carbon from the carboxyl group (i.e., ω2 or referred to as ω-1) of the chain series C_{11} to C_{19}. In most instances, these compounds represent intermediates in the ω-oxidation of FA by microorganisms.

Analogous to the isomer distribution of hydroxy FA, the major types containing a ketone functional group occur as 2-keto (α-keto), 3-keto (β-keto), and ω-1 keto acids. The chain length distribution for each type is C_6 to C_{20} or greater for 3-keto acids, and C_7 to C_{20} or greater for both 2-keto and ω-1 keto acids. In addition, 4-keto, 5-keto, and 6-keto acids are found in heated milk and pork (Deuel 1951).

1.3   ACYLGLYCEROLS

Acylglycerols consist of one, two, or three FA (acyl) residues esterified to the hydroxy residue(s) of a glycerol molecule (Figure 1.3). They are commonly referred to as complex lipids due to the presence of one or more individual acyl residues in the molecule that can be removed by chemical reaction. Acylglycerols represent the most common type of complex lipid in which FA are present. In the past, they have been referred to as neutral lipids, fats, or glycerides (mono-, di-, and triglycerides). They represent the most common lipid class present in foods. From natural sources, mono- and diacylglycerols are usually present as precursors to the formation of triacylglycerols (TAG) or phospholipids (PL). In foods, their presence may also indicate the degradation of TAG components by chemical or enzymatic deacylation.

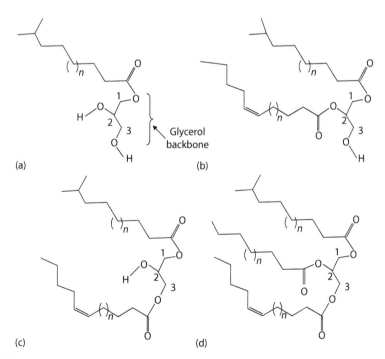

FIGURE 1.3   Schematic structures of the major acylglycerol classes: (a) 1-monoacyl-*sn*-glycerol, (b) 1,2-diacyl-*sn*-glycerol, (c) 1,3-diacyl-*sn*-glycerol, and (d) 1,2,3-triacyl-*sn*-glycerol.

The nomenclature of acylglycerols is based on the concept of substitution of the glycerol "backbone" by acyl residues in a similar manner to which substitutions along an FA carbon chain would be described. Acylglycerols can introduce the further complication of asymmetric substitution forming a chiral center at the central carbon of the glycerol backbone when one of the hydroxyl groups is substituted (monoacylglycerols) or when two or all three hydroxyl groups are substituted by different acyl residues [diacyl-, TAG, and PL]. To address this issue, the stereospecific numbering (*sn*) system is used to differentiate the carbinol groups. This defines a fixed numbering system for the glycerol backbone independent of the substituents. An example is given in Figure 1.4. Using a derived Fischer projection, the orientation of the secondary hydroxyl group to the left of the second carbon defines the upper carbon atom of the glycerol backbone as *sn*-1, and the lower carbon as *sn*-3. The reversal of the hydroxyl position at the second carbon reverses the numbering sequence. The prefix "*sn*-" denotes the use of stereospecific numbering. In previous literature, the carbon positions have been referred to as α, β, and α′ (Figure 1.4).

1.3.1 MONOACYLGLYCEROLS

Monoacylglycerols consist of a single substitution of one hydroxyl group of the glycerol molecule with an acyl residue via an ester linkage (Figure 1.3). Three positional isomers are therefore possible for the substitution: *sn*-1, the central carbon position *sn*-2, or the terminal carbon position *sn*-3. Unlike the *sn*-1 and *sn*-3 isomers, the *sn*-2 isomer retains its molecular symmetry and is therefore non-chiral.

1.3.2 DIACYLGLYCEROLS

Diacylglycerols occur where there is a substitution of two hydroxyl groups of the glycerol molecule with an acyl residue via an ester linkage (Figure 1.3). For diacylglycerols

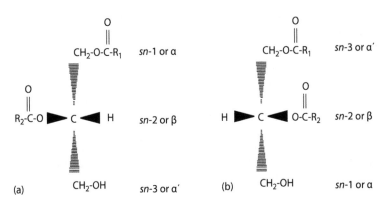

FIGURE 1.4 Stereochemical projection of a generic diacylglycerol illustrating the stereochemical numbering (*sn*) system and the chiral center at the *sn*-2 carbon. Each molecule, although identical in formula, is a stereoisomer (specifically an enantiomer, a nonsuperimposable mirror image). R_1 and R_2 represent substituent alkyl chains.

containing a single FA, three isomers are possible (*sn-1*, *sn-2*; *sn-1*, *sn-3*; and *sn-2*, *sn-3*). Where two different FA are involved, six positional isomers are possible.

1.3.3 TRIACYLGLYCEROLS

TAG contain full substitution of all three hydroxyl groups of the glycerol molecule with ester linked acyl residues (Figure 1.3). They have commonly been referred to as "oils" or "fats," depending on their melting point, and represent the depot lipids of both plants (in seeds) and animals (in adipose tissue). As such, they are common components of food systems. A large variety of positional isomers are possible, which depend on the diversity of component FA and their positional distribution within the molecule.

TAG derived from the seeds of tropical trees are unusual in possessing a large ratio of saturated to unsaturated FA. Hence, they form a solid at normal temperatures but have a very narrow melting range due to the limited positional isomerism exhibited. A good example of this is coca butter (Table 1.3). In general, oils derived from vegetables contain high proportions of unsaturated FA. The more unsaturated components are concentrated in the *sn-2* position with saturated FA more common at the *sn-1* and *sn-3* locations. However, TAG of plant origin may also be highly saturated, as demonstrated from coconut, which is dominated by short-chain saturated FA with specific isomeric distributions (Table 1.3).

TAG from animals are more dominated by C_{16} and C_{18} saturated FA, including significant proportions of fully saturated TAG giving rise to solid substances at room temperature (e.g., tallow, lard). In general, the saturated content of the *sn-2* position is greater than that from plants and 16:0 is concentrated at *sn-1*. Pork fat exists as a special case where 16:0 is concentrated at *sn-2*, 18:0 at *sn-1*, and a high proportion of monounsaturated FA at *sn-1* and *sn-3* (Table 1.3).

1.3.4 PHOSPHOLIPIDS

As is the case for TAG, PL contain full substitution of all three hydroxyl groups of the glycerol molecule. However, in PL, only two of these substitutions involve ester linked acyl residues. The third one is undertaken by a phosphate group together with variable alcohol functionality, referred to as the "head group." Biosynthetically, PL are derived from glycerol-3-phosphate rather than glycerol and therefore have the *sn-3* position occupied by the phosphate substituent and covalently bonded head group assembly. Positions *sn-1* and *sn-2* are therefore available for substitution by the wide diversity of FA, as is the case for other acylglycerols. However, due to their physiological role in animals and plants (see below), the FA composition of PL is tightly regulated. In general, the *sn-1* position is occupied by a more saturated acyl residue while the *sn-2* position may hold a more unsaturated FA.

The composition of the head group moiety increases the structural diversity of PL, which are grouped into a number of classes. The addition of serine as the head group defines the class of PL known as phosphatidylserines (PS) and, likewise, the

TABLE 1.3

FA Positional Distributions in TAG from Various Sources

Lipid Source	sn-Position	Fatty Acid (%)								
		8:0	10:0	12:0	14:0	16:0	18:0	18:1	18:2	18:3
Coconut	1	4	4	39	29	16	3	4	—	—
	2	2	5	78	8	1	1	3	2	—
	3	32	13	38	8	1	1	3	2	—
Coca butter	1	—	—	—	—	34	50	12	1	—
	2	—	—	—	—	2	2	87	9	—
	3	—	—	—	—	37	53	9		—
Corn	1	—	—	—	—	18	3	28	50	—
	2	—	—	—	—	2	—	27	70	—
	3	—	—	—	—	14	31	52	1	—
Soybean	1	—	—	—	—	14	6	23	48	9
	2	—	—	—	—	1	—	22	70	7
	3	—	—	—	—	13	6	28	45	8
Olive	1	—	—	—	—	13	3	72	10	1
	2	—	—	—	—	1	—	83	14	1
	3	—	—	—	—	17	4	74	5	1
Peanut	1	—	—	—	—	14	5	59	19	—
	2	—	—	—	—	2	—	59	39	—
	3	—	—	—	—	11	5	57	10	—
Beef fat	1	—	—	—	4	41	17	20	4	1
	2	—	—	—	9	17	9	41	5	1
	3	—	—	—	1	22	24	37	5	1
Pig (outer back)	1	—	—	—	1	10	30	51	6	—
	2	—	—	—	4	72	2	13	3	—
	3	—	—	—	—	—	7	73	18	—

Source: Data compiled from Deuel, H.J. Jr., *The Lipids, Their Chemistry and Biochemistry, Volume I: Chemistry*, Interscience Publishers Inc., New York, 1951.

addition of ethanolamine yields the phosphatidylethanolamines (PE) (Figure 1.5) and choline yields the phosphatidylcholines (PC). In addition, glycerol itself may also act as a head group, yielding the phosphatidylglycerol (PG) PL class. Finally, two PL molecules may be joined together by a glycerol head group to form the diphosphatidylglycerols (or cardiolipins).

PL molecules are unique amongst most lipids in being amphipathic structures. That is, they contain both a hydrophilic or polar region (the phosphate and head group) together with a hydrophobic or nonpolar region (the acyl residues and glycerol backbone). Their amphipathic nature, together with the resultant charge and steric qualities of the head group allow these molecules to self-assemble into various membrane structures. It is this role that they fulfill within plant and animal cells. Consequently, the PL content of most plant or animal sources is relatively low.

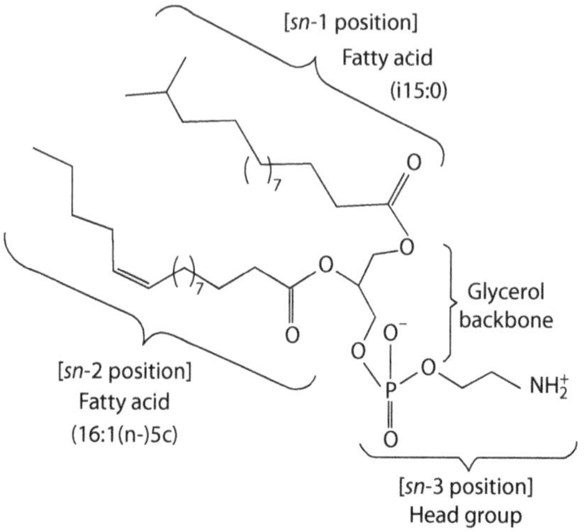

FIGURE 1.5 Structural representation of a PL molecule; in this case 1-14-methyltetradecanoyl-2-*cis*-hexadec-11-enoyl-*sn*-glycero-3-phosphoethanolamine.

1.4 WAXES

Waxes may be defined as FA esters of alcohols possessing a higher molecular weight. Waxes are chemically stable and insoluble in water and many organic solvents. Due to these properties, they are widely distributed in both plants and animals as protective coverings for tissues. Simple waxes are classed as monoesters of normal FA and normal long-chain alcohols. Complex waxes also exist where either the FA or alcohol components have complex structures in their own rights (e.g., vitamin esters or sterol esters, see Section 1.5.1).

The nomenclature of simple waxes is based on the stem name of the FA. The FA suffix "-ic acid" is replaced with "-ate." This stem term is then preceded by the name of the alcohol component with the suffix modified from the "-ol" of the free alcohol to "-yl." Many examples exist in bacteria and marine organisms where waxes serve as short-term storage lipids or as aids in buoyancy. Compounds with C_{21} to C_{44} have been reported from marine organisms, however the usual range from copepods or fish is C_{30} to C_{42}. Component alcohols are usually saturated or monounsaturated with 16:0 and 18:1 predominating. FA involved are more diverse and may include PUFA, although 16:1 and 20:1 are frequently major components.

1.5 ISOPRENOID LIPIDS

1.5.1 INTRODUCTION

Isoprenoid lipids represent a diverse collection of compound types, some of which are present in nearly every organism. The group includes the major classes of compounds such as terpenoids, steroids, and carotenoids. The latter two of these classes will be dealt with in more detail below. The great diversity of isoprenoid lipids shares a common structural origin. The basis of this is biosynthesis of the "backbone" structure from multiple, repeating, branched-chain unsaturated C_5 units called isoprene or isopentenyl units (Figure 1.6). Differing numbers of isoprene units may combine in a variety of ways, yielding a diversity of chain structures. Hence, isoprenoid compounds usually contain a carbon number divisible by five and are broadly classified in a terpenoid nomenclature based on this fact (Table 1.4). Indeed many isoprenoid

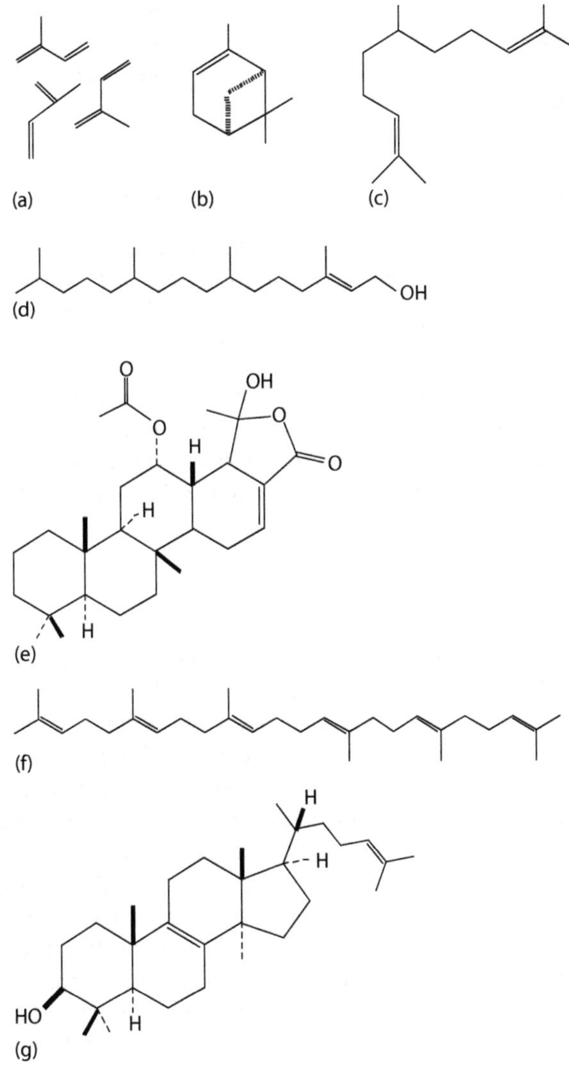

FIGURE 1.6 Examples of isoprenoid lipids illustrating the structural diversity of the group: (a) three isoprene units (2-methyl-1,3-butadiene) in various conformations, (b) α-pinene (2,6,6-trimethylbicyclo[3.1.1]hept-2-ene), (c) β-farnescene (3,7,11-trimethyl-1,3,10-dodecatriene), (d) phytol (3,7,11,15-tetramethyl-2-hexadecenol), (e) scalarin (12α-acetoxy-25α-hydroxyscalar-16-en-25,24-olide), (f) squalene (2,6,10,15,19,23-hexamethyl-2,6,10,14,18,22-tetracosahexaene), and (g) lanosterol (4,4,14α-trimethyl-5α-cholest-8,24-diene-3β-ol).

TABLE 1.4
Isoprenoid Lipids Classified in Terms of Terpenoid Class, with Common Examples (Refer to Figure 1.6 for Structures)

Terpenoid Class	Carbon Number	Examples (Common Name)
Hemiterpenes	5	Isoprene
Monoterpenes	10	α-Pinene
Sesquiterpenes	15	β-Farnesene
Diterpenes	20	Phytol
Sesterterpenes	25	Scalarin
Triterpenes	30	Squalene, Lanosterol
Polyterpenes	>30	Carotenes

Source: Data compiled from Kirk, D.N. and Marples, B.A., The structure and nomenclature of steroids, in *Steroid Analysis*, Makin, H.L.J. et al. (eds), Blackie Academic and Professional, London, U.K., 1995, pp. 1–24.

lipids can be easily identified by the ability to divide the compound structure neatly into its original isoprene synthetic units (the so-called isoprene rule). However, the subsequent cyclization and/or addition of further functional groups at sites of unsaturation result in the large diversity and complex nomenclature of isoprenoid lipids.

1.5.2 STEROIDS

1.5.2.1 Structural Nomenclature of Steroids

The steroids are a large group of compounds sharing a particular structural motif of C_{30} (triterpenoid) derived isoprenoid lipids. They include a number of ubiquitous compound classes such as sterols, adrenal steroids, sex hormones, bile acids, saprogenins, and others. As a group, they share the common feature of a tetracyclic ring system (three six membered and one five membered) derived from lanosterol, itself a cyclization product of the triterpenoid squalene (Figure 1.6). However, significant modification of the ring systems gives rise to the diversity of steroid classes. Due to the high degree of modification of the original ring systems, the majority of steroids do not conform to the isoprene rule. The use of common or trivial names for steroids is widespread and beyond the scope of this chapter (refer to Briggs and Brotherton 1970). The following sections will concentrate solely on an explanation of the IUPAC systematic nomenclature, which the reader will be able to relate to other descriptions.

The systematic nomenclature of steroids is complex due to the diversity of substituents and the stereochemistry of the ring system. The carbon numbering of the generic steroid ring system is shown in Figure 1.7. The carbon atoms are numbered in a cyclic fashion beginning with the combined rings *A* and *B* anticlockwise, ring *C* clockwise, and then ring *D* anticlockwise. Carbon substituents of ring junctions are then numbered, working backward from ring *D* to *A*. The isoprenoid side-chain at C-17 is numbered lastly in a specific fashion to accommodate the presence or absence of common

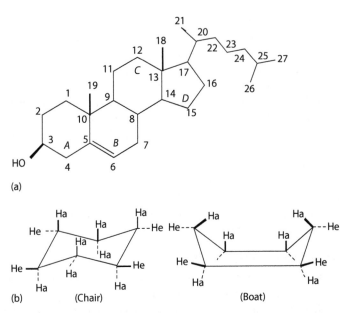

FIGURE 1.7 Steroids. (a) The carbon numbering system for steroids using cholesterol (cholest-5-en-3β-ol) as an example. The individual rings of the carbon skeleton are labeled *A* to *D*. (b) Representations of the "chair" and "boat" conformations for a six-membered carbon ring illustrating equatorial (He) and axial (Ha) hydrogen atoms.

substituents. Further carbon substitutions may also occur in some cases at C-24 (then numbered C-28, C-29) and C-4 (then numbered C-30, C-31) (e.g., scalarin, Figure 1.6).

A description of the molecular stereochemistry is also included in both the structural formula and the nomenclature. Each ring junction can exist in a *cis* or *trans* conformation, yielding six centers of asymmetry or chirality (C-5, 8, 9, 10, 13, 14). Hence, 64 stereoisomers are possible for the ring system alone. With the C-17 side chain forming a seventh site of asymmetry, the number of theoretical stereoisomers increases to 128. However, in practice, the isomeric possibilities are restricted by the overall conformational limitations of the ring system.

In isolation, each six-membered ring system may exist in either "chair" or "boat" conformations (Figure 1.7). While the chair conformation is favored in either case, the two hydrogen atoms attached to each carbon take up positions either in the general plane of the ring, termed equatorial (e), or perpendicular to the plane of the ring, termed axial (a). Equatorial or axial positions can be oriented above (β-configuration) or below (α-configuration) the general plane of the ring. By convention, β-configuration is represented by a heavy line and α-configuration by a dotted line in the structural formula. The junction between rings *A* and *B* can be in one of two orientations: with substituent hydrogens in a *trans* position (i.e., C-5α, C-10β configuration) or in a *cis* position (i.e., C-5β, C-10β configuration). Hence, the orientation of the C-5 hydrogen defines the type of *A/B* ring junction, a *trans* junction is defined by a 5α orientation (previously termed "allo") and a *cis* junction is defined by the 5β position (previously termed "normal").

In most naturally occurring steroids, sterols, and bile salts, the *C/D* ring junction is in the *trans* conformation. This is also the case for the *A/B* ring junction in most biologically active steroids. These junctions make the tetracyclic ring system planar in orientation. To aid in the deduction of stereochemistry from the structural formulas it is common to show the orientation of the C-5 hydrogen atom (e.g., Figures 1.6 and 1.7). The orientation of the remaining junctions can then be assumed in most cases. That is, unless otherwise indicated, substituent atoms at ring junctions are assumed to be C-8β, C-9α, C-10β, C-13β, C-14α, and the C-17 side chain in the β conformation.

Stereochemistry of side chain substituents is described by the sequence rule procedure, where any chiral centers formed by substituents at C-20 or above are assigned the (*R*) or (*S*) configuration. This designation, when employed, precedes the entire nomenclature name (see below). For brevity, side chain stereochemistry will not be described in further detail.

1.5.2.2 Named Nomenclature of Steroids

Steroids also possess their own set of nomenclature stem names based on the carbon number of the central ring system (Figure 1.8). These names are equivalent to the stem names for *n*-alkane hydrocarbons. Many of the standard IUPAC prefixes and suffixes are then applied directly to describe the related structural details, in a similar manner for other compounds (Table 1.5). However, there are some variations. For the following discussion, the parent ring system cholestane will be used to highlight examples.

1.5.2.2.1 Carboxylic Acids and Derivatives (Esters and Lactones)

The substitution of a carboxylic acid functionality to a methyl carbon employs the usage of the "-oic acid" suffix (e.g., 5β-cholest-26-oic acid, Figure 1.9). When substitution occurs on a methylene carbon, the suffix employed is "-carboxylic acid" (e.g., 5β-cholestane-24-carboxylic acid, Figure 1.9). Esters of either type of carboxylic acid are referred to in the same manner as other acids, by changing the "-oic acid" or "-ic acid" suffix to "-oate" or "-ate" and utilizing "ester" as the prefix. Hence, the methyl ester of the earlier examples becomes methyl 5β-cholest-26-oate and methyl 5β-cholestane-24-carboxylate, respectively. For lactone derivatives, the suffix "-ic acid" is replaced by "-lactone" and "carboxylic acid" changed to "-carbolactone." The suffix is preceded by the acid group location and then the hydroxyl group location (e.g., 5β-cholesto-26, 17α-lactone, Figure 1.9).

1.5.2.2.2 Aldehydes, Ketones, and Ethers

When a methyl group changes to an aldehyde, the suffix used is "-al." When an acid group is changed to an aldehyde, the suffix becomes "-aldehyde," but the name is derived from that of the acid (e.g., 5β-cholest-26-aldehyde). The prefix "oxo-" denotes the change of a methyl or methylene group to an aldehyde or ketone group, respectively. If additional carbon atoms are introduced as aldehyde groups, a separate nomenclature system is used. For ketone groups, the standard suffix "-one" and prefix "oxo-" are used as appropriate (e.g., 4-oxo-5β-cholest-26-oic acid, 5β-cholest-4-one). Ethers are named as alkoxy group prefixes (methoxy-, ethoxy-, etc.).

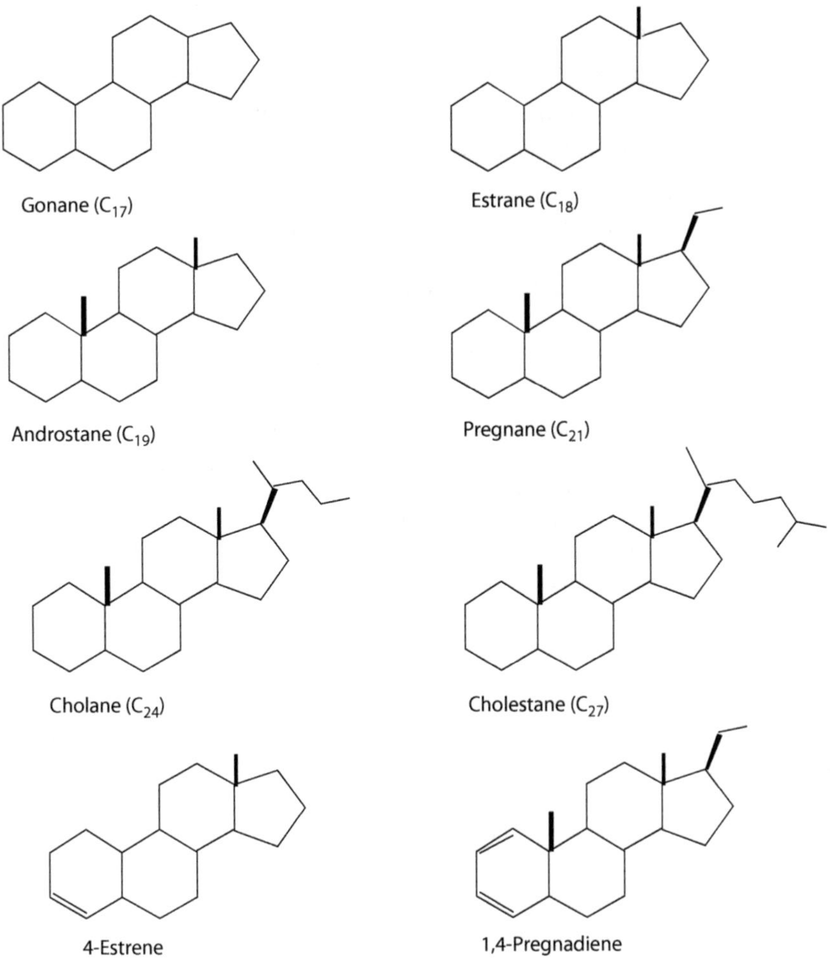

Gonane (C$_{17}$)

Estrane (C$_{18}$)

Androstane (C$_{19}$)

Pregnane (C$_{21}$)

Cholane (C$_{24}$)

Cholestane (C$_{27}$)

4-Estrene

1,4-Pregnadiene

FIGURE 1.8 Example generic carbon ring systems for saturated and unsaturated steroids of C$_{17}$ to C$_{27}$ with the systematic nomenclature stem names.

1.5.2.2.3 Alcohols and Derivatives (Esters and Ethers)

Alcohol groups are designated as for other lipid components, with the suffix "-ol" and prefix "hydroxy-" where appropriate (e.g., cholest-5-en-3β-ol, Figure 1.7). Esters are named by replacing the "-ol" suffix of the parent alcohol with "-yl" to generate the radical name. The acyloxy group is then denoted in anionic form following the main name (e.g., cholest-5-en-3β-yl acetate). As ester groups take naming precedence over oxo (aldehyde and ketone) groups, the systematic names of many steroid esters do not necessarily derive from the parent alcohol. An example is the addition of an acetyl group to the C-17 hydroxy group of 17β-hydroxyandrost-4-en-3-one (testosterone), which forms the compound named 3-oxo-androst-4-en-17β-yl acetate (Figure 1.9).

TABLE 1.5
Functional Groups in Order of Priority for the Selection of the Name Suffix in Systematic Steroid Nomenclature

Functional Group	Prefix	Suffix
Carboxylic acid	—	-oic acid/-carboxylic acid[a]
Lactone	—	-lactone/-carbolactone[a]
Ester or salt of acid	Alkyl group[a]	-oate
Aldehyde	oxo-	-al
Ketone	oxo-	-one
Hydroxyl	hydroxy-	-ol

Source: Compiled from Kirk, D.N. and Marples, B.A., The structure and nomenclature of steroids, in *Steroid Analysis*, Makin, H.L.J. et al. (eds.), Blackie Academic and Professional, London, U.K., 1995, 1–24.

[a] Refer to Section 1.5.2.

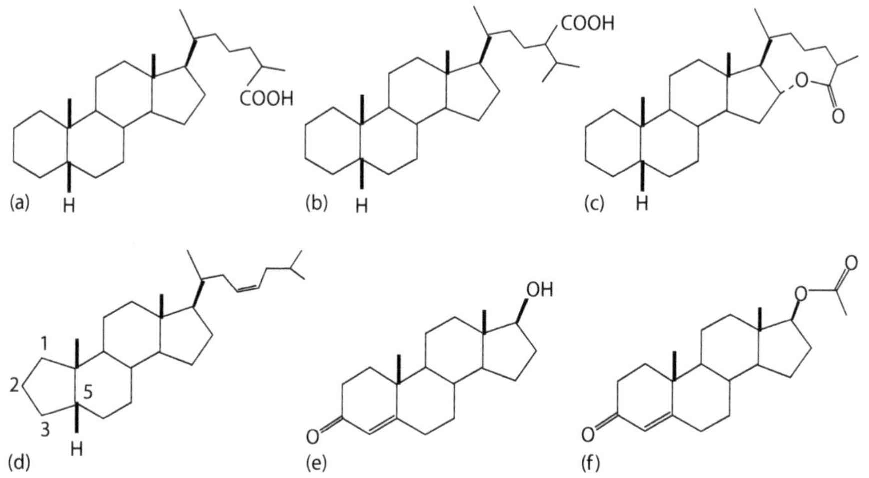

FIGURE 1.9 Example steroid structures with systematic nomenclature names: (a) 5β-cholest-26-oic acid, (b) 5β-cholestane-24-carboxylic acid, (c) 5β-cholesto-26,17α-lactone, (d) 23(Z)-4-nor-5β-cholest-23-ene, (e) 17β-hydroxyandrost-4-en-3-one, and (f) 3-oxo-androst-4-en-17β-yl acetate.

1.5.2.2.4 Unsaturation and Skeletal Modifications

The insertion of double bonds in the ring system does not require further nomenclature description as the geometry is fixed by the ring itself. For double bonds on the side chain, the older terms *cis* and *trans* used for FA nomenclature are recommended to be replaced with the more precise sequence rule terms (Z), usually equivalent to *cis*, and (*E*), usually equivalent to *trans*, respectively. Where the ring system has been modified by the loss of a carbon atom (either a methyl group or ring carbon), the prefix "nor-" is used preceded by the location of the missing carbon (Figure 1.9). Where additional carbons are added to the ring structure, the prefix "homo-" is used as above. Scission of a ring system is denoted by the prefix "seco-" with the location of the two carbon atoms where the break occurs.

1.5.3 CAROTENOIDS

Carotenoids consist of a group of pigments with an extremely wide distribution in natural products. The term is used to describe several groups of compounds derived from eight isoprene units in a manner so that the arrangement of the units is reversed at the center of the molecule. They are also characterized by containing a conjugated double bond system. All members of the carotenoids can be formally derived from the acyclic precursor lycopene (Figure 1.10) by reactions involving one or more options of hydrogenation, dehydrogenation, cyclization, insertion of oxygen, double bond migration, methyl migration, chain elongation, or chain shortening. Different groups of carotenoids are defined by these various modifications. The reader is referred to Goodwin (1980) and Britton et al. (2004) for further detail.

1.5.3.1 Carotenes

Carotenes are a specific group of carotenoids that retain their hydrocarbon character (i.e., without the addition of other functional groups) and can be described by the type of terminal chain structures (end groups) connected to their central $C_{22}H_{26}$ linear chain. End group designations for carotenes are given in Figure 1.10. Trivial names for carotenes are commonly employed. Again, we shall concentrate on the IUPAC systematic nomenclature. The stem name "carotene" is employed for this and the following groups, preceded by the designation of the end groups. Hence, lycopene is systematically described as ψ,ψ-carotene, α-carotene is β,ε-carotene, and β-carotene is β,β-carotene (Figure 1.10). When a carotene is referred to in most texts, the systematic name will be denoted in addition to the trivial name. However, following this, the compound is likely to be referred to only by its trivial name.

Carotenoids that contain an oxygenated functionality are referred to collectively as xanthophylls. They retain the systematic naming system of carotenes, with the additional functionalities named according to the usual nomenclature rules discussed earlier. For example, β,β-carotene-3,3′-diol (zeaxanthin) (Figure 1.11). Methoxy, carboxy, aldehyde, epoxy, and ketone derivates are also common.

In retro-carotenoids, all the single or double bonds of the conjugated polyene system have shifted by one position. The carbon atoms defining the new conjugated system are indicated in the nomenclature as a prefix prior to the term "-*retro*-." For example, 4′,5′-didehydro-4,5-*retro*-β,β-carotene-3,3′-diol (eschscholtzxanthin) (Figure 1.11).

Seco-carotenoids have undergone oxidative fission of a ring system without the loss of any carbon atoms. In these cases, the prefix "seco-" is utilized, preceded by

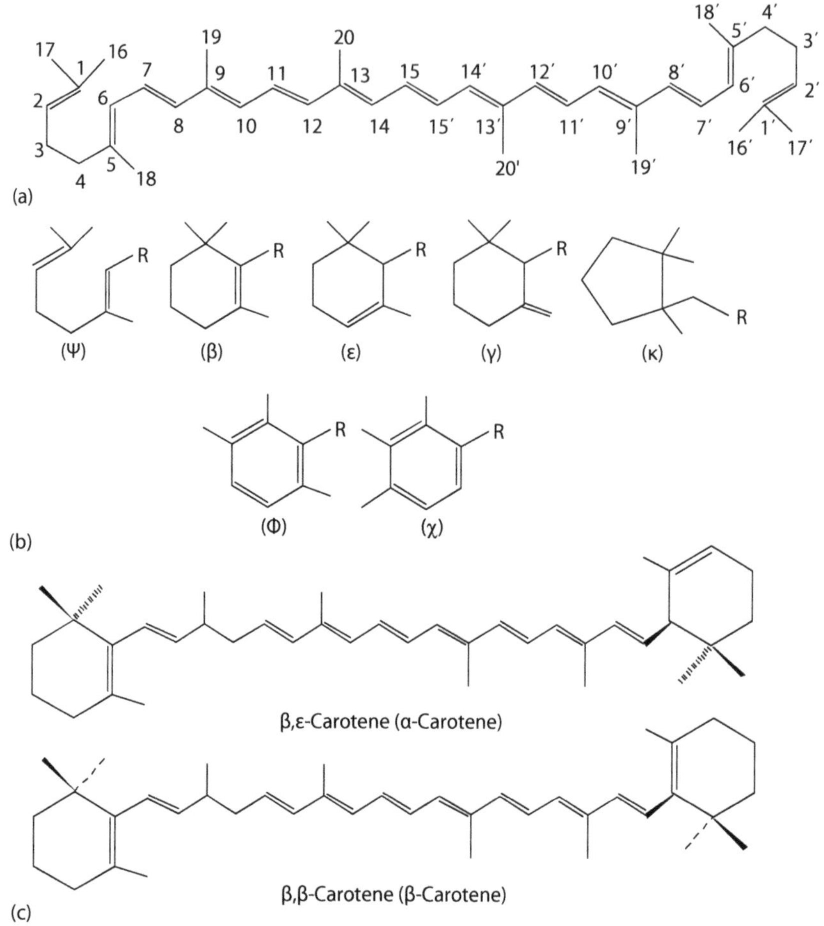

FIGURE 1.10 Examples of carotene structural diversity. (a) Lycopene (ψ,ψ-carotene) displaying the carbon number system for carotenes, (b) the systematic nomenclature end group designations for carotenes, and (c) α-carotene (β,ε-carotene) and β-carotene (β,β-carotene).

the numbers of the carbon atoms involved in the fission. The remainder of the stem name is taken from the description of the remaining molecule. For example, 5,6-seco-β-cartoen-5,6-dione (Figure 1.11). Where oxidative fission of a ring system results in the loss of any carbon atom, the resulting compound is termed as apo-carotenoid. The prefix "apo-" is preceded by the number of the carbon atoms from which the remainder of the molecule has been removed, with the stem name taken from the description of the remaining molecule (e.g., 3'-hydroxy-8-apo-β-cartoen-8-al (Figure 1.11)). Nor-carotenoids are compounds where carbon atoms have been removed by procedures other than the cleavage of carbon–carbon bonds. They are named along the normal standard system of nomenclature retaining the "carotene" stem name.

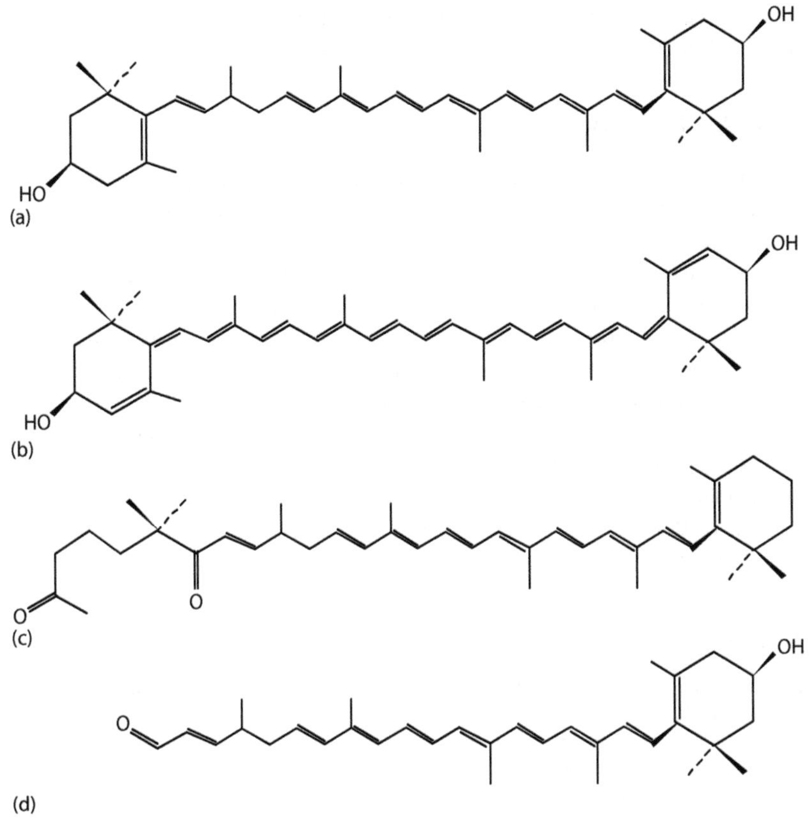

FIGURE 1.11 Examples of oxygenated carotenoids (xanthophylls): (a) zeaxanthin (β,β-carotene-3,3′-diol), (b) eschscholtzxanthin (4′,5′-didehydro-4,5′-*retro*- β,β-carotene-3,3′-diol), (c) 5,6-seco-β-caroten-5,6-dione, and (d) 3′-hydroxy-8-apo-β-caroten-8-al.

1.5.3.2 Stereochemistry of Carotenoids

The configuration of chiral centers is designated as described previously, using the (R) and (S) designations, together with locations, preceding the full systematic name. Stereochemistry also affects the carbon numbering of the end groups. When a potential chirality at C-1 is structured as in the β end group (Figure 1.10), with the polyene chain to the right of C-1, then the methyl group below the plane of the ring is numbered C-16 and the one above is numbered C-17. When the polyene chain is to the left of C-1, the numbering of the methyl groups is reversed. If the end group is acyclic (e.g., ψ end group, Figure 1.10), the methyl group, which is *trans* to the polyene chain, is numbered C-16 and the *cis* methyl group C-17.

The stem name "carotene" brings with it an implication of *trans* double-bond geometry for all bonds. Note that this is opposite to that of PEFA discussed earlier (see Section 1.2.4). If a *cis* double bond is present, this is distinguished in the structural name as either "cis-" or "Z-" following the appropriate carbon location. Compounds known to contain a *cis* double bond but where the location is unknown, or isomers of carotenes with an unlocated *cis* double bond are given the prefix "neo-" and an alphabetical suffix.

1.6   FINAL REMARKS

The IUPAC rules of lipid nomenclature have been developed to convey the maximum level of structural information within the name of the lipid. From this information, the reader should be able to reconstruct the component structure in a systematic fashion. For this reason, the established system of systematic nomenclature should be encouraged rather than the use of trivial or common names in most circumstances. However, exceptions to this generality can be justified in the cases of steroids and carotenoids. In these cases, systematic nomenclature is often unwieldy for everyday usage and common names are well established in the respective fields.

For the use of structural formulas, care should be taken in the interpretation and representation of isomeric and geometric features. This is particularly true for the stereochemistry of complex lipids such as acylglycerols and the ring systems and substituents of steroids. The development of a solid grounding in lipid nomenclature will stand the researcher or food technologist in good stead.

REFERENCES

Briggs, M.H. and Brotherton, J. 1970. *Steroid Biochemistry and Pharmacology*. London, U.K.: Academic Press.

Britton, G., Liasen-Jensen, S., and Pfander, H. 2004. *Carotenoids: Handbook*. Basel, Switzerland: Birkhauser.

Deuel, H.J. Jr. 1951. *The Lipids, Their Chemistry and Biochemistry, Volume I: Chemistry*. New York: Interscience Publishers Inc.

Goodwin, T.W. 1980. *The Biochemistry of the Carotenoids*. London, U.K.: Chapman and Hall.

Harwood, J.L. and Russell, N.J. 1984. *Lipids of Plants and Microbes*. London, U.K.: George Allen and Unwin.

IUPAC-IUB Commission on Biochemical Nomenclature. 1977. The nomenclature of lipids: Recommendations (1976). *Lipids*, 12: 455–468.

Kirk, D.N. and Marples, B.A. 1995. The structure and nomenclature of steroids. In *Steroid Analysis*, H.L.J. Makin, D.B. Gower, and D.N. Kirk (eds.), pp. 1–24. London, U.K.: Blackie Academic and Professional.

2 Chemical and Physical Properties of Lipids

David S. Nichols, Timothy B. Jordan, and Neil Kerr

CONTENTS

2.1 CHEMICAL PROPERTIES

2.1.1 INTRODUCTION

The chemical properties of lipids become evident during a chemical reaction and define a particular characteristic of the molecular structure. As such, the chemical properties of lipids may only be described in reference to a particular reaction. A number of lipid chemical properties are referred to as chemical indices and have been used historically to describe the characteristics of pure lipids or lipid mixtures.

2.1.2 MAIN CHEMICAL REACTIONS OF LIPIDS

2.1.2.1 Outlook

The reactions of lipids in foods are significant as they may alter the flavor, color, texture, nutritional value, or safety of the food product. The main chemical reactions of lipids in foods are influenced directly or indirectly by temperature and, in particular, the thermal processing that the food containing lipids may undergo. Most food produced today undergoes some form of heat treatment during its preparation, such as cooking, baking, boiling, toasting, roasting, canning, pasteurizing, or drying. The chemical and physical changes that occur in food lipids depend on the lipid composition and conditions of treatment. Heat treatment, particularly in the absence of oxygen, leads to direct thermal degradation (thermolytic) reactions. In the presence of air, additional oxidative reactions occur that are accelerated by the heat. In the case of lipids within the food context, the majority of chemical reactions involve the acyl residues of complex lipids. The following discussion will therefore concentrate on the reactions of acyl residues that are common throughout the differing classes of acylglycerols.

2.1.2.2 Thermolytic Reactions

In general, the unsaturated double bonds of lipids provide the main sites for thermolytic reactions to occur. Hence, for the major lipid class in foods (acylglycerols), it is the unsaturated acyl residues (FA) that represent the primary reactants. The major result of thermolytic reactions of unsaturated FA is the formation of dimeric and cyclic compounds by a number of mechanisms involving carbon–carbon bond formation within the same FA. The degree of unsaturation also increases the occurrence of thermolytic reactions.

In the case of acylglycerols, dimeric or cyclization reactions may also occur between acyl residues of the same molecule. Acylglycerols may also decompose thermolytically to release their component FA and other products. Quantitatively, this is the most common thermolytic reaction of triacylglycerols (TAG) and occurs with or without the presence of water, by different mechanisms.

2.1.2.3 Oxidative Reactions

Similar to thermolytic reactions, it is the unsaturated double bonds within lipids that make them vulnerable to oxidative reactions. However, since a direct reaction of molecular oxygen with the double bonds of lipids is thermodynamically unfavorable, oxidative reactions must be initiated via a radical intermediate that is generated by a separate mechanism. Once free radical formation is initiated, a chain reaction (autoxidation) may occur where the hydrogen atoms adjacent to carbon double bonds in lipids are removed by the radicals. The carbon atoms at these locations subsequently react with oxygen to form a peroxide radical ($ROO^•$), which in turn removes hydrogen atoms from further double bond sites to form peroxides ($ROOH$), which degrade to form alkyl radicals ($R^•$), which regenerate peroxide radicals by reacting with oxygen. Peroxides formed as the primary products of lipid autoxidation undergo further complex degradation and reactions with other molecules (refer to

Nawar 1986). Many specific peroxides are formed from the oxidation of polyenoic FA, with the major decomposition products of lipid autoxidation being characteristic of the FA composition of the acylglycerol lipids.

2.1.3 Chemical Indices of Food Lipid Quality

2.1.3.1 Introduction

Chemical indices are a measure of several specific chemical properties of lipids. Indices are widely used in the fat and oil industry as they are useful for describing the characteristics of lipid mixtures used in food manufacturing and processing, as well as to establish their purity. The composition of natural oils and fats are affected by cultivar and growing conditions as well as handling and processing. Chemical indices are a practical way of describing and quantifying this diversity by assigning commonly accepted values for certain animal and vegetable fats (Tables 2.1 and 2.2). They are, however, seldom used for the characterization of individual lipids as they do not in isolation provide proof of the identity or purity of an individual compound.

A large number of indices have been proposed and used by different sections of industry. These include the melting point, the refractive index, the specific gravity, the saponification value (SV), and the iodine value (IV), among others. Standard methods for assays have been published by the American Oil Chemists' Society (AOCS) http://www.aocs.org and the International Organisation for Standardization (ISO) http://www.iso.org among others. The following section focuses on indices particularly relevant to food processing and manufacture.

2.1.3.2 Iodine Value

The IV is an important parameter in trade specifications and may provide an indication of the adulteration of an oil or fat. It is also useful in determining the ability of a lipid to be hardened by hydrogenation and for monitoring hydrogenation process control.

Under certain conditions, iodine is absorbed quantitatively by unsaturated acyl residues of acylglycerols and provides a measure of the degree of unsaturation of a lipid (Table 2.1). The IV is defined as the number of grams of iodine absorbed by 100 g of a lipid. Lipids containing fully saturated acyl residues have a zero IV. The Wijs method, which involves reacting a sample with iodine monochloride in a solution of glacial acetic acid and detecting liberated iodine with standardized thiosulfate, is the standard chemical method, e.g., ISO 3961:1996; AOCS Cd 1d-92.

FAs containing acetylenic bonds, double bonds close to the carboxyl group, conjugated unsaturation, or other reducible (e.g., keto) groups rarely add the

TABLE 2.1

Effect of Unsaturation on the Iodine Value of Individual Fatty Acids

Fatty Acid	Iodine Value
16:1(n-)9c	99
18:1(n-)9c	89
18:2(n-)6	181
18:3(n-)3	273
20:4(n-)6	320

Source: Data compiled from Duel, H.J. Jr., *The Lipids, Their Chemistry and Biochemistry, Volume I: Chemistry,* Interscience Publishers Inc., New York, 1951.

TABLE 2.2
Physical and Chemical Properties of Lipids from Different Sources

Lipid Source	IV[a]	Sap[b]	SG[c] (20°C/20°C)	Viscosity (37.8°C)	Refractive Index (40°C)	Melting Point (°C)
Olive oil	75–94	184–196	0.910–0.916	43.2	1.468–1.471[d]	−3 to 0
Coconut oil	5–13	248–265	0.908–0.921[e]	29.8–31.6	1.448–1.450	23–26
Palm kernel oil	14–21	230–254	0.899–0.914[e]	47.8	1.448–1.452	24–26
Chicken fat	76–80	193–198	0.914–0924[f]	—	1.452–1.460	30–34
Tallow (mutton)	35–46	192–198	0.938–0.955[g]	—	1.452–1.458	44–51
Tallow (beef)	33–47	190–200	0.903–0.907[h]	62.1	1.450–1.458	45–48
Salmon oil	130–160	183–186	0.924–0.926[i]	42.3	1.472–1.477	—
Sunflower seed oil	118–145	188–194	0.918–0.923	49.1	1.467–1.469	−18 to −16
Rapeseed oil	94–120	168–181	0.910–0.920	—	1.465–1.469	−9
Canola oil	110–126	182–193	0.914–0.920	—	1.465–1.467	−20
Palm oil	49–55	190–209	0.891–0.899[j]	47.8	1.454–1.456	33–40
Castor oil	81–91	176–187	0.945–0.965[h]	259–325	1.466–1.473	−12 to −10
Pumpkin seed oil	116–133	174–197	0.903–0.926[i]	—	1.466–1.474	—
Maize oil	107–135	187–195	0.917–0.925	—	1.465–1.468	−12 to −10

Source: Data compiled from Abramovic, H. and Klofuta, C., *Acta Chim. Sloven.*, 45, 69, 1998; Anon., Official methods and recommended practice of the American Oil Chemists' Society, in *Physical and Chemical Characteristics of Oils, Fats and Waxes*, AOCS Press, Champaign, IL, 1997; Firestone, D., *Physical and Chemical Properties of Oils, Fats, and Waxes.*, AOCS Press, Champaign, IL, 1999.

[a] Iodine value.
[b] Saponification value.
[c] Specific gravity.
[d] 20°C.
[e] 40/20°C.
[f] 10/10°C.
[g] 55/55°C.
[h] 25/25°C.
[i] 15/15°C.
[j] 50/20°C.

theoretical amount of iodine. Samples containing significant proportions of these compounds will give lower than expected IV. The presence of conjugated double bond systems can be detected by ultraviolet spectroscopy. Hydroxyl groups in some lipids may interact with the Wijs reagent, giving a higher-than-expected IV. Likewise, the presence of a significant quantity of sterols in the lipid sample will also result in a higher than expected IV, as double bonds within the sterols absorb additional iodine.

Since the IV is a measure of the relative unsaturation of a compound or lipid sample, there also exists a linear relationship between this index and both the melting point and the refractive index, where one value may be used to estimate the other (see Sections 2.2.5 and 2.2.6). Other analytical techniques (e.g., gas chromatography of FA composition Cd 1c-85) can also be used to estimate the IV. Interest in recent years has focused on spectroscopic techniques for the rapid determination of IV. Fourier transform (FT) near infrared (IR), near IR, FT Raman, and ^{1}H and ^{13}C nuclear magnetic resonance spectroscopy have all been investigated (Ng and Gee 2001). The most promising results have been obtained with FT-near IR spectroscopy, which only takes a few minutes to determine the IV (Cox et al. 2000).

2.1.3.3 Acid Value or Neutralization Value

The neutralization or acid value (AV) of a lipid sample gives an indication of the unbound or free FA content. This is achieved by direct titration of the sample in an appropriate solvent with an alkali (e.g., ISO 660:1996). The AV is defined as the number of milligrams of KOH required to neutralize 1 g of the sample. Although refined oils are largely devoid of free FA, considerable amounts may be present in crude oils. Their presence may be an index of the oil purity. The degree of edibility of a fat is generally considered to be inversely proportional to the total amount of free FA.

2.1.3.4 Saponification Value

The SV gives a measure of the average length of the acyl residues present in complex lipid (acylglycerol) components. It is defined as the number of milligrams of KOH necessary to saponify 1 g of a lipid sample. The saponification number is inversely proportional to the molecular weight of a lipid. Samples of natural oils or fats, which are largely mixed acylglycerols containing a variety of FA, typically have SVs in the range 190–200. Higher values indicate the presence of increased amounts of longer chain FA (e.g., canola oil *ca.* 320), hydroxy acids (e.g., castor oil *ca.* 310), or unsaponifiable material. Lower values are characteristic of samples rich in shorter FA (e.g., palm oil 260–280 or butter fat 240–260). Table 2.2 lists some typical values.

2.2 PHYSICAL PROPERTIES

2.2.1 Introduction

The physical properties of lipids derive directly from their chemical structures and functional groups. Physical properties greatly influence the functions of lipids in foods and the methods required for manipulation and processing. As for the chemical reactions and chemical indices of food lipids, the physical properties of acylglycerol lipids are mediated by their FA composition. Describing the physical properties of differing FA types therefore applies equally to acylglycerols in which they may be components. Physical properties can be used to assess the purity or quality of lipid material in reference to known standards or preferred characteristics. Properties of particular interest to food scientists include density, surface tension, viscosity, polarity, melting point, and refractive index. Of these properties, melting point, density, and refractive index, in particular, vary progressively with the acyl residue chain length and degree of unsaturation.

2.2.2 DENSITY

The differences in density (or specific gravity) between different fats or oils are due to the differences in FA composition of the acylglycerol lipids. The density of acylglycerol lipids increases with the molecular weight of the component FA, and also with higher proportions of unsaturated or hydroxy FA (Table 2.2). It also tends to increase with the oxidation (rancidity) of a lipid. For fats that are solid at room temperature, the density is best determined well above the melting point, usually 40°C–50°C.

2.2.3 SURFACE TENSION AND VISCOSITY

Surface tension decreases with increasing temperature. This relationship is linear over a large part of the temperature range. The measurement of surface tension is markedly affected by the presence of impurities in the sample.

Viscosity is an important parameter for the design of industrial processes, e.g., it determines the rate at which an oil drains from a fried food. It is also an important factor affecting the stability of foods. Saturated FA generally have higher viscosities than unsaturated FA as their molecular structure enables close proximity of the carbon chains allowing intermolecular interactions such as van de Waals forces to establish. Conversely, the *cis* configuration of double bonds within FAs prevents the close alignment of molecules and hence results in weaker intermolecular interactions. For example, olive oil (10% unsaturated FA) has higher viscosity than sunflower oil (70% unsaturated FA). Some typical viscosity values are given in Table 2.2. Viscosity can also be used to evaluate the quality of fats used in frying, as these show an increased viscosity as they approach heat-induced breakdown (reviewed by Gertz 2000). The increased viscosity is coincident with polymerization, oxidation, gumming, and foaming tendencies.

2.2.4 POLARITY

The polarity of a lipid affects its volatility, solubility, and nonspecific binding to other polar compounds. Lipids are often functionally classified into neutral and polar on the basis of their mobility in thin-layer chromatography when a neutral or polar solvent system is used. Neutral lipids such as wax esters, steryl esters, ether lipids, and TAG are chemically neutral while other neutral lipids such as free FA, fatty alcohols, and monoacylglycerols are actually slightly polar due to the presence of hydroxyl and/or carboxyl groups. Table 2.3 lists the relative polarities of some common lipid classes.

TABLE 2.3

Examples of Lipid Classes Listed in Order from Most Polar to Least Polar

Sphingomyelin
Phosphatidylinositol
Phosphatidylserine
Sulfolipids
Monoacylglycerols
Diacylglycerols
Sterols
Triacylglycerols
Aldehydes
Hydrocarbons

Source: Data compiled from Hemming, F.W. and Hawthorne, J.N., *Lipid Analysis*, Bios Scientific Publishers, Oxford, U.K., 1996, Chapter 2.

2.2.5 MELTING POINT

Natural lipid mixtures do not have definite melting points as they are a mixture of various compounds. When heated, they soften due to the melting of individual components before becoming fully liquid. Some fats have a double melting point where they melt, then solidify before melting again. Therefore, empirical methods have been adopted to characterize the melting of mixtures such as TAG. Commonly employed methods include the drop point, the temperature at which the first drop of liquid falls from an open capillary containing the solid mixture, and the slip point, the temperature at which a solid mixture in an open capillary placed in a water bath begins to move upward. These methods are generally reproducible for homogeneous mixtures but can give quite variable results with substances formed from a number of lipid sources (Table 2.2).

Purified lipids have sharp and reproducible melting points. The melting points of acylglycerols closely approximate the FA residues that they contain. The melting points of saturated FA (and the acylglycerols containing them) do not increase uniformly with increased chain length but form two alternate series with even chain length FA having a higher melting point. These series converge as chain length increases (Figure 2.1). The introduction of a double bond leads to a decrease in the melting point. FA with a double bond in the *cis* configuration has a lower melting point than the corresponding isomer with a *trans* configuration. The position of the double bond also affects the melting point. Melting points are lowered as the double bond moves to the center of the carbon chain. This effect is much greater for double bonds in the *cis* rather than the *trans* configuration (Gunstone 1958). These acyl chain reactions have been interpreted in terms of an effective chain length model. Thus, for monoenoic acids where the double bond exists closer toward either end of the molecule, the acyl moieties behave as if they were saturated chains with an effective chain length identical with the length of the longer of the two chain segments separated by the double bond (Cevec 1991). Additional unsaturation lowers

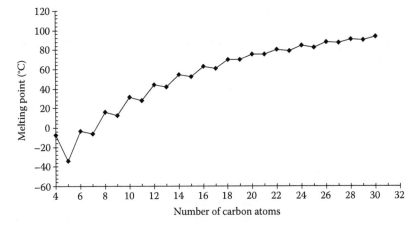

FIGURE 2.1 Melting points of saturated straight-chain fatty acids demonstrating the difference between even- and odd-chain length series.

the melting point further unless the double bonds are conjugated. The effectiveness of an unsaturated FA in lowering the melting point of an acylglycerol depends not only on the degree of unsaturation within the FA but also on the *sn* position of the FA within the acylglycerol. For example, natural cottonseed TAG oil can be changed from a liquid to a semisolid by chemical rearrangement (interesterification) of the FA *sn* positions (Duel 1951).

The substitution of the alkyl chain also affects the melting point. For saturated FA, hydroxyl groups raise the melting point while methyl groups lower it. *Iso*-branched FA generally have a higher melting point than the corresponding *anteiso*-branched FA, e.g., the melting points of 15:0, i15:0, and a15:0 are 52.3°C, 51.8°C, and 23.0°C, respectively.

2.2.6 REFRACTIVE INDEX

Refractive index is easily determined (e.g., ISO 6320:2000) and often used as a criterion of purity. A compound is considered to be optically active, or chiral, if linearly polarized monochromatic light is rotated when passed through it. The molecular basis of a chiral compound is a molecular structure that is not superimposable on its mirror image. Chirality can be caused by a number of factors such as restricted rotation about a single double bond, but is often indicative of an asymmetrical molecular structure.

The amount of optical activity can be measured using a polarimeter and is determined by the length of the path the light traverses, the concentration of the compound, the compound, and the wavelength of the light (normally a sodium light). Specific rotation, α, is defined as

$$[\alpha] = \frac{\alpha}{l \times d} \tag{2.1}$$

where
  α is the measured optical rotation in degrees
  l is the sample path length in decimeters
  d is the density if the sample is pure liquid, or the concentration if the sample is a solution expressed as g/cm^3

Generally, the refractive index varies with density (specific gravity) and, hence, is related to the molecular weight and degree of unsaturation of acyl residues within acylglycerols. The refractive index of saturated FAs have a linear increase with chain length when measured at a temperature above 40°C. Below 40°C, intramolecular forces may cause a nonrandom arrangement of molecules affecting the refractive index. The refractive index of unsaturated FAs increases with the degree of unsaturation, hence the refractive index of a lipid is also related to IV. Conjugated unsaturation has a marked increase in the refractive index. Some typical values are shown in Table 2.2.

Most mixtures of TAG show little optical activity, but castor oil and those unusual oils containing acylglycerols of cyclopentene acids (e.g., chaulmoogric and hydnocarpic acids present in the seed oils of *Flacourtiaceae*) are notable exceptions.

Samples containing sterols will display some optical activity due to the presence of these optically active compounds. Since most edible oils have low optical activity, a method for the detection of adulteration of edible oils with castor oil based on the optical activity of ricinoleic acid (1,2-dihydroxyoctadecenoic acid), a major constituent of castor oil, has been proposed (Babu et al. 1996).

TAG with FA differing greatly in chain length show small, but measurable, optical rotation. A saturated TAG with a longer acyl chain length at position *sn*-1 than at position *sn*-3 has a negative optical rotation. However, in a complex mixture of TAG, considerable fractionation would be required before analysis, as the optical rotations of the different components would partially cancel each other.

2.3 ABSORPTION SPECTRA

2.3.1 ULTRAVIOLET

Ultraviolet (UV) and visible light spectra of lipids result from electronic excitations and transitions in the molecule. Hence, functional groups with high electron density (e.g., carboxyl groups, double bonds) absorb strongly in the UV or visible spectrum with characteristic maximum wavelengths (λ_{max}) and extinction coefficients (ε_{max}). For lipids, a large portion of a relatively complex molecule may be transparent in the UV spectrum, yielding UV spectra similar to that of a much simpler molecule derived from the UV-absorbing functional groups. Values for common UV-absorbing functionalities are shown in Table 2.4. The conjugation of double bonds brings the absorption bands into the easily accessible region. Extinction coefficients rise with increased conjugation. With three or more conjugated double bonds, the absorption shows three peaks, a main one and subsidiary maxima on either side. The peaks are spread further apart as the number of double bonds increases.

UV spectra are particularly valuable for the analysis of carotenoids, due to their long conjugated double-bond systems resulting in the strong absorption of both UV and visible light. For carotenoids, both the position of λ_{max} and the shape of the

TABLE 2.4
Ultraviolet Absorption Values for Common Lipid Functionalities

Functionality		Absorption	Extinction
Name	Structure	Maxima (nm)	Coefficient
Double bond	—C=C—	177	12,600
Conjugated double bonds	—C=C–C=C—	217	20,900
Carboxyl	—COOH	208–210	32–50
Conjugated carboxyl	—CH=CH–COOH	206	13,500

Source: Data compiled from Hemming, F.W. and Hawthorne, J.N., *Lipid Analysis*, Bios Scientific Publishers, Oxford, U.K., 1996, Chapter 2.

absorption spectrum can yield valuable information regarding the structure of the carotenoid. An advantage of the limited absorption in the UV spectrum for most lipids is that the characteristic functional groups may be recognized in molecules of widely varying structure.

2.3.2 INFRARED

The stretching and bending vibrations of molecular bonds give rise to absorptions in the IR region of the electromagnetic spectrum, between visible light and microwaves. Absorption bands are expressed as a frequency (wavenumber) or the number of waves per centimeter. As every separate chemical bond within the molecule may absorb IR energy by stretching and bending, the IR spectrum from a given compound is complex and unique to that structure. A peak-by-peak correlation of an IR spectrum is an excellent identification tool. Although the IR spectrum of a given compound structure is unique in its entirety, it is also true that certain functional groups or structural arrangements of atoms also absorb at characteristic frequencies regardless of the rest of the molecule.

All FA show a strong absorption in the 2750–3000 cm^{-1} range because of the large number of CH_2 and CH_3 groups in the molecules (Table 2.5). Cyclopropane FA give characteristic absorption bands at 1020 and 3050 cm^{-1} due to the vibration of the

TABLE 2.5
Infrared Absorption Bands of Common Lipid Functionalities

Functional Group	Frequency (cm^{-1})	Intensity[a]
$-CH_3$	2962 and 2872 ± 10	S
	1450 ± 20	M
	1375 ± 5	M
$-OCH_3$	1430	M
$-CH_3$	2926 and 2853 ± 10	S
	1465 ± 10	M
	750–720	M
	3100–3000	M
	1025–1000	M
$-C(CH_3)_2$	1385 and 1365 ± 5	M
	1170 ± 5	M
$-C(CH_3)_3$	1395–1385	M
	1365	S
	1250–1200	S

Source: Data compiled from Hemming, F.W. and Hawthorne, J.N., *Lipid Analysis*, Bios Scientific Publishers, Oxford, U.K., 1996, Chapter 2.

[a] S, strong; M, medium.

methylene CH_2 and stretching of the C—H bonds in the cyclopropane ring, respectively. Unsaturation can be detected by the C—H stretch near $3020\,cm^{-1}$. *cis* and *trans* isomers show little difference, although *trans* double bonds are associated with C—H deformation at $950–1000\,cm^{-1}$. This can be used for the quantitative determination of *trans* double bonds as long as there is limited conjugation. Hydroxy FA has a characteristic C—O stretch at $1045\,cm^{-1}$.

REFERENCES

Abramovic, H. and Klofuta, C. 1998. The temperature dependence of dynamic viscosity for some vegetable oils. *Acta Chim. Sloven.*, 45, 69.

Anon., 1997. Official methods and recommended practice of the American Oil Chemists' Society. In: *Physical and Chemical Characteristics of Oils, Fats and Waxes*. AOCS Press, Champaign, IL.

Babu, S., Sudershan, R.V., Sharma, R.K., and Bhat, R.V. 1996. A simple and rapid polarimetric method for quantitative determination of castor oil. *J. Am. Oil Chem. Soc.*, 73, 397.

Cevec, G. 1991. How membrane chain-melting phase-transition temperature is affected by the lipid chain asymmetry and degree of unsaturation: An effective chain length model. *Biochemistry*, 30, 7186.

Cox, R., Lebrasseur, J., Michiels, E., Buijs, H., Li, H., van de Voort, F.R., Ismail, A.A., and Sedman, J. 2000. Determination of iodine value with a Fourier transform-near infra red based global calibration using disposable vials: An international collaborative study. *J. Am. Oil Chem. Soc.*, 77, 1229.

Duel, H.J. Jr. 1951. *The Lipids, Their Chemistry and Biochemistry, Volume I: Chemistry*. Interscience, New York, Chapters 2, 3.

Firestone, D. 1999. *Physical and Chemical Properties of Oils, Fats, and Waxes*. AOCS Press, Champaign, IL.

Gertz, C. 2000. Chemical and physical parameters as quality indicators for used frying fats. *Eur. J. Lipid Sci. Technol.*, 102, 566.

Gunstone, F.D. 1958. *An Introduction to the Chemistry of Fats and Lipids*. Chapman & Hall, Norwich, U.K.

Hemming, F.W. and Hawthorne, J.N. 1996. *Lipid Analysis*. Bios Scientific Publishers, Oxford, U.K., Chapter 2.

Nawar, W.W. 1986. Chemistry of thermal oxidation of lipids. In: *Flavor Chemistry of Fats and Oils*, Min, D.B. and Simouse, T.H. (Eds.). AOCS Press, Champaign, IL, pp. 39–60.

Ng, S. and Gee, P.T. 2001. Determination of iodine value of palm and palm kernel oil by carbon-13 nuclear magnetic resonance spectroscopy. *Eur. J. Lipid Sci. Technol.*, 103, 223.

3 Principles of Lipid Analysis

David S. Nichols, Timothy B. Jordan, and Neil Kerr

CONTENTS

3.1 INTRODUCTION

There are several key developments in the history of lipid analysis, which have shaped the current knowledge of lipids in food systems. The first of these is the development of efficient, quantitative solvent-extraction systems for the recovery of pure lipid extracts from a variety of food and natural sources. While taken somewhat for granted today, significant research was devoted to this area during the 1940s–1960s. Without the establishment of this fundamental area, much of what we know today would not have followed. Following from these discoveries came the techniques required for the chemical separation of lipid types based on the solubility of different lipid types in various solvent systems. These techniques continue to underpin the analysis of lipids today.

Second, major advances were achieved with the development of chromatographic techniques for the separation and identification of lipids. What can now be considered the scientific "field" of chromatography had many of its early advances in the development of lipid-separation techniques, from paper chromatography and thin layer chromatography (TLC) through to gas chromatography (GC) and high-performance liquid chromatography (HPLC).

Finally, over recent years, there have been continuing revolutions in instrument hardware, which have had profound influences on some areas of lipid analysis. The development of supercritical fluid extraction (SFE) of lipids now gives an alternative to wet solvent extractions, which have basically remained unchanged in principle and practice since the 1960s. Advances in mass spectrometry (MS) and analytical chromatography now allow the direct analysis and identification of complex lipids from total lipid extracts without the requirement for solvent fractionation.

3.2 EXTRACTION OF LIPIDS FROM FOODS

3.2.1 INTRODUCTION

The initial step in the analysis of any lipids from foods is to quantitatively separate the lipid components from other protein, amino acid, carbohydrate, and aqueous ingredients. This should also be accomplished in a manner that does not degrade complex lipids or introduce contamination. Three major strategies will be discussed for the extraction of lipid components: solvent extraction using various combinations of organic solvents and two more recent innovations—SFE and accelerated solvent extraction (ASE).

3.2.2 SOLVENT EXTRACTION

The systems of solvent extraction rely of the differing solubility of lipids from that of other food components. These differences in solubility relate mainly to issues of

TABLE 3.1
The General Gradation of Polarity
Among Common Lipid Classes and
Organic Solvents

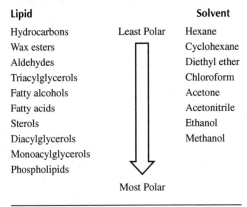

Lipid		Solvent
Hydrocarbons	Least Polar	Hexane
Wax esters		Cyclohexane
Aldehydes		Diethyl ether
Triacylglycerols		Chloroform
Fatty alcohols		Acetone
Fatty acids		Acetonitrile
Sterols		Ethanol
Diacylglycerols		Methanol
Monoacylglycerols		
Phospholipids		
	Most Polar	

charge or polarity, and the nature of the association between lipids and other components within the material to be extracted.

However, as lipids represent such a diverse range of compound types, there is also the issue of ensuring that the solvent extraction is capable of obtaining all the lipids present to yield a total lipid extract or total solvent extract. In essence, this issue is related to the relative polarity of the lipid types and extraction solvents (Table 3.1). Neutral or nonpolar lipids such as hydrocarbons, sterol esters, acylglycerols, and carotenoids can be bound through non-covalent interactions such as van der Waals forces and hydrophobic associations between hydrocarbon chains of lipids and hydrophobic protein domains within foods. Such lipids can be extracted with relatively nonpolar solvents such as diethyl ether or chloroform. Polar lipids such as phospholipids, glycolipids, and steroids can be bound through hydrogen bonding, electrostatic and hydrophobic interactions to protein components, and are usually membrane associated. The extraction of these lipids requires more polar solvents such as ethanol or methanol to disrupt these molecular interactions. Covalent associations can also exist between fatty acids (FA) (hydroxy or complex branched-chain), which are covalently bound as esters, amides, or glycosides to polysaccharide components. The extraction of bound lipid residues requires initial chemical cleavage through acid or alkaline hydrolysis.

Hence, a mixture of organic solvents, with a range of polarities, is necessary to ensure comprehensive extraction of non-covalently bound lipids. Many methods from the 1940s onward have utilized mixtures of chloroform and methanol for the quantitative extraction of lipids. However, the use of more polar organic solvents such as methanol also results in the extraction of unwanted contaminants in the form of sugars, amino acids, and salts. While these contaminants can be removed by purification or "washing" of the solvent extract, there exists a well-established extraction protocol that incorporates a partitioning of the solvent extract against an aqueous phase for the removal of contaminants.

The Bligh and Dyer extraction procedure is an example of a one-step solvent extraction and purification procedure to achieve a quantitative lipid extract. The procedure was originally developed for the extraction of fish tissue, but is equally applicable to the extraction of a wide range of materials with various modifications. The basis of the extraction utilizes a mixture of chloroform, methanol, and water, exploiting the one-to-two phase relationship of differing proportions of this mixture. It was determined experimentally that the solvent ratio of 2:1:0.8, (v/v/v) of methanol:chloroform:water was not only miscible in a single phase but gave near-quantitative extraction of lipids (around 94% from fish tissue, as good as any other method). The addition of a further 1:1 volumes of chloroform:water created a biphasic system, with a (predominantly) chloroform layer containing lipids and a (predominantly) methanol–water layer containing the non-lipid contaminants. Bligh and Dyer themselves noted that around 1% of the total lipid from fish tissues may be lost by partition into the methanol–water phase and that this fraction contained lipids of particularly high polarity. Recent revisiting of the solvent ratios demonstrated that a small increase in the yield of total lipid could be achieved by increasing the proportion of methanol remaining in the chloroform phase and that this was due predominantly to an increased yield of phospholipids.

Such solvent systems are calculated on the total volume of water in the extraction system, including any moisture content that may be present in the material to be extracted. It is therefore common to either dehydrate the sample prior to extraction or derive an estimate of the moisture content to appropriately adjust the volume of water added.

3.2.3 SUPERCRITICAL FLUID EXTRACTION

The manipulation of lipids with supercritical fluids (SF) is based on the discovery that, in general, SF can more easily dissolve compounds than liquid solvents. Until 1869, only three phases of matter were described. At this time, a critical phenomenon was discovered in the phase behavior or carbon dioxide. Under the combined extremes of pressure and temperature, a fourth phase was discovered, the SF. The generalized phase diagram for a pure substance is shown in Figure 3.1. The phase behavior we are usually used to follows the solid arrow, where at normal pressures, a substance will respond to increased temperatures by first crossing the equilibrium point between a solid and a liquid (line B-T) and then the equilibrium point between a liquid and a gas (line T-CP). It is also possible for a substance to pass directly from a solid to gas phase (sublimation; line A-T) under conditions of combined low pressure and temperature. However, under conditions of combined high pressure and temperature, a critical point (CP) is reached where the properties of the substance become intermediate between those of the liquid and gas phases. The SF possesses a density roughly two orders of magnitude higher than the gas phase and about one third that of the liquid, while the properties of the SF as a solvent are comparable to that of the liquid. Hence, while there is no decrease in solvation properties, an SF offers benefits in density (read viscosity and volume) over a liquid.

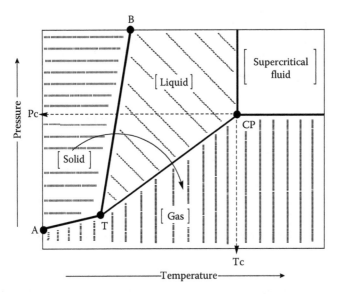

FIGURE 3.1 Idealized phase diagram for a pure compound describing the formation of an SF. Abbreviations: T, triple point; CP, critical point; Pc, critical point pressure; Tc, critical point temperature.

Several compounds have been investigated for SF extractions. However, when efficiency, cost, and safety issues are considered, the sole viable option is often concluded to be carbon dioxide. Since the critical point temperature (Tc) and critical point pressure (Pc) values for carbon dioxide are 31°C and 73 atm, respectively, a dedicated extraction instrument with high-pressure pumps and control valves is required for SF extractions. The benefits of SF procedures are the liberation of the extractions from (relatively) large volumes of chlorinated organic solvents. Increasing concerns of solvent toxicity and disposal have fueled the development of SF techniques. This has been particularly true in the industrial setting where the problems of organic solvents are magnified and the cost of dedicated high-pressure extraction equipment can be more easily justified.

The low polarity of carbon dioxide appears as a limitation to its use in extraction technologies requiring a total lipid extract, as more polar lipids may not be extracted. Many of the extraction protocols used for these applications rely on a "modified" solvent system based on the addition of small volumes of polar organic solvent (methanol or ethanol) to the extraction system. Such application for the extraction of lipids in the food industry is widespread, particularly with plant-based oils. However, the potential problems with the extraction of total lipids by SF have also yielded specific applications in the selective extraction or concentration of lipid components. Examples exist of the separation of FA from triacylglycerols (TAG) and squalene from sterols.

3.2.4 ACCELERATED SOLVENT EXTRACTION

The extraction of target analytes from samples has traditionally been the most labor- and time-consuming facet of lipid analysis. Recently, an automated liquid extraction

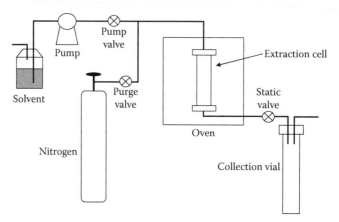

FIGURE 3.2   Schematic arrangement of the major components of an ASE system.

technique utilizing high pressure and temperatures has been commercialized. ASE was developed in an effort to reduce extraction times, increase extraction efficiencies, and reduce solvent consumption (Richter et al. 1996). Other names given to ASE include pressurized fluid extraction (PFE), pressurized liquid extraction (PLE), subcritical fluid extraction (as opposed to supercritical SFE), or pressurized hot-water extraction (PHWE). Figure 3.2 shows the basic components of an ASE system.

The principle of ASE is to pressurize and heat a solvent to enhance its extractive properties. Heating the solvent increases its capacity to dissolve analytes, while also increasing the rate of diffusion and disrupting solute–matrix interactions (such as van der Waals forces and hydrogen interactions). Higher temperatures also decrease solvent viscosity and surface tension, allowing it to penetrate more effectively into the matrix. These effects all contribute to an enhanced solvent–matrix interaction, allowing a rapid, more efficient extraction of analytes compared to other techniques (Richter et al. 1996). The second facet of ASE is the use of increased pressure. This has two main benefits: first, it allows the solvent to be heated above its boiling point while keeping it in the liquid phase, and second, it aids in forcing the solvent into matrix pores, potentially increasing extraction efficiency.

The ASE parameters most influential to the extraction of analytes are the properties of solvent, followed by the temperature of the extraction. As a general rule, a solvent that works well for traditional methods will work well with ASE. Some consideration can also be given to reducing the "strength" of the solvent, owing to the effects of temperature and pressure described above. Small amounts of apolar organic modifier (e.g., methanol) can also increase or selectively alter the extractive properties of the method. The effect of pressure has minor importance, so long as it is sufficient to keep the solvent in the liquid phase. Extraction time and the number of cycles may also contribute to the overall effectiveness of the method, both positively (greater extraction efficiencies through longer interaction between solvent and matrix and the use of fresh solvent) and negatively (increased time and potential for degradation). Typically, the shortest time required to achieve an appropriate level of recovery and precision is utilized.

TABLE 3.2

Example of Dispersants Used to Modify Samples for ASE and Their Functions

Function	Agent
Dispersant	Diamateceous Earth
	Celite-545
	Ottawa Sand
Adsorption	Silica gel—removal of nonpolar lipids
	Alumina—removal of nonpolar lipids, color
	Florisil—removal nonpolar lipids
	C18—removal of organics, polar compounds, lipids, color
	Carbon—removal of organics, color
	Copper—removal of sulfur
Drying	Sodium sulfate
	Hydromatrix

Source: Adapted from Dionex Application Note 210.

Sample preparation generally involves grinding or mixing the sample with an inert support matrix, achieving a consistent, free-flowing texture, which is then transferred to a stainless steel extraction cell containing a filter at the bottom. The support matrix can be simple bulking/dispersing agents, or they can be incorporated to perform other functions such as drying or adsorbing interferences (Table 3.2). The layering of these materials can also be incorporated into the cell to achieve different separating conditions (see In-cell fractionation, Section 3.3.3.4). The selection of solvent can also effect the retaining properties of the sorbents. Although ASE is predominantly used for extraction of solid samples, samples containing much water can be dried before extraction, or liquid samples can be mixed with a large proportion of dispersant.

There are three main advantages for using ASE over traditional methods such as Soxhlet: equal or greater efficiency, decreased time due to better extraction, and decreased solvent usage. Disadvantages include high cost of equipment, limited number of samples, and the requirement for a high-pressure gas connection (N_2/air). A major application of ASE has been in the determination of total lipids (fat) in foods. Total fat determinations of food generally require long and complex extraction procedures, employing acid hydrolysis or alkaline pretreatment followed by Soxhlet or sonication extraction. These methods are time consuming and use large volumes of solvent. The extraction of fats from dairy products (cheese, butter, milk, milk powder), meats, and chocolate by ASE has been shown to be equivalent in respect to yield to traditional methods employing Soxhlet or other schemes (Dionex Application Notes 340, 344, 345). In the case of cheese, the use of ASE removed the need to perform acid hydrolysis before extraction. Extraction times vary from 8–18 min, using 20–30 mL of solvent. Table 3.3 highlights examples of solvents and temperatures used for the determination of total fats in various food products. Lean

TABLE 3.3

Examples of Extraction Solvents and Temperatures Used for the Determination of Total Fats in Food Products by ASE

Product	Extraction Solvent	Extraction Temperature (°C)
Meat	Petroleum ether or hexane	125
Whole milk powder	Hexane:dichloromethane:methanol 5:2:1 (v/v)	80
Skim milk powder	Hexane:dichloromethane:methanol 3:2:1 (v/v)	80
Whey protein concentrate	Hexane:dichloromethane:methanol 2:3:3 (v/v)	80
Cheese	Hexane:isopropanol 3:2 (v/v)	110
Butter	Petroleum ether:acetone 3:2 (v/v)	100
Whole milk	Petroleum ether:isopropanol 2:1 (v/v)	120
Chocolate	Petroleum ether	125

Source: Adapted from Dionex Application Notes 340, 344, 345.

fish muscle tends to be dominated by polar phospholipids, which are not readily extracted by some traditional methods. A method for total lipid extraction from fish using ASE has been developed by Isaac et al. (2005). An optimization study using various solvents and temperatures found that isopropanol:hexane (65:35, v/v) at 115°C extracted the greatest amount of total lipids. A second step using hexane:diethyl ether (90:10, v/v) was also included, which improved the yield by 5%.

A recent development in the composition of the extraction cell and associated sample pathways has allowed the use of acidic or alkaline solvents to be used with ASE systems (Dionex Application Note 361). The Dionium™ components resist corrosion even when using acidic or alkaline solvents at the high temperatures and pressures used in ASE. This allows the direct transfer of acid hydrolyzed samples into the ASE cell, followed by a hexane extraction at 100°C. After extraction, direct gravimetric determination can be done, or the FA can be esterified for GC analysis. A comparison between the ASE and Mojonnier (AOAC Method 996.06) methods showed equivalence in the amount and precision of FA methyl esters (FAME) extracted from various foods (mayonnaise, corn chips, parmesan cheese).

3.3 PRINCIPLES OF LIPID CLASS SEPARATION

3.3.1 INTRODUCTION

Total lipid extracts from natural or food sources are likely to contain a diverse mixture of lipid classes. While one or two lipid classes may dominate on a proportional basis, it may often be the amount or composition of minor lipid classes that are of particular interest. Conversely, the presence of minor lipid classes may interfere with the determination of properties of interest from the major classes. In such cases, the fractionation of lipid classes is often required to enable individual analyses of the lipids present to be undertaken. The type of fractionation approach can also depend on the characteristics of the lipid extract. For example, those from animal sources are likely to have a high proportion of polar lipids with the remainder being neutral or nonpolar. Extracts or oils from plant seeds are likely to be the reverse. There are two main strategies for the separation of lipid classes. First, solvent fractionation based on the specific chemical reaction of certain lipid classes and resultant differences in solubility. Second, a suite of chromatographic techniques based on differences in lipid class polarity.

3.3.2 SOLVENT FRACTIONATION

There are two main procedures for the solvent fractionation of total lipid extracts (Figure 3.3). Both are primarily aimed at the isolation of total FA from the lipid extract for analysis. By reaction of lipid classes containing acyl residues, the liberated FA may be separated from the remaining neutral lipids (Figure 3.3, route A), or derivatized directly to a form, which may be analyzed together with neutral lipid components (Figure 3.3, route B).

Route A follows the protocol of an initial saponification. Here, a base catalyzed hydrolysis cleaves the ester linkages of acyl residues in phospholipids, acylglycerols,

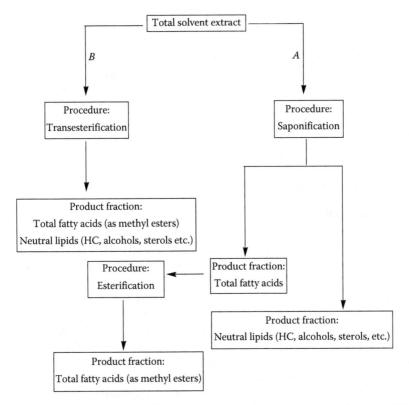

FIGURE 3.3 A flow diagram outlining two strategies for the solvent fractionation of lipid classes based on the hydrolytic reaction of O-acyl ester lipids.

and wax esters to yield an amalgam of total FA (as salts), long-chain alcohols (if wax esters were present), and remaining lipid classes, which do not contain ester linkages and are therefore unaffected by the reaction process (hydrocarbons, isoprenoids, ether linkages of O-alkyl residues). The high polarity of the FA allows them to be partitioned into an aqueous extract, leaving the isolated fraction of "nonsaponifiable" lipids. Following acidification of the FA fraction, they may be re-extracted back into an organic solvent and further derivatized for analysis (e.g., to FA methyl esters for GC analysis).

Route B does not strictly represent a fractionation protocol, but is worthy of highlighting in the instances where an alternative "rapid" procedure is preferred for the analysis of total FA. Here, an acid catalyzed transesterification reaction can be undertaken, converting total ester-linked acyl residues directly to their methyl esters. As by-products, nonsaponifiable lipids are also retained in this fraction (hydrocarbons, isoprenoids) and analyzed simultaneously by chromatographic techniques. While more rapid, the technique yields a more complex sample for analysis.

Figure 3.4 gives an example of possible results from solvent fractionation procedures. A total FA fraction has been isolated from the blue whale subcutaneous blubber sample and derivatized as methyl esters for GC (route A). The sample shows a well-separated profile of FA components and no other lipids. The dorsal blubber sample represents an example where the neutral lipid fraction has been collected. Here, nonsaponifiable lipids are retained in the analysis fraction, in this example, ether linked 1-O-alkyl-sn-glycerols derived from 1-O-alkyl-sn-2,3-diacylglycerols originally present in the blubber sample. However, there is also the carryover of some FA liberated from the ester linkages of complex lipids during the saponification procedure. While such carryover is uncommon in the fractionation of most lipid extracts, it may occur due to inefficient manipulation of the solvent fractions or where the proportion of FA liberated during saponification far exceeds that of nonsaponifiable components (in this example, the lipid extract consisted of 92% TAG, 5% FA, and only 1% 1-O-alkyl-sn-2,3-diacylglycerol).

3.3.3   CHROMATOGRAPHIC SEPARATION

3.3.3.1   Introduction

The basis of all chromatographic techniques may be considered as the interaction between lipid molecules and a solid matrix of specific character, termed the stationary phase. These interactions are further influenced by the medium in which the lipid sample is contained. This medium may be a liquid (for dissolved lipids) or a gas (for volatilized lipids) and is employed to move the lipid sample over or through the stationary phase. It is therefore known as the mobile phase.

The major characteristics of lipids that are exploited to chromatographically separate different classes are polarity and the degree of ionization. The polarity of a lipid significantly influences properties such as solubility, volatility, and nonspecific binding to other polar materials. The degree of ionization determines the amount of intermolecular interactions, which reduces compound volatility. In addition, such interactions with polar solvents increase compound solubility and can increase the strength of interactions between compounds and polar stationary phases. Hence, the

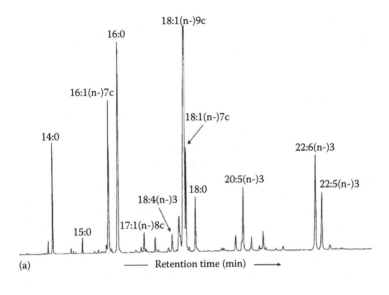

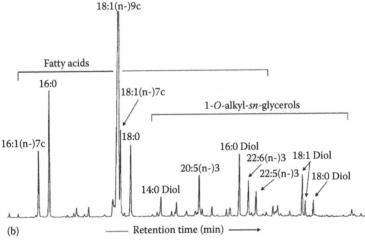

FIGURE 3.4 Partial capillary gas chromatograms of lipid fractions from blue whale (*Balaenoptera musculus*) lipid samples. (a) Total FA (analyzed as methyl esters) derived from the solvent fractionation of lipid classes from a subcutaneous blubber lipid extract by the hydrolytic reaction of *O*-acyl ester lipids. (b) The neutral lipid fraction, containing nonsaponifiable lipids (analyzed as bis(trimethylsilyl)trifluoroacetamide (OTMS) esters), derived from the solvent fractionation of lipid classes from a dorsal blubber lipid extract. In this case, FA have contaminated the neutral lipid fraction and have been detected as OTMS esters.

common rationale of all chromatography systems is to select an appropriate stationary and mobile phase to exploit the differences in polarity and ionization between lipid classes. The degree of lipid class separation can either be "simple," yielding polar and nonpolar fractions, or "complex," enabling the separation and isolation of acylglycerols, FA, sterols, glycolipids, and phospholipids.

Chromatography systems can be classified according to the physical arrangement of the phases that are employed. In each case, the stationary phase is held in place as a packing of small particles either in a column or as a layer on a flat surface. The mobile phase then moves through the column or over the surface layer.

3.3.3.2   Column Chromatography

Column chromatography falls into the category of liquid–solid chromatography, as the stationary phase consists of a solid matrix through which the liquid mobile phase passes, usually under the influence of gravity only. The process relies on the partitioning or adsorption of lipid classes onto the stationary phase. Lipid classes can then be washed (eluted) from the stationary phase with increasing polarity and strength of the mobile phase. This may involve a stepwise change in the solvents of the mobile phase (isocratic elution) or a gradually changing mixture of solvents (gradient elution). For this technique, the stationary phase is arranged as a tightly packed cylindrical column covered with the initial mobile phase (Figure 3.5). The lipid sample is placed on the surface of the solid phase, and as the continuous addition of mobile phase progresses, lipid components are separated by the differing rates at which they proceed through the column based on their interaction with both phases. Common stationary phases used for this technique include silica (silicic acid), alumina, and

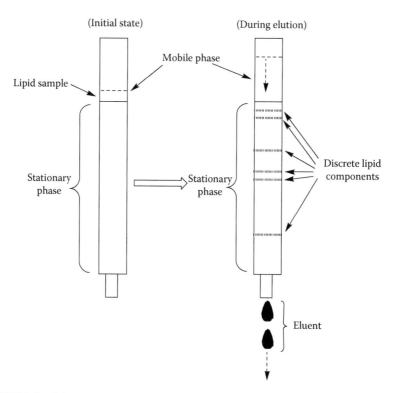

FIGURE 3.5   Schematic representation of the initial and intermediate stages of the column chromatography procedure. See Section 3.3.3.2.

ion-exchange resins. Well-established procedures using defined mobile phases are available for the separation of all lipid classes and relatively large amounts of material may be applied to such columns (Kates 1986).

The separated lipid classes can then be collected as discrete fractions in solvent. This is particularly useful either for the direct detection and quantitation of isolated lipid fractions or for further derivatization and analysis of lipid components. For example, acylglycerols, wax esters, and phospholipid classes may be separated by column chromatography and each fraction reacted as in Section 3.3 to analyze the FA present in each lipid class.

3.3.3.3 Thin-Layer Chromatography

TLC employs the same principles as those of column chromatography but yields separations more suited to analytical preparations and the identification of lipids through *in situ* reaction of separated lipid components. TLC can achieve a very high degree of separation efficiency and are often used for the separation of individual lipid components from within the same lipid class. For example, the separation and identification of individual phospholipid types from a polar lipid fraction, or the separation of FA based on molecular structure such as chain-length or the degree of unsaturation.

In TLC, the stationary phase usually consists of silica (or a derivative thereof) together with a binding agent that bonds the phase in a uniform thin layer to a glass or aluminum plate. After the application of the sample to one end of the plate, it is placed vertically in a reservoir of the mobile phase (Figure 3.6). The mobile phase remains in the same composition throughout the procedure and progresses across the stationary phase layer under capillary action; the plate being removed from the reservoir before the solvent front reaches the end of the plate (the plate is then said to be "developed"). Lipid components are thereby separated along the TLC plate in the direction of the mobile phase travel based on the same principles discussed above. However, unlike column chromatography, where separated components are eluted in solvent fractions, components remain deposited on the stationary phase.

This fact introduces two advantages and one disadvantage of TLC compared to column chromatography. The first advantage is the ability to redevelop the TLC plate in the same or different mobile phase, thereby improving the separation of components achieved by the initial development. This may be in the same dimension (referred to as dual or multi-development) or in a second dimension (referred to as two dimensional) to that of the first development using a second mobile phase (Figure 3.6). Two-dimensional TLC enables a high degree of separation to be achieved and is particularly useful for very complex mixtures of lipid types, or where the separation of very similar components within a lipid class is required.

The second advantage of bound components following separation is that this enables a further method of component identification based on the carefully measured position of spots on the plate following development. This is described in terms of the distance traveled by each component from the origin relative to the solvent front (Figure 3.6) and is called the R_f value. Characteristic R_f values have been established for lipid components under various conditions. To identify separated

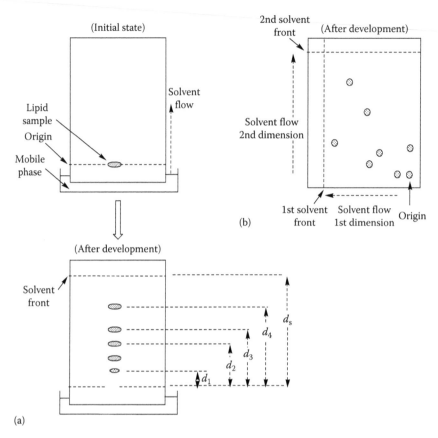

FIGURE 3.6   Schematic representation of the initial and end stages of the TLC procedure. (a) One-dimensional TLC. An example of the distances traveled for each separated component is shown. The R_f value for each component can therefore be calculated by (d_X/d_S), where X is 1, 2, 3 or 4; and (b) two-dimensional TLC. See Section 3.3.3.

components, lipids must be visualized using a stain. Lipid components can be visualized on TLC plates nondestructively through the use of staining compounds such as rhodamine-6G, iodine vapor, or 2′,7′-dichlorofluorescein. Many specific stains are also utilized to aid in the identification of lipid spots by initiating chemical reactions with specific functionalities (destructive stains). For example, there are stains specific for the phosphate group of phospholipids, for phospholipids containing an amino or choline group, and stains specific for glycolipids and sterols.

The disadvantage of TLC compared to column chromatography is one of component recovery for further manipulation. TLC involves the drying of lipids on the stationary phase, thereby risking the oxidation of some components. Many of the stains useful for identification are destructive and therefore decompose the sample by staining. Where nondestructive stains are employed, the spots identified can be scraped from the TLC plate and lipids re-extracted from the stationary phase, although this process is seldom quantitative.

When faced with the separation and identification of a complex lipid mixture, it is often a valuable strategy to employ chromatographic techniques in tandem. Column chromatography is adept at rapidly separating lipid classes. Specific lipid classes can then be analyzed by TLC for the separation and presumptive identification of individual components. Extensive TLC procedures are described by Stahl (1969).

3.3.3.4 Solid Phase Extraction

Solid phase extraction (SPE) can be viewed as an extension to column chromatography (see Section 3.3.3.2). While all the principles are the same, the inconvenience of preparing the columns has been largely eliminated by premade commercial products. The advantages are uniformity in sorbent chemistry, physical properties, size of cartridge and sorbent load, and the ability to extract small samples. In addition, they are generally usable "off the shelf." Overall, SPE gives intra- and inter-laboratory consistency and repeatability of results. The only disadvantage is that the cartridges do not lend themselves to large-scale separation of compounds, as the largest cartridges are usually 1 g of solid phase sorbent, or 10 g in special cases. Another advantage of SPE is that since the original development of silica and C18 products, there has been large expansion of sorbents available. Those used in the analysis of lipids have been silica, NH_2 (aminopropyl), CN (cyanopropyl), diol, C_{18}, C_8, C_2, phenyl, quaternary amine, propylbenzenesulfonate, propylsulfonate, and silver modified ion exchange. Much of this work covering the separation of lipid classes by SPE has been reviewed by Ruiz-Gutiérrez and Pérez-Camino (2000) and Aluyor et al. (2009). An example detailing the separation of unsaturated FAME is given in Figure 3.7.

In addition to a column configuration, SPE products are also produced as disks and 96 well plates for varying applications. Further options include polymeric phases, which replace silica. In addition, a solid phase micro-extraction (SPME) system may be used for certain GC applications. However, these SPE formats are not widely employed for lipid analysis, finding greater application in the removal of lipids from samples for the analysis of other components.

Poerschmann and Carlson (2006) developed a variation of SPE combined with ASE, which they termed "ASE in-cell fractionation." The technique relied upon placing an appropriate silica-based sorbent (solid phase) at the outlet end of the ASE extraction cell (Figure 3.8) and controlling the solvent composition and temperature of two sequential extractions (refer Section 3.2.4). The automated fractionation of neutral lipids from phospholipids was thereby possible. Silica gel and cyanopropyl-based silica sorbents were the most efficient at discriminating between the neutral lipids and phospholipids. The fractionation scheme eluted the neutral lipids using 9:1 *n*-hexane:acetone (v/v) while the phospholipids were subsequently eluted using 1:4 chloroform:methanol (v/v).

3.4 ANALYSIS OF LIPID CLASSES

3.4.1 Introduction

The analysis (separation, identification, and quantitation) of lipid classes from total lipid extracts is of prime importance to many food-industry applications. While the

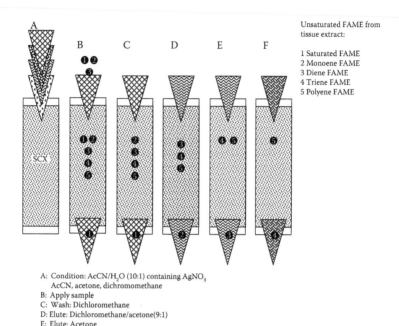

A: Condition: AcCN/H₂O (10:1) containing AgNO₃
AcCN, acetone, dichromomethane
B: Apply sample
C: Wash: Dichloromethane
D: Elute: Dichloromethane/acetone(9:1)
E: Elute: Acetone
F: Elute acetone/AcCN(97:3)
Higher unsaturated lipids can be eluted with different combinations of acetone/AcCN.

FIGURE 3.7 Schematic representation of a stepwise SPE procedure for the fractionation of FAME by their degree of unsaturation. (Adapted from Simpson, N and Van Horne, K.C., *Sorbent Extraction Technology Handbook*, Varian Sample Preparation Products, Harbor City, CA, 1993.)

chemical and chromatographic approaches outlined above are elegant, well established, and robust, they are relatively time consuming and rely on separate stages of separation, identification, and quantitation.

Direct analysis of lipid classes enables faster and more efficient usage of analytical resources and sample material. Many of the direct analysis techniques are derived from the basic principles of chromatography described above. However, they have evolved specific instrumentation to enable (in most cases) improved separation and, more importantly, one or more integrated methods of compound identification and quantitation. The marriage of these three aspects into a closely integrated process is the prime advantage of direct analysis techniques. The disadvantage is that to achieve this marriage, a rather expensive and complicated analytical instrumentation is required.

Below, three example techniques are described in general terms for the direct analysis of lipid classes. The first two are derived directly from the chromatographic techniques discussed above, while the third introduces the technique of MS for the identification of both simple and complex lipids.

3.4.2 Thin Layer Chromatography–Flame Ionization Detection

Thin layer chromatography–flame ionization detection (TLC-FID) is an analytical technique utilizing the separation characteristics of TLC combined with a rapid and

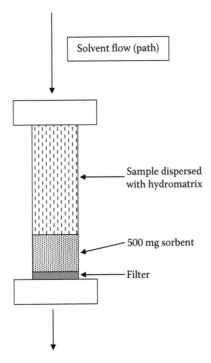

Solvent flow (path)

Sample dispersed
with hydromatrix

500 mg sorbent

Filter

FIGURE 3.8 Schematic representation of an ASE sample cell configured for in-cell fractionation of lipids by the incorporation of SPE sorbent at the bottom of the extraction cell.

sensitive method of quantitation, flame ionization detection. It is particularly useful for the rapid separation and quantitation of total lipid classes.

In this instrumentation, the stationary phase consists of a thin layer of silica or alumina bonded to a quartz rod (0.9 mm × 15 cm). Ten rods can be placed simultaneously in a metal frame and a sample applied to each rod. The rod assembly may then be developed in a solvent system as for a TLC plate. Once developed, the rod assembly is placed within an instrument, the iatroscan TH-10 (Iatron Laboratories, Tokyo, Japan). The basis of the iatroscan involves a mobile flame ionization detector (FID). Briefly, a FID consists of two platinum electrodes spaced across a flame of hydrogen burning in air. The FID is positioned in a moveable mounting so that it is able to pass along the length of each rod in sequence. Separated components along the rod are therefore combusted by the FID and a current signal generated. The signal is then displayed as a chromatogram, depicting the rod length and R_f of detected components from which identifications are based (Figure 3.9).

The advantages of the iatroscan system include the rapidity of the procedure to process multiple samples and a detector system that allows both identification of components (based on known R_f values) and a method for quantitation (see below). TLC-FID systems also offer a further refinement—the partial scanning of rods by the detector. With this option, certain lipid classes can be separated from a complex mixture by an initial rod development while leaving the remaining lipids at or near the origin. The FID can then be arranged to scan only the upper section of the rod

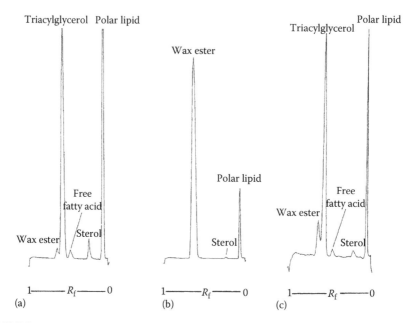

FIGURE 3.9   TLC-FID chromatograms of lipid extracts from the deep-sea fish orange roughy (*Hoplostethus atlanticus*). (a) Ovary, (b) muscle, and (c) liver. In this example, lipid classes were separated by a single development in a solvent system (mobile phase) of hexane:diethyl ether:acetic acid (60:17:0.2, v/v/v) with the rod (stationary phase) fully scanned. See Section 3.4.2.

where the initial lipid classes have been separated. Once scanned in this manner, the rods may be redeveloped in the same or different solvent system to further separate the remaining lipid classes together or in an individual series. Such a process in effect yields an "endless" stationary phase resulting in the possibility of resolving a large number of components.

The disadvantages of the iatroscan include the limitations of the FID detector. First, the ionization current generated by lipids of different classes is not the same. Hence, some lipid classes (e.g., sterols and phospholipids) generate a high ionization current per unit mass while others (e.g., hydrocarbons, TAG, wax esters) generate a much lower detector response per unit mass. This fact necessitates the calibration of the iatroscan system for each lipid class to be quantitated over the concentration range required. Further complications arise from analysis parameters such as hydrogen flow rate and scan speed also affecting ionization efficiency. Calibration curves should therefore be produced under the same instrumental conditions to be used for analyses. The material or standards used for lipid-class calibrations may also influence quantitation. For example, single species TAG standards yield higher detector responses than mixed species standards due to the production of sharper peaks. Ideally, the lipid standards used for calibration should be similar in composition to the samples analyzed.

3.4.3 High-Performance Liquid Chromatography

HPLC represents an advanced derivative of the column chromatography principles already described (see Section 3.3.3.2). Very small uniform particles of matrix are employed in a very tightly packed column. This guarantees an even flow of mobile phase through the column but requires high pressures to force the solvent through the matrix at a constant and reasonable rate. Coating the matrix particles is the stationary phase, which may be made from a variety of materials bonded to the matrix particles. Polar phases include quaternary ammonium, aminoalkyl, or cyanoalkyl groups. Unlike column chromatography, a range of nonpolar stationary phases have also been developed, the most common being octadecyl or octyl hydrocarbon groups. Use of these columns is referred to as "reverse-phase" HPLC.

The general components of an HPLC system are described in Figure 3.10. Rather than utilizing a premixed mobile phase, reservoirs of component solvents are provided and are mixed in the appropriate ratios under computer control. This allows the mobile phase composition to be altered during the elution sequence (gradient elution) and represents a significant advantage in improving chromatographic separations. Following the passage through the column, the eluent is passed to a detector unit. For many HPLC applications, the most common type is the UV detector, which monitors a fixed wavelength through the flow of eluent, commonly 200–210 nm. In such cases, the solvents used must be "UV transparent" and free of UV-absorbing impurities (e.g., "HPLC-grade" solvents). UV detectors are only of use with lipid components that absorb in this region. Some problems may arise with certain lipids. For example, FA do not absorb strongly in this region and must be derivatized with a UV-absorbing group to be quantitated. The requirement for such derivatization negates the major premise of "direct" analysis. Hence, in these situations, alternative techniques or an alternative detection system may be preferred. Other common types include refractive index detectors. Increasingly, mass spectrometric detectors are utilized in conjunction with HPLC systems due to the advantages offered from the definitive structural detail yielded by mass fragmentation spectra of lipid components (see Section 3.5.2).

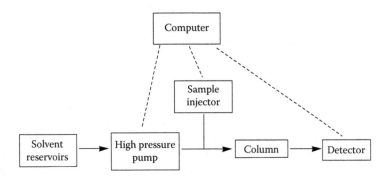

FIGURE 3.10 Schematic arrangement of a basic HPLC system. See Section 3.4.3.

The general application of reversed-phase HPLC can be employed directly in the separation and quantitation of total lipid classes. However, the high degree of resolution possible from HPLC columns is more often employed for the separation and analysis of individual lipid components from complex mixtures or individual lipid classes previously separated by other chromatographic procedures (e.g., column chromatography). Examples include HPLC analysis of acylglycerols (Kuksis 1994) and phospholipids (Porter and Weenen 1981).

3.4.4 MASS SPECTROMETRY

3.4.4.1 Introduction

MS as a technique yields a wealth of structural information for the identification of lipids. The degree of information obtained can vary from a simple molecular weight determination for a purified lipid to the one-step separation, molecular weight determination, and molecular structural determination for each component in a complex lipid mixture. Two factors influence this process, the physical design of the mass spectrometer and the type of ionization used to introduce the lipid sample into the mass spectrometer.

The advances in the production of compact, robust, and cheaper mass spectrometers have also seen their blossoming application in routine analysis procedures, for the stand-alone analysis of lipids and also in combination as "detectors" for other chromatographic systems such as GC-MS and LC-MS, *viz.*, HPLC-MS. GC-MS has particular advantages for the analysis of lipids with relatively high volatility (i.e., relatively low molecular weight and polarity) such as hydrocarbons, isoprenoids, and FA (Couderc 1995). The separation applications of HPLC for lipid components and lipid classes have been discussed above and may be linked on-line to the mass spectrometer through interfaces such as thermospray or electrospray ionization, usually used to remove the "large" volumes of solvent from entering the mass spectrometer.

While it is beyond the scope of this chapter to address the physical design and fundamentals of mass spectrometers, it is salient to briefly summarize the relevant methods of sample ionization and their differing application to lipid analysis and partner chromatographic systems. The reader is referred to Murphy (1993) for background regarding the various designs and operation of mass spectrometers.

The analysis of lipids by MS firstly requires the alteration of the component from its natural state to that of an ion (positive or negative) in the gas phase. A wide variety of techniques are available for the generation of gaseous ions dependent on the volatility of the lipid type. Lipid classes that are relatively volatile (i.e., those amenable to analysis by GC, hydrocarbons, FA, isoprenoids) are easily transformed to the gas phase by heating and can then utilize ionization techniques, which require a preprepared gaseous form, such as electron impact (EI) or chemical ionization. Other lipids that are relatively nonvolatile (e.g., phospholipids) require an ionization technique that incorporates a desorption/desolvation aspect, such as fast atom bombardment or thermospray/electrospray ionization. Each ionization technique yields its own degree of sensitivity, structure, and molecular weight information. The exact information or degree of identification required, along with the lipid properties, may also influence the type of ionization technique employed.

3.4.4.2 Electron Impact Ionization

EI (or bombardment) ionization has been the most widely applied method to date. The technique relies on the analyte being introduced into the ion source of the mass spectrometer (under vacuum) in a preprepared gaseous form. Within the ion source, the neutral analyte molecules come into association with a beam of energetically excited electrons. Transfer of energy from the electrons to the analyte molecule ultimately results in the loss of an electron from the parent molecule (forming a molecular ion, M+), which then decays by fragmentation into a series of further charged ions (Figure 3.11). The resulting fragmentation ions can be separated in a magnetic and/or electronic field based on their mass/charge (m/z) ratio and the resultant mass spectrum generated. Major information gained from this technique is usually the molecular weight of the analyte (represented by the parent molecular ion) and some structural detail related to the formation of specific fragment ions. Established databases for EI mass spectra of many lipids are readily available (e.g., McLafferty et al. 1991). Two examples of electron mass spectra of FA are given in Figure 3.11. Specific fragmentations give rise to characteristic ions of FA, which in most cases can be interpreted to give unequivocal identifications. However, a particular shortcoming of EI ionization in lipid analysis is also highlighted in Figure 3.11. The large amount of energy transferred to highly unsaturated FA from EI ionization results in a high degree of fragmentation and complex rearrangements. While a mass spectrum characteristic of polyenoic FA is produced, it is dominated by low molecular weight ions (Figure 3.11; ions m/z 67, 79, 91) and lacking in high molecular weight ions including the molecular ion. Hence, in isolation, an EI mass spectrum of a polyenoic FA is insufficient for the complete identification of the compound.

3.4.4.3 Chemical Ionization

Chemical ionization relies on the transfer of a charge between an introduced ion and analyte molecule in the gas phase. Gases such as methane or ammonia may be charged in a modified EI ion source and then used as an ion source for the sample. The resultant ionization of the analyte is "soft," i.e., it transfers little excess energy, which markedly decreases the degree of parent ion fragmentation, leaving a high proportion of analyte molecular ions. Hence, where EI ionization often results in many fragment ions with the molecular ion sometimes difficult to detect, chemical ionization usually yields a high proportion of the molecular ion allowing very accurate mass determination of the analyte but less structural information. Further details of chemical ionization procedures are given by Harrison (1992).

An important advancement in ionization techniques has been the development of atmospheric pressure chemical ionization (APCI). Under this design, the ion source is kept at atmospheric pressure rather than under high vacuum. Following ionization, the sample enters the high-vacuum region of the mass spectrometer through a small

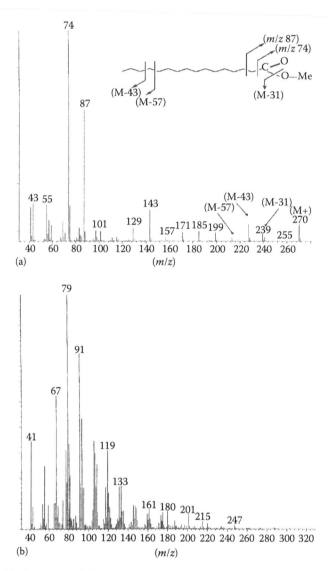

FIGURE 3.11 Examples of EI mass spectra for two FA (analyzed as methyl esters). (a) The mass spectrum of methyl hexadecanoate (the derivative of hexadecanoic acid, 16:0, molecular weight 270) demonstrating characteristic fragmentations and resultant ions. (b) The mass spectrum of methyl eicosapentaenoate (the derivative of eicosapentaenoic acid, 20:5(n-)3, molecular weight 316) demonstrating the characteristic ions of polyenoic FA but the lack of higher molecular weight ions.

orifice. In mass spectrometers associated with HPLC systems, APCI introduces an important benefit in removing the necessity to introduce large volumes of eluent into the high-vacuum area of a mass spectrometer. In APCI, the majority of the solvent is contained within the ion source region where it is utilized as the primary ion source (Henion and Lee 1990).

3.4.4.4 Fast Atom Bombardment

A significant proportion of complex lipids contain polar structural features, which engage in intermolecular associations resulting in a low volatility and difficulties in manipulating them into the gas phase necessary for EI or chemical ionization. Some polar groups can be chemically derivatized to increase the compound volatility; however, this increases the molecular weight and can alter the chemical structure of the analyte. Fast atom bombardment (FAB) represents a specific ionization technique for polar lipids or complex lipids of low volatility.

The FAB technique relies on the generation of a high velocity beam of heavy, neutral gas atoms (e.g., xenon or argon), which are directed against the lipid sample mixed in a "matrix" material (usually glycerol). The matrix absorbs most of the collisional energy but sufficient energy is transferred to the sample components for soft ionization events to occur. The ions of sample components formed (termed secondary ions) are emitted from the surface of the matrix (desorbed) and can then be analyzed as molecular ions. Where a beam of gaseous ions is used (usually caesium ions) rather than neutral atoms for the desorption of sample secondary ions, the technique is referred to as liquid secondary ion mass spectrometry (LSIMS). However, in practice, there is little difference between the outcomes of either technique. FAB MS has gained wide application to the analysis of phospholipids and other polar acylglycerols from isolated fractions or directly from complex lipid extracts (Kerwin 1999). A simplified example is shown in Figure 3.12 where the phospholipid molecular species from the total solvent extract of a bacterial culture have been analyzed by negative-ion FAB-MS. Based on the determination of phospholipid molecular

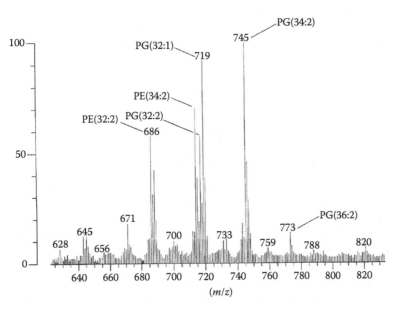

FIGURE 3.12 Example of a partial negative-ion FAB mass spectrum of the total solvent extract from a marine bacterium showing phospholipid molecular species. Acronyms: PE, phosphatidylethanolamine; PG, phosphatidylglycerol.

ion species, the phospholipid class and combined acyl chain composition of each component can be assigned. Such analyses, combined with the knowledge of the total FA composition of the sample, can allow specific component speciation to be inferred. For example, with the knowledge that the lipid sample analyzed in Figure 3.12 contains a mixture of saturated and monounsaturated FA from C14 to C18, the phosphatidylethanolamine component PE(32:2) could represent three possible molecular species: PE(16:1/16:1), PE(15:1/17:1), or PE (14:1/18:1). A further refinement to resolve these possibilities would be to utilize FAB-tandem MS to identify the acyl residue daughter ions released from the parent molecular species by secondary ionization.

3.4.4.5 Thermospray/Electrospray Ionization

Thermospray and, more recently, electrospray ionization have found wide application as an interface technology between HPLC and MS instruments. They represent powerful techniques for the analysis of complex lipids directly from solutions (Henion and Lee 1990, Murphy 1993). In most instances, the total HPLC eluent can be sent directly into the heated thermospray ion source. Here, the combination of heat and eluent velocity creates a plume of small diameter particles suspended in a vapor (nebulization). A strong electric charge forms on the surface of the liquid particles, and as the droplets evaporate, the increase in charge ionizes analyte molecules, which are discharged directly from the droplet into the gas phase. From here, they may enter the mass spectrometer directly.

Electrospray ionization is similar in effect to the thermospray technique and is useful for similar applications. The difference resides in the use of a high electric field to nebulize the sample solution (or sample and eluent), creating droplets with excess electric charge. As the droplet solvent evaporates during traverse of a desolvation chamber, charge transfers to the analyte molecules and these are released as gaseous ions. A further refinement in this technique is the use of electronic lenses to direct ions more efficiently into the mass spectrometer. As the analyte is not subject to heating, the possibility for the thermal decomposition of complex lipid components is less.

3.5 ANALYSIS OF LIPID COMPONENTS

3.5.1 Gas Chromatography–Mass Spectrometry

The development of GC arose from the substitution of a liquid mobile phase with an inert gas (e.g., helium or hydrogen). The introduction of a sample mixture in a volatile form through a heated zone then allowed gaseous compounds to interact chemically with the stationary phase of the GC column on a similar basis to that described for column chromatography (see Section 3.3.3.2). In addition, GC instruments contain the column unit within a variably heated zone, which may be temperature programmed to improve the efficiency of compound separation. The development of narrow bore capillary GC columns and the wide diversity of stationary phase types have led to many diverse applications of GC to lipid analysis. The diversity of these applications has been further enhanced by the differing types of detectors to which

GCs may be coupled. Common detectors include FID, electron capture detectors (ECDs), and, increasingly, mass spectrometers.

As discussed in Section 3.4.4, GC-MS has particular advantages for the analysis of lipids with relatively high volatility (i.e., relatively low molecular weight and polarity), such as hydrocarbons, isoprenoids, sterols, and FA, whether they are individual lipid components or derived from more complex lipids. However, even the presence of potentially polar functional groups within these lipids requires derivatization to less polar moieties prior to GC-MS analysis.

3.5.1.1 Sterols

The hydroxyl functional group of sterol compounds requires derivatization to a less polar state prior to GC-MS analysis. The formation of sterol-trimethylsilyl ethers (TMS ethers) by reaction of sterols with bis-(trimethylsilyl)trifluoroacetamide (BSTFA) is the most commonly used method. Sterol-TMS ethers produce characteristic MS fragmentations under EI ionization allowing the ready identification of components separated on a nonpolar GC capillary column (Jones et al. 1994). If GC-MS analysis is not available for sterol identification, the positive identification of sterol components is significantly more complicated requiring the analysis of relative retention times of components from multiple GC columns and the co-injection of known standards.

Phytosterols are increasingly important agricultural products for the health and nutrition industries. Likewise, cholesterol is an important constituent of animal food products where the potential formation of oxidation products is a cause of concern (Guardiola et al. 2004). Comprehensive GC-MS techniques are the method of choice for such analyses (e.g., Pizzoferrato et al. 1993) with continual improvements in methodologies also offering the application of MS/MS techniques (Ai 1997).

3.5.1.2 Fatty Acids

Similar to sterols, free FA or FA liberated from complex lipids must be derivatized to reduce the polarity of their carboxyl group prior to GC analysis. A variety of derivatives have been utilized for this purpose and each has specific advantages and disadvantages depending on the derivatization procedure or degree of interpretive information required. Section 3.4.4.2 has referred to the MS EI ionization analysis of FA as methyl esters and this remains probably the most popular derivatization technique for routine analysis. However, FA methyl esters are not always the most useful derivatives for the elucidation of structural information, particularly in the case of unsaturated FA.

Saturated FA methyl esters are readily identified by a prominent molecular ion (M^+) and other significant fragmentation ions of $(M-31)^+$, $(M-43)^+$ with intensity maxima at $(m/z = 74)$ and $(m/z = 87)$ (see Figure 3.13). In contrast, the EI mass spectra of unsaturated FA vary considerably from their saturated analogues and also change with the degree of unsaturation. While M^+ is still distinct, the significant fragmentation ions alter to $(M-32)^+$, $(M-74)^+$, and $(M-116)^+$. The intensity maxima (base peak) also changes to $(m/z = 55)$ (see Figure 3.14). As the degree of unsaturation of FA methyl esters increases, the higher molecular weight fragmentation ions progressively disappear and the base peaks progressively shift to higher masses

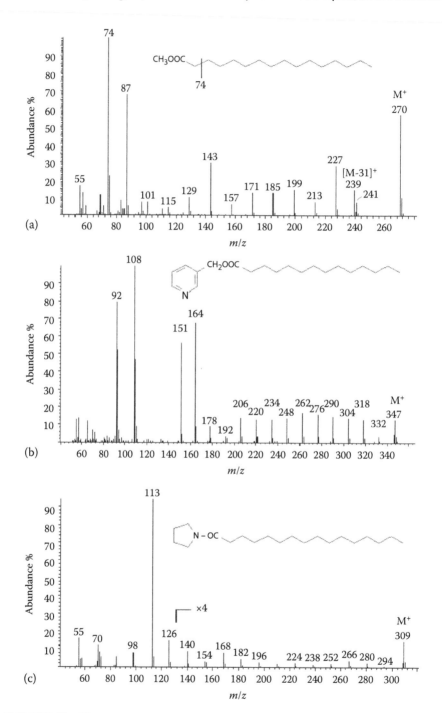

FIGURE 3.13   The mass spectra of hexadecanoic acid as various ester derivatives: (a) methyl ester, (b) piconyl ester, (c) pyrrolidine ester, and

(continued)

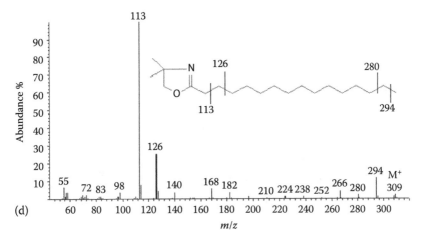

(d)

m/z

FIGURE 3.13 (continued)   (d) dimethyloxazoline ester. Spectra adapted from Christie (1989). (From Christie, W.W., *Gas Chromatography and Lipids*, The Oily Press Ltd., Bridgwater, U.K., 1989.)

of ($m/z=67$) for dienes and ($m/z=79$, 97, 109) for polyenes. The mass spectra of polyenoic FAME are therefore limited in determining the exact number of double bonds present, unless the weak M^+ may be discerned. Further, there are no ions to indicate the location or stereochemistry of double bonds, even in monounsaturated FAME, due to double bond migration occurring along the carbon chain during the formation of the molecular ion (Christie 1989).

The three remaining commonly employed FA esters for GC-MS analysis (pico-nyl esters, pyrrolidine esters, and dimethyloxazoline (DMOX) esters) all utilize a nitrogen-containing functionality. This imparts a particular advantage for the struc-tural determination of double-bond location, as the charge of the M^+ remains mostly located on the derivatization group rather than the double bond(s) of the acyl chain. Subsequently, double bonds do not tend to migrate along the acyl chain and their position is revealed by characteristic fragmentation ions.

Saturated FA piconyl esters display strong base peak ions at ($m/z=92$, 108, 151, 164) derived from the derivatization group. The M^+ is also easily distinguished, as it is the only odd-numbered fragmentation ion in the spectrum (see Figure 3.13). Within the remaining spectrum, fragmentation ions progress by loss of the methyl group $(M-15)^+$ and subsequently a series of ions separated by 14 amu representing succes-sive loss of the methylene groups. Unsaturated FA piconyl esters possess an enhanced M^+ (once again odd-numbered) and a disruption of the 14 amu ions pattern seen in saturated FA piconyl esters. The comparison of Figures 3.13 and 3.14 demonstrates an enhanced fragmentation ion pair 14 amu apart on the methyl side of the double bond location ($m/z=288$, 274; Figure 3.14). The position of the double bond itself is revealed by a change in the fragmentation ion separation from 14 to 26 amu ($m/z=234$ to 260; Figure 3.14) representing cleavage at either side of the double bond position.

Saturated FA pyrrolidine esters generate base peaks of ($m/z=113$, 126). Due to the presence of a nitrogen atom within the derivatization functional group, the M^+ is

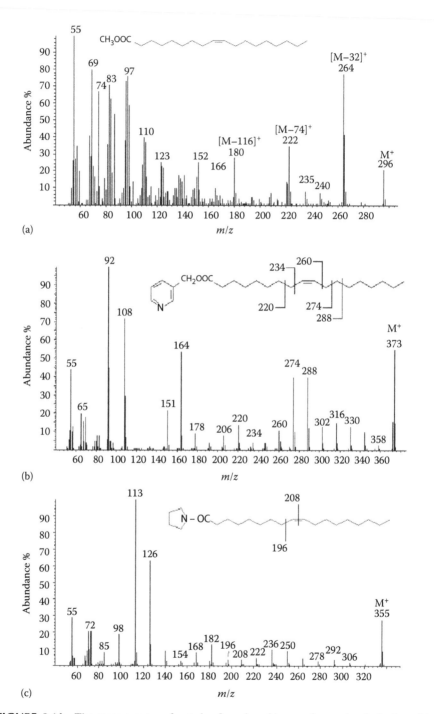

FIGURE 3.14 The mass spectra of octadec-9-enoic acid as various ester derivatives (a) methyl ester, (b) piconyl ester, (c) pyrrolidine ester, and

(*continued*)

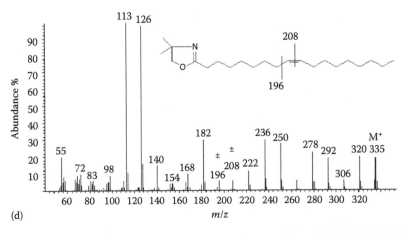

(d)

FIGURE 3.14 (continued) (d) DMOX ester. Spectra adapted from Christie (1989). (From Christie, W.W., *Gas Chromatography and Lipids*, The Oily Press Ltd., Bridgwater, U.K., 1989.)

also odd-numbered and therefore quite distinctive (Figure 3.13). Similar to saturated FA piconyl esters, saturated pyrrolidine esters display a series of fragmentation ions separated by 14 amu representing successive loss of the methylene groups, but at a lower intensity. The double bond position in unsaturated FA pyrrolidine esters is revealed by a change in the separation of the fragmentation series from 14 to 12 amu (Figure 3.14) representing cleavage across the double bond itself.

Closely related in function (if not chemistry) to pyrrolidine, FA esters are DMOX derivatives. Through coincidence, both saturated and monounsaturated FA DMOX esters give very similar mass spectra to that of pyrrolidine esters, and may be interpreted in the same manner (see Figure 3.14). DMOX esters exhibit improved chromatographic properties when compared to other nitrogen-containing derivatives and are in particular more easily interpreted in the case of polyunsaturated FA. However, the derivatives are less stable and readily degrade in contact with trace levels of moisture.

3.5.2 Liquid Chromatography–Mass Spectrometry

The coupling of MS to LC has been a relatively recent achievement when compared to GC-MS. Although LC-MS has been around since the 1970s, it was not until the commercialization of ESI and APCI interfaces in the 1990s that the technique has become mainstream (see Sections 3.4.4.2 and 3.4.4.4). LC is ideally suited to the analysis of "complex" lipids, such as TAG and phospholipid compounds, which are not amenable to GC analysis and the coupling of MS detection offers direct structural information and very high sensitivity. It is advantageous to use MS detection over UV as most lipids have little or no absorption at UV wavelengths, especially those containing saturated FA chains. While derivatization with benzoates, dinitrobenzoates, pentafluorobenzoates, and nicotinic acid can improve UV absorbance, these are labor-intensive methods, which use potentially dangerous chemicals (Peterson and Cummings 2006).

The two most common interfaces, ESI and APCI, can be viewed as complimentary in their capabilities. Nonpolar lipids are ideally suited to analysis by LC-APCI-MS, while ESI is suited toward the more polar lipids, such as phospholipids. A review of the literature shows that ESI is more commonly used for lipid analysis than APCI, and also tends to have greater sensitivity. Although both ESI and APCI are regarded as soft ionization techniques, APCI can produce "in-source" fragmentation that is useful for structural elucidation when using single MS; however, information about the molecular weight can be lost due to this fragmentation. In contrast, ESI produces very little fragmentation, with pseudo-molecular ions ([M+H]$^+$) or adduct ions (e.g., [M+Na]$^+$, [M+NH$_4$]$^+$) predominating.

3.5.2.1 Free Fatty Acids

Underivatized free FA can be separated using conventional reversed-phase HPLC on a C18 column with detection using negative ESI-MS (Chu et al. 2009). Deprotonated molecular ions [M–H]$^-$ are the predominant ions observed. The retention times of free FA are proportional to the chain length and inversely proportional to the number of double bonds. The position of double bonds in unsaturated FA can be deduced from ozonolysis, the products of which are an aldehyde and a carboxylic acid created from the position of the double bond (Thomas et al. 2007).

3.5.2.2 Triacylglycerols

The relatively nonpolar TAG are ideally suited to LC-APCI-MS analysis. The separation of different TAG can be achieved using reserved phase columns, usually with nonaqueous mobile phases (Holcapek et al. 2005). Solvent mixtures commonly used include isopropanol:acetonitrile:hexane, or acetonitrile paired with acetone, chloroform, or dichloromethane. As with other complex lipids, the retention times of TAG increase with increasing chain length and decrease with the number of double bonds. The separation of *cis*- and *trans*-isomers is also possible with appropriate choice of mobile phases. For example, the *cis*- and *trans*-isomers of the TAG containing three C18:1 chains have been separated using a C8 column and a mobile phase gradient consisting of water:isopropanol:*n*-butanol containing ammonium formate (McIntyre 2008). Ammonium adducts of TAG tend to be more stable than the protonated molecular ion, hence, the inclusion of ammonium formate in the previously mentioned study (McIntyre 2008). The degree of unsaturation can be easily calculated from the [M+H]$^+$ or [M+NH$_4$]$^+$ ions, with a reduction of 2 amu for each double bond present.

3.5.2.3 Phospholipids

Phospholipids can be separated using normal phase chromatography and elute in clusters, based on their polar head group (Peterson and Cummings 2006). The groups elute from a silica column in order of increasing polarity of the head group (i.e., glycerol, ethanolamine, inositol, serine, choline). The elution order can be changed somewhat using a diol-modified silica column (Karlsson et al. 1996), and various tandem MS techniques can be used to identify molecular species within each class (see Table 3.4). Reversed-phase chromatography can be used to separate individual molecular species, although it is recommended that this be undertaken following normal-phase separation of each class. As with other classes of lipids, ESI tends to

TABLE 3.4
Characteristic Mass Fragmentations Produced by Collision-Induced Dissociation (CID) of Phospholipids Using Different MS/MS Techniques

Class of Phospholipid	Common Mass Fragments from CID MS/MS	(+)ESI Neutral Loss Scan	(+)ESI Precursor Ion Scan	(−)ESI Precursor Ion Scan
Phosphatidylethanolamine	651, [M+H-144]$^+$	141		196
Phosphatidylglycerol	Loss of acyl chains	172		153
Phosphatidylserine	(M−87) loss of serine	185, 87		
Phosphatidylinositol	223, 241, 259, 279, 315	260		241
Phosphatidylcholine	(M−15), (M−60), (M−86) from fragmentation of choline group		184 (also for sphingomyelin)	
Sphingomyelin			184	168

Source: Pulfer, M. and Murphy, R.C. 2003. *Mass Spectrometry Reviews*, 22, 332, 2003; Peterson, B.L. and Cummings, B.S. 2006. *Biomedical Chromatography*, 20, 227, 2006.

be more sensitive toward phospholipids than APCI, although large proportions of sodium adducts tend to be formed.

3.5.2.4 Carotenoids

Carotenoids are large, nonvolatile compounds, which are ideally suited to analysis by HPLC. While they possess excellent absorbance in the 400–500 nm range, mass spectrometric methods can act as confirmatory methods or help elucidate the structure of unknown compounds. Separation can be achieved using a reversed-phase column (C18 or C30), and APCI operated in positive mode produces abundant [M+H]$^+$ ions (Clarke et al. 1996, Brydwell 2001). Carotenoids containing oxygen functionality also produce significant [M−H$_2$O+H]$^+$ fragment ions, which can be the base peak for some compounds. For example, β-carotene produces intense [M+H]$^+$ ions at *m/z* 537 with no fragmentation, while lutein has both [M+H]$^+$ and [M−H$_2$O+H]$^+$ ions at *m/z* 569 and 551, respectively.

3.5.2.5 Tandem Mass Spectrometry

Tandem MS (MS/MS) is very useful for identifying and characterizing novel species and for identifying molecular species in complex mixtures. Zehethofer and Pinto (2008) and Griffiths (2003) provide excellent overviews of MS/MS applied to lipid analysis, and the latter includes many details that are beyond the scope of this book. While it is possible to tentatively identify species from single MS alone, a definitive identification of an unknown requires MS/MS techniques. For example, a phospholipid that shows a mass corresponding to C28 and containing one double bond may be 14:0–14:1 or 12:0–16:1, as they both have the same molecular weight. MS/MS techniques would allow these two species to be differentiated.

Scanning the product ions produced from CID of a parent ion can allow unique identification from fragmentation pattern in many cases, and the position of double bonds can also be deduced in some cases. However, care must be taken to ensure artifacts from CID do not interfere with the identification, e.g., migration of double bonds. Multiple-reaction-monitoring (MRM) is a more sensitive MS/MS technique, which involves monitoring only selected fragment ions produced from CID rather than scanning a range of masses. The main benefit from this mode is a large increase in sensitivity; however, the method is highly specific, and knowledge of components and their precursor-product ion fragmentation must be known *a priori*. Novel or nontarget components will be missed when operating in product-ion scan or MRM mode, as these techniques are intended to target specific classes of compounds, or individual species, respectively.

Neutral-loss and precursor-ion scan modes are ideally suited to profiling specific classes of compounds, especially in complex mixtures. These modes are useful when specific groups of compounds either lose a common mass or produce a common product ion after CID, respectively. For example, a precursor ion scan at m/z 184 will detect the choline and sphingomyelin phospholipids. Table 3.4 shows examples of the common mass fragments used in MS/MS techniques to identify specific lipid classes (compiled from Pulfer and Murphy 2003, Peterson and Cummings 2006).

3.5.2.6 Quantitation

The quantitation of lipids in LC-MS in very problematic, and while many methods exist, there is no consensus as to which method is best (Brydwell 2001). The main problem arises from the fact that relatively few lipid species can be obtained as authentic standards of known concentration. In contrast, there are hundreds, even thousands of different molecular species that can be separated and identified using LC-MS. As the response of both ESI and APCI is linear over a wide dynamic range, the relative concentrations of different molecular species within a class of lipids can be determined simply from relative peak areas. This can be useful when comparing the relative concentrations of various analytes from different samples, although absolute quantitation leads to large errors. Relative response factors for a variety of TAG have been tabulated for APCI, ELSD, and UV, allowing at least some degree of quantitation relative to known standards (Holcapek et al. 2005). The sensitivity of compounds differs due to the chain length and degree of unsaturation, and other components can also have major effects. For example, the composition of the polar head group of phospholipids has a substantial influence on the detector signal (Pulfer and Murphy 2003). Ionization suppression due to matrix effects can be considerable with ESI, although APCI seems to be less affected.

3.6 FINAL REMARKS

Figure 3.15 illustrates the general approach to analytical lipid analysis described in this chapter, highlighting the major stages and the types of procedures that may be employed. Emphasis has been given to describe the principles of several widely applicable techniques focused on diverse outcomes. The objective has been to better inform the reader to choose an appropriate analysis strategy for their needs and resources.

The approach to the extraction of lipids from samples remains a fundamental area for the majority of analysis techniques. Extraction by organic solvents remains

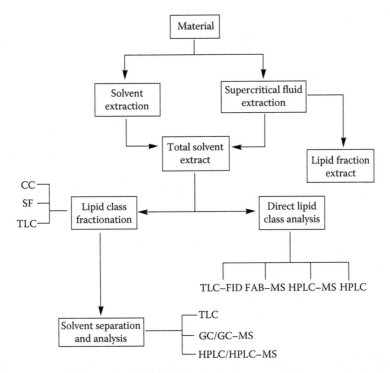

FIGURE 3.15 Flow chart of the general approach to analytical lipid analysis highlighting the major stages and types of procedures that may be employed.

the mainstay of routine procedures although new technologies such as SFE and ASE have appealing applications in industrial areas where large volumes of solvent usage is problematic or there are advantageous interactions between the extraction technique and certain lipids.

Certain lipid classes (e.g., FA, hydrocarbons) can be fractionated from lipid extracts by their differing solubilities in solvent systems, particularly following chemical reactions to release ester-linked FA from complex lipids. However, for the separation of complex lipid classes and/or individual components within lipid classes, chromatographic techniques afford robust and reliable methods. Column chromatography is useful for the collection of lipid fractions retained in solvent for further analysis while TLC yields separations more suited to the analytical identification of lipids.

Great advances have been achieved in the direct analysis of lipid extracts, i.e., those requiring minimal preparative fractionation, by the use of dedicated analytical instrumentation such as TLC-FID and HPLC. MS enables the combination of aspects of lipid separation by molecular weight and structural detail from techniques of ionization-induced fragmentation. Combined with chromatographic procedures, this represents the most powerful but expensive tool available for lipid analysis.

REFERENCES

Ai, J. 1997. Rapid measurement of free phytosterols in tobacco by short-column GC/MS/MS. *Journal of Agricultural and Food Chemistry* 45: 3932–3935.

Aluyor, E.O., Ozigagu, C.E., Oboh, O.I., and Aluyor, P. 2009. Chromatographic analysis of vegetable oils: A review. *Science Research Essays* 4: 191–197.

Brydwell, W.C. 2001. Atmospheric pressure chemical ionization mass spectrometry for analysis of lipids. *Lipids* 36: 327–346.

Christie, W.W. 1989. *Gas Chromatography and Lipids*. Bridgwater, U.K.: The Oily Press Ltd.

Chu, X., Zhao, T., Zhang, Y., Zhao, A., Zhou, M., Zheng, Z., Dan, M., and Jia, W. 2009. Determination of 13 free fatty acids in *Pheretima* using ultra-performance LC-ESI-MS. *Chromatographia* 69: 645–652.

Clarke, P.A., Barnes, K.A., Startin, J.R., Ibe, F.I., and Shepherd, M.J. 1996. High performance liquid chromatography/atmospheric pressure chemical ionization-mass spectrometry for the determination of carotenoids. *Rapid Communications in Mass Spectrometry* 10: 1781–1785.

Couderc, F. 1995. Gas chromatography tandem mass-spectrometry as an analytical tool for the identification of fatty acids. *Lipids* 30: 691–699.

Dionex Corporation. 2009. Application Notes. http://www.dionex.com

Griffiths, W.J. 2003. Tandem mass spectrometry in the study of fatty acids, bile acids, and steroids. *Mass Spectrometry Reviews* 22: 81–152.

Guardiola, F., Bou, R., Boatella, J., and Codony, R. 2004. Analysis of sterol oxidation products in foods. *Journal of AOAC International* 87: 441–466.

Harrison, A.G. 1992. *Chemical Ionisation Mass Spectrometry*, 2nd edn., Boca Raton, FL: CRC Press.

Henion, J. and Lee, E. 1990. Atmospheric pressure ionisation LC/MS for the analysis of biological samples. In *Mass Spectrometry of Biological Materials*, C.N. McEwan and B.S. Larsen (eds.), Chapter 15. New York: Marcel Decker.

Holcapek, M., Lisa, M., Jandera, P., and Kabatova, N. 2005. Quantitation of triacylglycerols in plant oils using HPLC with APCI-MS, evaporative light-scattering, and UV detection. *Journal of Separation Science* 28: 1315–1333.

Isaac, G., Waldback, M., Eriksson, U., Odham, G., and Markides, K.E. 2005. Total lipid extraction of homogenized and intact lean fish muscles using pressurized fluid extraction and batch extraction techniques. *Journal of Agricultural and Food Chemistry* 53: 5506–5512.

Jones, G.J., Nichols, P.D., and Shaw, P.M. 1994. Analysis of sterols and hopenoids. In *Chemical Methods in Prokaryotic Systematics*, M. Goodfellow, and A.G. O'Donnell (eds.), pp. 163–195. Chichester, U.K.: John Wiley & Sons.

Karlsson, A.A., Michelsen, P., Larsen, A., and Odham, G. 1996. Normal-phase liquid chromatography class separation and species determination of phospholipids utilizing electrospray mass spectrometry/tandem mass spectrometry. *Rapid Communications in Mass Spectrometry* 10: 775–780.

Kates, M. 1986. Techniques of lipidology: Isolation, analysis and identification of lipids. In *Laboratory Techniques in Biochemistry and Molecular Biology*, Vol. 3, R.H. Burdon and P.H. van Knippenberg (eds.). Amsterdam, the Netherlands: Elsevier.

Kerwin, J.L. 1999. Mass spectrometric characterisation of lipids, *Recent Research Developments in Lipids* 3: 205–217.

Kuksis, A. 1994. GLC and HPLC of neutral glycerolipids. In *Lipid Chromatographic Analysis*, T. Shibamoto (ed.), Chapter 5. New York: Marcel Decker Inc.

McIntyre, D. 2008. The analysis of triglycerides in edible oils by APCI LC/MS. Agilent Application. http://www.chem.agilent.com/Library/applications/5989–8441EN.pdf (accessed August 5, 2009).

McLafferty, F.W., Stauffer, D.B., Twiss-Brooks, A.B., and Loh, S.Y. 1991. An enlarged data base of electron-ionisation mass spectra. *Journal of the American Society for Mass Spectrometry* 2: 432–437.

Murphy, R.C. 1993. Mass spectrometry of lipids. In *Handbook of Lipid Research*, Vol. 7, F. Snyder (ed.). New York: Plenum Press.

Peterson, B.L. and Cummings, B.S. 2006. A review of chromatographic methods for the assessment of phospholipids in biological samples. *Biomedical Chromatography* 20: 227–243.

Pizzoferrato, L., Nicoli, S., and Lintas, C. 1993. GC-MS characterisation and quantification of sterols and cholesterol oxidation products. *Chromatographia* 35: 269–274.

Poerschmann, J. and Carlson, R. 2006. New fractionation scheme for lipid classes based on "in-cell fractionation" using sequential pressurized liquid extraction. *Journal of Chromatography A* 1127: 18–25.

Porter, N.A. and Weenen, H. 1981. High-performance liquid chromatography separations of phospholipids and phospholipid oxidation products. *Methods in Enzymology* 72: 34–40.

Pulfer, M. and Murphy, R.C. 2003. Electrospray mass spectrometry of phospholipids. *Mass Spectrometry Reviews* 22: 332–364.

Richter, B.E., Jones, B.A., Ezzell, J.L., Porter, N.L., Avdalovic, N., and Pohl, C. 1996. Accelerated solvent extraction: A technique for sample preparation. *Analytical Chemistry* 68: 1033–1039.

Ruiz-Gutiérrez, V. and Pérez-Camino, M.C. 2000. Update on solid-phase extraction for the analysis of lipid classes and related products. *Journal of Chromatography A* 885: 321–341.

Simpson, N. and Van Horne, K.C. 1993. *Sorbent Extraction Technology Handbook*. Harbor City, CA: Varian Sample Preparation Products.

Stahl, E. 1969. *Thin-Layer Chromatography. A Laboratory Handbook*. London, U.K.: Allen and Unwin Ltd.

Thomas, M.C., Mitchell, T.W., Harman, D.G., Deeley, J.M., Murphy, R.C., and Blanksby, S.J. 2007. Elucidation of double bond position in unsaturated lipids by ozone electrospray ionization mass spectrometry. *Analytical Chemistry* 79: 5013–5022.

Zehethofer, N. and Pinto, D.M. 2008. Recent developments in tandem mass spectrometry for lipodomic analysis. *Analytica Chimica Acta* 627: 62–70.

4 Lipids and Food Quality

Anna Kołakowska and Zdzisław E. Sikorski

CONTENTS

4.1 INTRODUCTION

Lipids contribute very significantly to the quality of almost all kinds of foods. They affect the structure of various raw materials and have a large impact on the nutritional value, safety, and the sensory properties of foods. They are the richest kind of energy source in the human diet, supply the organism with numerous substrates for the synthesis of biologically important metabolites and structures, and serve as a carrier of fat-soluble vitamins and other lipophilic substances needed in nutrition. However, they can also be a vehicle of harmful contaminants. They add their own attributes to different sensory properties of foods, and affect the rheological behavior of many products by having their characteristic structure and melting point and by interacting physically with other constituents. Furthermore, the products of enzymatic and chemical reactions of lipids contribute to the formation of pleasant or undesirable notes of color and flavor. These effects are caused in different degrees by various natural fats differing in chemical and functional properties, as well as by novel, tailor-made lipids produced by applying biotechnological and chemical processes. The effect on food quality is related predominantly to the contents, distribution in the food matrix, chemical composition, and reactivity of the lipids, as

71

well as to their physical changes due to processing, and the interactions with other components. A very large asset in the activities aimed at increasing the role of lipids in improving the quality of food is the rapidly growing potential of analytical methods (Sacchi and Paolillo 2007).

4.2   THE CONTENTS AND COMPOSITION OF LIPIDS IN FOODS

4.2.1   The Contents and Distribution

Almost all plants and animals used by humans for food contain lipids, although in different quantities. The fat content in various raw materials and products is given in published nutritional tables in percent of the wet weight of the edible parts of the food, as percent of dry matter, or in grams in one serving of the food. In wet weight, it ranges from close to null to about 70%. It depends on the species, genotype, breed, physiological stage, and nutritional status of the plant or animal, as well as on the location of the lipids in the grain, nut, fruit, tuber, or animal carcass (Table 4.1). The temperature and other conditions of vegetation or breeding are also important. Thus, the published data should be treated with caution, knowing that, e.g., a fillet of herring may contain 4% or 20% of fat, depending on the season of catch, and the fat content of cod muscle is about 0.7% while that of the liver of the same fish is about 10 times higher. Most vegetables and fruits are very poor in fat, contributing generally about

TABLE 4.1
Proximate Contents of Fat in Food Raw Materials and Products

Raw Materials	Fat Content % Wet Weight	Products	Fat Content % Wet Weight
Cow's milk	3.9	Kippered herring	13
Sheep's milk	7.2	Clipfish	1.5–2.5
Egg yolk	33	Tenderloin steak, lean	9
Beef chuck, medium	16	Liverwurst, smoked	27
Beef liver	3	Mortadella	25
Beef tongue	15	Ham, cooked, sliced	3.3
Lamb leg, medium	18	Chicken balls, Chinese	13.5
Crab meat	1	Chicken breast, roasted	6.5
Oyster	1.5	Yogurt	3.4
Sole fillet	2	Low-fat yogurt	0.2
Wheat grain	2.2	Halvah	29–34
Barley grain	2.1	Cheddar cheese	34
Rice grain	2.4	Cheddar cheese reduced fat	24
Oat grain	7.1	Ice cream	21
Cashew nut	42	Whole meal bread	2.9
Walnut	59	White bread	1.5
Potato	0.1	Twix, Mars	24
Paprika	0.2	Whole milk chocolate	28–32

0.3% of the total weight. A notable exception is avocado with about 20% of lipid in the edible part. The content of fat in muscle tissue is about 2% in lean beef, fish, white poultry meat, and shellfish; 3.7% in cow's milk and 2%–4% in grains; about 30% in fatty pork; 32% in an egg yolk; and up to 35% in fillets of some fatty fish. Oil-bearing nuts and pulses contain approximately from 20% fat in soybeans to 65% in walnuts.

In processed foods, the content of fat depends on the raw material specificity and the required sensory properties of the products, which are reflected in the recipe. It is also affected by the current recommendations of nutritionists. In numerous cases the fat content is controlled to fulfill the requirements of regulatory agencies. Thus, e.g., according to the EC regulation the minimum content of milk fat in sweet cream unsalted butter should be 82%.

4.2.2 LIPID CLASSES AFFECTING FOOD QUALITY

4.2.2.1 Structural Lipids and Depot Fats

The cell and organelle membranes are made of polar lipids—phospholipids (PL) and nonesterified cholesterol in animal organisms, whereas in plants of PL and glycolipids (see Chapter 1 for nomenclature). The latter are also found in the central nervous systems of some animals. The formation of the matrix of cellular membranes is a spontaneous process caused by self-association of the amphipathic lipid molecules, which is entropically driven by water. These membranes are responsible for compartmentalization of the cells and tissues (Van Meer et al. 2008). If a muscle tissue like that in lean fish contains only 0.3% w/w of lipids, they consist almost entirely of PL. Galactoglycerols and PL are important in nutrient and antioxidant delivery systems (Herslof 2000).

The storage lipids form the hydrophobic core of lipid droplets in the cells. They consist of triacylglycerols (TAG), steryl esters, and waxes in different proportions in various organisms. Waxes are characteristic for organisms of polar regions and deep-ocean fish. Due to the buoyancy wax, these animals may quickly change their vertical position. The storage lipids can be regarded as efficient, anhydrous reservoirs of energy and of fatty acid (FA) and sterol components, which are necessary for membrane biogenesis. In the animal carcass, they are present mainly in the subcutaneous adipose tissue, as kidney and groin fat, and as intramuscular fat responsible for the marbling of meat, i.e., interspacing of fat sheets within the lean tissue. In fatty fish, there is much fat both in the subcutaneous layer and in the muscles; however, its quantitative distribution depends on the species of the animal. In milk, dairy commodities, mayonnaise, and in comminuted meat and fish products the fat is dispersed in the form of globules of various sizes.

Depot fats have been exploited by humans for centuries. They can be partially removed or added during processing and are used as frying medium. Polar lipids, PL, mono- and diacylglycerols are usually used in smaller amounts as functional additives. The intake of PL with plant and animal foods can be regarded as the only benefit of the excessive consumption of invisible fats. However, most of the invisible fat consumed currently originates from added fats and frying fats absorbed by fried products.

4.2.2.2 Fatty Acids

FA composition determines the physical properties, stability, and nutritional value of lipids. All food lipids of natural origin contain acyls of saturated, monoenoic, and polyenoic FA (PEFA) in various proportions. Variations in FA composition make it possible to determine the origin of the lipids. The profile of FA, especially in PL, which due to their function are less prone to changes, is considered as a species "identity card" (Grahl-Nielsen 1999). The FA pattern of depot fat reflects the composition of lipids in the animal diet and the conditions of vegetation of the plant. This may be used to modify it in the meat of farm animals, fish, poultry, and eggs. In ruminants' feeding, special procedures are required to avoid the modifying influence of the rumen micro-flora on the fat composition. There are noticeable differences among hybrids in the sus-ceptibility to modification according to species, variety, and genus. Changes in the FA composition of pig muscle fat induced by diet may increase the susceptibility to oxida-tion to a greater extent than would be expected by the alterations in FA (Kołakowska et al. 1998). The composition of FA in plant lipids is generally altered at molecular level and is the subject of intensive research that has already led to successful modification of rape-seed oil to the low erucic, high erucic, high oleic, high linoleic, and high lauric acid oils. Genetic modifications and spontaneous mutations of soybeans have produced low linolenic acid oil, high oleic acid oils, and high saturated FA oils. Modifications in the FA composition of fats and oils have been presented in detail in Chapter 19.

FA distribution in TAG as well as in PL affects the physical properties, lipolytic and oxidative stability, and nutritional availability of the lipids. In many TAG the FA are nonrandomly distributed. In plants, monoenoic FA and PEFA are dominant at a sn-2 position (Orthoefer 1996). In pig depot fat and in cow's milk, the TAG sn-2 posi-tion is occupied by palmitic acid. In blubber seals, long-chain n-3 PEFA are esteri-fied rather in sn-1,3 positions, whereas in muscle TAG in the sn-2 position that is typical for lipid muscles of nearly all fish (Ackman 1994). The type of FA in the sn-2 position is critical for the bioavailability and digestibility of oils and fats. In palm oils and cocoa butter, the saturated FA are in sn-1 and sn-3 positions, while oleic acid is in sn-2; in olive oil, 80% oleic acid is in sn-2. In lard, the palmitic acid and $trans$-FA are in sn-2 position. Biotechnological syntheses of structured TAG lead to a tailor-made distribution of various acyls in the TAG molecules.

4.3 LIPIDS AFFECTING THE NUTRITIONAL VALUE AND SAFETY OF FOODS

4.3.1 NUTRITIONAL BENEFITS

Lipids play many essential roles in cell biology. In addition to what has been men-tioned in Section 1.1, they also serve as first and second messengers in the transducing of signals and in molecular recognition phenomena, provide a barrier for ion separa-tion, prevent cells from water loss, and act as temperature sensors as well as insulating and contour-building material. From the nutritionists' point of view, one of the most important functions of lipids in the diet is that they serve as substrates for eicosanoids, which cannot be synthesized by the human organism. Deficiencies in long-chain n-3 PEFA in the western diet and insufficient supply of sea fish lead the nutritionists' attention to atypical sources of plant oils, which, besides alfa linolenic acid, also contain stearidonic acid, like in the seeds of echium. The flax, borage, blackcurrant, and evening primrose oils are also treated as valuable sources of phytosterols and tocopherols. Special health-promoting effects are claimed as regards some milk fat FA, predominantly the short-chain FA (Stołyhwo and Rutkowska 2007). The relation of lipids in the diet and the human health has been presented in Chapters 11 through 13.

4.3.2 FOOD LIPIDS POSING HEALTH HAZARDS

Health hazards associated with the consumption of lipids may be caused by

- Improper quantity and composition of the consumed lipids
- Natural, harmful components of fats
- *trans*-Isomers of FA formed due to processing
- Lipid oxidation products
- Lipid polymerization products
- Effects of interactions with other food components, especially proteins
- Toxic substances accumulated from the environment and packaging materials
- Added synthetic antioxidants

Some unprocessed fats and oils contain different amounts of natural, harmful components. Crude cottonseed oil contains the toxic, phenolic compound gossipol, which is naturally present in some varieties of cottonseed in concentrations of up to 1%. Alkaline refining is effective in removing gossipol from the oil, since the sodium derivative of gossipol is water soluble.

The fats of dairy products and, in much lower extent, of meat from cattle and sheep contain acyls of several *trans*-FA in amounts of 2%–5%. These acids occur naturally, since they are produced in the rumen of ruminants by bio-hydrogenation of unsaturated FA. Partial hydrogenation of vegetable oils and heat isomerization of frying oils also lead to the formation of *trans*-FA. However, the quantitative distribution of the isomers in milk fat and in partially hydrogenated oils is different (Figure 4.1). The *trans*-FA present predominantly in milk fat, i.e., vaccenic acid $C_{18:1}(11t)$, which, in summer and autumn, may make up as much as about 80% of the total *trans*-FA, may be desaturated in the human organism to the conjugated linoleic acid (CLA), which has numerous beneficial biological effects (Yacoob et al. 2006).

Various products of lipid oxidation in foods may have unfavorable health consequences. There is some evidence pointing to toxic effects of hydroperoxides from the diet. The toxicity of thermooxidized frying fats on humans is better verified. Particular attention is paid to the toxicity of oxidized cholesterol, especially at temperatures higher than 100°C and in presence of unsaturated lipids, as well as to the role of oxy fitosterols.

Chemical reactions of oxidized lipids with proteins, especially during long-term storage and due to high-temperature processing lead to the formation of polymers of lowered digestibility. Furthermore, they cause loss in reactive, essential amino acids. Detailed information on the biological effects of the reactions of oxidized lipids in foods is given in Chapters 9 and 22.

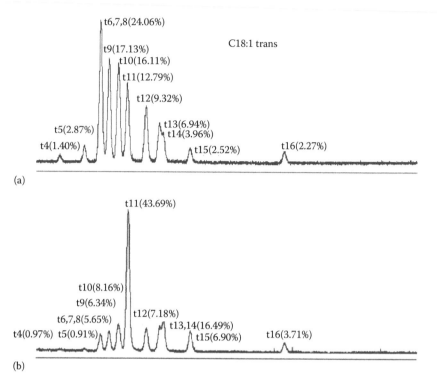

FIGURE 4.1 *trans*-Isomers of FA in milk fat (a) and in partially hydrogenated rapeseed oil (b). (Courtesy of Andrzej Stołyhwo, Warsaw University of Life Sciences, unpublished.)

The lipids of various food raw materials accumulate a large number of different hydrophobic components from the environment. Many of these compounds present serious health hazards to humans. The contamination of meat of some fatty fish, predominantly fish livers from polluted waters, may in same cases exceed the permissible limits of concentration of, e.g., polychlorinated pesticides. Fatty food products may become contaminated with fat-soluble low-molecular constituents of plastic materials like plasticizers, stabilizers, or slip additives, as well as monomeric or oligomeric components of the plastics (Piotrowska 2005). A spectacular case is the long-lasting problem of too-high migration of plasticizers from polyvinyl chloride gaskets in lids for glass jars. An in-depth treatment of the determination of toxic substances in food lipids.

4.4 THE EFFECTS OF LIPIDS ON THE SENSORY ATTRIBUTES OF FOODS

4.4.1 Introduction

Lipids generally have a desirable impact on the sensory properties of many foods by affecting the mouth feel, e.g., the richness of whole milk as opposed to the blank taste of skim milk, or the smoothness of high-quality ice cream. The particular sensory attributes of foods that are affected by lipids include mainly the color, rheological properties, and flavor. They depend on the content, chemical composition, physical changes due to processing, as well as chemical and enzymatic transformations of the lipids during storage of the food, and in manufacturing operations. The desirable contribution of fats, lipochromes, and lipid-soluble vitamins may be easily lost due to deteriorative reactions, predominantly oxidation and hydrolysis during storage at abuse conditions.

4.4.2 Lipids and the Color of Foods

Lipids are involved in color formation in foods by carrying different colored substances and by participating as substrates in reactions leading to generation or modifications of colored compounds. The surface pigmentation of marine animals is largely due to different carotenoproteins, that may be yellow, orange, red, purple, blue, or green, depending on the structure of the complexes, the kind of carotenoid, predominantly astaxanthin, cantaxanthin, and β-carotene, as well as the properties of the proteinaceous component. The dissociation of the protein moiety from the complex in bright light brings about the fading of colors of fish skin. During prolonged frozen storage, the released carotenoids diffuse into the lipid layer causing a yellowish discoloration. The boiling of crustaceans turns the blue or blue-gray carapace into bright red due to heat denaturation of the carotenoprotein crustacyanin with release of the free, red astaxanthin. Carotenoid pigments are also responsible for the color of the flesh oil of redfish (*Sebastes marinus*). Vegetable oils also contain different carotenoids, generally in concentrations about or below 0.1%. In palm oil, the carotenoid pigments, about 0.3%, are responsible for the orange color. Olive oil contains the green pigment chlorophyll in concentrations of up to several hundred micrograms per gram. Lipid oxidation products increase the rate of browning of meat hemoproteins. This shortens the case life of beef cuts in retail display. A measure to prevent the rapid loss of the desirable bright red color of the meat is to use suitable antioxidants in cattle feeding. The carbonyl compounds formed due to lipid oxidation react with protein amino groups leading to browning of processed and long-stored foods.

4.4.3 The Role of Lipids in Food Texture

The rheological properties are affected by fat in meat and meat products, in fishery products, in dairy commodities, in pastry, cakes, and mayonnaise. The desirable texture of culinary meat is, in part, due to the proper marbling of the muscles with thin fat layers; that of comminuted sausages is conditioned by adequate content of fat as well as by the optimal size and distribution of the lipid globules in the formulation. The effect of fat on the rheological properties of meats also depends on the FA composition of the feed given to the farm animals. In fish belonging to the fatty species, the meat texture depends very significantly on the seasonal changes in fat content. Baltic sprats caught in summer are unsuitable as raw material for the canned smoked Baltic sprats in oil because at fat content below 6%, the texture of the fish is too hard. Very high quality of hot smoked mackerel can be assured only by using raw material containing about 30% fat. The tender texture of the lightly salted maatjes is due to the abundance of lipids in the tissues of the immature, fatty herring. The great role of fat in affecting the sensory properties of some delicatessen foods has been praised by the renown German fish technologist Peter Biegler: "Ein mild gesalzener Schinken, oder mild gesalzener Kaviar oder mild gesalzener Lachs dürfte schon den Göttern gut gemundet haben" (Biegler 1960). No delicatessen product of this type can be made of too lean meat or fish. The cream for producing whipped cream without any whipping agents should contain about 30% fat. The desirable sensory sensation caused by the melting of chocolate in the mouth is due to the narrow range of melting temperature 28°C–36°C of the fat of cocoa butter. In bread and other baked goods, the lipids contribute to texture by interacting with proteins and polysaccharides. They have a shortening effect by disrupting the structure of the gluten network, stabilize the gas bubbles within the loaf, and as amylose–lipid complexes have the ability to form gels, thus contributing also to the texture of the products.

4.4.4 LIPIDS AND FOOD FLAVOR

Short-chain aldehydes and alcohols generated due to enzymatic lipid oxidation participate in the formation of the desirable "green aroma" of fresh melons, cucumbers, tomatoes, bananas, and mushrooms. Lipid degradation products in low concentration contribute to the mild, rather pleasant, plant-like, melon-like, sea-weedy aroma of the very fresh fish. Due to reactions catalyzed by endogenous lipoxygenases, hydroxyperoxide lyases, Z,E-enal isomerases, and alcohol dehydrogenases, the PEFA of fish lipids are degraded to aldehydes, ketones, and alcohols with 6, 8, and 9 carbon atoms. The gradual loss in the intensity of the fresh fish aroma is in part caused by microbial conversion of the carbonyl compounds into alcohols, which have higher aroma threshold values. During the storage of frozen fish, off-flavor develops due to the oxidation of lipids.

In producing the characteristic flavor of meat dishes, the lipids participate through reactions of their oxidized forms with various thermal degradation products, predominantly nitrogenous compounds. The secondary lipid oxidation products, especially carbonyl compounds, are also responsible for the off-flavor notes known as warmed-over flavor (see Chapter 17).

The flavor of cheese is created, in part, by the products of lipid hydrolysis and oxidation due to the activity of various enzymes of milk and the cheese microflora. Among the flavor compounds produced directly from lipids are short-chain FA, ketones, alcohols, and lactones. Furthermore, just as in the case of other foods, the lipid degradation products participate in the formation of other volatile products in reactions with a variety of other cheese components.

The reader interested in problems of food flavor may find more information in the forthcoming CRC book in the same series *Food Flavors: Chemical Properties and Sensory Characteristics*, edited by Henryk Jeleń.

4.5   INTERACTIONS OF LIPIDS WITH OTHER FOOD COMPONENTS

Many lipids found in nature are intimately bound to proteins and saccharides. The interactions within such structures are usually weak, but also covalent bonds may occur, e.g., lipid inclusion in amylose or some lipid fractions in fish muscle tissue. During the processing and storage of food, the lipids are released and new bonds may be formed. Interactions are promoted purposely or undergo spontaneously during the processing and storage of foods.

Both the original lipids, the products of their chemical and enzymatic alterations, and added fats and lipid surfactants participate in interactions with proteins and saccharides in different foods. As a result of the disintegration of the tissue structures due to mincing and by applying high shear forces in homogenization, the fats are emulsified in the food matrix in the presence of proteins and other surfactants, forming the desirable properties of mayonnaise and comminuted sausages. The interactions of lipids, proteins, and saccharides during mixing of a cake batter lead to the typical texture of the cake. In cereal products, various lipid–saccharide and lipid–protein interactions take place, although the effects caused by endogenous, native, neutral lipids, and by added fats and surfactants are different in several aspects. The interactions of lipids with starch affect the leaching of amylose out of the starch granules and swelling of the granules due to heating, decrease the rate of staling of bread, and improve the texture of the loaf (Eliasson 1998). Lipoprotein films formed in the bread dough by interactions of polar lipids with some wheat proteins are desirable in bread making, while the nonpolar fats and FA destabilize the protein foams (Marion et al. 1998).

The reactivity of fish lipids and proteins depends on the level of oxidation and reaction conditions. PEFA interact easier and form stronger bonds than other FA with proteins and starch. The interactions of lipids with other food components create new product attributes and also affect the properties of fats, particularly their extractability and availability in the human organism.

A comprehensive treatment of the subject of interactions of lipids in foods has been presented in Chapter 22.

REFERENCES

Ackman, R. G. 1994. Seafood lipids. In *Seafoods: Chemistry, Processing Technology and Quality*, F. Shahidi and J. R. Botta (eds.), pp. 34–48. London, U.K.: Chapman & Hall.

Biegler, P. 1960. Fischwaren-Technologie. Theorie und Praxis der Fabrikationsmethoden zur Konservierung von Fischen, In *Der Fisch. Mitteilungen für die Fischindustrie*, Band V. K. Baader (ed.), p. 81. Lübeck, Germany: Verlag Der Fisch – Clara Baader.

Eliasson, A. Ch. 1998. Lipid-carbohydrate interactions. In *Interactions: The Keys to Cereal Quality*, R. J. Hamer and R. C. Hoseney (eds.), pp. 47–79. St. Paul, MN: American Association of Cereal Chemists, Inc.

Grahl-Nielsen, O. 1999. Comment: Fatty acid signatures and classification trees: New tools for investigating the foraging ecology of seals. *Can. J. Fish. Aquat. Sci.*, 56: 2219–2223.

Herslof, B. G. 2000. From drug delivery to functional foods: A lipid story, *INFORM*, 11: 1109–1115.

Kołakowska, A., Zygadlik, B., and Szczygielski, M. 1998. Susceptibility of muscle lipids in pigs to oxidation depending on breed soybean or rapeseed diet. *Pol. J. Food Nutr. Sci.*, 7/48: 655–662.

Marion, D., Dubreil, L., Wilde, P. J., and Clark, D. C. 1998. Lipids, lipid-protein interactions and the quality of baked cereal products. In *Interactions: The Keys to Cereal Quality*, R. J. Hamer and R. C. Hoseney (eds.), pp. 131–167. St. Paul, MN: American Association of Cereal Chemists, Inc.

Orthoefer, F. T. 1996. Vegetable oils. In *Bailey's Industrial Oil and Fat Products, Edible Oil and Fat Products: General Applications*, Vol. 1, 5th edn., Y. H. Hui (ed.), pp. 19–44. New York: John Wiley & Sons, Inc.

Piotrowska, B. 2005. Toxic components of food packaging materials. In *Toxins in Food*, W. M. Dąbrowski and Z. E. Sikorski (eds.), pp. 313–333. Boca Raton, FL: CRC Press.

Sacchi, R. and Paolillo, L. 2007. NMR for food quality and traceability. In *Advances in Food Diagnostics*, L. M. L. Nollet, F. Toldra, and Y. H. Hui (eds.), pp. 101–117. Ames, IA: Blackwell Publishing.

Stołyhwo, A. and Rutkowska, J. 2007. Milk fat: Structure, composition and health-promoting properties. In *Food Chemistry*, Vol. 3, 5th edn., Z. E. Sikorski (ed.), pp. 39–89. Warsaw, Poland: Wydawnictwa Naukowo-Techniczne (in Polish).

Van Meer, G., Voelkar, D. R., and Feiebson, G. W. 2008. Membrane lipids: Where they are and how they behave. *Nat. Rev./Mol. Cell Biol.*, 9: 112–124.

Yacoob, P., Tricon, S., Burdge G. C., and Calder, P. C. 2006. Conjugated linoleic acids (CLAs) and health. In *Improving the Fat Content of Food*, Ch. Williams and J. Buttriss (eds.), pp. 182–209. Cambridge, U.K.: CRC Press.

5 Lipids in Food Structure

Wioletta Błaszczak and Józef Fornal

CONTENTS

5.1 PLANT TISSUE LIPIDS

5.1.1 CEREAL LIPIDS

The interest in food cereals has mainly concentrated on grains as a source of dietary fiber and phytochemicals. It is generally emphasized that whole grains are rich in fiber, vitamins, minerals, and phytochemicals including vitamin E, carotenoids, inulin, sterols, and phenolics.

The cereal lipids have also become a focus of interest as they exhibit nutritional and technological potential. Oat contains much more lipids than any other cereal (2%–12%), and owing to that it may also be considered as an excellent source of unsaturated fatty acids, including several essential fatty acids. Oat is an important source of glycolipids, providing a significant amount of digalactosyldiacylglycerol as well as phospholipids. It has been demonstrated that the major fatty acid residues present in the lipid fraction are palmitic, oleic, and linoleic ones (Zhou et al., 1999; Aro et al., 2007). The oat lipids are distributed in the whole grain, whereas in other cereal grains they are mostly present in the germ. They occur in the form of oil droplets, which can easily be detected under a microscope (Figure 5.1A, B, and D). However, the lipids in the starchy endosperm and in bran of cereals account for most of the total lipids since the embryonic axis and *scutellum* represent a relatively small proportion of the whole grain (Zhou et al., 1999). The oat lipids affect the functional properties of oat starch, but most of all they are responsible for the flavor/off-flavor attributes of oats. The typical flavor of oat products results from lipid oxidation and the formation of N-heterocyclic compounds upon heat processing (Zhou et al., 1999). During processing of the oat grain (rolling, flaking, and cooking), the loss of lipids is

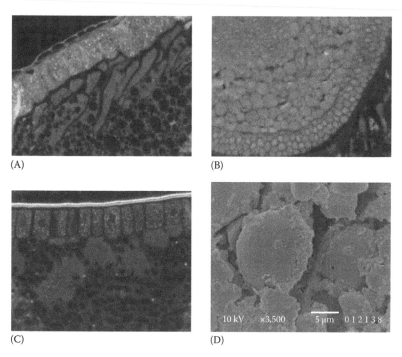

(A)

(B)

(C)

(D)

FIGURE 5.1 Fluorescence of lipid bodies in aleurone layer (A) and germ (B) of oat and wheat (C) kernel. Arrows show the lipid bodies on starch granule surface (D) (scanning electron microscopy (SEM) photo of oat endosperm).

rather insignificant (Yiu, 1986). Zhou and coauthors reported that water soaking of flours decreased significantly the content of linoleic acid, which was also linked with a selective loss of esterified polyenoic fatty acids upon processing. Wet fractionation of the flour from non-heated oat grain very often leads to hydrolysis of triacylglycerols of the endosperm fraction.

Wheat grain is an essential raw material for the production of semolina and pasta, and for bread making. Spring wheat flours have been found to manifest a higher concentration of total free lipids (1285 mg/100 g) as compared to the winter ones (1102 mg/100 g). In the case of spring wheat, the nonpolar lipids were mainly found in the endosperm part of the kernel (Konopka et al., 2006). The contribution of the nonpolar fraction may reach from 70% to 90% of the total lipid fraction for both wheat varieties. According to these authors, the nonpolar lipids in the flour form minute-like oil storage globules (spherosomes). The spherosomes are mainly localized in the germ and aleurone layer, yet in a far lower concentration in mature starch endosperm (Figure 5.1C). Spring wheat varieties contain also more glycolipids (134–215 mg/100 g of flour) than the winter ones (158–196 mg/100 g of flour) (Konopka et al., 2006). Glycolipids appear to be localized in the membrane of amyloplasts, and thus their content may be significantly affected by the size and number of starch granules. A significant relationship has also been found between endosperm hardness and its lipid composition. This fact may indicate that grain hardness is mainly correlated with the content of free glycolipids. Harder wheat varieties have been shown to display a higher content of oleic acid in the lipid fraction (Konopka et al., 2005). It is generally believed that lipids, especially the polar ones, significantly affect the quality of dough (viscoelastic properties), loaf volume, and texture of bread as well as its staling (Dubreil et al., 2002; Georgopoulos et al., 2006). In the kernel, polar lipids are mainly responsible for the formation of cellular membranes. Using confocal scanning laser microscopy (CLSM), it was shown that the lipids in wheat dough were generally located around gas/bubble cells and formed the structure of a protein–starch matrix. The presence of puro-indolines in bread was connected with defatting of the gas bubble surface and a decrease in the size of lipid droplets. Droplets embedded in the starch–protein matrix influence gas cells/bubbles expansion during fermentation and baking and determine porosity, one of the sensory attributes of bread (crumb texture) (Dubreil et al., 2002).

5.1.2 COTYLEDON LIPIDS—THEIR ROLE IN SEED STRUCTURE AND QUALITY

Structural investigations of soybean cotyledons have shown that the main subcellular structure is formed by protein bodies 5–20 μm in diameter, surrounded by a cytoplasmic protein network, which the lipid bodies are embedded in (Wolf and Baker, 1981). A detailed observation under scanning electron microscopy (SEM) demonstrated that lipid bodies revealed a spherical shape ranging in diameter from 0.05 to 1.2 μm. The cooking of soybeans at 115°C for 30 min elicits changes in lipid bodies, thus leading to their coalescence into large droplets. The influence of processing on soybean lipids was also studied (Wolf and Baker, 1981; Sarkar et al., 1996); however, under mild conditions of soaking and cooking of soybean, no striking difference was found in lipid content. Contrary to this, the lipid content was significantly affected by the fermentation of soybeans. The major fatty acids in unfermented and fermented soybeans were palmitic, stearic, oleic, linoleic, lino-lenic, and arachidic ones. Except for arachidic acid, all the identified lipid com-pounds were released upon fermentation by microbial lipases (Sarkar et al., 1996). The full-fat soy flour is a popular product obtained from steamed soybeans. The steaming process is used mainly to inactivate lipoxygenase and trypsin inhibitors and to modify the grassy/beany and bitter flavors of raw soybeans. The changes in the appearance of the lipid bodies under the microscope upon steaming were demonstrated by Wolf and Baker (1981), who illustrated the bursting-like and blurring of the lipid bodies on a protein surface.

The drying process of seeds with high moisture may significantly influence the quality of a seed constituent. A high quality of seeds is especially important in the case of oil production from the oil seeds. Commercial varieties of rapeseed may contain about 40% (by weight) of the lipid fraction, with neutral lipids (triacylglycer-ols) that may account for 95% of the total lipid in the seed. Rapeseed belongs to the dicotyledonous plants in which the embryo constitutes as much as 80% of the mature seed. The seed contains also a radicle and two conduplicate cotyledons. The embryo is formed by storage parenchyma cells, containing the major reserves of lipids and proteins. Transmission electron microscopy (TEM) and SEM studies of rapeseed

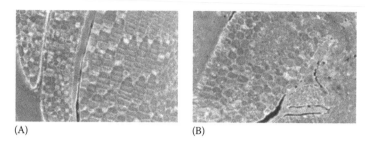

(A)   (B)

FIGURE 5.2   Light microscopy (LM) photo of rapeseed cotyledons: (A) native and (B) dried.

have illustrated lipid bodies appearing as individual oil droplets with a spherical shape and diameter ranging from 0.1 to 1.5 μm (Yiu et al., 1982).

Improper drying of rapeseed may evoke significant changes in seed structure, which in turn leads to difficulties during further processing (Fornal, 1998). Microscopic studies of raw (Figure 5.2A) and processed rapeseeds distinctly showed changes in protein bodies connected with their denaturation, damages in the lipoprotein membrane surrounding individual fat droplets, and concentration of free fat near the cell walls (Figure 5.2B). The cell wall damage upon processing has been shown to elicit significant alterations in the properties of the press cake obtained, due to changes in its adhesion and cohesion forces. This phenomenon results from poor percolation of a solvent or even its lack because of press cake powdering. It has been revealed that very poor percolation of hexane through the crushed press cake layer evokes high oil content of rapeseed meal, thus disturbing the technological process, and lowers the capacity of oil production, which is unacceptable from the economic point of view.

5.2   FOOD SYSTEM LIPIDS

5.2.1   FAT IN SPREADS

Plastic fats form a three-dimensional network structure of crystals, which the liquid oil is trapped in. Fats contain triacylglycerols that are solid at room temperature and may also contain a variety of acylglycerols that reveal slightly different melting points and demonstrate the ability to form crystals i.e., upon cooling (Figure 5.3A and B). Long-chain compounds, including triacylglycerols, show polymorphism upon crystallization. There are three main polymorphic forms of fat crystals known, named alpha, beta-prime, and beta, in order of increasing stability. The form alpha is hexagonal and is characterized by a low-density structure with a cross-sectional area of about $0.2 \, nm^2$. It is the least organized crystal structure as compared to the other forms. This form may be obtained when melted fat is subjected to rapid cooling. The beta-prime structure is orthorhombic and is more common for many natural fats. All the beta crystals have the chain axes oriented in one direction, demonstrating the triclinic form with a cross-sectional area of about $0.185 \, nm^2$ (deMan, 1982).

Shortenings are composed of liquid oil and fat crystals, whereas margarine and butter may additionally contain around 16% of water (Heertje, 1998). The

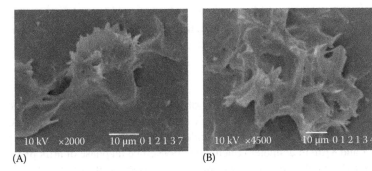

(A)   (B)

FIGURE 5.3   SEM photo of fat crystals formed in commercial margarine. The initial (A) and an advanced (B) stage.

solid/liquid ratio of the fat phase typically accounts for 1/4 to 1/3 for margarine and butter, respectively, at room temperature. The texture properties, i.e., consistency or mouthfeel, of these products are closely related to the organization of fat crystals. To study the microstructure of fat spreads and differences in the polymorphism of fat crystals by SEM or by polarized light microscopy (PLM), it is necessary to remove the liquid oil completely. Otherwise, the amorphous surface layer coming from the oil phase may disturb the original crystallization pattern. The SEM studies of commercial shortenings and margarines confirmed the above-mentioned suggestion that beta and beta-prime forms did occur in the products analyzed (Chawla et al., 1990). However, the predominance of one form over another is more dependent on the processing conditions than on the type of the fat spread examined. Under SEM, margarine demonstrates the prevalence of beta form crystals. The fat crystals may appear as bundles, with the crystals being more or less parallel to each other. When analyzed under SEM, some other commercial margarines display individual, rod-shaped crystals uniform in size and shape. The microstructure of beta crystals can be visible as platelets and/or spherulites with random arrangement, appearing as single crystals or as aggregates ranging 20–30 μm in size. Beta prime crystals are likely to form nonuniform in size (5–7 μm), single, needle-shaped units. When the transformation of fat crystals occurs i.e., from beta-prime to beta form, the crystals change their structure from initially small in size single units to aggregates and grow in size along with storage time, yielding a sandy texture not acceptable by consumers.

Water droplets play an important role in the structure and stability of margarine. They are being formed upon intense mixing of fat and water phase during the production of margarine and, as shown by the cryo-SEM (CSEM), are just a few μm in diameter. The orientation of crystals at the water droplet surface results in the stabilization of the latter in the structure of margarine. The nature of the fat crystalline network as well as the size, shape, and aggregation of the fat crystals varies in respect of margarine texture, which in turn depends on the product's application (creaming, cake, or puff pastry) (Heertje, 1998).

In opposite to margarine, butter is substantially different in its microstructure. In the microstructure of butter, the fat globules remaining after the churning process of cream form a discontinuous structure with a fat crystalline matrix. The limited interaction of fat globules with the rest of the matrix, or the lack of it, affect such properties as hardness, spreadability, mouthfeel, emulsion stability, and salt release (Heertje, 1998).

5.2.2 Fat in Dairy Products

The lipid phase of milk consists of 98% of triacylglycerols, whereas phospho- and glycolipids or mono- and diacylglycerols were found to be the minor constituents of milk fat. The milk fat was visualized under a microscope in the form of globules between 1 and 8 μm in diameter (Buchheim, 1998). The fat phase of milk plays a significant role during the processing of raw milk. Upon the milk cooling, the fat phase starts to slowly crystallize, and the distribution of the fat crystals in milk fat globules may be significantly different in each individual globule (Figure 5.4A). When milk is subjected to shear forces during agitation at low temperature, the fat globules are deformed, which in turn leads to the phenomenon of fat globule membranes' coalescence. The structure of the milk fat globule membrane is highly complex and very sensitive to any kind of treatment. The major constituents of the membrane material are glycoproteins, phospholipids, and glycolipids. Denaturation of glyco-proteins of the milk fat globule membrane during milk heating does not impair the

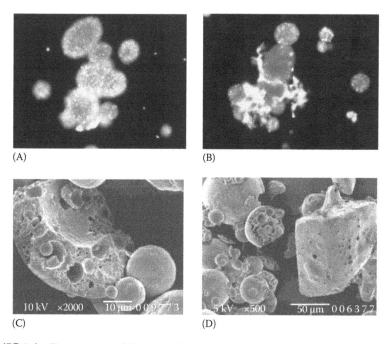

(A) (B)

(C) (D)

FIGURE 5.4 Fluorescence of fat crystallites on protein surface in powdered milk (A) and phenomenon of fat globules clustering (B). SEM photo of spray-dried milk (C) and milk whey (D) (pores after removing of the interior lipids are visible). (Photos, were presented in the work of Blaszczak, W., Journal entitled: Application of microscopy methods in food analysis, published by Institute of Animal Reproduction and Food Research of Polish Academy of Sciences, Olsztyn, 2008.)

emulsion stability but may evoke irreversible changes in the structure of the membranes (Buchheim, 1998). Due to the changes in the protein/fat phase, the ultrahigh temperature (UHT) treated or sterilized milk does not manifest any tendency for fat creaming during longtime storage. A process that strongly influences the fat phase is high-pressure homogenization with pressure up to 250 bar and temperature between 60°C and 80°C. This process enables reducing the size of fat droplets, resulting in 10- to 15-fold increase in the surface area per unit volume of fat and the formation of new types of interfacial layers, thus making the fat phase more stable. For UHT-treated or sterilized milk products or evaporated milk, the size of the fat globules should be around 0.5 μm. The high-pressure homogenized (HPH) cream (10%–20% of fat content) exhibits a tendency for the clustering of fat globules (Figure 5.4B), which in turn may lead to not only fat creaming upon storage but also a significant increase in the cream viscosity. The phenomenon of fat globule clustering should not be observed in UHT-treated and/or sterilized milks, creams, or evaporated milk. The fat globule clustering resulting from separation (creaming and sedimentation) or heat-induced coagulation may also occur in condensed milk products as a result of improperly selected processing conditions (Buchheim, 1998). The microstructure of condensed unsweetened milk is formed by aggregates consisting of protein (casein, denatured whey protein) and fat droplets with an average diameter around 0.5 μm. The size of the protein aggregates and the protein/fat ratio within these aggregates may significantly influence the product's stability.

The surface properties of powdered milk are affected by, among other things, fat globules. Fat globules of powdered dairy products (whole milk powder, powdered cheeses or cream) are responsible not only for surface structure formation, but they appear in the interior of micelle particles as capillary lipids (Figure 5.4C). The interior lipids can be encapsulated by unbroken particles where they often associate with milk proteins. Such complexes are resistant to oxidation and additionally make fat resistant to extraction (Kalab et al., 1989). The observation of powdered milk under SEM indicated its porous surface. The porous structure is typical of spray-dried products and results from rapid water evaporation during treatment from the material associated with the surface of the spherical and granular particles of the powder (remnants of lactose, protein, lipid) (Figure 5.4D) (Sherestha et al., 2007). Fat globules present in whole milk powder with a homogenized fat phase may reach 0.5 μm in diameter. A cream or cheese powder manifests a similar appearance to the microstructure of whole milk powder, except that the fat phase constitutes a higher proportion of the sample analyzed. Numerous LM observations of cheese powder have distinctly shown fat globules unequally distributed on the surface of the casein micelles. Moreover, the microstructure of powdered cheese appears to be more clustered as compared to that of powdered milk. This fact may be related to strong aggregation of fat and its association with the protein matrix (Błaszczak and Fornal, 2008).

Soft, mould-ripened, and hard cheeses are the most popular types of cheese. The ripening process of soft cheeses involves the same biochemical changes that occur in hard cheeses. The most complex and important changes occur during proteolysis. The variability of milk proteinases and coagulants, starters, as well as all the microorganisms present in the cheese potentially contributes to the changes connected with proteolysis. The biochemical changes undergone during proteolysis are

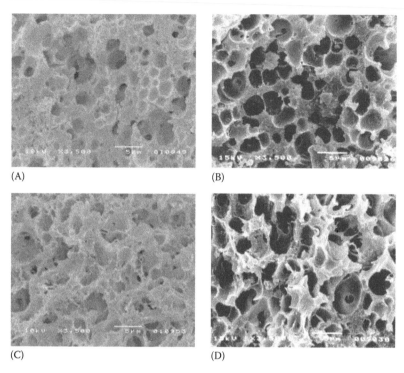

(A)   (B)

(C)   (D)

FIGURE 5.5   SEM photo of fresh (A,B) and ripened (C,D) hard cheeses. A,B—Swiss-type, and C,D—Dutch-type cheeses.

significantly responsible for the texture formation of the cheese curd (Sousa and McSweeney, 2001). The curd is composed mostly of casein micelles (~24%) that aggregate in milk during cheese making and in fat globules (~34%) entrapped in the coagulum. The changes in the microstructure of a protein matrix proceeding during cheese ripening were demonstrated on an example of Swiss- and Dutch-type cheeses (Figure 5.5) (Błaszczak and Fornal, 2008). While the Swiss type of cheese demonstrates a compact and dense protein matrix (Figure 5.5A and B), the Dutch cheese shows an open and strongly porous structure (Figure 5.5C and D). These differences can be related to the fact that in the case of the Swiss cheese, the majority of lipid cells exhibit a minute-like character and are unequal in size and shape. Such a variety in lipid cells morphology results very often in inhomogeneous fat distribution within the protein matrix and additionally affects its microstructure. The lipid cells present in the protein matrix of the Dutch cheese become more uniform and manifest a delicate structure with very thin walls. The hydrolytic processes undergone in the cheese matrix upon ripening (12 weeks of storage) result in a decrease in the pH value of the cheese, which in turn changes the contents of nitrogen components. In the SEM pictures, these changes reveal mostly an alteration in the fat structure from small droplets to a more amorphous and fibrous structure. The lipid constituents of the cheese structure become bigger and more irregular as compared to the "fresh" cheeses. These alterations in the form of lipids are more significant in the case of the

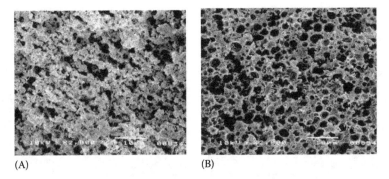

(A)   (B)

FIGURE 5.6   SEM photo of fresh (A) and ripened (B) Camembert-type cheese.

Swiss cheese than in the Dutch one. The microstructure of hard cheeses can also be significantly affected by starter cultures used in their production, as well as by different processing parameters.

Changes in the microstructure of Camembert-type cheese (soft type) during its ripening are demonstrated in Figure 5.6 (Błaszczak and Fornal, 2008). Camembert is characterized by the complex structure of its matrix that is composed of the outer (peel) and inner part. In soft cheeses, the ripening proceeds from the peel (periphery) toward the center. The inner part of fresh Camembert is formed by a heterogeneous protein matrix consisting of minute-like, granular (spherical) structures combined together into a porous network (Figure 5.6A). The ripening process evokes alterations in the microstructure of casein micelles, thus changing their shape from the spherical particles to fibrous-like structures. It leads to the formation of a continuous protein network (Figure 5.6B). A microscopic examination of Camembert cheese indicated that the surface zone was completely devoid of fat globules. On the contrary, the inner part of the cheese was rich in small fat globules. The concentration of fat globules decreased and their size increased gradually from the center toward the surface zone (Yiu, 1985).

Food manufactures responding to steadily increasing demands of consumers for a healthy diet have introduced low-fat food to the market. In such a context, soft- and hard-type cheeses with a reduced fat content have also appeared on the market. It is generally known that a reduction of fat content, especially in the products naturally rich in fat, very often leads to undesirable modification of their texture properties. The textural properties of cheese are strongly influenced by the structural properties of the protein matrix and fat droplets embedded in it. The reduction of fat content affects protein components in a way that more non-interrupted protein zones constitute the cheese structure thereafter. Such a phenomenon makes the three-dimensional protein matrix more resistant to deformation (Lobato-Calleros et al., 2006).

Modified starches and/or other stabilizers of polysaccharide origin are used in cheese production since they interact or embed in the protein matrix, thus hampering protein aggregation (Figure 5.7). The polysaccharide ingredients may affect the melting properties and moisture retention of the low-/reduced-fat products. Low-fat dairy products are very often identified by consumers as rubbery, bland, firm, and unpalatable. Yet, the application of resistant starch in low-fat imitation cheeses was

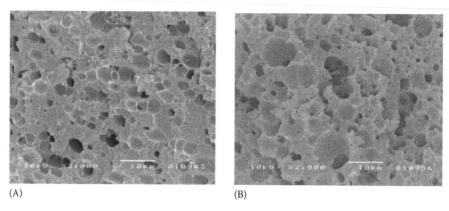

(A)   (B)

FIGURE 5.7   SEM photo of cheese-like products: (A) low-fat cheese and (B) cheese containing vegetable oil (pores after removing of the fat globules are visible).

not found to affect the end product's functionality (Noronha et al., 2008). The microstructure of low-fat cheese with 10% of native corn starch is formed by a protein matrix that contains fat globules with diameters ranging from 2 to 12 μm (Noronha et al., 2008). The addition of pre-gelatinized starch to the cheese at the same concentration affected the character of the fat globules. They seemed to be less numerous and larger in size (~9–20 μm) as compared to those in the native starch-containing cheese. According to these authors, such changes are probably connected with the competition between the pre-gelatinized starch and casein for cold water.

Another example of reduced-fat cheese is a cheese-like product/analog containing emulsified vegetable oils that replace saturated milk fat. Fats are commonly found as oil-in-water or water-in-oil emulsions, but also as water-in-oil-in-water multiple emulsions. The latter is taken into consideration as suitable for the production of reduced-fat food products (Lobato-Calleros et al., 2008). The milk fat globules in natural cheese are stabilized by natural surfactants that are milk proteins, whereas in cheese-like products different hydrocolloids stabilizing the emulsion droplets are added. Moreover, the spatial distribution of the emulsion droplets and interactions with the protein matrix in cheese-like products as well as cheese microstructure were found to be significantly dependent on the type of hydrocolloids added to the formulated milk. The substitution of milk fat globules for oil-in-water emulsion droplets in a cheese-like product evokes a decrease in its hardness, springiness, and chewiness; however, an increase in cohesiveness has also been found as compared to that of the white fresh cheese (Lobato-Calleros et al., 2002).

5.2.3   Fat in Food Emulsions

Typical food emulsions were defined as a dispersed system of two immiscible liquids stabilized by emulsifier(s) and/or stabilizer(s). According to another definition, liquid droplets and/or liquid crystals are dispersed in a liquid. Both oil droplets and/or fat crystals have a significant impact on the structure of an emulsion (morphology) and its properties. The fat crystal network of palm oil, observed under PLM, demonstrated some structural elements ranging from 1 to 140 μm in size that were forming the crystal network. The lower structures were related to crystallites, while the upper ones to the clusters of crystallites (Aguilera et al., 2000). Emulsions based on dairy products are characterized by a fat phase in a partially solid state, which in turn results in some modification of emulsion microstructure and properties (deMan, 1982). The milk proteins are found to be natural emulsifiers very often used in food emulsions. Oils or fats may be emulsified in milk serum (skim milk) with the formation of a stable emulsifier layer. Milk fat and protein (skim milk powder), caseinate or whey protein concentrates, and low-molecular-weight (LMW) emulsifiers are used in the production of aerated products. The LMW emulsifiers are often used in combination with milk proteins for their competitive adsorption to fat–serum or air–serum interfaces (Relkin et al., 2006). Mayonnaise (Figure 5.8A) and salad dressings can be considered as semisolid oil-in-water emulsions. The analyses of a diluted sample of mayonnaise under TEM enable visualizing an interfacial film originating from the egg yolk, surrounding the oil droplets (Tung and Jones, 1981). The interfacial film observed under TEM in diluted mayonnaise was explained in literature as a phenomenon that may be formed of low-density lipoproteins and microparticles of egg yolk granules. The droplet diameter in mayonnaise varies significantly from 1 up to 40 μm and has been found to depend on product type and oil content. Under CLSM, even smaller droplet forms of 0.2 μm were found in high-fat (80%) mayonnaise. The internal structure of

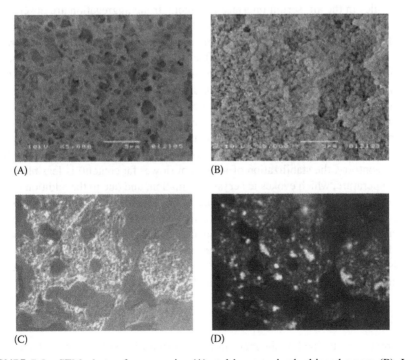

(A)

(B)

(C)

(D)

FIGURE 5.8   SEM photo of mayonnaise (A) and homogenized whipped cream (B). LM photo of fat particles in pork batter under *Nomarski* (C) and fluorescence (D).

the oil droplets varied in character and appeared to be completely amorphous or partially crystalline, or containing much smaller droplets inside (Langton et al., 1999). The degree of crystalline appearance varied in the oil droplets and depended on the type of product. Both soybean and rapeseed oil-based mayonnaises demonstrated crystalline regions in the interior of the oil droplets. Fat crystals formation within fat globules was also observed in ice cream products as well as in whipped cream or butter examined under PLM at 5°C (deMan, 1982). Interactions of emulsifiers with aqueous and lipid phases resulted in the formation of not only liquid crystalline structures but also some mesomorphic phases. The major mesomorphic structures were characterized as lamellar, hexagonal, and cubic, and their geometric forms were found to be affected by processing. The lamellar structures are formed upon the heating of the emulsifier with water, which in turn results in the alteration in the bimolecular layers of lipid that are separated by layers of water. The hexagonal phase consists of two different cylindrical structures—hexagonal I and II. The former structure contains lipophilic hydrocarbon chains inside the core of the cylinders, whereas the latter demonstrates the hydrocarbon chains outside of the cylinders. Monoacylglycerols may also form micellar aggregates that demonstrate an inner structure of the lamellar type and appear in PLM as rod-shaped "batonnets." The lamellar structure displays threadlike, striated networks (deMan, 1982).

The microstructure of whipped frozen emulsions stabilized by different proteins is characterized by partial replacement of native whey proteins by the pre-denatured ones and casein introduced into the oil-in-water emulsion. A high concentration of fat globules in the air–serum interface and their strong aggregation are observed as well. A more heterogeneous distribution of proteins at the fat interface results, in turn, from the formation of protein aggregates. Fat globule microstructure in frozen aerated products may be significantly influenced by the treatment of proteins before homogenization (Relkin et al., 2006). The aerated food emulsions were found to be a very complex system since it contains emulsified droplets composed of partly crystalline and partly liquid fat. The incorporated gas cells are stabilized by an adsorbed layer of partly coalescent fat globules (Allen et al., 2008). A high content of fat in whipped cream affected its clumped structure during aeration (Figure 5.8B), which in turn facilitates rapid and complete stabilization of the incorporated air bubbles. On the contrary, the stabilization of ice cream (lower fat content) is feasible due to low temperature, which evokes ice crystal formation, and due to the addition of sugars, emulsifiers, and hydrocolloid thickeners. The stabilization of whipped homogenized dairy cream may be obtained not only by partial coalescence but also by the association of casein molecules on the surface of fat globules via calcium bridges. As it has been shown on model foam, the stabilization by bridging the aggregation of adsorbed milk protein on the surface of an emulsion droplet can be elicited by acidification (Allen et al., 2006; Relkin et al., 2006).

5.2.4 EMULSIFIED MEAT PRODUCTS

Meat products obtained by the incorporation of small fat particles into a lean meat-salt-water mixture are defined as meat emulsions, or alternatively as meat batters (Gordon and Barbut, 1990). The differences in the definition between the meat emulsions and batters resulted from the theories of stabilization of finely comminuted products. One of them refers to the stabilization of the emulsion structure, indicating the role of an interfacial protein film around fat bodies that stabilizes them upon processing. The second theory is connected with gel formation by the myofibrillar proteins as the main structural element determining the stability of meat batter. There is strong evidence that the physical properties of the fat applied as well as the gelation of meat proteins are critical to the fat and water holding as well as to the texture in processed meat batters.

The microstructure of beef emulsion is regarded as a homogeneous mixture of lipid droplets of various size surrounded by the protein matrix (Carroll and Lee, 1981). It is affected by such factors as type of meat, fat and ingredients, level of fat, moisture and salts, as well as the comminution process and conditions of cooking (Lee, 1985). Dispersion of fat and comminution of muscle tissue was found to be less uniform in the batter prepared from beef than that prepared from a beef–pork combination. Finer fat particles in the case of pork (Figure 5.8C and D) can relate to its softer consistency as compared to the hard consistency of beef fat. When the batter is obtained from chicken, the tissue is completely comminuted and forms a continuous matrix. However, the fat may not be finely dispersed as it is usually observed in the case of beef and pork. The differences in fat dispersion may be influenced by the different types of a chopping machine used in the production, chopping time, and the shear rate. The hardness and melting properties of fat also influence its dispersion and significantly affect thermal stability of the emulsion. Destabilized emulsion reveals a discontinuity of a protein matrix, resulting from losses not only in fat but also in moisture. These changes lead to nonuniform dispersion of fat, and they may also influence the water binding properties of protein in the meat system. The microscopic studies of the meat emulsions allowed to distinguish two groups of fat particles on the basis of their size. While the fat particles ranging in diameter from 1 to $20\,\mu m$ formed the first group, those having diameters over $20\,\mu m$ represented the second group. The first group formed spherical globules remaining in a suspension in the so-called "true emulsions." The latter group formed angular structures physically entrapped in the protein matrix (Lee, 1985). The majority of fat particles appeared in an angular form and reached more than $20\,\mu m$ in diameter. The greater importance of the larger fat particles than those with diameters in the range, for example, of $12–15\,\mu m$ is due to the fact that the former were found to be most responsible for the stability of the emulsion.

REFERENCES

Aguilera, J.M., Stanley, D.W., and Baker, K.W. 2000. New dimensions in microstructure of food products. *Food Sci. Technol.*, 11:3–9.

Allen, K.E., Dickinson, E., and Murray, B.S. 2006. Acidified sodium caseinate emulsion foams containing liquid fat: A comparison with whipped cream. *LWT-Food Sci. Technol.*, 39:225–234.

Allen, K.E., Murray, B.S., and Dickinson, E. 2008. Development of a model whipped cream: Effects of emulsion droplet liquid/solid character and added hydrocolloid. *Food Hydrocoll.*, 22:690–699.

Aro, H., Jarvenpaa, E., Konko, K., Huopalahti, R., and Hietaniemi, V. 2007. The characterization of oat lipids produced by supercritical fluid technologies. *J. Cereal Sci.*, 45:116–119.

Błaszczak, W. and Fornal, J. 2008. Application of microscopy methods in food analysis. *Pol. J. Food Nutr. Sci.*, 58(2):183–198.

Buchheim, W. 1998. The structure of dairy food. *Pol. J. Food Nutr. Sci.*, 7/48 (3(S)):24–36.

Carroll, R.J. and Lee, C.M. 1981. Meat emulsions-fine structure relationships and stability. In *Studies of Food Microstructure*, D.N. Holcomb and M. Kalab (eds.), pp. 105–110. Chicago, IL: Scanning Electron Microscopy, Inc.

Chawla, P., deMan, J.M., and Smith, A.K. 1990. Crystal morphology of shortenings and margarines. *Food Struct.*, 9(4):329–336.

deMan, J.M. 1982. Microscopy in the study of fats and emulsions. *Food Microstruct.*, 1:209–222.

Dubreil, L., Biswas, S.C., and Marion, D. 2002. Localization of puroindoline-a and lipids in bread dough using confocal scanning laser microscopy. *J. Agric. Food Chem.*, 50(21):6078–6085.

Fornal, J. 1998. The changes of plant materials microstructure during processing. *Pol. J. Food Nutr. Sci.*, 7/48, No 3(S):9–23.

Georgopoulos, T., Larsson, H., and Eliasson, A.C. 2006. Influence of native lipids on the rheological properties of wheat flour dough and gluten. *J. Texture Stud.*, 37(1):49–62.

Gordon, A. and Barbut, S. 1990. The microstructure of raw meat batters prepared with monovalent and divalent chloride salts. *Food Struct.*, 9(4):279–296.

Heertje, I. 1998. Fat crystals, emulsifiers and liquid crystals. From structure to functionality. *Pol. J. Food Nutr. Sci.*, 7/48 (2(S)):7–18.

Kalab, M., Caric, M., Zaher, M., and Harwalkar, V.R. 1989. Composition and some properties of spray-dried retentates obtained by the ultrafiltration of milk. *Food Struct.*, 8(1):225–235.

Konopka, I., Rotkiewicz, D., and Tanska, M. 2005. Wheat endosperm hardness. Part II. Relationships to content and composition of flour lipids. *Eur. Food Res. Technol.*, 220(1):20–24.

Konopka, I., Czaplicki, S., and Rotkiewicz, D. 2006. Differences in content and composition of free lipids and carotenoids in flour of spring and winter wheat cultivated in Poland. *Food Chem.*, 95:290–300.

Langton, M., Jordansson, E., Altskar, A., Sorensen, Ch., and Hermansson, A.M. 1999. Microstructure and image analysis of mayonnaises. *Food Hydrocoll.*, 13:113–125.

Lee, C.M. 1985. Microstructure of meat emulsions in relation to fat stabilization. *Food Microstruct.*, 4(1):63–72.

Lobato-Calleros, C., Ramirez-Santiago, C., Osorio-Santiago, V.J., and Vernon-Carter, E.J. 2002. Microstructure and texture of manchego cheese-like products made with canola oil, lipophilic, and hydrophilic emulsifiers. *J. Texture Stud.*, 33:165–182.

Lobato-Calleros, C., Rodriguez, E., Sandoval-Castilla, O., Vernon-Carter, E., and Alvarez-Ramirez, J. 2006. Reduced-fat white fresh cheese-like products obtained from W1/O/W2 multiple emulsions: Viscoelastic and high-resolution image analysis. *Food Res. Int.*, 39:678–685.

Lobato-Calleros, C., Sosa-Perez, A., Rodriguez-Tafoya, J., Sandoval-Castilla, O., Perez-Alonso, C., and Vernon-Carter, E.J. 2008. Structural and textural characteristics of reduced-fat cheese-like products made from $W_1/O/W_2$ emulsions and skim milk. *LWT-Food Sci. Technol.*, 41:1847–1856.

Noronha, N., Duggan, E., Ziegler, G.R., Stapleton, J.J., O'Riordan, E.D., and O'Sullivan, M. 2008. Comparison of microscopy techniques for the examination of the microstructure of starch-containing imitation cheeses. *Food Res. Int.*, 41:472–479.

Relkin, P., Sourdet, S., Smith, A.K., Goff, H.D., and Cuvelier, G. 2006. Effects of whey protein aggregation on fat globule microstructure in whipped-frozen emulsions. *Food Hydrocoll.*, 20:1050–1056.

Sarkar, P.K., Jones, L.J., Gore, W., Craven, G.S., and Somerset, S.M. 1996. Changes in soy bean lipid profiles during kinema production. *J. Sci. Food Agric.*, 71(3):321–328.

Sherestha, A.K., Howes, T., Adhikari, B.P., Wood, B.J., and Bhandari, B.R. 2007. Effect of protein concentration on the surface composition water sorption and glass transition temperature of spray-dried skim milk powders. *Food Chem.*, 104:1436–1444.

Sousa, M.J. and McSweeney, P.L.H. 2001. Studies on the ripening of Cooleeney an Irish farmhouse Camembert-type cheese. *Irish J. Agric. Food Res.*, 40:83–95.

Tung, M.A. and Jones, L.J. 1981. Microstructure of mayonnaise and salad dressing. In *Studies of Food Microstructure*, D.N. Holcomb and M. Kalab (eds.), pp. 231–238. Chicago, IL: Scanning Electron Microscopy, Inc.

Wolf, W.J. and Baker, F.L. 1981. Scanning electron microscopy of soybeans and soybean protein products. In *Studies of Food Microstructure*, D.N. Holcomb and M. Kalab (eds.), pp. 239–252. Chicago, IL: Scanning Electron Microscopy, Inc.

Yiu, S.H. 1985. A fluorescence microscopic study of cheese. *Food Microstruct.*, 4(1):99–106.

Yiu, S.H. 1986. Effects of processing and cooking on the structural and microchemical composition of oats. *Food Struct.*, 5(2):219–224.

Yiu, S.H., Poon, H., Fulcher, R.G., and Altosaar, I. 1982. The microscopic structure and chemistry of rapeseed and its products. *Food Struct.*, 1(2):135–144.

Zhou, M., Robards, K., Glennie-Holmes, M., and Helliwell, S. 1999. Oat lipids. *JAOCS*, 76(2):159–169.

6 Phospholipids

Jan Pokorný and Štefan Schmidt

CONTENTS

6.1 INTRODUCTION

In addition to the rather nonpolar triacylglycerols (TAG), several classes of relatively polar lipids are also present in food materials. The most important components are phospholipids, containing a phosphate or (rarely) a phosphonate group. Another class includes glycolipids, which contain bound sugar. A large, physiologically important group contains sialic acid or closely related derivatives. Lipids containing sulfur are less frequent. Phenolic acids are bound in another group of polar lipids. Lipids containing sugars, sialic acid, or sulfates may also contain a phospho group. As their chemical structures are rather complicated, they are usually called by their trivial or semisystematic names, and only very rarely by their systematic names (the IUPAC/ IUB terminology is used in such cases).

6.2   CHEMICAL STRUCTURES OF PHOSPHOLIPIDS, GLYCOLIPIDS, AND GANGLIOSIDES

6.2.1   CHEMICAL STRUCTURE OF PHOSPHOLIPIDS

The most common phospholipids are glycerophospholipids, earlier called phosphatides. They are derived from glycerol, where sn-1 and sn-2 positions are substituted
with acyls, and the sn-3 position is esterified with phosphoric acids. This group is
called phosphatidyl. If only sn-1-position is substituted with an acyl and the sn-2
position remains free, a lysophosphatidyl is produced.

The bound phosphoric acid may remain free, and such compounds are called
phosphatidic acids (Figure 6.1), or a hydrogen atom of the bound phosphoric acid
may form inner salt or be substituted by another group. After the substitution with
choline, phosphatidylcholine is formed (Figure 6.2), which may exist as an inner salt
(Figure 6.3). Phosphatidyl substituted with ethanolamine is called phosphatidylethanolamine and, in an analogous way, phosphatidylserine is produced (Figure 6.4). A
different situation is produced by the substitution with inositol, as inositol has free
hydroxyl groups (Figure 6.5) that are often substituted by another molecule of phosphoric acid or two molecules.

A specific form of phospholipids are those that are formed when the phosphoric
acid group is substituted by glycerol (phosphatidylglycerol, Figure 6.6). If two
phosphatidyl residues are substituted with diacylglycerol, phosphatidyldiacylglyc-
erol results. Glycerol substituted with a phosphatidyl in the sn-1 and sn-3 positions

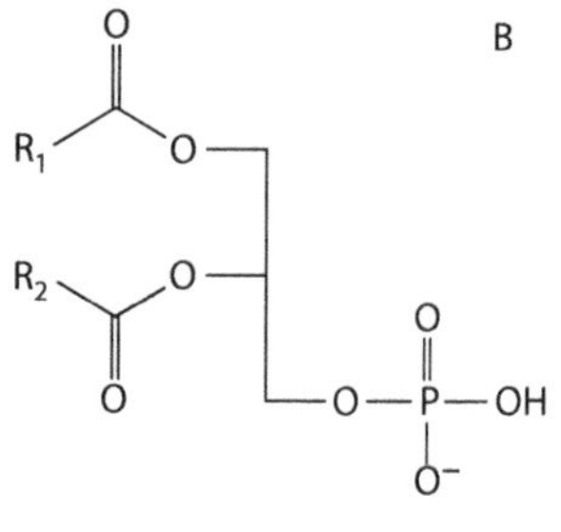

FIGURE 6.1   Chemical structure of phosphatidic acid.

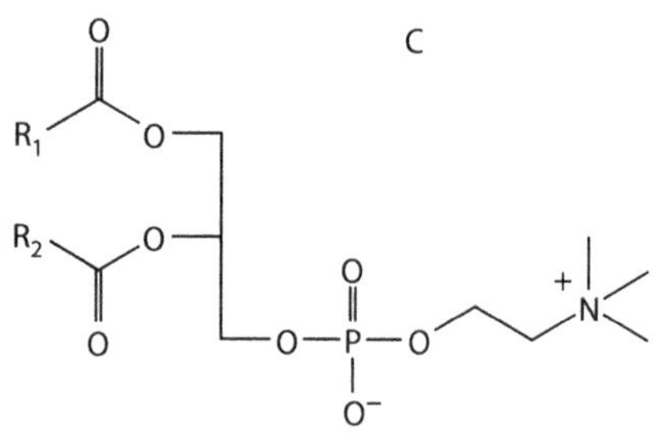

FIGURE 6.2   Formation of inner salts by reaction of phosphatidic acid with hydroxyamine.

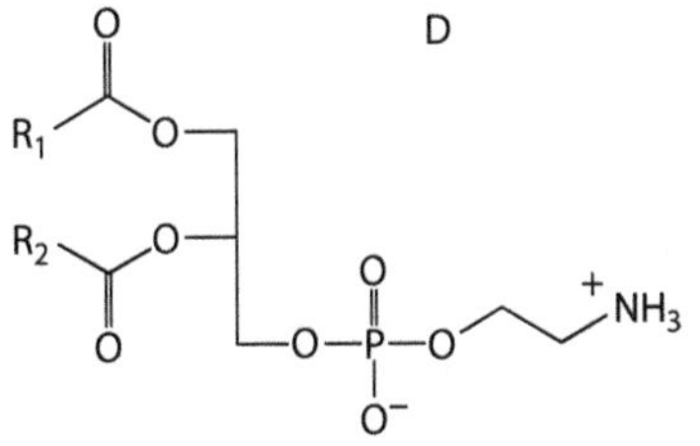

FIGURE 6.3   Chemical structure of phosphatidylcholine.

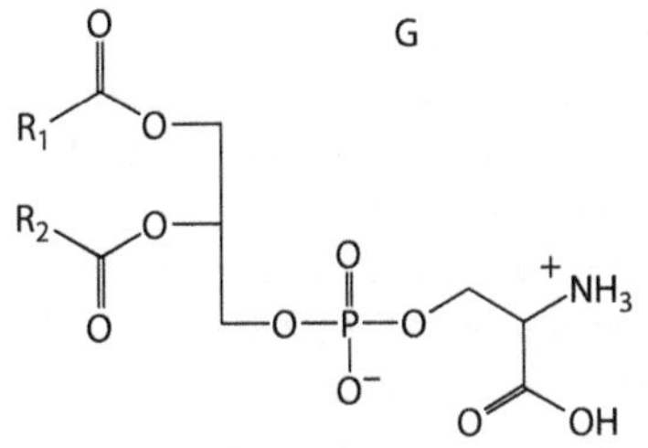

FIGURE 6.4   Chemical structure of phosphatidylserine.

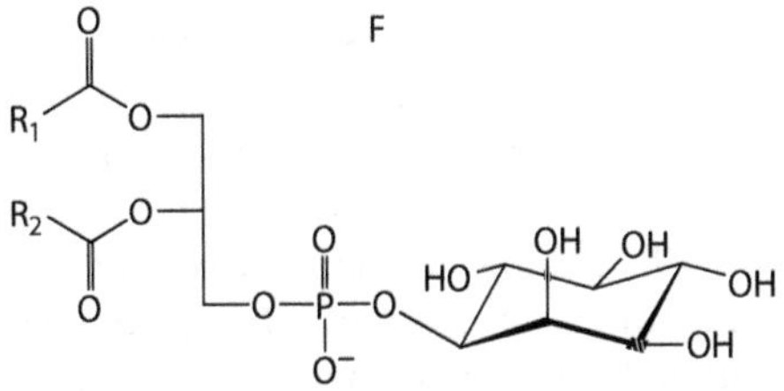

FIGURE 6.5   Chemical structure of phosphatidylinositol.

(the *sn*-2 position remains free) is called cardiolipin, which is a biologically active compound. All the other above-mentioned glycerophospholipids could, naturally, also exist in a lyso form (Figure 6.7). All phospholipases are thermolabile.

Phospholipases are present in foods, which cleave glycerophospholipids at different places. They are differentiated by adding letters (A, B, C, D). It is evident from Figure 6.8 which phospholipase may cleave a particular bond.

Aldehydes may exist in two forms:

$$O=CH-CH_2-CH_2-R$$

or

$$HO-CH_2-CH=CH-R \text{ (1-alken-1-ol)}$$

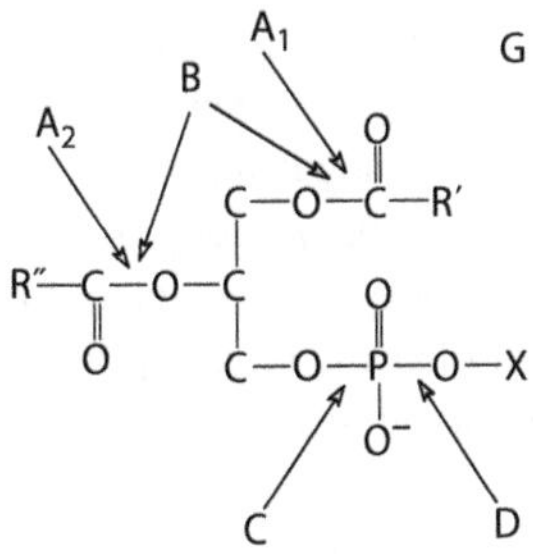

FIGURE 6.6   Location of points of attack of phospholipases of glycerollipids.

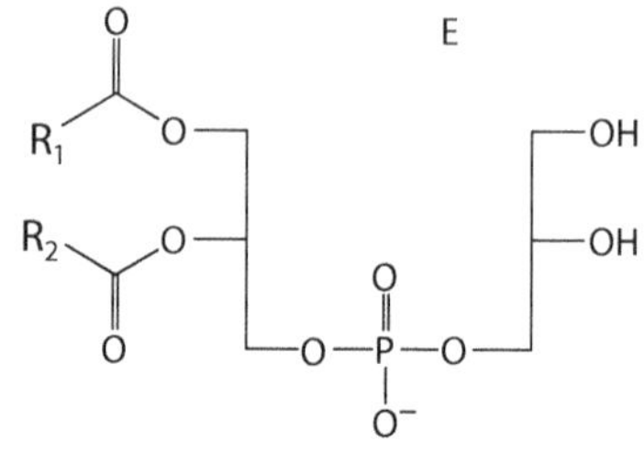

FIGURE 6.7 Chemical structure of phosphatidylglycerol.

The latter can form ethers with a hydroxyl group of glycerol. Phosphatidic acids, where the *sn*-1 position, if etherified with an aldehyde, are called plasmenic acid. The plasmenylcholine is derived in a similar way.

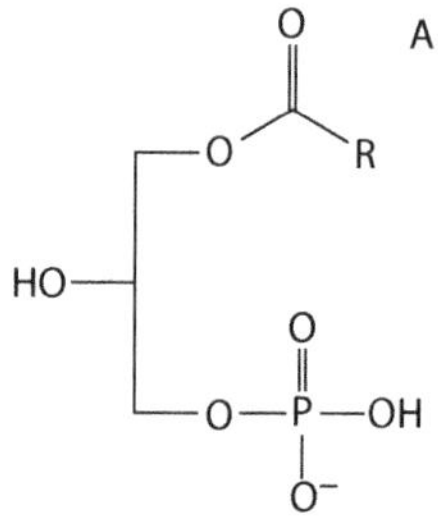

Instead of one phosphoric acid group in phosphatidylglycerol, a diphospho group can be substituted, such as in 1,2-diacyl-*sn*-glycerol-3-diphosphocytidine.

In addition to glycerolphospholipids, other alcoholic bases are possible. The most widely occurring ones are sphingo-lipids, which are formed by substitution of sphingenin (systematic name: (2-S, 3-S, 4-E)-2-amino-4-octadecene-1,3-diol. The 2-position may be esterified with a

FIGURE 6.8 Chemical structure of a lysophospholipid.

fatty acid, forming a ceramide. If the 1-hydroxyl is esterified with a molecule of phosphoric acid, ceramide phosphate is produced. The phosphoric acid can be substituted with a choline molecule, producing sphingomyelin (*N*-acyl-4-D-hydrosphingenine phosphocholine, Figure 6.9). Sphingenin derivatives exist in animal tissues, the related (2-S, 3-S, 4-R)-2-amino-1,3,4-octanetriol (D-hydroxysphinganine) is found in plants. Sphingolipids, ceramides, and sphingomyelins may also contain bound sugars.

6.2.2 CHEMICAL STRUCTURE OF GLYCOLIPIDS

Glycolipids, more correctly glycosyl lipids, contain both a lipid moiety and bound sugars, most often D-galactose, arabinose, or fucose. They often contain a phospho group too. A simple glycolipid is psychosin, which belongs to a group of sphingosine glycosides. Their more correct collective name would be 1-monoglycosylsphingolipids. Glycosides of ceramides are called cerebrosides. In this group of glycosphingolipids, several sugar units may be bound one to another, for example, a tetragalactoside.

Glycoglycerolipids contain bound glycerol, such as 1,2-diacyl-3-beta-D-galactosyl-*sn*-glycerol, or a related digalactosides, which were detected in plants such as spinach leaves. Glycolipids may also contain bound phenolic acids, which are active as antioxidants.

6.2.3 STRUCTURE OF GANGLIOSIDES

Gangliosides (formerly mucolipids) belong to a group of polar lipids of high biological interest. They are acidic glycosphingolipids or glycolipids, which contain either

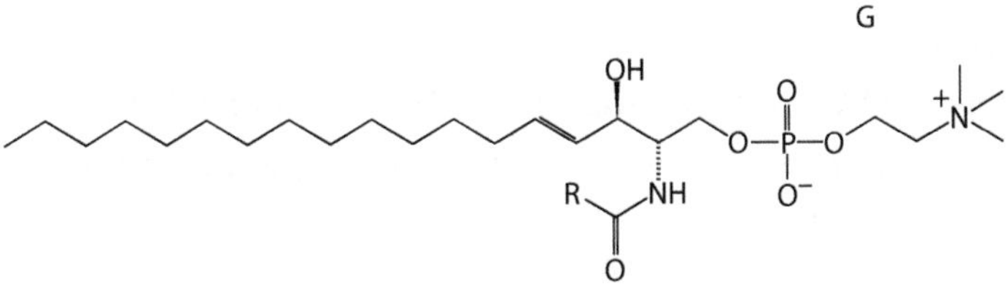

FIGURE 6.9 Chemical structure of sphingomyelin.

bound sialic acids or sulfuric acid. The sugar moiety is sulfated in sulfoglycosphingolipids, for example, in glycosylceramide sulfates.

Glycolipids that contain bound sialic acid residues (one or more) are called gangliosides. Sialic acid is a common name of *N*-acetyl or *N*-glycoloyl neuraminic acid (5-amino-3,5-dideoxy-D-glycero-D-galactononusonic acid), which is bound in ceramides. Because of their complicated chemical structures, they are most often expressed by a sequence of codes, not using chemical structures. More details are found in special literature.

6.3 COMPOSITION OF NATURAL PHOSPHOLIPIDS

6.3.1 OCCURRENCE OF PRIMARY PHOSPHOLIPID CLASSES IN TOTAL PHOSPHOLIPIDS

The composition of a lipid fraction in a lipid extract from food is rather variable, not only between two different species, but also within the same species (Table 6.1). Phosphatidylcholine is the main phospholipid component in most phospholipid concentrates. In egg yolk, the phosphatidylcholine content is particularly high in comparison with that of egg white, cereals, vegetables, legumes, and mushrooms

TABLE 6.1
Distribution of Lipid Classes in Different Phospholipid Concentrates

Phospholipid Class	Egg Yolk	Milk Fat	Liver	Muscle	Soybean	Rapeseed
PC	68–86	10–26	46–52	47–63	18–32	18–26
PE	8–24	25–45	24–28	18–24	6–17	14–31
PS	tr-2	tr-6	2–5	2–5	tr-2	1–2
PI	tr	tr-4	6–9	4–9	17–24	6–14
SPH	1–5	10–30	3–5	tr-1	tr-1	tr-1
Lyso	2–6	tr-1	tr-1	tr-1	tr-1	2–12
PA	tr-1	tr-1	2–4	tr-1	tr-6	tr-6

Notes: PC, phosphatylcholine; PE, phosphatidylethanolamine; PS, phosphatidylserine; PI, phosphatidylinositol; SPH, sphingomyelins; Lyso, lysophospholipids; PA, phosphatidic acid; tr, traces.

TABLE 6.2
Example of Fatty Acid Distribution in Two Phospholipid Classes

Material	Phospholipid	Saturated Acids	Monoenoic Acids	Polyenoic Acids
Egg yolk	PC	53–54	27–28	18–19
	PE	49–50	19–20	30–31
Rapeseed	PC	8–12	60–66	22–30
	PE	11–12	51–56	32–38

Note: PC, phosphatidylcholine; PE, phosphatidylethanolamine; results in %.

(Table 6.2). However, differences exist between phospholipid classes from the same species, depending on variety, degree of ripeness, feed composition, climatic conditions, and soil composition. For example, in pig adipose tissue, phosphatidylethanolamines contain less saturated fatty acids and more polyunsaturated acids (PUFA) than phosphatidylcholines (Body, 1988). The content of eicosatetraenoic acid is particularly high in the phosphatidylethanolamine fraction, and fish phosphatidylethanolamines are also rich in docosahexaenoic acid (Bandarra et al., 2001).

6.3.2 Fatty Acid Composition of Phospholipids

The composition of fatty acids in *sn*-1 position is also somewhat different from that in the position *sn*-2 (Table 6.3). In the former case, saturated fatty acids prevail, while the latter contains more unsaturated fatty acids. The unsaturated fatty acid content bound in the position *sn*-2 was 97.3%, 91.4%, and 95.9% in egg, cottonseed, and groundnut phospholipids, respectively (Vijayalakshmi and Rao, 1972).

TABLE 6.3
Composition of Major Fatty Acids in 1- and 2-Positions in Phosphatidylcholine (%)

Fatty Acids in Position	Fatty Acid	Chicken Muscle	Egg Yolk
Position 1	Palmitic acid	30.2	38.2
	Stearic acid	1.1	9.3
	Monoenoic acids	3.0	3.0
Position 2	Palmitic	20.0	21.8
	Stearic	4.0	11.2
	Monoenoic	20.0	1.5
	Polyenoic	5.5	9.3

6.4 OCCURRENCE OF PHOSPHOLIPIDS IN FOODS

6.4.1 PHOSPHOLIPIDS CONTENT IN FOODS

Phospholipid contents are very similar (about 1%–2% dry matter) in microbial, plant, and animal tissue, showing their importance among structural lipids. If the content of neutral lipids is low, phospholipids may account for 20%–40% of lipid extract (e.g., in marine invertebrates or cereals). In egg yolk, 23% of the total lipids are phospholipids and other polar lipids (Kuksis, 1985), because phospholipids are necessary for the growing avian organism in the egg. On the contrary, in adipose tissue or in oilseeds, the content of phospholipids varies between 1% and −3% of total lipids. In oilseeds rich in oil (such as in rapeseed or coconut oil), the content of phospholipids is lower than in oilseeds with lower oil content (such as soybeans or corn germ), when the results are expressed on the basis of the respective oil content. It is much the same if the content is expressed in terms of total dry matter of the oilseed.

Phospholipids are mainly extracted (especially on the plant scale) by nonpolar solvents, together with other lipids, and are present in crude oils. However, in the original material, phospholipids are primarily bound to proteins (as lipoproteins), e.g., in membranes or may be bound in other tissue components. For example, phospholipids interact with chlorophyll pigments, where they form complexes between the central magnesium ion of the chlorophyll molecule and the phospho group of the phospholipids.

If the material is extracted by nonpolar solvents, the yield of phospholipids is relatively low, but the bonds between proteins and phospholipids are destroyed by heat denaturation of the protein moiety (best carried out by application of steam). Another way is the application of alcohols.

6.4.2 FUNCTIONS OF PHOSPHOLIPIDS IN ORIGINAL FOOD MATERIALS

In both animal and plant tissues, phospholipids are primarily bound in cellular membranes. Phospholipids are very important emulsifiers in living tissues, where they are bound as lipoproteins. They help to transport nonpolar lipids in blood and other intercellular fluids (Chapter 11).

The scope of the functions of phospholipids can be broadened by their transesterification (Figure 6.10). The acyl, located in the position *sn*-2, can be replaced by another acyl.

In foods, phospholipids are often present at the interface of emulsions or they cooperate in forming films on the surface of solid particles. The best emulsion stability is achieved when neutral phospholipids, such as phosphatidylcholine, are added to negatively charged lipids. It is sometimes difficult, as the majority of lipid fraction is neutral. In technological practice, it is suitable to select common phospholipid concentrates, in which a certain amount of negatively charged phospholipids is present.

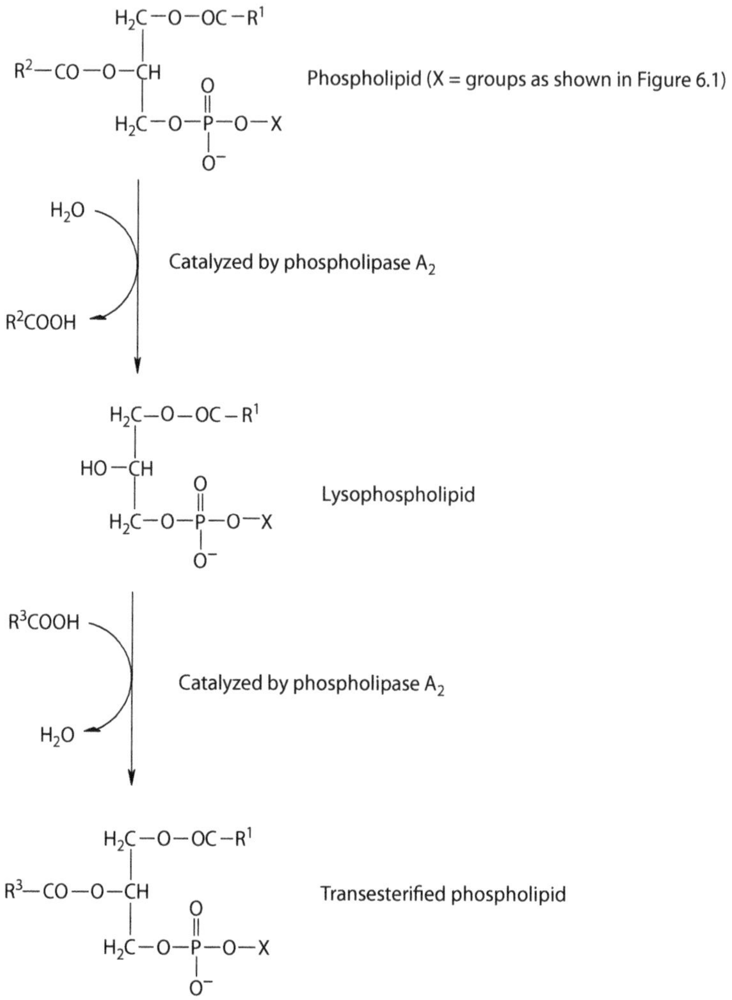

H₂C—O—OC—R¹ (structure diagram)

Phospholipid (X = groups as shown in Figure 6.1)

Catalyzed by phospholipase A₂

Lysophospholipid

Catalyzed by phospholipase A₂

Transesterified phospholipid

FIGURE 6.10 Reaction of transesterification of phospholipids.

6.5 PROPERTIES AND APPLICATIONS OF PHOSPHOLIPIDS

6.5.1 MANUFACTURE OF PHOSPHOLIPID CONCENTRATES

The most important phospholipid concentrates are those resulting from oilseed processing. Oilseeds are ground, heated to high temperature (over 100°C) with application of steam. Lipoproteins and cell membranes are damaged by these operations, and phospholipids are liberated from their complexes with proteins and similar substances, such as polypeptides. They are then expeller pressed to obtain crude oil. However, only small amounts of phospholipids are obtained in this form. The cakes still contain about 5%–10% oil. The next operation is the extraction of expeller cakes. Pentane gives rather low yields of phospholipids, but larger yields are

TABLE 6.4
Specification of Commercial Lecithin

Type of Lecithin	Plastic Lecithin	Fluid Lecithin	Food Grade
Acetone insolubles (%)	≤65	≤62	≤50
Moisture (%)	Max. 1	Max. 1	Max. 0.3
Benzene insolubles (%)	Max. 0.3	Max. 0.3	Max. 1.5
Acid value (mg/g)	Max. 30	Max. 32	Max. 36

obtained with hexane (which is the most widely used solvent, and the largest with heptane (or isohexane): 0.06%, 0.22%, and 0.31%, respectively. The effect of solvents is, however, selective as high yields of phosphatidic acids and phosphatidylethanol-amines were obtained with hexane, and higher contents of phosphatidylcholines and phosphatidylinositols with heptane.

Crude oils contain 1%–3% phospholipids, which belong to minor lipids, and accompanying substances, the presence of which makes oil unsuitable for human nutrition. The acceptance of crude oils depends on the quality of processed seeds and the technology. Virgin oils are an exception, as they are acceptable under certain conditions, which will be discussed later.

Phospholipids are removed in the first step, called degumming. They are treated with hot water, which precipitates them, but unhydratable phospholipids, bound mostly as salts, are not precipitated. Therefore, phosphoric acid or citric acid is applied to decrease the phospholipid content in degummed oil. The acids cleave metal ions from phospholipids. Most phospholipids are hydrated, so that they become insoluble in neutral oils and precipitate. The phospholipid concentrates are then washed, water is removed by distillation. The preparations containing 50%–70% phospholipids are thus obtained. They are called lecithin.

Most lecithins obtained during the degumming of edible oils are unsuitable for human use. They are added back to the extracted meal and used as feed. Only soybean lecithin has a quality sufficient for human food and industrial uses. Properties required for soybean lecithin are shown in Table 6.4 (adapted after Procise, 1985). The phospholipid fraction of soybean lecithin contains about 30%–32% phosphatidylcholine and 20%–29% other substances. The fatty acid composition of soybean lecithin is shown in Table 6.5. The technical soybean lecithin obtained by extraction with a hydrocarbon solvent is very different from the phospholipid fraction obtained by extraction with chloroform or diethyl ether and methanol (after Folch); however, it is rather more expensive.

6.5.2 Modifications of Technical Phospholipid Concentrates

For many purposes, technical soybean lecithin must be modified. The dark brown color is improved by oxidative bleaching with hydrogen peroxide. Lecithin obtained in this way contains additional hydroxyl groups, which increase its activity as an emulsifier. Fluid lecithins are obtained by adding fatty acids or by several other methods.

TABLE 6.5
Fatty Acid Composition of Soybean Lecithin Determined after Extraction with Different Solvents (% Total Phospholipids)

Phospholipid Class	Hexane Extract	Chloroform- Methanol
Phosphatidylcholine	14.5	36.7
Phosphatidylethanolamine	20.1	13.3
Phosphatidyl serine	3.2	2.9
Phosphatidylinositol	11.9	19.3
Phosphatidic acid	30.2	10.6

Lecithins prepared from soybean oil refining contain about 30% neutral lipids. For some purposes, the material is fractionated with acetone, because phospholipids are insoluble in acetone while neutral lipids are soluble without limitations. They are then obtained in powder form, which should be protected against moisture. They can be fractionated with ethanol as well. Another method of phospholipid modification is hydrogenation under pressure.

Phosphatidylcholines are the most important fraction of soybean lecithin. Their content may be increased by transesterification with choline hydrochloride, catalyzed by phospholipase D. Phosphatidyl choline content may be thus increased from 30% to 60%–70%, and from 75%–80% to more than 90% (Jumeja et al., 1989). Similarly, phosphatidylserine can be produced from phosphatidylcholine by enzyme-catalyzed interesterification (Yaqoob et al., 2001). Another modification of lecithins is the interesterification of lysophosphatidylcholine with fish PUFA under catalytic action of phospholipase A2 (Na et al., 1990).

Aldehydes react with lysine, and unsaturated imines are produced by their dehydration (Figure 6.11), which are very reactive. (The reaction can be applied to phospholipids as a source of amine groups, too.)

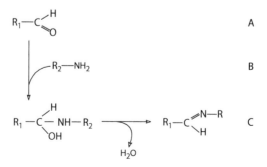

FIGURE 6.11 Reaction of an amino compound with an aldehyde.

6.5.3 Applications of Phospholipids in the Food Industry

Phospholipids from animal sources, mainly egg yolk phospholipids, can be used similarly as soybean lecithin; however, these are very expensive. For this reason, they are used only for pharmaceutical products or for biochemical research.

Phospholipids from plant sources are obtained on the industrial scale almost exclusively as a by-product of crude oil refining (Section 6.4.2). Naturally, the production of high quality soy lecithin is much lower than needed. Therefore, it is sometimes mixed with other raw materials. The most suitable lecithin substitute is produced by phosphorylation of a mixture of diacylglycerols and monoacylglycerols. It is still better to use purified diacylglycerols, obtained after removal of monoacylglycerols under very low pressure distillation, and the neutralization of the phosphorylated product with ammonia.

The applications of phospholipids in the food industry are manifold. In the bakery industry, phospholipids increase the volume of bakery goods, improve the fat dispersion, and possess antistaling properties, too. They also add to the nutritional value (Schäfer, 1998).

In the chocolate industry, lecithin increases the fluidity of the chocolate mass and prevents the crystallization of high melting TAG. Particles of some foods, such as dried milk or dry soups, are covered with a film of fat, which prevents a rapid and efficient contact with water. Phospholipids act as wetting and dispersion agents so that they enable fast dissolution in the aqueous phase or rapid dispersion (in instant foods). There are numerous other uses of lecithins for edible purposes.

Most lecithin is utilized for nonfood purposes, as emulsifiers, dispersion agents, adhesives, or lubricants. Lecithin is added to coatings, such as paints, waxes, and polishes. Its use in the cosmetic and pharmaceutical industries is also important (Wendel, 2001). The double bonds of an unsaturated phospholipids may be treated with epoxides (Figure 6.12), forming either hydroxylated secondary amine of a hydroxylated phospholipids (Figure 6.13), which has higher surface active properties or can be substituted by other polar substituents.

Lipid hydroperoxide reacts with phosphotidylethanolamine with the formation of an unsaturated imine, which is very reactive (Figure 6.14) as they may polymerize into melanoidins (Figure 6.15). In a similar reaction with phosphatidylcholine, the phospholipid molecule is cleaved with the formation of trimethylamine oxide (Figure 6.16).

6.6 ROLE OF PHOSPHOLIPIDS IN FOOD QUALITY

The main positive functions of phospholipids in foods are their surface-active properties. They act as emulsifiers and stabilizers of emulsions or suspensions (Van Nieuwenhuyzen and Szuhaj, 1998). They facilitate the dispersion of solid particles in water phase and improve the texture of multiphase food materials. Phospholipids make the texture smooth and improve the pleasantness by increasing the viscosity of morsel during chewing. The ingested food tastes full and homogenous. The bitter taste is also suppressed, but other tastes are not affected.

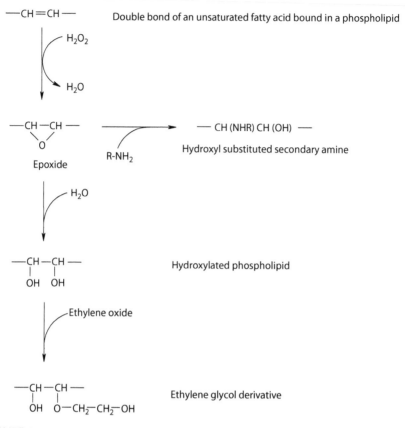

FIGURE 6.12   Reaction of unsaturated phospholipids with epoxides.

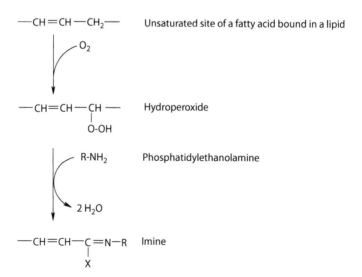

FIGURE 6.13   Reaction of a lipid hydroperoxide with phosphatidylethanolamine.

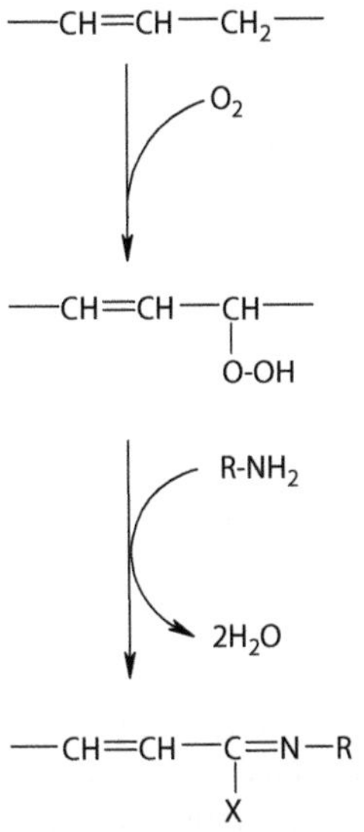

$$-CH=CH-CH_2-$$

$$\downarrow \; O_2$$

$$-CH=CH-CH-$$
$$\quad\quad\quad\quad | $$
$$\quad\quad\quad\quad O\text{-}OH$$

$$\downarrow \; R\text{-}NH_2$$

$$\downarrow \quad 2H_2O$$

$$-CH=CH-C=N-R$$
$$\quad\quad\quad\quad | $$
$$\quad\quad\quad\quad X$$

FIGURE 6.14 Reaction of phosphatidylethanolamine with a lipid hydroperoxide.

Phospholipids can increase the nutritional value of food and can be applied to functional foods (Schneider, 2001).

Phospholipids increase the oxidative stability of fats and oils and fatty foods, in that, they act as synergists of tocopherols (Khan and Shahidi, 2001), and other natural antioxidants, such as flavonoids. They stabilize even polyunsaturated oils (Kouřimská et al., 1994) and fish oils. Phosphatidylcholine reacts with peroxy radicals to yield trimethylammonium oxides. Phosphatidylcholines also react with lipid hydroperoxides in the nonradical way to produce imines (Figure 6.13). Phospholipids can also bind heavy metal traces, which otherwise would act as prooxidants, to produce inactive, undissociated salts.

Bulk phospholipids are relatively stable on storage (Réblová et al., 1991), but are less stable when dissolved in oils. They slowly change during the storage of lipid foods. In unheated food materials, phospholipases are active so that hydrolysis occurs, accompanied by formation of phosphatidic acids. In addition to hydrolysis, oxidation reactions also occur. The oxidation of PUFA, detected in phospholipids, proceeds in the way similar to that of the neutral lipid oxidation, only the peroxy free radicals of the hydroperoxides formed from them can react with amine groups in a nonradical way.

6.7 CHANGES OF PHOSPHOLIPIDS DURING FOOD PROCESSING

One process, which is very important from the standpoint of industrial production of phospholipid concentrates, is the removal of phospholipids from crude vegetable oils (Section 6.4.1). Refined oils contain only traces of phospholipids.

During the storage of raw materials or of unheated foods, phospholipids become partially hydrolyzed. The reaction is catalyzed by different phospholipases. The hydrolyzed products affect food texture, mainly of increased surface activity. Lysophospholipids are formed during cheese ripening.

Acidic phospholipids easily form salts with free amino acids or bound basic amino acids, such as lysine, or with metals ions always present in foods (most easily with calcium and magnesium ions).

Glycerophospholipids contain PUFA in the position *sn*-2. They are oxidized by similar mechanisms as PUFA bound in TAG.

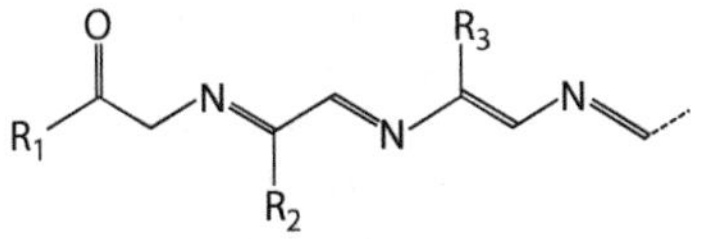

FIGURE 6.15 Chemical structure of melanoidin.

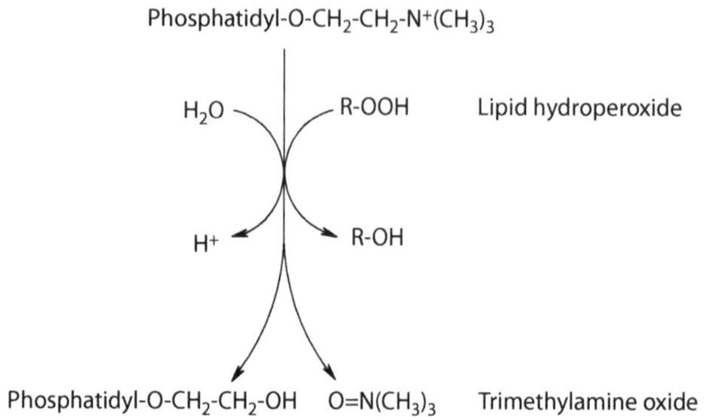

Phosphatidyl-O-CH$_2$-CH$_2$-N$^+$(CH$_3$)$_3$

H$_2$O   R-OOH   Lipid hydroperoxide

H$^+$ R-OH

Phosphatidyl-O-CH$_2$-CH$_2$-OH   O=N(CH$_3$)$_3$ Trimethylamine oxide

FIGURE 6.16 Oxidation of phosphatidylcholine by lipid hydroperoxide.

Amine groups of phospholipids, particularly of phosphatidylethanolamine or serine, react with reducing sugars, osones, and other products of sugar degradation to form intermediary products, which are polymerized into brown melanoidins (Figure 6.14). The pathway via Amadori rearrangement is similar to the way of amino acids (Utzmann and Lederer, 2001). They thus contribute to the darkening of phospholipid concentrates on storage, which turn brown by other mechanisms (oxidation, with production of melanoids). Melanoidins are partially bleached by lipid hydroperoxides as the peroxide destroys the system of imino bonds.

Changes of phospholipids during cooking, baking, or frying are similar to those of TAG, only the browning reactions are more intensive. Phosphatidylethanolamines are decomposed more easily during microwave heating than other phospholipids, because of the presence of a primary amine group (Yoshida et al., 2001). Similar degradation was observed during heating of pork. Phospholipids were decomposed during microwave heating of soybeans following the order:

phosphatidylethanolamine > phosphatidylcholine > phosphatidylinositol (Takagi and Yoshida, 1999).

Because phospholipids enhance the decomposition of lipid hydroperoxides, they stimulate the formation of flavor-active volatiles in meat and meat products (Mottram, 1999).

Among other interaction reactions, binding of phospholipids to phytin or other phospho derivatives via calcium or magnesium salt is very important. Amine groups of phospholipids may also form complexes with polyphenolic compounds and their oxidation products, particularly with quinones.

REFERENCES

Amate, L., Ramírez, M., and Gil, A. 1999. Positional analysis of triglycerides and phospholipids rich in long-chain polyunsaturated fatty acids, *Lipids*, 14, 865–871.

Bandarra, N. M., Batista, I., Nuñes, M. I., and Empis, J. M. 2001. Seasonal variation in chemical composition of horse mackerel, *Eur. Food Res. Technol.*, 212, 535–539.

Body, D. R. 1988. The lipid composition of adipose tissue, *Prog. Lipid Res.*, 27, 31–60.

Devos, M., Poisson, L., Ergan, F., and Pencrea'h, G. 2006. Enzymic hydrolysis of phospholipids from *Isochyris albama* for DHA enrichment, *Enzyme Microb. Technol.*, 39, 548–555.

Ghop, S.-H., Khox, H. T., and Gel, P. T. 1982. Phospholipids of palm oil, *JAOCS*, 59, 296–299.

Hamaguchi, N., Yokoyama, Y., Kasahara, Y., Hayashi, A., Sono, R., Tebayashi, S., Chul-Sa, K., and Hon-Sik, K. 2006. Heat deterioration of phospholipids. IV. Thermally deteriorated products from phosphatidylethanolamine and several sugars, *J. Oleo Sci.*, 55, 607–613.

Hara, S., Hasuo, H., Nakasato, M., Higaki, Y., and Totani, Y. 2002. Modification of soybean phospholipids by enzymatic transacylation, *J. Oleo Sci.*, 51, 417–421.

Hidalgo, F. J., Nogales, K., and Zamora, R. 2008. The role of aminophospholipids in the removal of cyto- and geno-toxic aldehydes products during lipid oxidation, *Food Chem. Toxicol.*, 46, 43–48.

Holmbäck, J., Karlsson, A. A., and Arnoldsson, K. C. 2001. Characterization of N-acylphosphatidylethanolamine and acylphosphatidylglycerol in oats, *Lipids*, 36, 153–155.

Hon, L., Jie, J., Chang-hu, X., Bim, Z., and Jia-chao, X. 2003. Seasonal changes in phospholipids of mussel (*Mytilus edulis* L.), *J. Sci. Food Agric.*, 83, 133–135.

Jianming, L. and Blank, I. 2009. Odorants generated by thermally induced degradation of phospholipids, *J. Agric. Food Chem.*, 51, 4364–4369.

Jumeja, L. R., Yamana, T., and Shimizu, S. 1989. Enzymatic method of increasing phosphatidylcholine content of lecithin, *JAOCS*, 66, 714–717.

Khan, M. A. and Shahidi, F. 2001. Tocopherols and phospholipids enhance the oxidative stability of borage and evening primrose triacylglycerols, *J. Food Lipids*, 7, 143–150.

Kouřimská, L., Pokorný, J., and Réblová, Z. 1994. Phospholipids as inhibitors of oxidation during food storage and frying, *Prehr.-Technol. Biotechnol. Rev.*, 32, 91–94.

Kuksis, A. 1985. Animal lecithin, in *Lecithin*, Szuhaj, B. F. and List, G. R., Eds., AOCS Press, Champaign, IL, pp. 105–162.

Lange, R., Engst, W., Elsner, A., and Brücker, J. 1994. Composition, preparation, and properties of rapeseed phospholipids, *Fat Sci. Technol.*, 96, 169–174.

Mottram, D. S. 1999. Flavour formation in meat and meat products, *Food Chem.*, 62, 415–424.

Na, A., Eriksson, S.-G., Österberg, E., and Holmberg, K. 1990. Synthesis of phosphatidylglycerol with n-3 fatty acids by phospholipase A2 in microemulsion, *JAOCS*, 67, 766–770.

Pokorný, J. 2006. Production, separation and modification of phospholipids for use in food, in *Modifying Lipids for Use in Food*, Gunstone, F. D., Ed., Woodhead Publishing Ltd., Cambridge, U.K., pp. 369–390.

Procise, W. E. 1985. Commercial lecithin products, in *Lecithin*, Szuhaj, B. F. and List, G. R., Eds., AOCS Press, Champaign, IL, pp. 163–182.

Réblová, Z., Pokorný, J., and Pánek, J. 1991. Oxidation of stored soybean lecithin, *Nahrung*, 35, 665–666.

Schäfer, W. 1998. Lecithin: More than an additive, *Getreide, Mehl, Brot*, 26–30.

Schmid, H. H. O., Schmid, P. C., and Natarajan, Y. 1990. N-Acetylated glycerophospholipids and their derivatives, *Prog. Lipid Res.*, 29, 1–43.

Schmitt, H. 2009. Phospholipids: Making life healthier, *Wellness Foods Eur.*, 2, 36–41.

Schneider, M. 2001. Phospholipids for functional foods, *Eur. J. Lipid Sci. Technol.*, 103, 98–101.

Soma, D. and Bhattacharyya, D. K. 2006. Preparation of surface-active properties of hydroxy and epoxy fatty acid-containing soy phospholipids, *JAOCS*, 83, 1015–1020.

Sono, R., Sakamoto, S., Hamaguchi, M., Tebayashi, S., Hen-Sik, K., and Horiike, M. 2002. Heat deterioration of phospholipids, II. Isolation and identification of new thermally deteriorated products from soybean lecithin, *J. Oleo Sci.*, 51, 191–202.

Takagi, S. and Yoshida, H. 1999. Microwave heating influences on fatty acid distribution of triacylglycerols and phospholipids of soybeans, *Food Chem.*, 66, 345–351.

Utzmann, C. M. and Lederer, M. D. 2000. Independent synthesis of aminophospholipid-linked Maillard products, *Carbohydr. Res.*, 325, 157–168.

Van Nieuwenhuyzen, W. and Szuhaj, B. F. 1998. Effects of lecithin and proteins on the stability of emulsions, *Fett/Lipid*, 100, 282–291.

Van Nieuwenhuyzen, W. and Tomas, M. C. 2008. Update of vegetable lecithin and phospholipid technology, *Eur. J. Lipid Sci. Technol.*, 110, 472–486.

Venkatesh, M., Kakali, M., Kuntal, M., and Mukherjee, P. K. 2009. Enhanced oral bioavailability and antioxidant profile of ellagic acid, *J. Agric. Food Chem.*, 57, 4559–4565.

Vijayalakshmi, B., and Rao, S. V. 1972. Fatty acid composition of phospholipids in seed oils containing unusual acids, *Chem. Phys. Lipids*, 9, 82–86.

Wang, W. and Wang, S.-S. 2007. Preparation of condensed soybean phospholipid and its feeding value, *Food Sci. Technol.*, 1, 194–195.

Wang, T., Hammond, E. G., Cornett, J. L., and Fehr, W. R. 1999. Positional analysis of triacylglycerols and phospholipids rich in long-chain polyunsaturated fatty acids, *Lipids*, 14, 865–871.

Wendel, A. 2001. Lecithin in the first 150 years. II. Evolution of a global pharmaceutical industry, *Inform*, 11, 992–997.

Yaqoob, M., Nabi, A., and Masoon-Yasinzai, M. 2001. Bioconversion of phosphatidylcholine or phosphatidylserine using immobilized enzyme mini columns, *Process Biochem.*, 36, 1181–1185.

Yoshida, H., Hirakawa, Y., and Abe, S. 2001. Influence of microwave roasting on positional distribution of fatty acids of triacylglycerols and phospholipids in sunflower seeds, *Eur. J. Lipid Sci. Technol.*, 103, 201–207.

Yoshida, H., Tomiyama, Y., Tanaka, M., and Mizushina, Y. 2007. Distribution of fatty acids in triacylglycerols and phospholipids from peas (*Pisum sativum* L.), *J. Sci. Food Agric.*, 87, 2709–2714.

Yoshida, H., Saiki, M., Tomiyama, Y., and Mizushima, Y. 2008. Positional distribution of fatty acids in triacylglycerols and phospholipids from Adzuku beans (*Vigna anguleris*), *Eur. J. Lipid Sci. Technol.*, 110, 158–163.

Youngjin, Y. and Eunok, C. 2009. Lipid oxidation and stability of tocopherols and phospholipids in soy-added fried products during storage in the dark, *Food Sci. Biotechnol.*, 18, 356–361.

Zhang, H., Tian, H., and Sun, B. 2007. Effects of free phospholipids and sulfur-containing volatile components in Maillard model systems, *Food Sci. China*, 28, 8–85.

Zhao, L. C. 2006. Research on production technology of powdered phosphatide, *Food Sci. Technol.*, 8, 81–83.

7 Cholesterol and Phytosterols

Erwin Wąsowicz and Magdalena Rudzińska

CONTENTS

7.1 INTRODUCTION

Sterols constitute a large group of compounds with a broad range of biological activities and physical properties. They are derivatives of steroids and are common in eukaryotic cells but rare in prokaryotes. The natural occurring sterols usually have a 1,2-cyclopentano-phenanthrene skeleton and 27–30 carbon atoms with a hydroxyl group at C-3 and side chain of at least seven carbon at C-17. Sterols are minor constituents present in the fat of the human diet. They comprise the major portion of the unsaponifiable fraction of most edible fats and oil. Sterols can be classified according to their origin as animal sterols or as plant sterols. The latter can be subdivided into phytosterols (higher plant sterols) and mycosterols (lower plant sterols present in the lipid fraction of yeast and fungi). Cholesterol is the main animal sterol, while sitosterol, campesterol, and stigmasterol are the major plant sterols. These phytosterols and cholesterol are all 4-desmethylsterols that share identical ring structures. The various sterols differ only in their side chain. Surprisingly, these minor differences result in major changes in biological functions. They are widely distributed in nature, and occur both in the free form or, most frequently, as esters of higher aliphatic acids and glycosides. Sterols, sterol esters, and sterol glycosides are, to different degrees, soluble in fats but completely insoluble in water.

Since the discovery of cholesterol, sterols have continued to be the focus of research activities of many chemists, biochemists, and clinicians, as attested to by the fact that three Nobel Prizes awarded between 1910 and 1985 were associated with work on sterols. The continued interest of food scientists in dietary oxidized cholesterol is related to atherosclerosis. In this chapter, the recent developments in the area of occurrence and biological effect of phytosterols, especially in the lowering of blood cholesterol, and their presence in diet is extended in relation to the first edition of this title [1].

7.2 CHOLESTEROL

7.2.1 Structure and Occurrence

Cholesterol, with a C27 carbon skeleton (Figure 7.1), is synthesized and utilized by animals. Today, it is most widely known for its association with atherosclerotic heart disease. However, despite its negative reputation, cholesterol is a necessary constituent of all human cells and fulfills important functions, as an essential component of mammalian cell membranes, a precursor for steroid hormones and bile acids, as well as being involved in various cell signaling pathways [2]. Cholesterol can be present in the free form or esterified at the hydroxyl group with fatty acids of various chain length and saturation. It also occurs in plants, usually in very small quantities, and marine algae. The content of cholesterol in some foods is presented in Table 7.1.

There exists a strong positive correlation between increased serum cholesterol concentrations and the risk of coronary heart disease (CHD)—the reason why most consumers are concerned about excessive intake of cholesterol [3–5]. Although the role of dietary cholesterol in human health has not yet been fully understood, factors

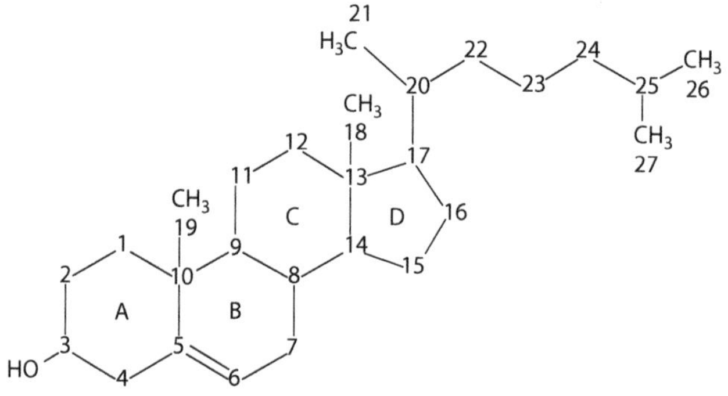

FIGURE 7.1 Structure of cholesterol showing the carbon atom numbering.

TABLE 7.1

Cholesterol and Phytosterol Content in Selected Food Products

Product	Cholesterol (mg/100 g)	Product	Phytosterols (mg/100 g)
Skim milk	1.8	Refined oils	
Whole milk	13.6	Corn	768–1200
Curd cheeses	5–37	Olive	140–220
Processed and hard cheeses	51–99	Soybean	300–440
Cream and sweet cream	35–106	Rapeseed	680–880
Butter	183–248	Sunflower	320–350
Pork	72–100	Flax	310–320
Lard	92	Rice bran	600–2800
Beef	65–82	Peanuts	600–1600
Tallow	109	Sesame seeds	200–250
Polish sausage (different types)	27–83	Almonds	160–270
Chicken, whole	75	Pecans	140–270
Turkey, light meat	60	Kidney beans	30–160
Liver	300–360	Soybeans, mature	100–120
Raw whole egg	450	Wheat grain	70–124
Raw egg yolk	1260	Bananas	31
Tuna	38	Carrots	12
Cod	73	Tomatoes	7
Lobster	95	Potatoes	5
Shrimp	152	Strawberries	1

that raise serum cholesterol, such as dietary cholesterol, are generally considered to be unfavorable. Dietary cholesterol intake is variable, but is often less than 300 mg/day, and currently 200 mg/day is recommended [6]. Approximately 25% of the serum cholesterol production rate is due to absorbed dietary cholesterol, while 75% is accumulated for by endogenously synthesized cholesterol. Dietary and endogenous cholesterols are inversely correlated, suggesting that they are co-regulated. The average cholesterol gastrointestinal absorption is only 56%, and varies between individuals.

7.2.2 Methods of Reducing Cholesterol in Foods

Various physical, chemical, and biological methods have been proposed for reducing cholesterol in foods. These include blending with vegetable oils, extraction with organic solvent, absorption with saponin to form cholesterol complexes, vacuum distillation, degradation by cholesterol oxidase, and removal by supercritical carbon dioxide. Most of these methods are relatively nonselective. The removal of cholesterol from dairy products, lard, and egg yolk was most effectively achieved by powdered β-cyclodextrin (β-CD) [7]. β-CD is not toxic, edible, nonhygroscopic, chemically stable, and easy to separate; these properties are positive for cholesterol removal. Crosslinked β-CD (obtained with adipic acid) is very efficient for removal of cholesterol and has effective recycling efficiency. Using β-CD, cholesterol reduction reached more than 90% and no significant changes were found in most physicochemical and sensory properties in products after β-CD treatment. Alonso et al. [8] have shown on a commercial scale a very effective process for cholesterol removal from pasteurized nonhomogenized milk at 4°C with 0.6% β-CD without changing fatty acid and triacylglycerol composition.

On an industrial scale, frying fat, trade name "Appetize," was introduced in the United States [9]. Appetize contains neither cholesterol nor *trans* fatty acids and develops good fried flavor associated with frying in beef tallow. Appetize is obtained by blending cholesterol-stripped animal fats (80%) with vegetable oils. Michicich et al. [10] have shown that the butter users preferred Appetize Lard over margarine, suggesting that Appetize Lard offers an appealing alternative to margarine for cholesterol-conscious consumers.

Another possible way of lowering cholesterol in food is the application of special diets for feeding animals. Many experiments with varying feeding conditions have been preformed. Precht [11] has shown that the cholesterol content in milk fat can be lowered by 8%–13% with special feeding conditions using rapeseed. A study by Fatouh [12] has shown that supercritical carbon dioxide is a useful tool for producing buffalo oil fractions that differ markedly in their properties. Cholesterol was concentrated in fraction one, and with increasing the fluid density, it decreased by more than 50% in fraction four.

7.2.3 Cholesterol and Its Relationship to Atherosclerosis

Atherosclerosis in human can have severe clinical sequelae, including heart attack, stroke, and peripheral vascular disease. In developed countries, atherosclerosis is responsible for more morbidity and mortality than any other single degenerative disease. Research over the past few decades has led to a new consensus on the sequence of events that initiate atherosclerotic lesions. Many of these events, including the accumulation of cholesterol in macrophages (foam cells), are accelerated by low-density lipoprotein (LDL) that has undergone oxidative modification. The accumulation of lipoprotein cholesterol is clearly central to the initiation of the "fatty streak"—the first anatomically defined lesion in atherosclerosis.

The casual relationship between elevated serum cholesterol and atherosclerosis has been established beyond doubt by several lines of evidence [13,14]. Approximately 20% of adults between the ages of 20 and 74 years have serum total cholesterol levels in the "high risk" category, that is, total cholesterol above $240\,mg/cm^3$ and LDL cholesterol greater than $160\,mg/cm$ [15].

Kanner [16] in his review stressed the association between LDL cholesterol and atherogenesis based in part on atherosclerotic complications and the observation that cholesterol-lowering therapy greatly diminished the clinical manifestations of atherosclerosis. Despite the association between atherosclerosis and LDL cholesterol, LDL particles did not appear to be atherogenic in themselves, but become so only after minimal modification. Oxidation is a process that leads to a biological modification of LDL particles. Oxidized lipoproteins have been identified in atherosclerotic lesions in both animals and humans. However, the origin of oxidized lipoproteins *in vivo* is not clear. Atherosclerosis may result, at least partly, from processes that occur following food ingestion. Staprans et al. [17] have shown that oxidized dietary lipids such as oxidized fatty acids and oxidized cholesterol in the diet are absorbed by the intestine and are incorporated into serum lipoproteins in both animals and humans and thereby contribute to the formation of oxidized lipoproteins in the circulation. In animals, oxidized lipoproteins contribute to fatty streak formation in the aorta and atherosclerosis. It is well established that typical diet in Western countries contains large quantities of oxidized fatty acids, oxidized cholesterol, cytotoxic aldehydes, and phospholipids.

7.2.4 OXIDATION OF CHOLESTEROL

The expressions "cholesterol oxidation products" (COPs) and "oxysterols" refer to a group of sterols similar in structure to cholesterol but containing an additional hydroxyl, ketone, or epoxide group, on the sterol nucleus or a hydroxyl group on the side chain of the molecule. Table 7.2 presents the names of most prominent COPs formed in foods, plasma, and tissues.

Foods of animal origin are suspected to contain some amount of COPs formed by autoxidation. Cholesterol autoxidation is a well-established free radical process that involves the same chemistry that occurs for the oxidation of unsaturated lipids. Cholesterol contains one double bond at the carbon-5 position; therefore, the weakest points in the structure are at the carbon-7 and carbon-4 positions. However, due to the possible influence of the hydroxyl group at carbon-3 and the tertiary atom at carbon C-5, the C-4 position is rarely attacked by molecular oxygen, and therefore the abstraction of an allylic hydrogen predominantly occurs at C-7 and gives rise to a series of A- and B-ring oxidation products. In the chain reaction, usually initiated by free radicals, epimeric hydroperoxides of cholesterol and cholesterol epoxides are formed. The presence of tertiary atoms at C-20 and C-25 in side chain adds to the center's sensitivity to oxidation, forming oxysterols (usually called side-chain oxysterols) (Figure 7.2) [18,19].

TABLE 7.2
Nomenclature of Some COPs

Systematic Name	Common Name	Abbreviated Name
Cholest-5-en-3β,7α-diol	7α-Hydroxycholesterol	7α-HC
Cholest-5-en-3β,7β-diol	7β-Hydroxycholesterol	7β-HC
5-Cholestane-3β,5α,6β-triol	Cholestanetriol	CT
Cholest-5-en-3β-ol-7-one	7-Ketocholesterol	7-kC
5,6α-Epoxy-5β-cholestan-3β-ol	Cholesterol-α-epoxide	α-CE
5,6β-Epoxy-5β-cholestan-3β-ol	Cholesterol-β-epoxide	β-CE
Cholest-5-en-3β,20α-diol	20α-Hydroxycholesterol	20-HC
Cholest-5-en-3β,22-diol	22-Hydroxycholesterol	22-HC
Cholest-5-en-3β,25-diol	25-Hydroxycholesterol	25-HC
Cholest-5-en-3β,26-diol	26-Hydroxycholesterol	26-HC

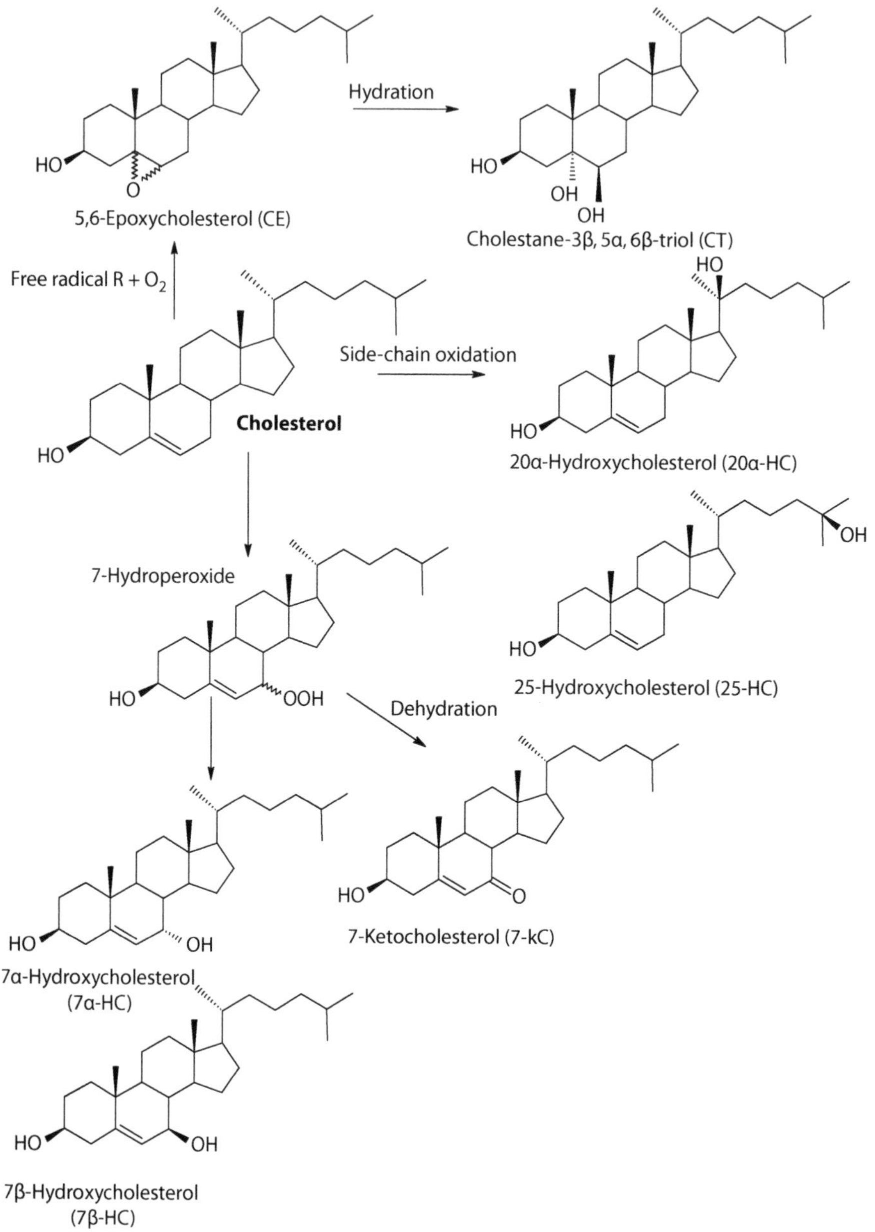

FIGURE 7.2 Cholesterol oxidation pathway.

In animal and human tissues, COP can be formed either enzymatically or by nonenzymatic oxidation. Enzymatic oxidation of cholesterol occurs by a number of cytochrome P-450 enzymes of the liver. Most prominent, cholesterol-7α-hydroxylase (EC1.14.13.17) and 26-hyroxylase (EC1.14.13.15) are implicated in bile acid biosynthesis. Notably, 24- and 25-hydroxylase activities were also found in mammalian liver. Outside the liver, 26-hydroxylase activity is detectable in a variety of cells, including fibroblasts, vascular endothelial cells, macrophages, and brain and kidney cells. Biosynthesis of C-21-steroid hormones requires 20- and 22-hydroxylation of cholesterol, forming 20α-HC, 22-HC, and cholest-5-ene-3β,20,22-triol. Further, the enzymatic origin of COPs is also possible through the action of dehydrogenases, but its significance for COPs formation in humans is not well documented [20].

Nonenzymatic formation of COPs from cholesterol *in vivo* can be directly induced by the action of reactive oxygen species like $^1O_2$, $^\bullet O_2^-$, ROO$^\bullet$, RO$^\bullet$, and HO$^\bullet$ [19]. However, much more attention has been paid to another mechanism, one in which polyunsaturated fatty acyl esters of cholesterol are subjected to oxidation, initially yielding cholesterol esters of fatty acid hydroperoxides with further reaction to 7-hydroperoxycholesterol acyl esters. The epimeric forms of 7-hydroperoxycholesterol are reduced to give 7α-HC and 7β-HC, or are dehydrated to give 7-kC. During cholesterol ester autoxidation, the Δ^5 double is also involved, forming α-CE and β-CE and further reacting to give CT. In addition, nonenzymatic side-chain oxidation of cholesterol occurs, yielding, for example, 25-HC.

Cholesterol oxides are present in our diet and have been identified in foods high in cholesterol [21–23]. As a rule, fresh foods contain very low level of cholesterol oxides. Storage, cooking, and processing tend to increase the COP content. Their concentration is particularly high (10–150 μg/g dry weight) in dried egg, milk powders, heated butter (ghee), precooked meat and poultry products, and heated tallow. The amount of COPs in ghee exceeds 12% of cholesterol in butter [24]. Currently, Soto-Rodriguez et al. [25] have found high extent of cholesterol oxidation (14.5%) in traditional Mexican sun-dried shrimp. Derewiaka and Obiedziński [26] have shown the content of COPs in commercial and thermally processed meats and meat products consumed in Poland. The amounts of COPs in heat-treated products ranged from 17.5 to 34.9 μg/g, and were statistically higher than before frying (2.2–10.7 μg/g). Processed meat products contained COP that equaled from 1% to 8.3% of cholesterol content.

The results from Zaborowska et al. [27] indicate that COPs contents in Polish sausage were between 4.4 and 36.5 μg/g. The content of COP in infant formulas after production ranged from 5.4 to 10.6 μg/g lipid extract [28]. The storage of model infant formulas in aluminum foil in a nitrogen atmosphere or vacuum at temperatures lower than 18°C up to 1 year does not increase the COPs content.

The formation of COPs in animal products can be minimized by the application of low processing temperatures (i.e., through minimal processing), by the use of oxygen-proof packaging and a protective atmosphere as well as by low-temperature and light-free storage, and by the dietary antioxidants in animal feed or antioxidants addition to foods. Conchillo et al. [29] have proved that vacuum packaging was particularly efficient in decreasing the rate of oxidation during frozen storage of cooked chicken breast. Flaczyk et al. [30] and Rudzińska et al. [31] have shown that butylated hydroxytoluene (BHT) was a poorer antioxidant than the crackling hydrolysates against cholesterol oxidation in meatballs during 7 days of refrigerated storage or storage at −18°C for 360 days.

Although there are many methods for determining COPs in foods, no generally accepted standardized methods are available. The quantification of cholesterol oxides in foods is difficult because their isolation is frequently hindered by the large amounts of interfering cholesterol, triacylglycerols (TAG), phospholipids, and other lipids present in the food. Analytical errors may result from artifacts of cholesterol oxidation, losses during processing, lability of certain oxysterols, poorly chosen internal standards, interference from contaminants, and insufficient sensitivity of detection. The identity of the separated compounds needs to be confirmed by mass spectrometry and, if possible, verified by labeled cholesterol [22,32,33].

7.2.5 Biological Effects of Oxysterols

COPs have several *in vitro* and *in vivo* biological effects at the cellular level including cytotoxicity, atherogenesis, mutagenesis, carcinogenesis, changes in cellular membrane properties, and inhibition of 3-hydroxy-3-methylglutaryl coenzyme A reductase. However, there are also indications that oxysterols may represent normal constituents of physiological processes, and serve as physiological mediators in many cholesterol-induced metabolic effects and as potential chemotherapeutic agents. These biological functions of COPs were reviewed by Schroepfer [34], Björkhem et al. [35], Garcia-Cruset et al. [36], Osada [37], Javitt [38], Gill et al. [2], and Lordan et al. [39].

The deleterious biological effects of various COPs have led to great interest at present because both exogenous and endogenous COPs may be related to the initiation or progression of various diseases. A considerable amount of research has focused on possible involvement of COP in the pathogenesis of atherosclerosis.

There is no direct evidence that oxidized cholesterol is atherogenic in humans. Hundreds of *in vitro* studies have demonstrated several activities of COPs related to the atherosclerotic process, indicating that these compounds may play an important role in the initiation and progression of atherosclerosis. It is widely accepted that administration of COPs to animals induces the formation of atherosclerotic lesions. Iuliano et al. [40] revealed that atherosclerotic plaques contained 45 times greater amounts of 7β-HC and 7-kC compared to normal human arteries. The presence of COP in plaques is interpreted as a consequence of LDL oxidation. Studies such as these have implicated COP in the development of atherosclerosis. Garcia-Cruset et al. [36] indicated that much more extensive research is required to assess the relative contribution of each source (dietary and endogenous) to the COPs levels found in animals and humans. However, apart from proatherogenic properties displayed by COP, several new lines of evidence support the view that they may also be involved in the pathogenesis of degenerative diseases such as Alzheimer's disease and age-related macular degeneration [39]. Consequently, taking into account adverse biological effects of COPs, it seems essential to control their presence in the diet.

7.3 PHYTOSTEROLS

7.3.1 Structure and Occurrence

Plant sterols are triterpenes similar in structure to cholesterol, containing the four-ring steroid molecule with 3β-hydroxyl group on the ring A and 5,6-double bond on ring B. Most phytosterols have a side chain composed of 9 or 10 carbon atoms and in the side chain often contain a double bond (Figure 7.1). The main physiological function of phytosterols is functionality and stabilization of cell membranes, and these components are synthesized exclusively by plants.

Plant sterols are classified into three groups:

1. 4-Desmethylsterols (cholestane series)
2. 4-Monomethylsterols (4α-methylcholestane series)
3. 4,4′-Dimethylsterols (lanostane series, also known as triterpene alcohols)

The most common representatives of phytosterols are sitosterol (Figure 7.3A), stigmasterol (Figure 7.3B) and campesterol (Figure 7.3C), brassicasterol (Figure 7.3D, specific mainly for *Cruciferae* family), and avenasterol (Figure 7.3E). 4-Methylsterols are intermediates in sterol biosynthesis and are always present in fats and oils accompanying 4-desmethylsterols. The predominant 4-methylsterols are citrostadienol, obtusifoliol, and gramisterol (Figure 7.4). 4,4′-Dimethylsterols (e.g., cycloartenol and 24-methylene-cycloartanol, Figure 7.4) are present in vegetable oils in minor amounts. Cycloartenol is the first cyclic product formed during biosynthesis of plant sterols in higher plants and algae. The conversion of cycloartenol to other plant sterols proceeds in three steps: the alkylation of the side chain, demethylation of the C-4 and C-14 methyl groups, and desaturation to form a double bond. The details of plant sterol biosynthesis are not clear, but it is likely that they are formed

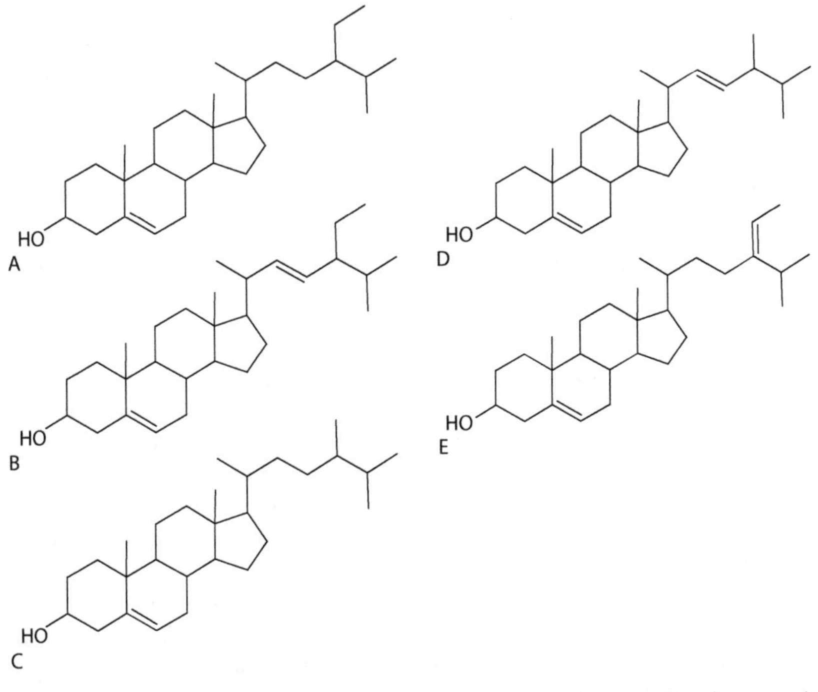

FIGURE 7.3 Chemical structure of 4-desmethylsterols: A, sitosterol; B, stigmasterol; C, campesterol; D, brassicasterol; E, avenasterol.

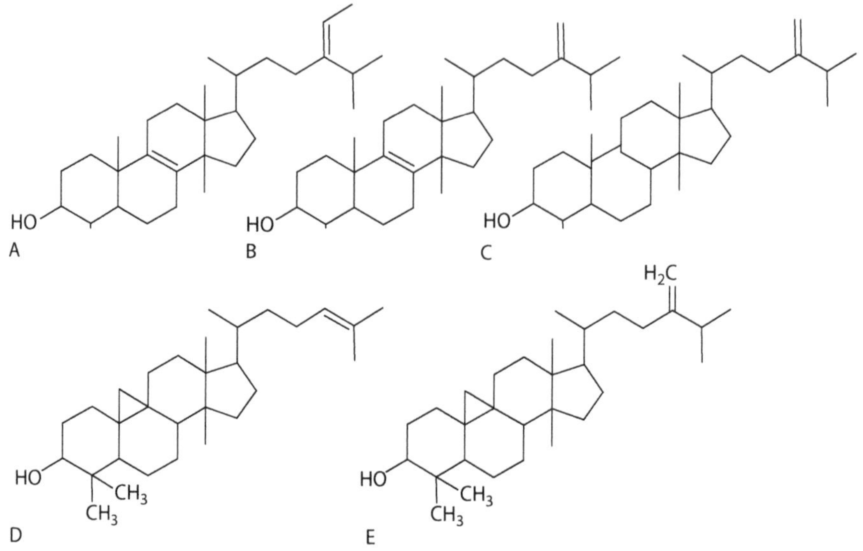

FIGURE 7.4 Structure of 4-monomethyl- and 4,4'-dimethylsterols: A, citrostadienol; B, obtusifoliol; C, gramisterol; D, cycloartenol; E, 24-methylene-cycloartanol.

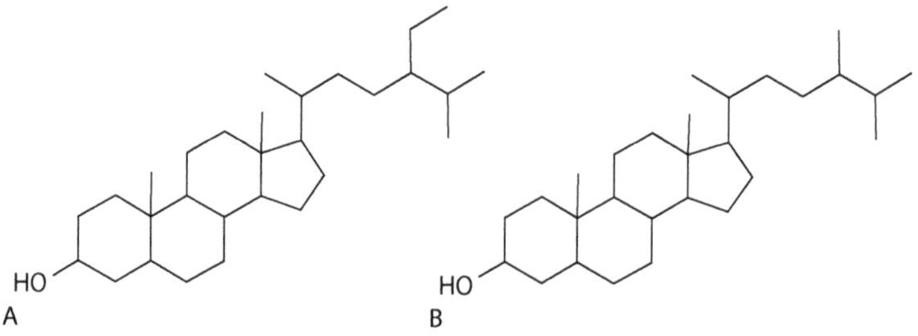

FIGURE 7.5 Chemical structure of saturated sterols (phytostanols): A, sitostanol; B, campestanol.

similarly to animal sterols [41]. Saturated plant sterols, stanols such as sitostanol and campestanol (Figure 7.5), occur usually in small amounts in some plants, e.g., wheat, rye [42], and corn kernels [43]. In all plant tissues, phytosterols exist as the free sterols, esters of fatty acids, steryl glycosides, and acylated steryl glycosides (Figure 7.6). Corn, rice, and other grains contain esters of hydroxycinnamic acid, where the sterol 3β-hydroxyl group is esterified with ferulic or p-coumaric acids (Figure 7.6) [44,45].

Phytosterols are poorly absorbed (0.6%–28%) in the human digestive system, however effectively hamper absorption of cholesterol [46]. The level of total phytosterols in plasma of healthy subjects is less than 1.0 mg/dL; however, in vegetarians, the observed amount was a few times higher. Patients with sitosterolemia have plasma phytosterol levels ranging from 12 to 40 mg/dL [47].

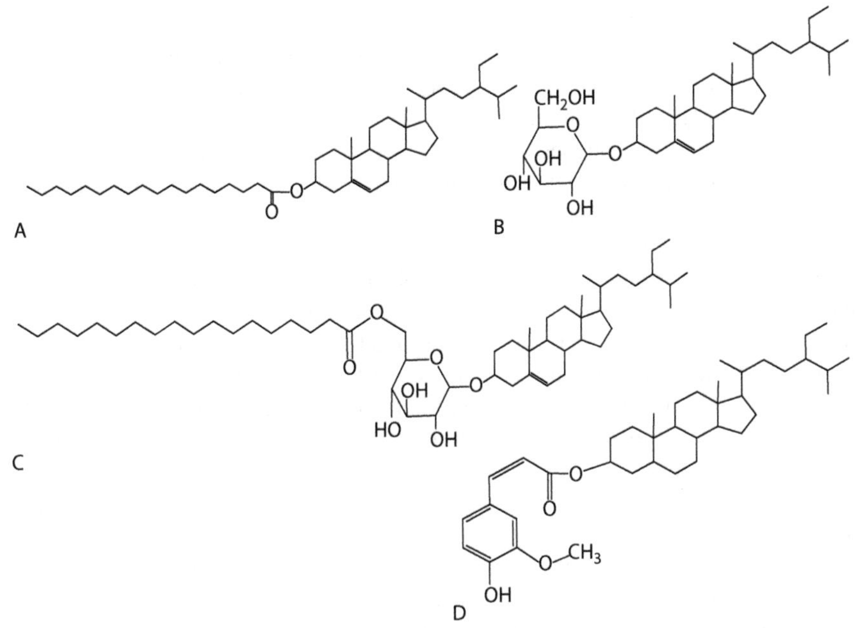

FIGURE 7.6 Chemical structure of steryl esters: A, sitosteryl stearate; B, sitosteryl β-D-glucoside; C, sitosteryl (6′-O-steaoryl) β-D-glucoside; D, hydroxycinnamate steryl ester.

7.3.1.1 Plant Sterols in Food Products

Plant sterols are found in all foods of plant origin. They make up the largest proportion of the unsaponifiable fraction of lipids. The sterol contents in common and typical foods and vegetable oils are listed in Table 7.1. The most important natural sources of phytosterols in human diets are oils and margarines. The content of phytosterols in most vegetable oils ranges from 1.0 to 5.0 mg/g of oil. Cold-pressed sunflower, soybean, and flax oils contained on average 3.2 mg of phytosterols per gram of oil, while crude soybean oil contained 3.0–4.4 mg/g of oil. Corn and rapeseed oils contain much higher amounts of phytosterol, 7.8–11.1 mg/g and 6.8–8.8 mg/g of oil, respectively. Specialty oil such as wheat germ oil contains 17–26 mg/g of phytosterols. Much lower amounts of phytosterols are found in palm oil (0.7–0.8 mg/g), coconut oil (0.7 mg/g), and olive oil (1.4–1.9 mg/g) [43,48–52].

During oil processing, the content of phytosterols decreases to the extend affected by the applied conditions. In expeller rapeseed oil, the content of plant sterols ranged from 8.4 to 8.7 mg/g, while in solvent-extracted oil the amount was 25% higher. During refining the amount of phytosterols decreased by 20%.

Among edible oils, rice bran oil (RBO) is a rich source of phytosterols, which are present in significant amount as γ-oryzanols, esters with ferulic acid (Figure 7.6). The amounts of γ-oryzanol in RBO is affected by genetic and environmental factors, and is within 9.0–29.0 mg/g [53]. The wild rice endogenous to North America contained 0.5–0.7 mg γ-oryzanols per gram of oil [54].

Nowadays, unconventional sources of edible oils are gaining attention owing to their health benefits, which are linked to unusual fatty acid composition and often high content of phytosterols. Oils from walnut and evening primrose contained 0.9 and 8.5 mg/g of phytosterols. The content of phytosterols in other specialty oils were as follows: pistachio oil, 5 mg/g; walnut oil, 1.8–3.0 mg/g; almond oil, 2.2–2.7 mg/g; hazelnut, 1.2 mg/g; and pecan oil, 2.7 mg/g [55–57].

Seeds of grapes and berries are by-products of juice production, and can be a source of oils rich in valuable compounds. Phytosterol content in selected berry seed oils varied from 4.0 to 6.9 mg/g [58], while in seed lipids from wild Canadian prairie fruits ranged from 3.6 to 8.6 mg/g [59].

Cereal products are important source of plant sterols that together with vegetable oils contribute up to 40% of daily intake of plant sterols. The sterol content in Finland rye, wheat, barley, and oat were 1.0, 0.7, 0.8, and 0.4 mg/g, respectively [60]. Authors also reported that analyzed grains contained significant amounts of phytostanols such as sitostanol and campestanol (Figure 7.5) in the total phytosterol fraction. Seitz [61] reported phytosterol ferulate esters in corn, wheat, rye, rice, and triticale. However, the amount of phytosterols in cereals is affected by genetic factors and growing conditions [62,63]. Harrabi et al. [43] demonstrated that each section of corn kernel contains characteristic phytosterols and phytostanols. The presence of phytosterols in bakery products is affected by the type of flour and the amount and type of fat used for dough formulation. The total phytosterol content in bread ranged from 0.4 to 0.9 mg/g, where the highest amounts were observed in bread baked with wholegrain flour [60].

Vegetable oils and cereals are considered the best natural sources of dietary plant sterols; however, consumption of raw and cooked vegetables may also contribute substantially to the total sterol intake. In fresh vegetables, the plant sterol contents are from low to moderate, ranging from 0.05 mg/g of in fresh potatoes to 0.4 mg/g in Brussels sprouts [64]. The content of phytosterols in uncooked broad bean and cauliflower was 0.3 mg/g; in courgette, 0.2 mg/g; and in carrot and cabbage, 0.1 mg/g [65].

Most nuts are regarded as good sources of phytosterols. The content of phytosterols in unshelled peanuts was 0.6–1.6 mg/g; in shelled peanuts, 0.6–1.3 mg/g; and in roasted peanuts, 0.6–1.1 mg/g [66].

Dietary phytosterol intakes are usually in the range of 100–450 mg/day and are affected by many factors like food tradition and major food sources [55,67]. The average intakes of phytosterols are as follows: Finland, 140–360 mg/day from nonenriched food sources [51]; United Kingdom, 163 mg/day [68]; and Ireland, 254 mg/day [69].

7.3.1.2 Food Products Enriched in Phytosterols and Phytostanols

During the last decade, increased interest in phytosterols as cholesterol-lowering food components was observed. Evidence of this phenomenon includes more than 40 patents on phytosterol products and more than 10 commercial phytosterol products being marketed in many parts of the world. The first phytosterol product was Cytellin, marketed by Eli Lilly & Co. from 1957 to 1982. The active ingredient in Cytellin was mainly free β-sitosterol with lower amounts of other phytosterols. The first food product enriched in esters of phytostanols was Benecol spread, which appeared in Finland in 1995. Since then there has been a significant progress in production and consumption of food products enriched in phytosterols, and currently they have spread to more than 20 countries. Food products enriched in phytosterols currently marketed in the world, besides margarines, are cream cheese spreads, milk, mayonnaise, pasta, cheese, yoghurts, meat products, snack bars, oils, beverages, bakery products, and salad dressings [70,71].

Tall oil and deodorizer distillate are the best sources of phytosterols for food products. Tall oil is a by-product of the wood pulp industry that contains 3%–7% by weight total sterols, mostly in esterified form. A deodorizer distillate may account for up to 18% (w/w) of phytosterols. They are recovered by direct extraction from these sources using supercritical fluid technology [72].

7.3.2 BIOACTIVITY OF PHYTOSTEROLS

Elevated level of LDL (carrier of LDL cholesterol) is an established major factor that increases the risk of CHD. The latter is the leading cause of mortality in Western countries and is rapidly increasing in the developing countries [73]. The cholesterol-lowering effect of phytosterols and phytostanols has been demonstrated in both humans and animals. Phytosterols and phytostanols, either in their free form or as fatty acid esters, inhibit the uptake of dietary and endogenous cholesterol from the gut, causing a decrease in serum total cholesterol and LDL levels [46,74]. High-density lipoprotein (HDL), often called good cholesterol, and TAG do not appear to be affected by dietary phytosterol consumption. An average daily dose of 2.8 g of phytosterols in normo- and middle-hypercholesterolemic subjects caused 10.9% reduction of LDL, whereas, on average, on every gram of phytosterols consumed, a reduction of 4.9% of LDL was observed [46]. Phytosterols appear not only to play an important role in the regulation of cholesterol metabolism but also exhibit anticancerogenic properties [47,74]. It is speculated that phytosterols may act through multiple mechanisms such as inhibition of carcinogens formation, cancer cell development, angiogenesis, invasion and metastasis, and through the promotion of apoptosis of cancerous cells [75]. A side effect associated with the consumption of phytosterols is a reduction of fat-soluble vitamins in the blood. But this can be compensated by supplements or foods rich in those vitamins.

Reduction of LDL concentrations by 10%–15% requires a phytosterol intake of 1.5–3.0 g/day, and this amount can only be provided by food products enriched in phytosterols or phytostanols. However, absorption of plant sterols in healthy individuals is low and ranges from 0.6% to 28% of unsaturated and from 0.044% to 12.5% of saturated sterols, whereas doses higher than 3.0 g/day only slightly improve phytosterol benefits [46].

7.3.3 CHANGES OF PHYTOSTEROLS DURING FOOD PROCESSING

7.3.3.1 Oxidation

The process of sterol oxidation has been described as being similar to free radical mechanism of fatty acid oxidation. Phytosterol oxidation products (POPs) are formed during autoxidation, the same process that was described for cholesterol oxidation pathway (Figure 7.2). Grandgirard et al. [76] reported that POPs were absorbed and accumulated in the plasma, liver, aorta, kidneys, and heart of hamsters. These products may have systemic effects *in vivo*, and therefore the potential to modulate human metabolism; indeed, oxyphytosterols have been reported to exert, *in vitro*, cytotoxicity [77].

The content and distribution of POP in food is dependent on numerous factors including the level of phytosterols in the food, temperature, time, sterol structure and form, degree of unsaturation, and water content of the lipid matrix [71]. In good quality fresh rapeseeds 12–15 µg/g of POP were observed, which can be related to enzymatic modification of sterols during maturation of seeds. The total content of POP in rapeseed oil increased during processing to 42–48 µg/g in expeller oil and to 52–59 µg/g in extracted crude oil. Further increase in the amount of POP was observed in refined oil to 100–110 µg/g [49]. Generally, thermal treatment of oil is the main source of POPs; Lampi et al. [78] reported 266–1098 µg/g of these components in rapeseed oil heated at 180°C for 24 h. When rapeseed, sunflower, soybean, and olive oils were exposed to sunlight for 30 days, 2421, 1007, 895, and 676 µg/g of POPs, respectively, were found, which can be related to the amount of sterols in these oils and the degree of unsaturation of fatty acids [79]. When sunflower and olive oils were heated at 150°C for 1 h, 241 and 37 µg/g of oxyphytosterols were observed, respectively. When the temperature was elevated to 200°C, the amount of POPs increased to 815 and 365 µg/g, respectively [80]. A blend of rapeseed oil and palm oils, sunflower, and high-oleic sunflower oils contained 41, 40, and 46 ppm of sterol oxides, respectively. After 2 days of frying, the amounts of these components increased to 60, 57, and 56 ppm, respectively. Total sterol oxide amounts found in French fries lipids fried at 200°C in the above-mentioned oils were 32, 37, and 54 ppm, respectively. The levels of POP in chips industrially fried in different vegetable oils and vacuum packaged was studied. The amounts of POPs in the chips lipids fried in palm oil increased from 6 to 9 µg/g after storage for 10 and 25 weeks, respectively [81,82].

POPs are also formed during storage of food, where time and temperature of storage have a significant influence on their type and quantity [82–84]. Rapeseeds after harvest were dried at elevated temperature and stored at typical conditions for 12 months. Seeds dried at elevated temperatures contained higher amounts of POP, and storage of these seeds at ambient temperature cause further oxidation of sterols.

Two liquid infant formulas made of milk and cereals were stored at 25°C for up to 9 months, and the content of sterol oxidation products increased to 1.0 mg/kg where 7-ketositosterol was a dominating derivative. In the same product, extent of stigmasterol oxidation was higher than that of cholesterol and β-sitosterol. The type and quality of raw material, as well as the processing conditions, seem to greatly influence oxysterols formation and accumulation in infant foods [84].

The effectiveness of different natural and synthetic antioxidants on the phytosterol oxidation was evaluated. Ethanolic extracts of raspberry, black currant, and tomato seeds protected sterols from oxidative degradation in peanuts, of which the black currant seed extract was the most effective [85]. Rudzińska et al. [86] analyzed the effectiveness of BHT, α-tocopherol, ethanolic extracts of rosemary, and green tea on stigmasterol oxidative degradation in sunflower purified TAG. α-Tocopherol offered the best protection shown in the smallest amount of stigmasterol oxidation products.

Phytostanols are considered to be less prone to oxidative degradation due to lack of double bond in the structure. However, these molecules have tertiary carbon atoms where oxidative attack is the most probable and 3-keto derivatives formed [87]. Aringer and Nordström [88] reported chromatographic and mass spectral data for several stanols hydroxyl derivatives and C-24 ethyl analogs of these oxides. Soupas et al. [89]

characterized sitostanol oxides formed during heating of sitostanol at 180°C for 3 h where nine sitostanol oxides were characterized, including epimers 7-hydroxysitostanol and 7-ketositostanol. Also, these authors have established that formation of sitostanol oxides in sitostanol enriched rapeseed oil and tripalmitin were very slow.

7.3.3.2 Polymerization

When thermo-oxidative changes of pure cholesterol were studied, compounds with molecular mass higher than sterol monomers were observed [90,91]. Authors suggested that cholesterol and oxysterols at elevated temperature might form oligomers. Lercker and Rodriguez-Estrada [92] identified 3,3′-dicholesterol ether when cholesterol standard was heated at 170°C. Rudzińska et al. [93] demonstrated that in heated phytosterol, significant amounts of oligomers were formed, and their contribution increased when elevated temperatures and longer times were applied. Some of the oligomers detected were dimers, trimers, and tetramers—products of condensation and/or polymerization of oxysterols (Figure 7.7). The dimers were the dominant oligomers present in heated sterol standards. Lampi et al. [94] detected 7% of dimers, 12% of polymers, and 11% of nonpolar monomers during heating stigmasterol at 180°C for 3 h. Oligomeric steroids are produced in natural plant products and are emerging as a significant pharmaceutical source for these compounds. Bounded ring to ring dimers and C-19 dimers of steroids are components present in membrane bilayers, micelle formation, and detergents [95].

7.3.3.3 Thermal Degradation

Phytosterols during thermo-oxidation undergo several modes of chemical reactions that are classified as carbon–carbon bond scission, and those that do not involve

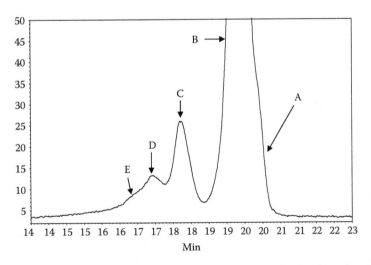

FIGURE 7.7 SEC-HPLC chromatogram of oligomers formed after heating of model phytosterol mixture at 180°C for 24 h. Standard mixture content: brassicasterol 4%, campesterol 16%, stigmasterol 5%, sitosterol 69%, avenasterol 6%. A, Fragmented sterol molecules; B, sterol monomers; C, dimers; D, trimers; E, tetramers.

such scissions [18]. Subsequent products are derived from the initial autoxidation, and include hydroperoxides, alcohols, ketones, aldehydes, and unfunctionalized alkyl chains or B-ring. Also, from transformations of side-chain hydroperoxides, volatile compounds are derived [18]. The well-known rancid off-odor of oxidized cholesterol samples was related to fourteen identified volatile compounds ranging from two to six carbon atom derivatives [96]. Among the identified volatiles were alcohols, aldehydes, ketones, and short-chain fatty acids. Rudzińska et al. [93] identified volatile compounds formed during thermo-oxidative degradation of sitosterol, campesterol, and their stanols. Standard heating at 60°C produced four hydrocarbons, two ketones, and acetaldehyde; among them acetone and 3-methyl-2-pentanone were the most abundant. With increase of temperature, the amount of volatiles increased accordingly, and in phytosterols heated at 120°C, 31 volatile compounds were identified. Among these compounds, hydrocarbons and ketones were the most abundant, where acetone dominated. In samples heated at 180°C, 30 major volatiles were identified, among them were 8 hydrocarbons, 13 ketones, 4 aldehydes, and 2 acids. When a standard phytosterol mix, consisting of brassicasterol, campesterol, stigmasterol, sitosterol, and avenasterol, was heated at 180°C for 24 h, volatile compounds were identified (Table 7.3) (Rudzińska et al., unpublished). Proposition of fragmented sterol molecules and volatile compound formation from the ring side of sitosterol after thermo-oxidation is presented in Figure 7.8, and from side chain in Figure 7.9. The mechanism of volatile compound formation from side chain of molecule was demonstrated by van Lier et al. [96].

TABLE 7.3
Composition of Volatile Compounds Identified as Products of Phytosterol[a] Decomposition during Heating at 180°C for 24 h (w/w, %)

Hydrocarbons	Percentage Contribution	Aldehydes	Percentage Contribution
4-Methyl-2-pentene	13.7	Acetaldehyde	1.3
2-Methyl-pentane	1.6	2-Ethyl-3-methyl-butanol	1.6
3,3-Dimethyl-1-octene	1.5	4-Methyl-benzaldehyde	1.9
Ketones		**Alcohols**	
Acetone	6.4	Ethanol	1.7
3-Methyl-2-butanone	2.1	2-Methyl-3-pentanol	1.7
2-Heptanone	8.7	**Acids**	
2-Nonanone	8.2	Formic acid	1.9
4-Decanone	34.6		
2-Cyclohexen-1-one	2.7	**Others**	
4,4,5,5-Tetramethyl-dihydro-2-furanone	3.8	Carbon dioxide	2.0
		Ethylene oxide	2.2

[a] Model phytosterol mixture content: brassicasterol 4%, campesterol 16%, stigmasterol 5%, sitosterol 69%, avenasterol 6%.

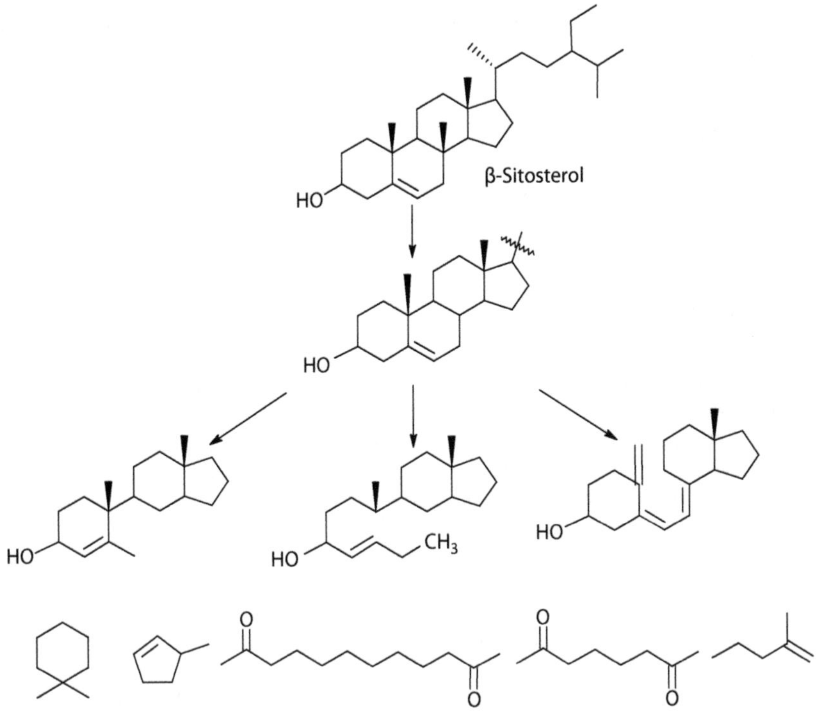

FIGURE 7.8 Formation of fragmented sterol molecules and volatile compounds from ring side of sitosterol after thermo-oxidation. (Adapted from van Lier, J.E. et al., *Chem. Phys. Lipids*, 14, 327, 1975; Smith, L.L., *Cholesterol Autoxidation*, Plenum Press, New York, 1981; Rudziňska, M. et al., *J. Am. Oil Chem. Soc.*, 86, 651, 2009.)

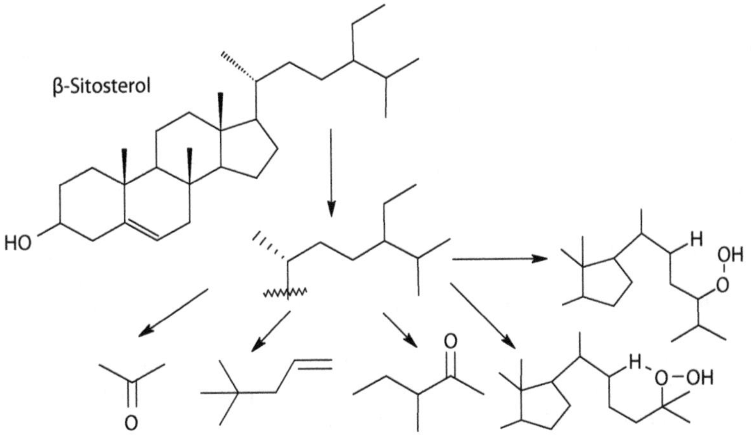

FIGURE 7.9 Formation of fragmented sterol molecules and volatile compounds from side chain of sitosterol after thermo-oxidation. (Adapted from van Lier, J.E. et al., *Chem. Phys. Lipids*, 14, 327, 1975; Smith, L.L., *Cholesterol Autoxidation*, Plenum Press, New York, 1981; Rudziňska, M. et al., *J. Am. Oil Chem. Soc.*, 86, 651, 2009.)

Presented results are preliminary and need developed analytical methods for quality and quantity detection.

ACKNOWLEDGMENT

Supported in part by the State Committee for Scientific Research, grant # N312 071 32/3209.

REFERENCES

1. Wąsowicz, E. (2002). Cholesterol and phytosterols, in *Chemical and Functional Properties of Food Lipids*, Sikorski, Z.E. and Kołakowska, A. (Eds.), CRC Press, Boca Raton, FL, p. 93.
2. Gill, S., Chow, R., and Brown, A.J. (2008). Sterol regulators of cholesterol homeostasis and beyond: The oxysterol hypothesis revisited and revised, *Prog. Lipid Res.*, 47, 391.
3. Grundy, S.M. et al. (1982). Rational of the diet heart statement of the American Heart Association. Report of the Nutrition Committee, *Circulation*, 65, 839A.
4. Gurr, M.I. (1992). Dietary lipids and coronary disease: Old evidence, new perspectives and progress, *Lipids Res.*, 31, 195.
5. Law, M.R. et al. (1994). Systematic underestimation of association between serum cholesterol concentration and ischemic heart disease in observational studies: Data from BUPA study, *Br. Med. J.*, 308, 363.
6. Ostlund, R.E. (2007). Phytosterols, cholesterol absorption and healthy diets, *Lipids*, 42, 41.
7. Astray, G. et al. (2009). A review on the use of cyclodextrins in foods, *Food Hydrocoll.*, 23, 1631.
8. Alonso, L. et al. (2009). Use of β-cyclodextrin to decrease the level of cholesterol in milk fat, *J. Dairy Sci.*, 92, 863.
9. Haumann, B.F. (1996). The goal: Faster and "healthier" fried food, *Inform*, 7, 320.
10. Michicich, M. et al. (1999). Consumer acceptance, consumption and sensory attributes of spreads made from designer fats, *Food Qual. Prefer.*, 10, 147.
11. Precht, D. (2001). Cholesterol content in European Bovine milk fats, *Nahrung/Food*, 45, 2.
12. Fatouh, A.E. (2007). Fractionation of buffalo butter oil by supercritical carbon dioxide, *LWT-Food Sci. Technol.*, 40, 1687.
13. McMillan, G.C. (1995). Historical review of research on atherosclerosis, in *Nutrition and Biotechnology in Heart Disease and Cancer*, Longenecker, J.B., Kritchevsky, D., and Drezner, M.K. (Eds.), Plenum Press, New York, p. 1.
14. Steinberg, D. (1995). Role of oxidized LDL, and antioxidants in atherosclerosis, in *Nutrition and Biotechnology in Heart Disease and Cancer*, Longenecker, J.B., Kritchevsky, D., and Drezner, M.K. (Eds.), Plenum Press, New York and London, U.K., pp. 39.
15. Sempos, C.T. et al. (1993). Prevalence of high blood cholesterol among U.S. adults. An update based on guidelines from the second Report of the National Cholesterol Education Program Adult Treatment Panel, *JAMA*, 269, 3009.
16. Kanner, J. (2007). Dietary advanced lipid oxidation endproducts are risk factors to human health, *Mol. Nutr. Food Res.*, 51, 1094.
17. Staprans, J. et al. (2005). The role of the dietary oxidized fatty acids in the development of atherosclerosis, *Mol. Nutr. Food Res.*, 49, 1075.

18. Smith, L.L. (1981). *Cholesterol Autoxidation*, Plenum Press, New York.
19. Smith, L.L. (1996). Review of progress in sterol oxidation: 1987–1995, *Lipids*, 31, 453.
20. Linseisen, J. and Wolfram, G. (1998). Origin, metabolism, and adverse health effects of cholesterol oxidation products, *Fett/Lipid*, 100, 211.
21. Savage, G.P., Dutta, P.C., and Rodriguez-Estrada, M.T. (2002). Cholesterol oxides: Their occurrence and methods to prevent their generation in foods, *Asia Pac. J. Clin. Nutr.*, 11, 72.
22. Sieber, R. (2005). Oxidized cholesterol in milk and dairy products, *Int. Dairy J.*, 15, 191.
23. Hur, S.J., Park, G.B., and Joo, S.T. (2007). Formation of cholesterol oxidation products (COP) in animal products, *Food Control*, 18, 939.
24. Jacobson, M.S. (1987). Cholesterol oxides in Indian ghae: Possible cause of unexpected high risk of atherosclerosis in Indian immigrant populations, *Lancet*, 1, 656.
25. Soto-Rodriguez, J. et al. (2008). Cholesterol oxidation in traditional Mexican dried and deep-fried food products, *J. Food Comp. Anal.*, 21, 489.
26. Derewiaka, D. and Obiedziñski, M. (2009). Oxysterol content in selected meats and meats products, *Acta Sci. Pol. Technol. Aliment.*, 8, 5.
27. Zaborowska, Z. et al. (2002). Cholesterol and cholesterol oxidation products in Polish commercial sausages, *Electron. J. Pol. Agric. Univ., Food Sci. Technol.*, 5, 2.
28. Przygoñski, K., Jeleñ, H., and Wąsowicz, E. (2000). Determination of cholesterol oxidation products in milk powder and infant formulas by gas chromatography and mass spectrometry, *Nahrung*, 44, 122.
29. Conchillo, A., Ansorena, D., and Astiasarán, I. (2005). Intensity of lipid oxidation and formation of cholesterol oxidation products during frozen storage of raw and cooked chicken, *J. Sci. Food Agric.*, 85, 141.
30. Flaczyk, E. et al. (2006). Effects of cracklings hydrolysates on oxidative stability of pork meatballs fat, *Food Res. Int.*, 39, 924.
31. Rudziñska, M. et al. (2007). Antioxidative effect of crackling hydrolysates during frozen storage of cooked pork meatballs, *Eur. Food Res. Technol.*, 224, 293.
32. Shan, H. et al. (2003). Chromatographic behavior of oxygenated derivatives of cholesterol, *Steroids*, 68, 221.
33. Appelqvist, L.Å. (2004). Harmonization of methods for analysis of cholesterol oxides in foods—The first portion of a long road toward standardization: Interlaboratory study, *J. AOAC Int.*, 87, 511.
34. Schroepfer, G.J., Jr. (2000). Oxysterols: Modulators of cholesterol metabolism and other processes, *Physiol. Rev.*, 80, 361.
35. Björkhem, J., Meaney, S., and Diczfalusy U. (2002). Oxysterols in human circulation: Which role do they hart? *Curr. Opin. Lipidol.*, 13, 247.
36. Garcia-Cruset, S. et al. (2002). Cholesterol oxidation products and atherosclerosis, in *Cholesterol and Phytosterol Oxidation Products: Analysis, Occurrence, and Biological Effects*, Guardiola, F., Dutta, P.C., Savage, G.P., and Colony, R. (Eds.), AOCS Press, Champaigne, IL, p. 241.
37. Osada, K. (2002). Cholesterol oxidation products: Other biological effects, in *Cholesterol and Phytosterol Oxidation Products: Analysis, Occurrence, and Biological Effects*, Guardiola F., Dutta P.C., Codony R., and Savage G.P. (Eds.), AOCS Press, Champaigne, IL, p. 278.
38. Javitt, N.B. (2008). Oxysterols: Novel biologic roles for the 21st century, *Steroids*, 73, 149.
39. Lordan, S., Mackrill, J.J., and O'Brien N.M. (2009). Oxysterol and mechanisms of apoptotic signaling: Implication in the pathology of degeneration diseases, *J. Nutr. Biochem.*, 20, 321.
40. Iuliano, L. et al. (2003). Measurement of oxysterols and α-tocopherol in plasma and tissue samples as indices of oxidant stress status, *Anal. Biochem.*, 312, 217.

41. Parish, E.J., Li, S., and Bell, A.D. (2008). Chemistry of waxes and sterols, in *Food Lipids: Chemistry, Nutrition, and Biotechnology*, Akoh, C.C. and Min, D.B. (Eds.), CRC Press, London, U.K., p. 99.

42. Dutta, P.C. and Appelqvist, L.Å. (1996). Saturated sterols (stanols) in unhydrogenated and hydrogenated edible vegetable oils and cereal lipids, *J. Sci. Food Agric.*, 71, 383.

43. Harrabi, S. et al. (2008). Phytostanols and phytosterols distribution in corn kernel, *Food Chem.*, 111, 115.

44. Moreau, R.A., Whitaker, B.D., and Hicks, K.B. (2002). Phytosterols, phytostanols, and their conjugates in foods: Structural diversity, quantitative analysis, and health-promoting uses, *Prog. Lipid Res.*, 41, 457.

45. Phillips, K.M. et al. (2002). Free and esterified sterol composition of edible oils and fats, *J. Food Compos. Anal.*, 15, 123.

46. Normén, L., Frohlich, J., and Trautwein, E. (2004). Role of plant sterols in cholesterol lowering, in *Phytosterols as Functional Food Components and Nutraceuticals*, Dutta, P.C. (Ed.), Marcel Dekker, Inc., New York, p. 243.

47. Calpe-Berdiel, L., Escolá-Gil, J.C., and Blanco-Vaca, F. (2009). New insights into the molecular actions of plant sterols and stanols in cholesterol metabolism, *Atherosclerosis*, 203, 18.

48. Rudzińska, M., Kazuś, T., and Wąsowicz, E. (2001). Sterols and their oxidized derivatives in refined and cold pressed seed oils, *Oilseed Crops* 22, 477.

49. Rudzińska, M., Uchmann, W., and Wąsowicz, E. (2005). Plant sterols in food technology, *Acta Sci. Pol. Technol. Aliment.*, 4, 147.

50. Verleyen, T. et al. (2002). Analysis of free and esterified sterols in vegetable oils, *J. Am. Oil Chem. Soc.*, 79, 117.

51. Piironen, V. and Lampi, A.-M. (2004). Occurrence and levels of phytosterols in foods, in *Phytosterols as Functional Food Components and Nutraceuticals*, Dutta, P.C. (Ed.), Marcel Dekker, Inc., New York, p. 1.

52. Eisenmenger, M. and Dunford, N.T. (2008). Bioactive components of commercial and supercritical carbon dioxide processed wheat germ oil. *J. Am. Oil Chem. Soc.*, 85, 55.

53. Patel, M. and Naik, S.N. (2004). γ-Oryzanol from RBO: A review, *J. Sci. Nutr. Res.*, 63, 569.

54. Przybylski, R. et al. (2009). Lipid components of North American Wild rice (*Zizania palustris*), *J. Am. Oil Chem. Soc.*, 86, 553.

55. Maguire, L.S. et al. (2004). Fatty acid profile, tocopherol, squalene and phytosterol content in walnuts, almonds, peanuts, hazelnuts and the macadamia nut, *Int. J. Food Sci. Nutr.*, 55, 171.

56. Miraliakbari, H. and Shahidi, F. (2008). Lipid class composition, tocopherols and sterols of tree nut oils extracted with different solvents, *J. Food Lipids*, 15, 81.

57. Ogrodowska, D. et al. (2009). Bioactive compounds of bio-oils unsaponificable fraction, *Lipids, Fats and Oils From Knowledge to Application*, 7th Euro Fed Lipid Congress, Graz, Austria, October 18–21, p. 234.

58. van Hoed, V. et al. (2009). Berry seeds: Source of specialty oils with high content of bioactives and nutritional value, *J. Food Lipids*, 16, 33.

59. Anwar, F. et al. (2008). Fatty acids, tocopherols and sterols profiles of seed lipids of selected Canadian prairie fruits, *J. Am. Oil Chem. Soc.*, 85, 953.

60. Piironen, V., Toivo, J., and Lampi, A.-M. (2002). Plant sterols in cereals and cereal products, *Cereal Chem.*, 79, 148.

61. Seitz, L.M. (1989). Stanol and sterol esters of ferulic and *p*-coumaric acids in wheat, corn, rye, and triticale, *J. Agric. Food Chem.*, 37, 662.

62. Määttä, K. et al. (1999). Phytosterol content in seven oat cultivars grown at three locations in Sweden, *J. Sci. Food Agric.*, 79, 1021.

63. Moreau, R.A., Singh, V., and Hicks, K.B. (2001). Comparison of oil and phytosterol levels in germplasm accessions of corn, teosinte, and Job's tears, *J. Agric. Food Chem.*, 49, 3793.
64. Piironen, V. et al. (2003). Plant sterols in vegetables, fruits and berries, *J. Sci. Food Agric.*, 83, 330.
65. Kaloustian, J. et al. (2008). Effect of water cooking on free phytosterol levels in beans and vegetables, *Food Chem.*, 107, 1379.
66. Awad, A.B. et al. (2000). Peanuts as source of β-sitosterol, a sterol with anticancer properties, *Nutr. Cancer.*, 36, 238.
67. Berger, A., Jones, P.J.H., and Abumweis, S.S. (2004). Plant sterols: Factors affecting their efficacy and safety as functional food ingredients, *Lipids Health Dis.*, 3, 5.
68. Morton, G.M. et al. (1995). Intakes and major sources of cholesterol and phytosterols in the British diet, *J. Hum., Nutr. Diet.*, 8, 429.
69. Hearty, A.P., Duffy, E., and Gibney, M.J. (2009). Intake estimates of naturally occurring phytosterols using deterministic and probabilistic methods in a representative Irish population, *Int. J. Food Sci. Nutr.*, 60, 533.
70. Moreau, R.A. (2004). Plant sterols in functional food, in *Phytosterols as Functional Food Components and Nutraceuticals*, Dutta, P.C. (Ed.), Marcel Dekker, Inc., New York, p. 317.
71. Ryan, E. et al. (2009). Phytosterol oxidation products: Their formation, occurrence, and biological effects, *Food Rev. Int.*, 25, 157.
72. Fernandes, P. and Cabral, J.M.S. (2007). Phytosterols: Applications and recovery methods, *Bioresour. Technol.* 98, 2335.
73. Trautwein, E.A. et al. (2003). Proposed mechanisms of cholesterol-lowering action of plant sterols, *Eur. J. Lipid Sci. Technol.*, 105, 171.
74. Jones, P.J.H. and Abumweis, S.S. (2009). Phytosterols as functional food ingredients: Linkages to cardiovascular disease and cancer, *Curr. Opin. Clin. Nutr. Metab. Care*, 12, 147.
75. Woyengo, T.A., Ramprasath, V.R., and Jones, P.J.H. (2009). Anticancer effects of phytosterols, *Eur. J. Clin. Nutr.*, 63, 813.
76. Grandgirard, A. et al. (2004). Incorporation of oxyphytosterols in tissues of hamster, *Reprod. Nutr. Dev.*, 44, 609.
77. Hovekamp, E. et al. (2008). Biological effects of oxidized phytosterols: A review of the current knowledge, *Prog. Lipid Res.*, 47, 37.
78. Lampi, A.-M. et al. (2002). Determination of thermo-oxidation products of plant sterols, *J. Chromatogr. B*, 777, 83.
79. Zhang, X. et al. (2006). Quantitative analysis of β-sitosterol oxides induced in vegetable oils by natural sunlight; artificially generated light; and irradiation, *J. Agric. Food Chem.*, 54, 5410.
80. Grandgirard, A. (2004a). Gas chromatographic separation and mass spectrometric identification of mixtures of oxyphytosterols and oxycholesterol derivatives. Application to a phytosterol-enriched food, *J. Chromatogr. A*, 1040, 239.
81. Dutta, P.C. (1997). Studies on phytosterol oxides. II: Content in some vegetable oils and in French fries prepared in these oils, *J. Am. Oil Chem. Soc.*, 74, 659.
82. Dutta, P.C. and Appelqvist, L.Å. (1997). Studies on phytosterol oxides. I. Effect of storage on the content in potato chips in different vegetable oils, *J. Am. Oil Chem. Soc.*, 74, 647.
83. Rudzińska, M., Gawrysiak-Witulska, M., and Przybylski, R. (2007a). Drying rapeseed affects phytosterol oxidative stability, *Oils, Fats and Lipids: From Science to Application*, 5th Euro Fed Lipid Congress, Gothenburg, Sweden, September 16–19, p. 311.
84. Garcia-Llatas, G. et al. (2008). Sterol oxidation in ready-to-eat infant foods during storage, *J. Agric. Food Chem.*, 56, 469.

85. Małecka, M. et al. (2003). The effect of raspberry, black currant and tomato seed extracts on oxyphytosterols formation in peanuts, *Pol. J. Food Nutr. Sci.*, 12/53, 49.
86. Rudzińska, M. et al. (2004). Inhibition of stigmasterol oxidation by antioxidants in purified sunflower oil, *J. AOAC Int.*, 87, 499.
87. Dutta, P.C. (2004). Chemistry, analysis, and occurrence of phytosterol oxidation products in food, in *Phytosterols as Functional Food Components and Nutraceuticals*, 1st edn., Dutta, P. (Ed.), Marcel Dekker, Inc., New York, p. 397.
88. Aringer, L. and Nordström, L. (1981). Chromatographic properties and mass spectrometric fragmentation of deoxygenated C_{27}-, C_{28}-, and C_{29}-steroids, *Biomed. Mass Spectrom.*, 8, 183.
89. Soupas, L. et al. (2004). Effects of sterol structure, temperature, and lipid medium on phytosterol oxidation, *J. Agric. Food Chem.*, 52, 6485.
90. Nawar, W.W. et al. (1991). Measurement of oxidative interaction of cholesterol, *J. Am. Oil Chem. Soc.*, 68, 496.
91. Kim, S.K. and Nawar, W.W. (1993). Parameters influencing cholesterol oxidation, *Lipids*, 28, 917.
92. Lercker, G. and Rodriguez-Estrada, M.T. (2002). Cholesterol oxidation mechanism, in *Cholesterol and Phytosterol Oxidation Products*, Guardiola, F., Dutta, P.C., Codony, R., and Savage, G.P. (Eds.), AOCS Press, Champaigne, IL, p. 1.
93. Rudzińska, M., Przybylski, R., and Wąsowicz, E. (2009). Products formed during thermo-oxidative degradation of phytosterols, *J. Am. Oil Chem. Soc.*, 86, 651.
94. Lampi, A.-M. et al. (2009). Distribution of monomeric, dimeric and polymeric products of stigmasterol during thermo-oxidation, *Eur. J. Lipid Sci. Technol.*, 111, 1027.
95. Li, Y. and Dias J.R. (1997). Dimeric and oligomeric steroids. *Chem. Res.*, 97, 283.
96. van Lier, J.E., da Costa, A.L., and Smith, L.L. (1975). Cholesterol autoxidation: Identification of the volatile fragments, *Chem. Phys. Lipids*, 14, 327.

8 Lipophilic Vitamins

Małgorzata Nogala-Kałucka

CONTENTS

8.1 INTRODUCTION

A hundred years ago, in 1909, the first lipophilic vitamin, vitamin A, was discovered; then, 10 years later, vitamin D_3 was identified, and in 1922 Evans and Bishop discovered vitamin E. The last of the fat-soluble vitamins, vitamin K_1, was identified in the 1930s by Dam and Doisy who achieved a worldwide recognition of their studies on vitamins and received the Nobel Prize for these efforts.

Vitamins do not constitute any source of energy for the human body, nor are they considered as structural material. But in cells and tissues, they have various regulatory functions that determine the development, physical efficiency, and health status of humans.

The main characteristic shared by lipophilic vitamins is the similarity in their occurrence in the human body and the effect exerted upon biological processes. Some vitamins can be generated in the human organism to a certain extent, but such a biosynthesis is usually insufficient to cover the physiological demand. This includes: vitamin A originating in the human body from β-carotene (provitamin A) and some other carotenoids as a result of bioconversion and vitamin D_3 being synthesized in the skin under the influence of the UV rays, to vitamin K, which can be produced in the gut by intestinal microflora. On the other hand, vitamin E is totally exogenous for humans and animals. The lipid-soluble vitamins, (A, D, E, and K) differ from other vitamins by lacking nitrogen in their molecular structures.

These vitamins can be accumulated in some human tissues, and thus it is not necessary to consume their required quantity every day. Their bioavailability depends on the presence of fat in foods, and the extent to which they are released in the gastrointestinal tract and then absorbed and distributed to tissues and organs. The biological activity of vitamins is important from the point of view of human physiology. It determines the way in which a compound is being assimilated from food and transformed into an active form of a vitamin thus preventing the signs of deficiency. Numerous intra- and extrasomatic factors affect the biological activity of vitamins, for example, their chemical structure, source, and intake, as well as the efficiency of the mechanisms of intestinal absorption. In case of vitamins A, D, E, and K, the presence of fat is indispensable, as the appropriate level of biliary and pancreatic secretion and also the presence of appropriate enzymes (e.g., carboxylic esterase) bring about decomposition of the ester forms of vitamins. The nutritional status of the organism and the related somatic reserves are of great significance. From a nutritional point of view vitamins A and E are of particular importance because they are supplied solely by consumption of the appropriate foodstuffs. Their biological activities can be affected by metabolic disorders connected with various diseases and administration of drugs. Drugs can interact with vitamins and act synergistically or antagonistically, thus drastically affecting the vitamin level. It was calculated that a properly composed daily diet supplies sufficient quantities of lipophilic vitamins. Considering physiological possibilities of their storage in some organs supplementation is not always necessary because their excess can lead to acute and chronic symptoms of toxicity, which was proved and described in case of vitamins A and D.

8.2 VITAMIN A AND CAROTENOIDS

8.2.1 STRUCTURE

A group of compounds described as vitamin A (retinol) includes those that, within their molecules, possess the characteristic β-ionone ring and the isoprene chain. They are capable of forming molecules of polyene alcohol, aldehyde, acid, and ester (retinal, retinoic acid, and retinyl palmitate) (Figure 8.1). They can occur in many isomeric forms as groups of compounds possessing diversified properties that are revealed by their various functions in metabolic and physiological processes. The

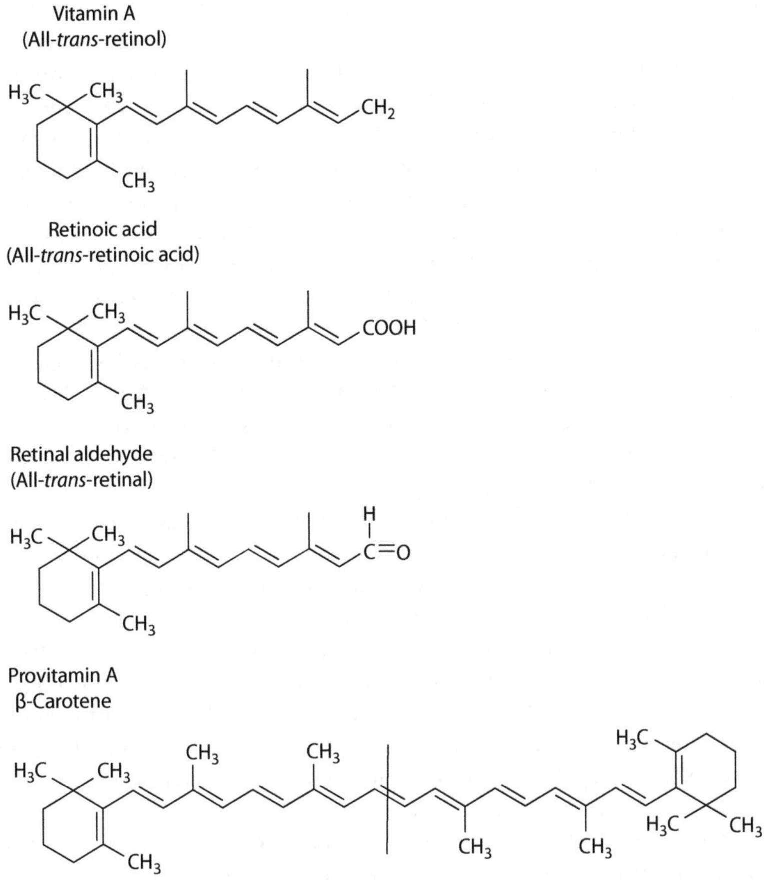

Vitamin A
(All-*trans*-retinol)

Retinoic acid
(All-*trans*-retinoic acid)

Retinal aldehyde
(All-*trans*-retinal)

Provitamin A
β-Carotene

FIGURE 8.1   Vitamin A: chemical structures.

vitamin A family includes several plant pigments identified in nature—carotenoids—described as provitamin A because they are converted to vitamin A in the human and animal organisms. These compounds are composed of two β-ionone rings and most often they have isomeric forms—α, β, and/or γ isomers. The most valuable is β-carotene, which displays the highest biological activity. It is composed of two β-ionone rings linked by the isoprene chain. Bioconversion comprises splitting of the symmetric double bond between C_{15} and $C_{15'}$, causing simultaneous formation of two molecules of retinal, which is reduced to retinol. The α and γ isomers form only one molecule of retinal. The rate of conversion of β-carotene to vitamin A depends on vitamin C, zinc, and the level of thyroid hormones.

8.2.2   METABOLISM AND FUNCTIONS

Vitamin A in foods is present in its ester form and, after enzyme-catalyzed hydrolysis, it is absorbed into the gastrointestinal tract. After forming an emulsion in the presence of bile salts and pancreatic juice, it is incorporated into molecules forming chylomicrons (Olson, 1996). Other forms of carotenoids are oxidatively converted to retinal in the intestinal mucosa. Then, via the lymphatic vessels, it is transported from the bloodstream to the liver. The majority of retinoids and carotenoids contained in chylomicrons is trapped by the liver, but some 25%–30% is taken up by other tissues. Vitamin A is stored mainly in the liver, but it can also be found in the ester form within the lipocytes. Transport from the liver depends on the level of retinol binding protein (RBP) (Wolf and Phil, 1991). About 95% RBP is related to the protein transporting a thyroid hormone—transthyretin (TTR). The intensity of the transfer of retinol from the retinol/RBP complex to the cells depends on the level of proteins binding the retinol inside the cell. Retinol binding proteins belong to the family of small proteins binding the hydrophobic ligands. Retinoids cause lowering of the level of thyroid hormones in blood serum. Vitamin A is well absorbed from the alimentary tract—its absorption efficiency is assessed as 80%. Retinol level should exceed $300 \mu g/dm^3$ of blood plasma (Gerstner, 1997). Total concentration of retinol in liver ranges from 20 to $300 \mu g/g$ tissue.

The oldest and best-defined function of vitamin A in humans is its participation in vision. As a light-sensitive factor, it is a component of rhodopsin. In the eye, light induces decomposition of rhodopsin with simultaneous splitting off of 11-*cis* retinal and opsin. During a series of conformational reactions in the dark regeneration of visual purple takes place. Long-lasting vitamin A deficiency leads to disturbed functioning of retina (i.e., night blindness, xerophthalmia); and at a later stage, keratomalacia and necrosis resulting in loss of sight. The second function of vitamin A is cell differentiation; its presence allows regulation of new cell generation processes, particularly those in epithelium, mucous membranes, and bone tissue. Vitamin A participates in synthesis of the adrenocortical and thyroid hormones and also in the metabolism of the steroid hormones. Retinol controls ubiquinone formation in the liver and probably serves as a coenzyme in the synthesis of glycoproteins and mucopolysaccharides. Vitamin A also plays a role in normal growth regulations, especially in young organisms (Gerstner, 1997). It affects spermatogenesis, and it has been discovered—although not yet fully clarified—that retinol and retinoic acid are essential for embryonic development. It has been suggested that vitamin A plays a significant role in immune response and inhibition of neoplastic cell development (Johnson, 2001) as well as in appetite and taste regulation. The physiological effects of the retinoid activity depend mainly on oxidation, isomerization, and the level of nuclear receptors; for example, the retinoid X receptor (RXR) is involved in the morphogenesis of heart and eyes.

8.2.3 Deficiency and Excess

Vitamin A deficiency may manifest itself by decreased immunity to infections, growth inhibition, and reproduction disorders. Typical pathological signs of vitamin A avitaminosis include lesions of the mucous membranes of the gastrointestinal tract, air passages, and urinary system. Skin keratosis can be seen, especially around outlets of the hair sheaths (phrynoderma). Vitamin A can be toxic when ingested in overly large doses. The most common sign of excessive vitamin A intake is the so-called dermal paralysis manifested by changes in skin pigmentation and dryness, accompanied by brittleness of nails and hair loss.

Vitamin A excess can be the reason for joint and bone pains as well as the weight loss (Eckhoff and Nau, 1990). Enlargement of the liver and spleen usually occurs (hepatosplenomegaly). Teratogenic activity has also been stated, notable birth defects (Hathcock, 1997). Symptoms of acute toxicity are headaches, nausea and vomiting, convulsions, and light intolerance.

8.3 VITAMIN D

8.3.1 STRUCTURE

Vitamin D is considered a hormone-like compound. Ergosterol is a vitamin D_2 precursor, a phytosterol that occurs mainly in plants. Ergocalciferol results from UV irradiation of ergosterol (Figure 8.2).

7-Dehydrocholesterol can be converted into the most active form—vitamin D_3 (cholecalciferol). Ergosterol and 7-dehydrocholesterol are clearly identified as provitamin D, because after irradiation of the skin with UV light or exposure to sunlight they can be converted into active vitamins D_2 and D_3.

The structures of compounds known as vitamin D are based on a system of three rings called the residue of phenanthrene; the fourth ring is the residue of cyclopentane. This general structure undergoes transformation by breaking one of the rings, and therefore there are only three aromatic rings in the molecules of vitamin D_2

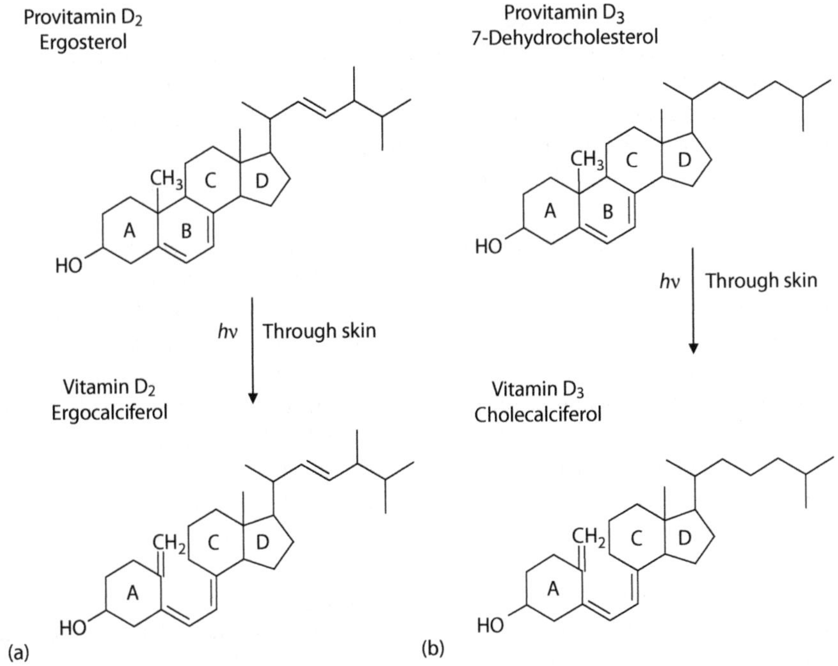

FIGURE 8.2 Vitamin D: chemical structures.

and D_3 (Figure 8.2). Photoisomerization alters molecules, particularly the biological activity. This alteration depends on the chemical structure of the isomer being generated. Evidence has also suggested the possibility of transforming cholesterol, *via* enzymatic and chemical dehydrogenation, into the precursor of vitamin D_3. Vitamin D is endogenously produced in the human and animal organisms, and is also supplied with food (Collins and Norman, 1990).

8.3.2 METABOLISM AND FUNCTIONS

The vitamin synthesized in the skin (D_3) and those supplied with food undergo the same transformations finally leading to generation of their active metabolites. They are deposited in the liver, lungs, and kidneys as well as in the fatty tissue. The amount of vitamin D in the organism depends on the diet and exposure to sunshine, and it ranges from 65 to 165 IU/100 mL blood. All forms of this vitamin are absorbed in the small intestine with bile salt, fatty acids, and acylglycerols present, and then delivered to the liver bound to a specific vitamin D-binding protein. From the liver the vitamin is transported via chylomicrons to the bloodstream. The molecules of vitamin D_2 and D_3 are activated in the liver by hydroxylation; the OH group appears at carbon-25 to give 25-hydroxy-D_3[25-(OH)]D. Further reaction occurs in the kidneys, bone, cartilage, and/or in placenta under the influence of the specific D_3-1-hydroxylase. The next two stages produce the most potent and physiologically active molecule, 1,25-dihydroxy vitamin D_3 [1,25-$(OH)_2D_3$]. According to the most recent nomenclature it is called calcitriol. Calcitriol acts through nuclear receptors and also through a still little known mechanism initiated by binding the calcitriol to the receptor in the cell membrane.

25-(OH)-D gets activated under the influence of the specific D_3-24-hydroxylase at carbon-24 [24,25-$(OH)_2D_3$]. Both circulate in the bloodstream and their main function as active metabolites is maintaining the homeostasis of calcium and phosphorus, which is vital for many body functions (Henry and Norman, 1984).

In the human organism, the vitamin may be stored in the fatty tissue for several years.

Active molecules of vitamin D work in conjunction with parathyroid hormone and calcitonin. The structure and mode of action of the vitamin D metabolites resemble those of typical steroid hormones, such as estrogen or testosterone, but their function in the endocrine system remains the subject of numerous studies and discussions (Bouillon et al., 1995). Vitamin D regulates specific gene expression following interaction with its intracellular receptor or target genes in order to stimulate or to suppress specific transcriptions (DeLuca and Zierold, 1998). Vitamin D is involved in normal cell growth and maturation. Several clinical studies have revealed that cell growth inhibition may protect the body against some kinds of neoplasms, for example, human leukemia, and colon, prostate, and breast cancer (Gann et al., 1996; Lips, 2006).

The action of vitamin D in human and animal organisms is primarily related to calcium metabolism and mineralization of bones and teeth, although the mechanism of these processes has not yet been univocally confirmed. The cooperation of vitamin A and K is necessary for proper functioning of vitamin D.

Vitamin D stimulates differentiation of bone-forming osteoblasts and inhibits formation of bone-decomposing osteoclasts. It has been assumed that vitamin D affects the blood sugar level by appropriate secretion of insulin by the pancreas (DeLuca and Zierold, 1998). Vitamin D is also thought to regulate the level of phosphorus (P) in kidneys and to facilitate transition of P from the organic to inorganic form, the junctions of the two being necessary for bone tissue generation (Arnaud and Sanchez, 1996). Vitamin D regulates calcium levels for normal nerve impulse transmission and muscle contraction, and also influences immune system regulation (white blood cells—monocytes and lymphocytes).

8.3.3 DEFICIENCY AND EXCESS

In human and animal organisms, deficiency of vitamin D can result from its decreased synthesis in the skin or disturbances in its metabolism related to malabsorption in the alimentary tract (Devqun et al., 1981). Certain liver and kidney disorders may result in excessive excretion of vitamin D in urine and feces, thus leading to vitamin D deficiency. If the calcium and phosphorus homeostasis is disturbed, bone changes quickly follow (Suda et al., 1992). The main symptom of vitamin D deficiency in newborn children is rickets; in adults, it is osteomalacia. In children, the disease primarily affects the chest, skull bones, and legs. Development of baby teeth is also slowed down. Changes in the bone tissue, becoming more intense, are also related to metabolic disturbances, and deficiencies in dietary calcium and phosphorus may lead to osteoporosis (bones become lighter, less dense, and prone to fractures) (Boonen et al., 2006). Clinical studies reveal that vitamin D deficiency plays a role in tuberculosis, stroke, high blood pressure, and inflammatory bowel disease. Low vitamin D level could be a risk of arteriosclerosis, rheumatoid arthritis or osteoarthritis (Thomas et al., 1998). Also, vitamin D deficiency has been recently associated with the metabolic syndrome of morbid obesity and type 1 and 2 diabetes mellitus (Botella-Carretero, 2007; Chatfield et al., 2007; Palomer et al., 2008). Clinical studies showed that vitamin D had influence on a number of immunological effects (Wijst and Hypponen, 2007).

Ingestion of excessive quantities of vitamin D may be toxic and can result in bone and/or pain, diffuse demineralization of bones, muscles fatigue, loss of appetite, thirst, sore eyes, itching skin, vomiting, diarrhea, urinary urgency, and abnormal calcium level in urine (indicative of kidney stones). High doses of vitamin D cause the build-up of calcium in the soft tissues and in such organs as liver, lungs, heart, kidneys, and muscles (Watson et al., 1997). An optimal dose of vitamin D should raise serum concentrations of 25-(OH)-D to desirable level of at least 75 nmol/L (Bischoff-Ferrari, 2007).

8.4 VITAMIN E

8.4.1 STRUCTURE

Vitamin E represents a group of lipids with isoprene structure and it is the only one possessing eight homologous forms. These are multimolecular phenolic compounds consisting of the ring system 2-methyl-6-chromanol with a hydroxyl group at C-6 and linked at C-2 to a 16-carbon saturated, isoprenoid chain derived from plant diterpenoic alcohol—phytyl. RRR-α-tocopherol has the R-configuration at each chiral center.

Built in this way, a molecule forms tocol structures. If a carbon chain of the same length, but with three double bonds, is linked to the ring system, it generates a tocotrienol structure. Binding a –CH₃ group to the C-5, C-7, and/or C-8 of the tocol structure results in the formation of four homologous tocopherols (–T); that is α-T, β-T, γ-T, and δ-T. If the –CH₃ is substituted in the tocotrienol structure then four corresponding tocotrienols (-T-3) are formed: α-, β-, γ-, and δ-T-3 (Figure 8.3). All these compounds have been given the common name of vitamin E, and they occur as enantiomers (D) and (L)—(D) α-T being predominant in nature. They can occur in free or bound forms, most often as esters that, before absorption, are hydrolyzed by pancreatic esterases. In comparison to other homologues α-T displays full biological activity. The biological activity of the remaining forms was determined as: β-T has 40% of α-T activity, γ-T has 10%, and δ-T has only slight activity. The greatest activity among tocotrienols was determined for α-T-3 (it was described as 30% of α-T activity) (Sokol, 1996). At the end of the twentieth century and at the beginning of the twenty-first century, further, naturally occurring forms of tocochromanols were discovered in food and in animal and human tissues: DM-T-3, DDM-T-3, and α-tocopheryl phosphate (Figure 8.3) (Ogru et al., 2003; Gianello et al., 2005). The latest techniques of isolation and analytical possibilities permitted to identify the structure and determine the antioxidant properties of δ-tocomonoenol occurring in kiwi fruit (*Actinidia chinensis*). The new structure of the T analog,

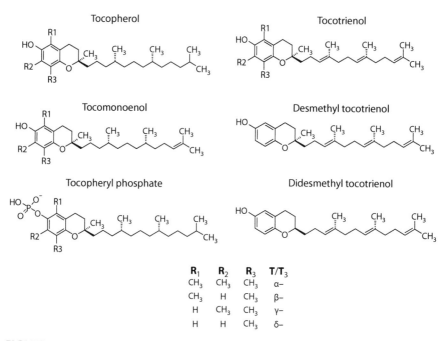

R₁	R₂	R₃	T/T₃
CH₃	CH₃	CH₃	α–
CH₃	H	CH₃	β–
H	CH₃	CH₃	γ–
H	H	CH₃	δ–

FIGURE 8.3 Vitamin E: chemical structures.

2,8-diethyl-2(4,8,12-trimethyltridec-11-enyl)chroman-6-ol, was elucidated on the basis of the electron impact mass spectra (EIMS), refractive index detector (RID), and nuclear magnetic resonance (NMR) spectra (Fiorention et al., 2009). Also, α- and δ-T were identified in the pulp and peel extracts of kiwi fruit.

8.4.2 METABOLISM AND FUNCTIONS

Vitamin E, like other lipids, is digested and absorbed only in the presence of bile. Bile functions as the lipid emulsifier and makes absorption of lipids in the intestinal epithelium possible. Vitamin E can be absorbed in 30%–90% while its assimilability amounts to about 30%, the rest being excreted from the organism. In adults, the concentration of α-T in plasma ranges should exceed 30 µmol/L (Biesalski, 1997). A diet rich in vitamin E allows its accumulation in the organism, for example, in the liver, pituitary gland, and adrenals as well as in the adipose and muscular tissues. Vitamin E takes part in various metabolic processes in human and animal organisms. Its primary role is thought to be that of determining the proper functioning of reproductive organs. Research has shown that vitamin E actively participates in tissue respiration and also in the synthesis of some hormones. Vitamin E protects against macrocytic anemia in children and occurrence of dermal changes. In concurrence with vitamin A, carotenoids, and vitamin C, it protects against neoplastic lesions (Johnson, 2001). It conditions proper structure and permeability of cell membranes. Being a natural antioxidant, vitamin E displays protective properties in relation to unsaturated fatty acids and β-carotene, reacting with radicals generated in metabolic processes and effectively quenching singlet oxygen, free radicals, thus preventing its reaction with other polyenoic fatty acids present in phospholipids of cell membranes (Bramley et al., 2000). T are considered to be secondary antioxidants; alone, they can quickly react with peroxide radicals, thus deactivating them (Porter et al., 1995; Kamal-Eldin and Appelqvist, 2001). In this way, they influence the inhibition of radical chain reactions. The antioxidant activity is largely affected by the hydrogen from the hydroxyl group and the related ability to form radicals of T, which are further capable of reacting among themselves forming dimers and trimers (Figure 8.4). At the end, radicals are deactivated, and the generated molecules possess antioxidant properties less potent than the mother T (Niki et al., 1984). Radical form of T can be regenerated by the system in which the ascorbic acid and glutathione are present. Interestingly, the regeneration of the α-T radical takes place at the junction of water and lipid phases (Packer and Kagan, 1993). Alpha-T and γ-T and their metabolites are involved in modulation of eicosanoid biosynthesis causing formation of compounds with anti-inflammatory properties (Reiter et al., 2007). In the investigation, products of the transformation of T in human organism were determined and the formed and expelled compounds were considered tocochromanol derivatives in the reaction of oxidation—they were called Simon metabolites. The amount of Simon metabolites was considered the indicator of the use of T in the *in vivo* transformations (Brigelius-Flohe and Traber, 1999).

Studies on tocochromanols revealed significant α-T-3 activity in humans. This was explained in terms of their presence in low-density lipoproteins (LDL) and very-low-density-lipoproteins (VLDL), which are the main means of transport of T to

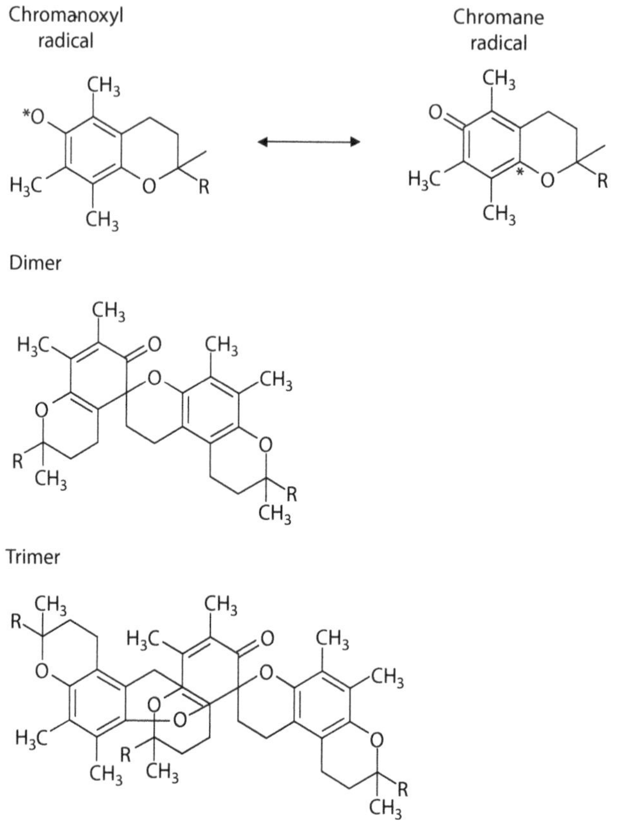

FIGURE 8.4 Tocopherol radicals, dimer, and trimer: chemical structures.

target cells (Pearce et al., 1992; Theriault et al., 1999). The phenyl side chain of tocotrienol has been postulated to be responsible for the differential membrane distribution and metabolism (Theriault et al., 1999). Studies conducted in vitro proved that antioxidant activity of α-T-3 against lipid peroxides generated from residues of the polyenoic fatty acids is significantly higher than that of α-T (Serbinova et al., 1991). The ability to prevent cardiovascular disease, coronary heart disease, stroke-associated brain damage, and cancer by arresting radical damage has been also stated (Hodis et al., 1995; Meydani, 1995). Free radicals may play a role in the pathogenesis of neurological disorders including Alzheimer and Parkinson diseases. Several clinical trials have shown that vitamin E consumption decreases the risk of lung, colon, and prostate cancers (Woodson et al., 1999; Giovannucci, 2000) However, although there is no unequivocal agreement some scientists believe that, but in an early stage, α-T can protect neurons from amyloid beta-protein toxicity (Behl et al., 1992; Fahn, 1992; Behl, 1999). Vitamin E is also indispensable for immune defense (Beharka, et al., 1997). It has been suggested that T and tocotrienols, acting as hormones or as secondary transmitters of genetic information control the expression of some genes (Munteanu and Zingg, 2007).

8.4.3 Deficiency and Excess

An inappropriate diet that does not ensure a proper supply of vitamin E results in a vitamin deficiency manifested by characteristic symptoms. It is very difficult to determine when the vitamin E deficiency occurs. In clinical practice, most often the concentration of vitamin E in blood plasma, in which the homolog α-T constitutes 80%–90% of total T content, is determined (Hosomi et al., 1997). Low vitamin E level usually accompanies infections and affects the immune and nervous systems. An insufficient quantity of this vitamin accelerates the breakdown of red blood cells, makes red cells susceptible to hemolysis, and may also be the reason for the decreased hemoglobin synthesis, which—particularly in infants and children—may result in anemia. The efficiency of the α-T activity depends on the equilibrium between its pro-oxidant and antioxidant effects. Vitamin E deficiency causes damage to cellular membranes resulting from oxidation of the unsaturated fatty acids in phospholipid tails. Greater permeability of cellular membranes brings about visible skin keratosis and ageing. Vitamin E deficiency can also manifest itself as muscular pain and progressing muscular dystrophy. In cooperation with vitamin A, vitamin E can affect vision (Jacques, 1999). It was also demonstrated that with the chronic oxidation stress present and the lack of the so-called co-antioxidants (i.e., the vitamin E–regenerating substances), this vitamin alone can bring about a pro-oxidant effect (Biesalski et al., 1997).

No signs of toxicity related to the vitamin E hypervitaminosis were observed. However, there are some symptoms linked with its excessive intake. Supplying an adult for a longer time with a dose over 1000 mg α-T acetate can result in considerable headache, vision disorders, and muscular weakness (Bendich and Machlin, 1993). It was also demonstrated that large doses of α-T act antagonistically to vitamin K, thus causing decrease in blood clotting capability (Corrigan 1982; Elmadfa and Bosse, 1985).

8.5 VITAMIN K

8.5.1 Structure

Vitamin K comprises derivatives of 1,4-naphthoquinone. Naturally occurring forms are equipped with structures possessing the unsaturated isoprenoid side chain linked to naphthoquinone at carbon-3 (Figure 8.5). Vitamin K_1, phylloquinone (2 methyl-3 phytyl-1,4 naphthoquinone), is one of the natural forms. Like compounds belonging to the vitamin E family, those also included in the vitamin K group have the isoprenoid side chain consisting of 20 carbon atoms with one double bond. This chain is derived from the plant diterpenoic alcohol ($C_{20}H_{39}OH$). These compounds have been detected in alfalfa oil.

Vitamin K_2 has about 75% of the potency of K_1, and is synthesized by bacteria in the intestinal tract of humans and various animals. This vitamin can also be found in putrefied fish meat. Compounds of this family differ from phylloquinone in the numbers of isoprene units in the side chain and also in the degree of their unsaturation. Hence the general term applied to them is menaquinones (Figure 8.5). Unlike vitamin K_1, they are incapable of optical light rotation. Maximum activity was observed for the *trans* configuration of the side chain double bonds.

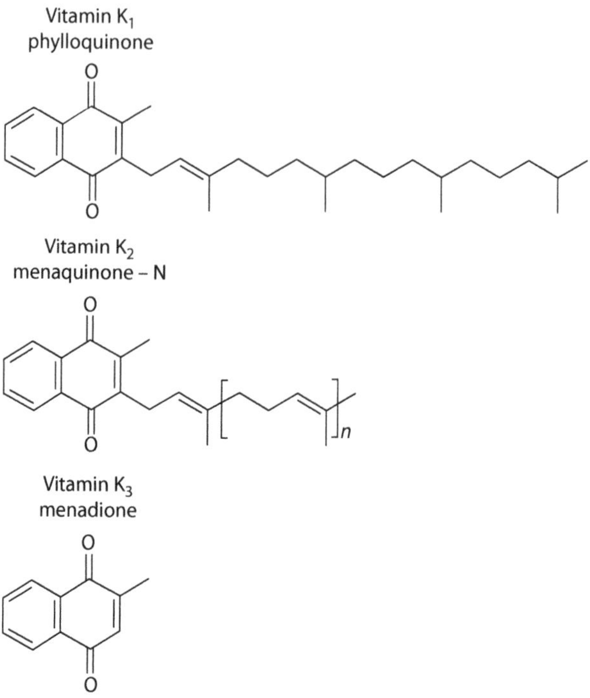

FIGURE 8.5 Vitamin K: chemical structures.

Vitamin K_3—menadione—is the only form isolated from *Staphylococcus aureus*, and also chemically synthesized. It is a synthetic compound which can be converted into K_2 in the gastrointestinal tract. Unlike other lipid-soluble vitamins obtained by chemical synthesis, vitamin K_3 is characterized by high biological activity just like the naturally occurring ones.

8.5.2 METABOLISM AND FUNCTIONS

The rate of vitamin K absorption in humans depends on the kind of fats included in the diet. Long-chain polyenoic fatty acids facilitate absorption of vitamin K_1 in the lymphatic vessels. The efficiency of this process is affected also by the presence of bile salt and pancreatic juices, the form of vitamin K, as well as the site in the gastrointestinal tract. Vitamin K_1 is primarily absorbed in the jejunum and ileum; only small amounts are absorbed in the colon. From the lymphatic system, vitamin K is transported to the circulatory system and, in chylomicrons, to the liver from which it is distributed to target tissues.

Intestinal anaerobes such as *Escherichia coli* and *Bacillus fragilis* are also capable of synthesizing vitamin K_2 (Holzapfel et al., 1998). The gastrointestinal bacterial flora of human and animal organisms appears to be responsible for a substantial supply of vitamins, especially vitamin K. It was demonstrated that about half of the total vitamin K amount present in humans is of dietary origin, and the remainder is synthesized by the intestinal bacteria of the large intestine; conversion of vitamin K_1 to K_2 takes place *via* the so-called shikimic pathway; however, this refers only to healthy individuals. The greatest concentrations of various forms of vitamin K were detected in the liver. Presence of phylloquinone and menaquinones of varying chain lengths (up to 13 prenyl units, $n = 13$, can occur) confirms their diversified origin, while heterogeneity of menaquinones depends on the kind of microorganisms capable of their synthesis (Shearer, 1992). Several studies of the total vitamin K content in liver, adrenals, kidneys, lungs, and bones prove that the possibility of its absorption is but slight. The concentration of phylloquinone in plasma ranges from 0.3 to $2.6\,nmol/dm^3$ (Sadowski et al., 1989). In the liver, the content of phylloquinone was determined as ca. 10% ($2–20\,\mu g$) of the total vitamin K stored, and the remainder is primarily composed of menaquinone-7 and menaquinone-8 (Hodges et al., 1990; Usi et al., 1990; Uchida and Komeno, 1988).

Since its discovery, the antihemorrhagic action of vitamin K has been its best-known function. Vitamin K is located in hepatic microsomes where the vitamin K-dependent carboxylation in prothrombin synthesis occurs (Bell, 1978). The plasma clotting factors (II, VII, IX, and X) are produced by the liver in precursor forms and are converted to functional proteins by vitamin K-dependent reaction (Corrigan, 1982). They also depend on vitamin K for their synthesis and contain γ-carboxyglutamic acid (Gla) residues (Stenflo and Suttie 1977; Suttie 1992), because Gla is part of a protein that controls binding of calcium (Price, 1988; Knapen et al., 1989). Vitamin K participates in the process of carboxylation, which gives the proteins "claws" so they can hold calcium ions. Vitamin K participates in protein carboxylation, during which the protein containing three residues of Gla is formed—osteocalcin. This protein is probably synthesized in cells similar to osteoblasts. Osteocalcin is active in calcium binding making it possible for the organic and inorganic matters to unite in the bone tissue. Its presence prevents osteoporosis in joint action with vitamin D_3 (Shea and Booth, 2007).

Many studies confirmed the activity of vitamin K in inhibiting generation of interleukin-6 (Feskanich et al., 1999) and its activity, and as a potential protector against hepatocellular carcinoma (HCC) (Sarin et al., 2006). People with disordered connection between apolipoprotein E and level of vitamin K in the brain are particularly prone to Alzheimer's disease. This has been explained in terms of vitamin K regulating the level of building calcium not only in bones, but also in the cardiovascular system and brain (Jie et al., 1993).

Vitamin K displays antibacterial and antimycotic properties, as well as analgesic and anti-inflammatory activity. Studies of the influence of vitamin K on blood sugar level are being continued because its presence in pancreas was observed; it is suggested that it may affect the insulin secretion. Other investigations have shown that vitamin K can be more powerful than vitamin E and coenzyme Q_{10} in scavenging radicals (Malorni et al., 1993).

8.5.3 DEFICIENCY AND EXCESS

In the elderly, when absorption of vitamin K and the efficiency of its synthesis decrease, lowered prothrombin level in blood as well as liver dysfunctions may occur

(Hodges et al., 1991). Blood clotting time is extended, resulting in hemorrhages from the nose, or urinary or gastrointestinal systems. In newborn infants, one symptom of decreased vitamin K level is the hemorrhagic syndrome (Lane and Hathaway, 1985). Substances such as dicoumarol and its derivatives decrease the utilization of vitamin K-dependent clotting factors because they are its antagonists. Several antibiotics, various intestinal illnesses, mineral oil, and irradiation can inhibit absorption of this vitamin. In large doses, vitamin E can enhance the anticoagulant action of vitamin K (Booth, 2000). Some physicians, by virtue of observations and clinical studies, suppose that vitamin K can act like a hormone, but such a role of this vitamin is not clearly understood.

The toxicity associated with excessive amounts of vitamin K in humans has not been univocally defined, although it has been suggested that excessive doses of one of its forms, menadione, can contribute to occurrence of hemolytic anemia, jaundice, kernicterus, and hyperbilirubinemia (Worthington-Roberts, 1988).

Vitamin K belongs to a group of compounds, the properties of which are still under investigation, and its function in humans has yet to be fully determined.

8.6 NATURAL SOURCES OF LIPOPHILIC VITAMINS AND THE EFFECT OF FOOD PROCESSING

8.6.1 INTRODUCTION

Lipophilic vitamins are widespread in raw materials and foodstuffs of plant and animal origin. However, they occur in various compositions and quantities. Improved analytical techniques employed for separation of vitamins permitted to identify new, unknown homologues of these compounds in raw materials and products in which they have not been identified so far.

Therefore, no products rich in all vitamins are encountered in the diet. Becoming familiar with the sources of lipid-soluble vitamins makes it possible for people to satisfy the needs of man's organism as far as substances indispensable for its correct functioning are concerned.

8.6.2 OCCURRENCE

Several plants and animal products serve as excellent sources of lipophilic vitamins (Table 8.1). In products of animal origin, vitamin A occurs as pure retinol and its derivatives, and as carotenoids in products of plant origin—pure retinol is more easily assimilated than its precursors. For this reason, the content of vitamin A in food products is determined by stating the content of retinol, β-carotene, and other carotenoids and then converting to retinol equivalents (RE). Common dietary sources of vitamins A and D are livers, especially the livers of marine fish, and oils obtained from the livers of shark, halibut, and tuna. Fish livers are used as raw material for obtaining oils with high concentration of these vitamins. The most efficient method for obtaining concentrates, or vitamins A and D alone, is to apply molecular distillation accompanied by simultaneous refining of the raw oil; this method allows for minimal thermal and oxidative losses of vitamin A. Other rich sources of

TABLE 8.1
Natural Sources of Vitamin A, D, E, and K[a]

Content (per 100 g of Edible Portion)

Food Category	Vitamin A Retinol Equivalent (µg)	Vitamin D (µg)	Vitamin E Tocopherol Equivalent (mg)	Vitamin K (µg)
Animal				
Beef meat—lean	20.0	0.3	0.5	12.5
Beef—liver	15,300	0.6	0.7	74.5
Pork meat—lean	6.0	b	0.3	18.0
Pork—liver	39,100	1.0	0.2	56.0
Chicken—hearts	9.0	b	1.2	720.0
Chicken—livers	12,800	1.3	0.4	80.0
Fish Products				
Herring	38.0/40.0	26.7/27.0	1.5/2.9	b
Mackerel	100.0	4.0	1.6	5.0
Sardine	20.2/60.0	10.7	3.8	b
Tuna	450.0	4.5	0.5	b
Salmon	15.0	16.3	0.9/2.0	b
Oil from cod liver	b	250	22.0	b
Oil from herring liver	b	3.5 mg	b	b
Dairy Products				
Milk 5.5% fat	31.0	0.2	0.1	4.0
Powder milk	253	1.2		55.0
Butter	653	1.2	2.2	60.0
Cottage cheese 30% fat d.m.	99.0	0.2	0.3	50.0
Cheddar 50% fat d.m.	360	0.3	1.0	3.0
Eggs				
Egg	190	2.9	0.8	47.5
Egg yolk	550/590	5.6	2.1	147.0
Oils				
Soybean	4.0	—	50.0	3.0
Sunflower	B	—	38.8	7.5
Sesame	583	—	29.0	10.0
Olive	9,400	—	24.5	49.6
Corn germ oil	608.0	—	13.6	60.0
Cereals				
Wheat germs	10.0	—	24.7	131.0
Wheat ground grain	Traces	—	2.1	30.0
Wheat brans	Traces	—	2.7	81.5
Oat flakes	B	—	1.5	63.0

(*continued*)

TABLE 8.1 (continued)
Natural Sources of Vitamin A, D, E, and K[a]

Content (per 100 g of Edible Portion)

Food Category	Vitamin A Retinol Equivalent (µg)	Vitamin D (µg)	Vitamin E Tocopherol Equivalent (mg)	Vitamin K (µg)
		Vegetables, fruits		
Tomato	114	—	0.8	8.0
Cauliflower	2.0	—	0.1	167
Broccoli	50.0	—	0.6	174
Green Peas	60.0	—		33.4
Parsley—leaves	1000.0	—	3.7	620
Lettuce	200.0	—	0.6	130.0
Kale	1400.0	—	1.7	817
Apple	6.0	—	0.5	20.2
Avocado	7.0	—	1.3	20.3
Kiwi	7.0	—	2.1	28.5
Orange	3.0	—	0.3	3.7

[a] On the basis of Belitz et al. (2004); Eitenmiller and Junsoo (2004); Elmadfa et al. (1997); Fiorentino et al. (2009).

[b] No data available.

vitamin A include various dairy products, whereas the main sources of carotenoids are dark-green leafy vegetables in which chlorophyll masks the yellow color of the carotenoids.

For vitamin E, the richest dietary sources are plant oils and products made from these oils, such as margarine, shortening, and salad dressings. Ample amounts of this vitamin occur in wheat germs and almost as much in grain meals. Green vegetables, such as broccoli, cauliflower, spinach, and parsley leaves well complement the diet with vitamins E and K, because they contain large amounts of phylloquinone. Sufficient quantity of vitamin K is also found in pluck, especially in giblets.

The content of lipophilic vitamins in foodstuffs is unstable, and it varies considerably. In raw materials and consequently in plant products the content of lipid-soluble vitamins depends upon the agro-technologic and climatic conditions, ripeness of seeds, the way they are harvested and stored, and finally upon the technology of food processing. Varying concentration of these vitamins in animal raw materials depends on the breed, age, and seasonal feeding of the animal. The biological activity of vitamins is affected by numerous extrinsic and intrinsic factors which influence their assimilation from food products. The variety of compounds displaying varying degree of biological activity of vitamins, especially vitamins A and E, has made it necessary to adopt a uniform system of expressing the organism's demand for these vitamins and their content in foods, taking into consideration differences in the bioactivity. The action of vitamin preparations in the form of esters or salts is weakened by fragments that do not show any vitamin activity apart from the active vitamin released in the alimentary tract (e.g., retinol). The biological activity of a compound is inversely proportional to participation of such fragments in its molecule. At first, a notion of an international unit (IU) was introduced to express the vitamin activity:

- IU vitamin A = 0.3 µg trans retinol or 0.6 µg β-carotene
- IU vitamin D = 0.025 mg cholecalciferol or ergocalciferol
- IU vitamin E = 1 mg tocopherol acetate

For the sake of greater analytic precision and as a result of numerous tests conducted on laboratory animals (e.g., rats and chickens), equivalents (E) have been proposed relating the quantity of a given compound to the form displaying the greatest activity of a given vitamin. For vitamin A it is the RE

1 µg RE = 1 µg retinol = 6 µg β-carotene = 12 µg of other provitamin A carotenoids

For vitamin E the T equivalent is being applied (TE):

1 mg TE = 1 mg α-T = 2 mg β-T = 4 mg γ-T = 5 mg α-T-3 = 25 mg β-T-3 or γ-T-3

It is sometimes difficult to calculate the RE because in order to determine the actual content of vitamin A in food products, meals or daily food allowance, it is necessary to first determine the content of retinol, β-carotene, and other carotenoids, and then calculate the RE. Currently, recommended dietary allowances (RDA) are being proposed to ensure that there are appropriate quantities of individual vitamins in the diet to cover the organism's demand. The recommended doses are subject to constant modification and take into consideration the changing life conditions and dietary patterns. In addition, increasing knowledge of human nutrition makes it necessary to update them from time to time (Russell, 1997).

8.6.3 FOOD PROCESSING

The vitamin content in foods, due to the constant demand for vitamins by human and animal organisms, is a significant criterion in the assessment of the effect of technological processes on the quality and biological/nutritive value of food products. The most significant losses of vitamins occur in raw materials during storage and as a consequence of handling and food processing. Decrease in vitamin level results from physical and chemical, or biochemical, changes and depend on various factors. Lipophilic vitamins in raw materials and foodstuffs can be affected by physical factors (temperature, sunlight, and UV light, oxygen/air), chemical factors (radicals, peroxides, metal ions, e.g., Cu^{2+}, Fe^{3+}), and enzymes, mainly oxidases such as lipoxygenase. Lipid-soluble vitamins are particularly susceptible to oxidation due to all of these factors, and the process is further accelerated by the presence of oxidized fats. These factors, acting jointly, could cause even greater vitamin losses in foodstuffs during technological processes, and afterwards during storage.

Vitamin A, in the retinol form, as well β-carotene in raw materials are sensitive to light, storage temperature, and atmosphere. Also plant oxidative enzymes could accelerate the changes and decomposition of vitamin A. The factor which also influences decomposition of lipid-soluble vitamins is pH of the environment.

In vegetables and fruits, during storage, β-carotene decays under the influence of the sunlight causing photo degradation of carotenoids manifested by discoloration, or bleaching. An investigation of the effect of light and temperature on tomato and carrot juices revealed faster decomposition of α-and β-forms than other carotenoids. Carotenoids losses were higher in blanched than in unblanched freeze-dried carrots. During tomato juice production, a 20% loss of β-carotene was observed; further loss (another 20%) occurred during 7 months storage at 22°C (Rodriguez-Amaya, 1993). Blanching of vegetables can bring about greater vitamin destruction than further storage of deep-frozen products and dishes (Gross, 1991). Research on the effects of storage of other foodstuffs, such as vacuum-packed rainbow trout fillets stored up to 6 months at −20°C, confirmed that no significant losses of carotenoids were noted (No and Storebakken, 1991). Beta-carotene, added to various plant oils heated at 120°C had some influence only in the refined corn oil, in which the addition of 0.05%–0.2% β-carotene slightly prolonged the oil stability during heating. Addition of 0.01%–0.2% β-carotene to olive and rapeseed oils displayed pro-oxidative properties (Wagner and Elmadfa, 1999).

The effects of several methods of food processing on the retention of carotenoids and vitamin A were studied. Results showed that canning at 121°C for 30 min destroyed carotenoids in carrot juice more than high temperature short time sterilization at 120°C, 110°C, for 30 s. Destruction of vitamin A during canning amounted to 55.7%. In other processes, vitamin A content was observed to decrease with increasing temperature and time of heating (Chen et al., 1995). In the carotenoid powder obtained from the carrot pulp waste, the stability of carotenoid was studied under light and dark storage at various temperatures (ranging from 4°C to 45°C) for 12 weeks. The experiment proved that the content of α- and β-carotene decreased with prolongation of the time of storage and light exposure; this was confirmed by high correlation between the change of color and the carotenoid content in the carrot powder (Chen and Tang, 1998). In model studies on the decomposition of β-carotene in Bickoff test and the influence of the homologous T added and exposed to light at 20°C, there was a significant decrease of decomposition dynamics—it was from 4.2 to 5 times slower in the case of α-, γ-, and δ-T than in the control sample (Nogala-Kalucka and Zabielski, 1999).

Irradiation significantly affects vitamin losses, depending on the applied dose and the environment conditions. The decomposition of homologous T after γ-irradiation of various edible oils (soybean, sunflower, palm, linseed, corn, and wheat germ oils) was investigated. The authors stated that with the range 2.5–20 kGy losses of T varied; at 20 kGy the loss in—for example—soybean oil amounted to 37%, 61% and 44% for α-, γ-, and δ-T, respectively; and in linseed oil—30% and 55% for α- and γ-T, respectively. Above 2.5 kGy rapid increase of peroxides occurred, which resulted in considerable acceleration of autoxidation (Gogolewski et al., 1996, 1997). A study on the changes in T in irradiated rice bran during storage confirmed the above results: 50%–82% loss of -T took place immediately after irradiation, and further decomposition (amounting to 10%–35%) was recorded during storage up to 52 weeks (Shin and Godberg, 1996). The tilapia and Spanish mackerel fillets were subjected to γ-irradiation (range 1.5–10 kGy) and then stored for 20 days. In both fish, decomposition of homologous T increased with the increasing irradiation dose. The 3.0 kGy dose was optimum for maintaining the T level, and doses higher than that brought about greater losses of α-T and γ-T. The 20-day storage only slightly influenced further changes in the homologous T content (Al-Kahtani et al., 1996).

Deep-frying of potato chips in rapeseed oil at 162°C have been investigated; α-T decomposed more rapidly than β-T and γ-T. However, the addition of rosemary extract or ascorbyl palmitate to the oil decreased the decomposition of T during frying (Gordon and Kourimska, 1995). Model studies of the antioxidant properties of T in linoleic acid and its methyl esters proved that at higher temperatures (37°C and 47°C) γ-T was a better antioxidant than α-T (Gottstein and Grosch, 1990). It is difficult to unequivocally determine the decomposition of, for example, β-carotene, as well as T, because they can migrate to the oil used for frying or, together with that oil, be absorbed by the product being fried (Pokorny, 1999; Pokorny and Schmidt, 2001).

Application of microwave heating for preparing dishes also results in losses of vitamin E. In microwave processing, the decomposition of T depended on the kind of oil and duration of heating (Yoshida et al., 1991a,b). Total loss of T varied, depending on the kind of oil being heated, it decreased in the following order: olive oil, rapeseed and corn oils, and soybean oil. Other studies upon changes of oxidation stability and T in soybean, corn, sunflower, rapeseed, and olive oils showed that, after 15 min the decrease of the T content was 5%, 14%, 43%, 32%, and 61%, respectively. After 25 min of heating (maximum heating time), α-T was completely destroyed in all oil samples, while in the soybean, corn, and rapeseed oils the presence of γ- and δ-T was noted (Marinova et al., 2001).

In industrial food processing, vitamin E loss is primarily caused by the oil refining process, which consists of several stages in which T are being partially removed. This significantly influences the change in the content of homologous T. Many studies on various oils (e.g., rapeseed, sunflower, and soybean) confirmed total loss amounting up to 30% T in the refined oil in comparison with the crude one. The largest amount of T, two-third, are removed during deodorization and one-third during degumming and bleaching (Nogala-Kalucka et al., 1993; Shahidi et al., 1997; Cmolik et al., 2000).

The physical refining caused higher losses in the total and individual T contents when compared to chemical refining (Tasan and Demirci, 2005).

T were determined also during storage of the rapeseed and soybean oils, crude and refined, at 20°C, with respect to the Lea number. At the Lea number 10 the T decomposition was the following: δ-T > γ-T > α-T for the rapeseed oil, and γ-T > α-T for the soybean oil. Generation of T dimers was noted, the quantities of which were smaller than the respective T being decomposed. The dimers, due to their antioxidant properties, can also inhibit the autoxidation of fats in which they occur. In margarine stored at 4°C and 20°C, after 136 days, the greatest decomposition observed concerned α-T amounting to 12%. Decomposition of γ-T and δ-T amounted to 8% each, at the lower temperature, and to 50%, 47%, and 36% for α-T, γ-T, and δ-T, at 20°C (Nogala-Kalucka and Gogolewski, 1995, 2000). On the basis of numerous studies on the preservation of oils and other edible plant fats, one can state that T losses depend primarily on the kind of refined plant oil or margarine (especially the composition of polyenoic fatty acids), storage temperature, packaging, access of light, and air/oxygen. Frying test showed that stability of oils depended on the lipid matrix and frying time and temperature, but most important was the type and quantity of T or tocotrienol homologue present (Rossi et al., 2007).

Not much literature data is devoted to changes of vitamin D and K in raw materials and foodstuffs. Both vitamins are light sensitive and acidity or alkalinity of the environment can additionally activate their decomposition. The effect of adding menadione to the stored oils upon T and formation of their dimers was studied. Commercial rapeseed and soybean oils were stored in transparent and brown bottles at 20°C. Addition of vitamin K_3 to oil samples depended on the T content in the tested oil (0.05 mM K_3: 0.5 mM T). In both rapeseed and soybean oils, stored in transparent bottles, decomposition of T was accompanied by simultaneous formation of their dimers. After 71 and 94 days, respectively, complete destruction of T and their dimers was noted. In brown bottles, the time of T decomposition was longer and amounted to 105 days for the rapeseed oil and 167 days for the soybean oil. In comparison with control samples, the decomposition of T was enhanced by addition of vitamin K_3 and exposure to light (transparent bottles) (Kupczyk and Gogolewski, 2001). The same authors, in a model study upon irradiation of T (range 2.5–10 kGy) and storage for two months at 4°C, noted that the decomposition of all irradiated T increased during storage and depended on irradiation doses and the reaction medium (samples were dissolved in benzene and ethanol and they were also irradiated "in substantia"), with the highest decomposition taking place in polar solvent. Destruction of δ-T irradiated "in substantia" was smaller than other homologues, which corroborates data concerning the influence of the environment on decomposition of T and also influence of the accompanying substances having, for example, an antagonistic effect (Kupczyk and Gogolewski, 1999).

Processing of raw materials brings about losses of bioactive food ingredients; therefore, fortification of some selected products is carried out in many countries worldwide. Typical foods fortified with lipophilic vitamins, especially with vitamins D and A, are dairy products, breakfast cereals, cookies, and plant margarine (Lamberg-Allardt, 2006; Butt et al., 2007; Kazmi et al., 2007). Nutritional studies carried out in many highly industrialized countries have shown inadequate intake of recommended vitamins by some population groups. Therefore, the primary aim of elaborating new food processing technologies is to maintain the maximum amount of vitamins of which a given product is a natural source.

8.7 ANTIOXIDANT CONCURRENCY

Proper daily diet is one of the preconditions for maintaining good health and it directly affects the general condition of the human organism. The increased supply of bioactive components—including vitamins—in daily food rations is called as a prophylactic and therapeutic factor in civilization diseases such as tumors, ischemic heart disease or diabetes. Apart from the already known factors triggering off these diseases there are also some significant endogenous factors such as reactive forms of oxygen or free radicals generated initially by chronic inflammatory conditions, environmental pollution, and ingested medicaments (Nogala-Kałucka, 2007). Formation of free radicals is a form of response to the attack of pathogens and a basic way of fighting infections and repairing damaged tissues. Therefore, the most important task to the human organism, or a cell—the basic system—is to keep the balance between the rate of formation of, for example, reactive forms of oxygen and free radicals and the antioxidant potential limiting their activity to physiological functions thus protecting the cells against destruction. Peroxidation of lipids in cell membranes is the best-known process. It can impair the properties of these membranes causing their increased permeability and inhibiting activity of membrane enzymes. The final effect is loss of the integrity of the cytoplasmic membrane and the intracellular membranes resulting in the cell's death. Relation between the occurrence of disease symptoms and—among other things—quality and quantity of ingested lipophilic and hydrophilic vitamins has been studied for many years (Fairfield and Fletcher, 2002). Clinical studies carried out on large populations regarding prevention of diseases brought about by the oxidative stress still remain one of the most important scientific trends.

Intensive studies on lipophilic vitamins permitted to precisely determine the effects of their deficiency in human and animal organisms. However, it was only during the last two decades that the mechanisms of the biofeedback of metabolism, activity, and regulation of the homeostasis of these vitamins, pointing to the participation of exogenous molecules functioning in the organism, their conditioning of gene expression and activity of already existing proteins were discovered (Lips, 2006; Zingg, 2007). Deficiency of pancreatic enzymes results in intestinal malabsorption of fat and consequently of the lipophilic vitamins.

Some vitamins, apart from their other functions in human and animal organisms, also participate in functioning of antioxidant defense mechanisms (Di Mascio et al., 1991). Carotenoids, vitamin A, E, D, and K, supplied with food, protect against free radicals being generated, reactive oxygen species, and peroxides of various types (Halliwell, 1996). Disturbance of the pro- and antioxidant equilibrium of the organism in favor of oxidation is the reason for many diseases. Dietary antioxidants—lipophilic vitamins—play an important role in relieving oxidation damages that occur in man's organism (Biesalski et al., 1997). Vitamin D_3 may act as a terminator of lipid peroxidation chain reaction (Lin et al., 2005). Also, it can be helpful in prevention of numerous kinds of cancer by regulating cell growth and decreasing the risk of malignant forms generation (Holick, 2006). Laboratory studies on animals as well as the clinical trials show that chronic deficiency of vitamins with antioxidative potential cause numerous biochemical changes resulting in development of, for instance, atherosclerosis, cardiovascular disorders, carcinogenic, and inflammatory processes (Albanes et al., 1995; Omenn et al., 1996).

Antioxidant vitamins protect physiologic functions of arterial endothelium, inhibit thrombocyte aggregation, relieve infections, and reduce progression of atherosclerotic changes. It is supposed that at an early stage they are capable of inhibiting development of AIDS (Gerstner, 1997; Anstead, 1998). Combined effect of lipophilic vitamins (A and E) in treatment of allergic rhinitis and prevention or treatment of acute and chronic skin disorders (atopic dermatitis) was found (Thiele and Ekanayake-Mudiyanselage, 2007; Wijst and Hypponen, 2007). Numerous studies emphasize molecular activity of vitamins E and C in cells as effective biological antioxidants (Munteanu and Zingg, 2007; Villacorta et al., 2007).

Development of analytic techniques and studies at the cell's molecular level and also utilization of the achievements of nutrigenomics and pharmacogenomic regarding combined effect of lipophilic vitamins with respect to the oxidative stress can facilitate formulation of unambiguous replies to many difficult questions and permit to establish appropriate prophylactic procedures, thus improving health of millions of people worldwide.

REFERENCES

Albanes, D. et al., Effects of alpha-tocopherol and beta-carotene supplements on cancer incidence in the alpha-tocopherol, beta-carotene prevention study, *Am. J. Clin. Nutr.*, 62, 1427S, 1995.

Al-Kahtani, H.A. et al., Chemical changes after irradiation and post-irradiation storage in tilapia and Spanish mackerel, *J. Food Sci.*, 61, 729, 1996.

Anstead, G.M., Steroids, retinoids and wound healing, *Adv. Wound Care*, 11, 277–285, 1998.

Arnaud, C.D. and Sanchez, S.D., Calcium and phosphorus, in *Present Knowledge in Nutrition*, E.E. Ziegle and L.J. Filer, Eds., ILSI Press, Washington, DC, 1996, p. 245.

Beharka, A., Redican, S., Leka, L., and Meydani, S.N., Vitamin E status and immune function, *Methods Enzymol.*, 282, 247, 1997.

Behl, C., Vitamin E and other antioxidants in neuroprotection, oxidative reactions in Parkinson's disease, *Neurology*, 40, 32, 1999.

Behl, C., Davids, J., Cole, G.M., and Schubert, D., Vitamin E protects nerve cells from amyloid beta-protein toxicity, *Biochem. Biophys. Red. Commun.*, 186, 994, 1992.

Belitz, H.-D., Grosch, W., Schieberle, P., and Burghagen, M., *Food Chemistry*, 3rd edn., Springer, Berlin, Germany/Tokyo, Japan, 2004.

Bell, R.G., Metabolism of vitamin K and prothrombin synthesis: Anticoagulants and the vitamin K—Epoxide cycle (review), *Fed. Proc.*, 37, 2599–2604, 1978.

Bendich, A. and Machlin, L.J., The safety of oral intake of vitamin E; date from clinical studies from 1986–1991, in *Vitamin E in Health and Disease*, L. Packer and J. Fuchs, Eds., Marcel Dekker, New York, 1993, p. 411.

Biesalski, H.K., Consensus statements. Antioxidant vitamins in prevention, *Clin. Nutr.*, 16, 151, 1997.

Bischoff-Ferrari, H.A., How to select the doses of vitamin D in the management of osteoporosis, *Osteoporos. Int.*, 18, 401–407, 2007.

Boonen, S. et al., Calcium and vitamin D in prevention and treatment of osteoporosis—Clinical update, *J. Int. Med.*, 259, 539–552, 2006.

Booth, S.L., Dietary intake and adequacy of vitamin K, *J. Nutr.*, 130, 785, 2000.

Botella-Carretero, J.I. et al., Vitamin D deficiency is associated with the metabolic syndrome in morbid obesity, *Clinical Nutr.*, 26, 573–580, 2007.

Bouillon, R., Okamura, W.H., and Norman, A.W., Structure–function relationships in the vitamin D endocrine system, *Endocr. Rev.*, 16, 200, 1995.

Bramley, P.M. et al., Review—Vitamin E, *J. Sci. Food Agric.*, 80, 913, 2000.

Brigelius-Flohe, R. and Traber, M., Vitamin E: Function and metabolism, *FASEB J.*, 13, 1145–1155, 1999.

Butt, M. et al., Bioavailability and storage stability of vitamin A fortificant (retinyl acetate) in fortified cookies. *Food Res. Int.*, 40, 1212–1219, 2007.

Chatfield, S. et al., Vitamin D deficiency in general medical inpatients in summer and winter, *Intern. Med. J.*, 37, 377–382, 2007.

Chen, B.H. and Tang, Y.C., Processing and stability of carotenoid powder from carrot pulp waste, *J. Agric. Food Chem.*, 46, 2312, 1998.

Chen, B.H., Peng, H.Y., and Chen, H.E., Changes of carotenoids, color and vitamin A contents during processing of carrot juice, *J. Agric. Food Chem.*, 43, 1912, 1995.

Cmolik, J. et al., Effects of plant-scale alkali refining and physical refining on the quality of rapeseed oil, *Eur. J. Lipid Sci., Technol.*, 102, 15, 2000.

Collins, E.D. and Norman, A.W., Vitamin D, in *Handbook of Vitamins*, L.J. Machlin, Ed., Marcel Dekker, New York, 1990, p. 59.

Corrigan, J.J. Jr., The effect of vitamin E on warfarin-induced vitamin K deficiency, *Ann. NY Acad. Sci.*, 393, 361–368, 1982.

DeLuca, H.F. and Zierold, C., Mechanisms and functions of vitamin D, *Nutr. Rev.*, 56, S4, 1998.

Devqun, M.S. et al., Vitamin D nutrition in relation to season and occupation, *Am. J. Clin. Nutr.*, 34, 1501, 1981.

Di Mascio, P., Murphy, M.E., and Sies, H., Antioxidants defense systems: The role of carotenoids, tocopherols and thiols, *Am. J. Clin. Nutr.*, 53, 194S, 1991.

Eckhoff, C. and Nau, H., Vitamin A supplementation increases levels of retinoic acid compounds in human plasma: Possible implications for teratogenesis, *Arch. Toxicol.*, 64, 502, 1990.

Eitenmiller, R. and Junsoo, L., *Vitamin E: Food Chemistry, Composition and Analysis*. Marcel Dekker, New York, 2004.

Elmadfa, I. and Bosse, W., Vitamin E—Eingenschaften Wirkungsweise und therapeutische Bedeutung, WVG, Stuttqart, Germany, 1985, p. 89.

Elmadfa, I. et al., Die grosse GU Nährwert – und Kalorien-Tabelle, Gräfe und Unzer Verlag GmbH, München, Germany, 1997, p. 7.

Fahn, S.A., A pilot trial of high-dose alpha-tocopherol and ascorbate in early Parkinson's disease, *Ann. Neurol.*, 32, 128, 1992.

Fairfield, K. and Fletcher, R., Vitamin for chronic disease prevention in adults: Scientific review, *JAMA*, 287, 23, 2002.

Feskanich, D. et al., Vitamin K intake and hip fractures in woman: A prospective study, *Am. J. Clin. Nutr.*, 69, 7479, 1999.

Fiorention, A. et al., δ-Tocomonoenol: A new vitamin E from kiwi (*Actinidia chinensis*) fruit, *Food Chem.*, 115, 187–192, 2009.

Gann, P.H. et al., Circulating vitamin D metabolites in relation to subsequent development of prostate cancer, *Cancer Epidemiol. Prev.*, 5, 121, 1996.

Gerstner, H., Vitamin A–functions, dietary requirements and safety in humans, *Int. J. Vitam. Nutr. Res.*, 67, 71, 1997.

Gianello, R. et al., α-Tocopheryl phosphate: A novel natural form of vitamin E, *Free Radic. Biol. Med.*, 39(7), 970–976, 2005.

Giovannucci, E., γ-Tocopherol: A new player in prostate cancer prevention? *J. Natl. Cancer Inst.*, 92, 1966–1967, 2000.

Gogolewski, M. et al., Effect of ionizing radiation on quality of some edible oils—Part I, (in Polish), *Brom. i Chem. Toksyk.*, 29, 63, 1996.

Gogolewski, M. et al., Effect of ionizing radiation on quality of some edible oils—Part II, (in Polish), *Brom. i Chem. Toksyk.*, 30, 149, 1997.

Gordon, M.H. and Kourimska, L., Effect of antioxidants on losses of tocopherols during deep-fat frying, *Food Chem.*, 52, 175, 1995.

Gottstein, T. and Grosch, W., Model study of different autoxidation properties of α-and γ-tocopherol in fats, *Fat Sci. Technol.*, 92, 139–144, 1990.

Gross, J., Carotenoids in vegetables, in *Pigments in Vegetables—Chlorophyll and Carotenoids*, J. Gross, Eds., AVI Book, New York, 1991, p. 136.

Halliwell, B., Antioxidants, in *Present Knowledge in Nutrition*, 7th edn., E.E. Ziegle and L.J. Filer, Eds., ILSI Press, Washington, DC, 1996, p. 596.

Hathcock, J.N. Vitamins and minerals efficacy and safety, *Am. J. Clin. Nutr.*, 66, 427, 1997.

Henry, H.L. and Norman, A.W., Vitamin D: Metabolism and biological action, *Ann. Rev. Nutr.*, 4, 493, 1984.

Hodges, S.J. et al., Age–related changes in the circulating levels of congeners of vitamin K_2, menaquinone—7 and menaquinone-8, *Clin. Sci.*, 78, 63, 1990.

Hodges, S.J. et al., Depressed levels of circulating menaquinones in patients with osteoporotic fractures of the spine and femoral neck, *Bone*, 12, 387, 1991.

Hodis, H.N. et al., Serial coronary angiographic evidence that antioxidant vitamin intake reduces progression of coronary artery atherosclerosis, *JAMA*, 273, 1849, 1995.

Holick, M., Vitamin D: Its role in cancer prevention and treatment, *Prog. Biophys. Mol. Biol.*, 92, 49–59, 2006.

Holzapfel, W.H. et al., Overview of gut flora and probiotics, *Int. J. Food Microbiol.*, 41, 85, 1998.

Hosomi, A. et al., Affinity for alpha-tocopherol transfer protein as a determinant of the biological activities of vitamin E analogs, *FEBS Lett.*, 409, 105, 1997.

Jacques, P.F., The potential preventative effects of vitamins for cataract and age-related macular degeneration, *Int. J. Vitam. Nutr. Res.*, 69, 198, 1999.

Jie, K.S. et al., Effects of vitamin K and oral anticoagulants on urinary calcium excretion, *Br. J. Haematol.*, 83, 100, 1993.

Johnson, I.T., Antioxidants and antitumor properties, in *Antioxidants in Food*, J. Pokorny, N. Yanishleva, and M. Gordon, Eds., CRC Press, Cambridge, U.K., 2001, p. 100.

Kamal-Eldin, A. and Appelqvist, L.A., The chemistry and antioxidant properties tocopherols and tocotrienols, *Lipids*, 31, 671, 2001.

Kazmi, S., Veith, R., and Rousseau, D., Vitamin D_3 fortification and quantification in processed dairy products, *Int. Dairy J.*, 17, 753–759, 2007.

Knapen, M.H., Hamulyak, K., and Vermeer, C., The effect of vitamin K supplementation on circulating osteocalcin (bone Gla protein) and urinary calcium excretion, *Ann. Intern. Med.*, 111, 1001, 1989.

Kupczyk, B. and Gogolewski, M., Effects of gamma irradiation and menadione (vit. K_3) on dissolution and dimerization of homologous tocopherols. Effect of storage time, *Food Sci. Technol.*, 3, 39, 1999.

Kupczyk, B. and Gogolewski, M., Influence of added menadione (vit. K_3) on dissolution and dimerization of tocopherols and autoxidation of triacylglycerols during storage of plant oils, *Nahrung/Food*, 45, 9–14, 2001.

Lamberg-Allardt, Ch., Vitamin D in foods and as supplements, *Prog. Biophys. Mol. Biol.*, 92, 33–38, 2006.

Lane, P.A. and Hathaway, W.E., Vitamin K in infancy, *J. Pediatr.*, 16, 351–359, 1985.

Lin, A., Chen, K., and Chao, P., Antioxidative effect of vitamin D_3 on zinc-induced oxidative stress in CNS, *Ann. N.Y. Acad. Sci.*, 1053, 319–329, 2005.

Lips, P., Vitamin D physiology, *Prog. Biophys. Mol. Biol.*, 92, 4–8, 2006.

Malorni, W. et al., Menadione—inducted oxidative stress leads to a rapid down—Modulation of transferring receptor recycling, *J. Cell. Sci.*, 106, 309, 1993.

Marinova, E. et al., Changes in the oxidation stability and tocopherol content in oils during microwave heating, in *Lipids, Fats and Oil: Reality and Public Perception*-abstracts, 24th World Congress and Exhibition of the ISF, Cognis, Berlin, Germany, 2001, p. 51.

Meydani, M., Vitamin E, *Lancet*, 345, 170, 1995.

Munteanu, A. and Zingg, J.-M., Cellular, molecular and clinical aspects of vitamin E on atherosclerosis prevention, *Mol. Aspects Med.*, 28, 538–590, 2007.

Niki, E. et al., Inhibition of oxidation of methyl linoleate in solution by vitamin E and vitamin C, *J. Biol. Chem.*, 259, 4177, 1984.

No, H.K. and Storebakken, T., Color stability of rainbow trout fillets during frozen storage, *J. Food Sci.*, 56, 969, 1991.

Nogala-Kalucka, M., Antioxidants in food. Healthy, technological, molecular and analytical aspects. (in Polish), Grajek W. Ed., *WNT*, Warszawa, 2007.

Nogala-Kalucka, M. and Gogolewski, M., Quantitative and qualitative changes in tocopherols and their dimers during storage of rapeseed and soybean oils—Crude and refined, *Ernährung/Nutrition*, 19, 537, 1995.

Nogala-Kalucka, M. and Gogolewski, M., Alternation of fatty acid composition, tocopherol content and peroxide value in margarine during storage at various temperature, *Nahrung*, 44, 431, 2000.

Nogala-Kalucka, M. and Zabielski, J. Model in vitro studies on the protective activity of tocochromanols with respect to β-carotene, in *Natural Antioxidants and Anticarcinogens in Nutrition, Health and Disease*, J.T. Kumpulainen and J.T. Salonen, Eds., Royal Society of·Chemistry, Cambridge, U.K., 1999, p. 334.

Nogala-Kalucka, M., Gogolewski, M., and Swiatkiewicz, E., Changes in the composition of tocopherols and fatty acids in postdeodorisation condensates during refining of various oils, *Fat Sci. Technol.*, 95, 144, 1993.

Ogru, E. et al. (Ed.), Vitamin E phosphate: An endogenous form of vitamin E, *Medimonod S.r. I*, 127–132, 2003.

Olson, J.A., Vitamin A, in *Present Knowledge in Nutrition*, E.E. Ziegle and L.J. Filer, Eds., ILSI Press, Washington, DC, 1996, p. 109–119.

Omenn, G.S. et al., Effects of a combination of beta-carotene and vitamin A on lung cancer and cardiovascular disease, *N. Engl. J. Med.*, 334, 1150, 1996.

Packer, L. and Kagan, V.E., Vitamin E the antioxidant harvesting center of membranes and lipoproteins, in *Vitamin E in Health and Disease*, L. Packer and J. Fuchs, Eds., Marcel Dekker, New York, 1993, p. 176.

Palomer, X. et al., Role of vitamin D in the pathogenesis of type 2 diabetes mellitus, *Diabetes Obes. Metab.*, 10, 185–197, 2008.

Pearce, B.C. et al., Hypocholesterolemic activity of synthetic and natural tocotrienols, *J. Med. Chem.*, 35, 3595, 1992.

Pokorny, J., Changes of nutrients at frying temperatures, in *Frying of Food*, D. Boskou and I. Elmadfa, Eds., Technomic, Lancaster, PA, 1999, p. 69.

Pokorny, J. and Schmidt, S., Natural antioxidant functionality during food processing, in *Antioxidants in Food*, J. Pokorny, N. Yanishlieva, and M. Gordon, Eds., CRC Press, Cambridge, U.K., 2001, p. 331.

Porter, N.A., Caldwell, S.E., and Mills, K.A., Mechanisms of free radical oxidation of unsaturated lipids, *Lipids*, 30, 277–290, 1995.

Price, P.A., Role of vitamin K-dependent proteins in bone metabolism, *Ann. Rev. Nutr.*, 8, 565, 1988.

Reiter, E., Jiang, Q., and Christen, S., Anti-inflammatory properties of α- and γ-tocopherol. *Mol. Aspects Med.*, 28, 668–691, 2007.

Rodriguez-Amaya, D.B., Stability of carotenoids during the storage of foods, in *Shelf-Life Studies of Foods and Beverages*, G. Charalambous, Ed., Elsevier, Amsterdam, the Netherlands, 1993, p. 591.

Rossi, M., Alamprese, C., and Ratti, S., Tocopherols and tocotrienols as free radical-scavenger in refined vegetable oils and their stability during deep-fat frying, *Food Chem.*, 102, 812–817, 2007.

Russell, R.M., New views on the RDAs for older adults, *J. Am. Diet. Assoc.*, 97, 515, 1997.

Sadowski, J.A. et al., Phylloquinone in plasma from elderly and young adults: Factors influencing its concentration, *Am. J. Clin. Nutr.*, 50, 100, 1989.

Sarin, S. et al., High dose vitamin K_3 infusion in advanced hepatocellular carcinoma. *J. Gastroenterol. Hepatol.*, 21, 1478–1482, 2006.

Serbinova, E. et al., Free radical recycling and intramembrane mobility in the antioxidant properties of α-tocopherol and α-tocotrienol, *Free Radic. Biol. Med.*, 10, 263, 1991.

Shahidi, F., Wanasundara, P.K., and Wanasundara, U.N., Changes in edible fats and oils during processing, *J. Food Lipids*, 4, 199, 1997.

Shea, M. and Booth S., Role of vitamin K in the regulation of calcification, *Int. Cong. Series*, 1297, 165–178, 2007.

Shearer, M.J., Vitamin K metabolism and nutritive, *Blood Rev.*, 6, 92, 1992.

Shin, T.S. and Godber, J.S., Changes of endogenous antioxidants and fatty acid composition on irradiated rice bran during storage, *J. Agric. Food Chem.*, 44, 567, 1996.

Sokol, R.J., Vitamin E, in *Present Knowledge in Nutrition*, E.E. Ziegle and L.J. Filer, Eds., ILSI Press, Washington, DC, 1996, p. 130.

Stenflo, J. and Suttie, J.W., Vitamin K-dependent formation γ-carboxyglutamic acid, *Annu. Rev. Biochem.*, 46, 157, 1977.

Suda, T., Takahashi, N., and Abe, E., Role of vitamin D in bone resorption, *J. Cell Biochem.*, 49, 53, 1992.

Suttie, J.W., Vitamin K and human nutrition, *J. Pediatr.*, 106, 351–359, 1992.

Tasan, M. and Demirci, M., Total and individual tocopherol contents of sunflower oil at different steps of refining, *Eur. Food Res. Technol.*, 220, 251–254, 2005.

Theriault, A. et al., Tocotrienol: A review of its therapeutic potential, *Clin. Biochem.*, 32, 309,1999.

Thiele, J. and Ekanayake-Mudiyanselage, S., Vitamin E in human skin: Organ-specific physiology and considerations for its use in dermatology, *Mol. Aspects Med.*, 28, 646–667, 2007.

Thomas, M.K. et al., Hypovitaminosis D in medical inpatients, *N. Engl. J. Med.*, 338, 777, 1998.

Uchida, K. and Komeno, T., Relationships between dietary and intestinal vitamin K, clotting factor levels, plasma vitamin K and urinary Gla, in *Current Advances in Vitamin K Research*, J.W. Suttie, Ed., Elsevier Science, New York, 1988, p. 477.

Usi, Y., Tamimura, H., and Nashimura N., Vitamin K, concentration in the plasma and liver of surgical patients, *Am. J. Clin. Nutr.*, 51, 846, 1990.

Villacorta, L., Azzi, A., and Zingg, J.-M., Regulatory role of vitamins E and C on the extracellular matrix components of the vascular system. *Mol. Aspects Med.*, 28, 507–537, 2007.

Wagner, K.-H. and Elmadfa, I., Nutrient antioxidant and stability of frying oils: Tocochromanols, β-carotene, phylloquinone, ubiquinone 50, in *Frying of Food*, D. Boskou and I. Elmadfa, Eds., Technomic, Lancaster, PA, 1999, p. 69.

Watson, K.E. et al., Active serum vitamin D levels are inversely correlated with coronary calcification, *Circulation*, 96, 1755, 1997.

Wijst, M. and Hypponen, E., Vitamin D serum levels and allergic rhinitis, *Allergy*, 62, 1085–1086, 2007.

Wolf, G. and Phil, D., The intracellular vitamin A binding proteins: An overview of their function, *Nutr. Rev.*, 49, 1, 1991.

Woodson, K. et al., Serum alpha-tocopherol and subsequent risk of lung cancer among male smokers, *J. Nutr. Cancer Inst.*, 91, 1738–1743, 1999.

Worthington-Roberts, B.S., Maternal nutrition and the course and outcome of pregnancy: Nutrient functions and needs, in *Nutrition Throughout the Life Cycle*, S. Rodwell Williams and Worthington-Roberts, B.S., Eds., Mosby College Publishing, St. Louis, MO, 1988, p. 96.

Yoshida, H., Hirooka, N., and Kajimoto, G., Microwave heating effects on relative stabilities of tocopherols in oils, *J. Food Sci.*, 56, 1042, 1991a.

Yoshida, H., Tatsumi, M., and Kajimoto, G., Relationship between oxidative stability of vitamin E and production of fatty acids in oils during microwave heating, *JAOCS*, 68, 566, 1991b.

Zingg, J.-M., Vitamin E: An overview of major research directions, *Mol. Aspects Med.*, 28, 400–422, 2007.

9 Lipid Oxidation in Food Systems

Grzegorz Bartosz and Anna Kołakowska

CONTENTS

9.1   INTRODUCTION

Lipid oxidation in food systems is a detrimental process. It deteriorates the sensory quality and nutritive value of a product, poses health hazards, and presents a number of analytical problems (Figure 9.1).

Lipid oxidation is affected by numerous internal and external factors such as fatty acid (FA) composition, content and activity of pro- and antioxidants, irradiation, temperature, oxygen pressure, surface area in contact with oxygen, and water activity (a_w). Because lipids are only a part of a food product, it is difficult to find a food component that would not be capable of affecting lipid oxidation.

9.2   FREE RADICALS AND REACTIVE OXYGEN SPECIES

Molecular oxygen (which in the ground state is a biradical, its molecule $^\bullet O{-}O^\bullet$ having two unpaired electrons) is not very reactive with respect to nonradical substrates. However, oxygen forms reactive oxygen species (ROS) whose ability to react with organic molecules, including food constituents, is much higher. The collective term ROS includes both free radicals (molecules having an odd electron) and

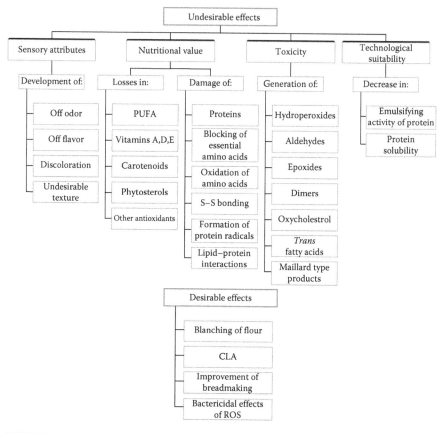

FIGURE 9.1 Effects of lipid oxidation on the quality of food. (Adapted from Kołakowska, A., Lipid oxidation in food systems, in *Chemical and Functional Properties of Food Lipids*, Sikorski, Z.E. and Kołakowska, A. (eds.), CRC Press, Boca Raton, FL, 2003, pp. 133–166.)

species that are not free radicals (like H_2O_2, singlet oxygen $^1O_2$, O_3, hypochlorite, and peroxynitrite). ROS are formed in every living cell. Their potentially damaging effects of their uncontrolled reactions with cell constituents are counteracted by antioxidative enzymes and low-molecular-mass antioxidants. Formation of ROS is not only an inevitable evil of aerobic metabolism; they play an important role in the defense against invading pathogens and in intra- and intercellular signaling as autocrine and paracrine factors (Bartosz, 2009). However, shifting the balance between the production of ROS and activities of antioxidants in the prooxidative direction (excess ROS production or insufficient antioxidant defense) leads to a situation referred to as *oxidative stress* where ROS inflict more or less massive damage on biological systems.

Oxidative stress accompanies many age-related degenerative diseases including atherosclerosis, cancer, trauma, stroke, asthma, arthritis, coronary heart disease, cataract, retinal damage, and hepatitis, and may contribute to their development. Oxidative stress contributes also to aging of cells and organisms.

While excessive ROS are predominantly implicated in causing cell damage, they also play a major physiological role in several aspects of intracellular signaling and regulation. Depending on their concentration, ROS elicit either a positive response (cell proliferation) or a negative cell response (growth arrest or cell death). While high concentrations of ROS cause cell death or even necrosis, low concentrations of superoxide radical and hydrogen peroxide stimulate proliferation and enhance survival in a wide variety of cell types. ROS can thus play a very important physiological role as secondary messengers. Other examples include regulation of the cytosolic calcium concentration (which itself regulates the above-mentioned biological activities), regulation of protein phosphorylation, and activation of certain transcription factors such as NF-κB and the AP-1 family factors.

One-electron reduction of the oxygen molecule produces *superoxide radical anion* ($O_2^{\bullet-}$). The main source of superoxide in living cells is the one-electron leakage of the mitochondrial respiratory chain and, in plant cells, of the chloroplasts redox system. Reduced nicotinamide adenine dinucleotide phosphate (NADPH) oxidases are enzymes that produce superoxide for the sake of defense (phagocytes) or signaling. Superoxide is also produced by other enzymes and by autoxidation of reduced forms of thiols, flavins, pteridines, diphenols, metalloproteins (e.g., hemoglobin, myoglobin, and ferredoxin), and xenobiotics. In food (especially meat), the same sources produce superoxide in the fresh tissue but soon their activities cease. Some enzymes, like xanthine oxidase, may act longer, the more that its substrates (xanthine and hypoxanthine) may be formed as a result of catabolism of purine nucleotides (Choe and Min, 2006a,b).

At low pH, $O_2^{\bullet-}$ can be protonated to a *perhydroxyl radical* $HO_2^{\bullet}$, which is a better oxidant than $O_2^{\bullet-}$:

$$O_2^{\bullet-} + H^+ \leftrightarrow HO_2^{\bullet-}$$

pK_a for this reaction is 4.88, so higher oxidative activity of the radical pair: $O_2^{\bullet-}$ radical/perhydroxyl radical can be expected at lower pH.

Two-electron reduction of molecular oxygen, one-electron reduction of $O_2^{\bullet-}$, and nonenzymatic or enzymatic dismutation of $O_2^{\bullet-}$ (reaction below) produce H_2O_2:

$$O_2^{\bullet-} + O_2^{\bullet-} + 2\,H^+ \rightarrow H_2O_2 + O_2$$

Hydrogen peroxide is generally less reactive than superoxide. However, its reactions with protein thiol groups may be important, especially in the context of signal transduction. In contrast to the electrically charged superoxide radical anion, this neutral molecule easily penetrates cellular membranes and other barriers, and may reach sites where it can be made more reactive.

Reaction of hydrogen peroxide with reduced forms of redox-active metal ions, especially Fe^{2+}, leads to formation of the extremely reactive *hydroxyl radical* $^{\bullet}OH$ in the Fenton reaction. Hydroxyl radical may also be formed by decomposition of peroxynitrite and in the reaction between hypochlorite and hydrogen peroxide:

$$^-OCl + H_2O_2 \rightarrow {}^{\bullet}OCl + {}^{\bullet}OH + {}^-OH$$

Another source of hydroxyl radical is ionizing radiation

$$H_2O \text{ (ionizing radiation)} \rightarrow H_2O^* \rightarrow H^{\bullet} + {}^{\bullet}OH$$

where H_2O^* represents an excited water molecule (possessing excess energy deposited by ionizing radiation). Hydroxyl radical may also be produced by UV-induced homolytic decomposition of H_2O_2:

$$H_2O_2 \text{ (UV)} \rightarrow 2{}^{\bullet}OH$$

and, albeit with much lower yield, during sonication of water and water-containing substances (Choe and Min, 2006).

Hydroxyl radical is the most reactive ROS. It has very high redox potential for one-electron reactions (about 2.3 V), so it can react with any organic molecule with very high reaction rate constants (10^9–10^{10} M^{-1} s^{-1}).

Nitric oxide NO${}^{\bullet}$ is also a free radical, playing mainly a signaling role in the body, being responsible, first of all, for smooth muscle relaxation and lowering blood pressure but also for retrograde signaling in the nervous system, having bactericidal action at high (micromolar) concentrations. It is produced mainly enzymatically, by NO${}^{\bullet}$ synthases. *In vivo*, nonenzymic production of NO${}^{\bullet}$ plays a minor role but may be more important in food. Nonenzymic sources of NO include reduction of nitrite NO_2^- under acidic conditions (pH < 4) by various reductants, by carotenoids exposed to light, or by strong reducing agents such as ascorbate. Cured muscle food products such as sausages and ham have their characteristic color from the stable product of reaction between myoglobin and NO${}^{\bullet}$.

Reaction of $O_2^{{\bullet}-}$ anion with NO${}^{\bullet}$ produces *peroxynitrite:*

$$O_2^{{\bullet}-} + NO^{\bullet} \rightarrow ONOO^-$$

Peroxynitrite is a strong oxidant capable of performing one- or two-electron oxidations of various compounds including antioxidants, lipids, and constituents of proteins and nucleic acids, and also a nitrating and nitrosylating agent. Its protonated form, *peroxynitrous acid* HOONO, decomposes

$$HOONO \rightarrow {}^{\bullet}OH + NO_2^{\bullet}$$

or isomerizes rapidly (half-life time of about 1 s at neutral pH). Isomerization to nitric acid leads to loss of reactivity. Decomposition to ${}^{\bullet}OH$ and $NO_2^{\bullet}$ is believed to represent *ca.* 30% of the reactivity of peroxynitrous acid.

Reactions of peroxynitrite lead to nitrosylation of –SH groups (formation of nitrosothiols –SNO) and nitration (introduction of the $-NO_2$ group), especially of tyrosine and tryptophan residues in proteins and guanine in nucleic acids, and also of other compounds (e.g., γ-tocopherol). The nitration reaction is favored, and oxidation reactions hampered, by the presence of carbon dioxide, which reacts with peroxynitrite to form 1-carboxylato-2-nitrosodioxidane $ONOOCO_2^-$. This compound in turn undergoes homolysis: $ONOO^- + CO_2 \rightarrow ONOOCO_2^- \rightarrow CO_3^{\bullet-} + NO_2^{\bullet}$ (Brannan et al., 2001).

Enzymatically catalyzed oxidation of chloride ions by hydrogen peroxide leads to the formation of *hypochlorous acid/hypochlorite*

$$H_2O_2 + Cl^- \rightarrow HOCl + HO^-$$

mainly by myeloperoxidase and lactoperoxidase. As the pK_a value of hypochlorous acid is about 7.5, comparable amounts of hypochlorous acid and its anion are present at near-neutral pH. Hypochlorite is a strong bactericidal agent and is produced for defensive purposes mainly by granulocytes. It is a chlorinating and oxidizing agent, forming 3-chlorotyrosine residues in proteins and initiating lipid peroxidation.

Singlet oxygen is an excited form of the oxygen molecule formed by energization of ground-state (triplet) oxygen molecule with light in the presence of photosensitizers (Cullen et al., 2009).

Ozone is another ROS, a minor constituent of the stratosphere but very important since it absorbs UV radiation. Ozone is also formed in the atmosphere mainly by photochemical reactions. Ozone is used for food decontamination. In the United States, gaseous ozone treatment or washing with ozonated water is used for solid food processing, and recently ozonation has been applied for the treatment of fruit juices (Gómez-López et al., 2009).

Another ROS that may be present if food is *chlorine dioxide* (ClO_2), a reactive gas used for pump bleaching in the paper industry and water disinfection, applied for food decontamination (inactivation of many microorganisms and viruses) in some countries (e.g., in the United States and Russia). In the European Union, its use is not legalized. Chlorine dioxide preserves also the color of vegetables by inhibiting polyphenol oxidase, responsible for oxidation of phenols which polymerize subsequently into melanins. In contrast to chlorine and hypochlorite, which react via both oxidation and electrophilic substitution, ClO_2 is essentially an oxidizing agent, producing no organochlorine compounds (Morrissey et al., 1998).

Reactions of the above-mentioned ROS with organic molecules lead to the formation of secondary ROS such as alkyl radicals $R^{\bullet}$, alkoxyl radicals $RO^{\bullet}$, peroxyl radicals $ROO^{\bullet}$, and hydroperoxides. Peroxyl and alkoxyl radicals, having one-electron redox potential above 1 V, are also quite efficient oxidants.

While ROS play multiple, both beneficial and deleterious roles in living organisms, their reactions in food are generally harmful, leading to lipid peroxidation and damage to other food components, which causes deterioration of food quality, for example, rancid flavor, unacceptable taste, shortening of shelf life, and even generation of toxic compounds. They can damage proteins and deplete the food in antioxidants, cause discoloration, and produce toxic compounds. The main reaction induced by ROS in food components is lipid peroxidation since this chain reaction can be propagated once initiated (Bartosz, 2003; Halliwell and Guteridge, 2007).

9.3 MECHANISMS OF LIPID PEROXIDATION

9.3.1 FREE RADICAL OXIDATION

Free radical induced lipid peroxidation is a chain reaction consisting of three distinctive steps: initiation, propagation, and termination. In the initiation step, a hydrogen atom is removed from a molecule LH and a free radical form of lipid (alkyl radical) $L^{\bullet}$ is formed:

$$LH \rightarrow L^{\bullet} + H^{\bullet}$$

The susceptibility of food to lipid oxidation depends on a number of factors, one of them being the level of polyunsaturated fatty acids (PUFA) present, another the levels of pro- and antioxidants. The hydrogen bound to the carbon atom separating the C=C bond is the easiest to be detached; therefore, PUFA or phospholipids (PLs) containing such residues are most prone to peroxidation. Linoleate was found to be 40 times more prone to peroxidation than oleate, and linolenate 2.4 times more prone than linoleate. Heat, UV, visible light, and metal catalysts can accelerate this step. It is usually initiated by hydroxyl, peroxyl, or hydroperoxyl radicals (but not superoxide or hydrogen peroxide) or hemoproteins activated to the ferryl state.

In the propagation step, alkyl radical reacts with oxygen to form peroxyl radical, which, in turn, can abstract hydrogen atom from another lipid molecule:

$$L^{\bullet} + O_2 \rightarrow LOO^{\bullet}$$

The reaction proceeds rapidly, as characterized by the rate constant $k = 3 \times 10^8 \ M^{-1} \ s^{-1}$.

$$LOO^{\bullet} + LH \rightarrow LOOH + L^{\bullet}$$

This reaction is relatively slow (rate constant of $10 \ M^{-1} \ s^{-1}$).

Reaction between two radicals terminates the reaction:

$$LOO^{\bullet} + L^{\bullet} \rightarrow LOOL$$

$$L^{\bullet} + L^{\bullet} \rightarrow LL$$

The reaction can also be inhibited if peroxyl radical reacts with an antioxidant to form an unreactive free radical, which does not abstract hydrogen from another lipid molecule. Such a role is most often played by vitamin E (tocopherol, TOH):

$$LOO^{\bullet} + TOH \rightarrow LOOH + TO^{\bullet}$$

The reaction rate constant for the reaction of TOH with the peroxyl radical $(k = 8 \times 10^4 \ M^{-1} \ s^{-1})$ is about four orders of magnitude higher than that for hydrogen abstraction from FA by the peroxyl radical. Thus, relatively small amounts of TOH may be sufficient to effectively inhibit lipid peroxidation. The tocopheryl radical can be reduced back by ascorbate (AH) and perhaps also by other antioxidants to regenerate TOH:

$$TO^\bullet + AH \rightarrow TOH + A^\bullet$$

where $A^\bullet$ represents AH radical. Otherwise, reaction between tocopheryl radicals may lead to a nonradical final product. However, under some conditions, TOH may also act as a prooxidant as the tocopheryl radical can (though with a low rate constant) abstract a hydrogen atom from FA or FA residues in acylglycerols (Porter et al., 1995; Kanner, 2007). Other lipid-soluble antioxidants important in the prevention of lipid peroxidation are vitamin A, carotenoids, ubiquinol, and bilirubin. Peroxidation may lead to the consumption of antioxidants present in the food and thus impair its nutritional quality (Hur et al., 2007).

Lipid hydroperoxides are relatively stable; edible oils may contain significant amounts of hydroperoxides, up to 10 mM. It has been estimated that daily uptake of hydroperoxides in humans may reach 1.4 mmol. Hydroperoxides may be decomposed to alkoxyl radicals by cleavage of the oxygen–oxygen bond, which has lower energy (184 kJ mol^{-1}) than the bond between oxygen and hydrogen (377 kJ mol^{-1}). However, they may be decomposed, especially by transition metal ions. This leads to the formation of secondary free radicals and reinitiation of lipid peroxidation (see below).

Also cholesterol is oxidized by ROS although its oxidation rate is relatively slow compared to that of PL containing PUFA residues. Hydroxyl radical abstracts hydrogen from cholesterol, mainly the C-7 hydrogen, producing free radicals. Main cholesterol oxidation products found in food are 7-ketocholesterol, β-epoxycholesterol, and α-epoxycholesterol. While fresh food contains none or undetectable levels of cholesterol oxidation products, their amount can reach frequently 1% of total cholesterol and occasionally even about 10% of total cholesterol in stored, processed, and heat-treated food. In meat, the rate of cholesterol oxidation is accelerated during storage following cooking. Cholesterol oxidation products have been shown to be more injurious to arterial cells than pure cholesterol and more likely to induce atherosclerosis and coronary heart disease. They also affect cell membrane functions, especially permeability (Choe and Min, 2006).

9.3.2 PHOTOOXIDATION

UV radiation and visible light may initiate lipid peroxidation in two ways. First, they may directly induce dissociation of *bis*-allylic hydrogen, leading to the formation of alkyl radical L$^\bullet$. However, the main effect of light is via photosensitized reactions, that is, light-induced formation of ROS. There are two main pathways of photosensitized reactions. Both require a photosensitizer, a molecule that, in the singlet state (all electrons paired) ^1S, absorbs the energy of the light to reach a singlet excited state $^1S^*$, and undergoes intersystem crossing to become an excited triplet-state molecule $^3S^*$ (with two unpaired electrons):

$$^1S + light \rightarrow {^1S^*} \rightarrow {^3S^*}$$

The excited triplet-state sensitizer reacts with ground-state oxygen (which is in the triplet state), and the excitation energy is transferred to the oxygen molecule singlet-state sensitizer and singlet oxygen $^1O_2$ is formed (Type II mechanism):

$$^3S* + {}^3O_2 \rightarrow {}^1S + {}^1O_2$$

When oxygen in the system becomes depleted, a shift from Type II to Type I photosensitized reactions is favored. In this case, superoxide anion radical is formed instead of $^1O_2$

$$^3S* + {}^3O_2 \rightarrow S^+ + O_2^{\bullet -}$$

Type III reaction, in which the excited sensitizer induces a direct damage to a target molecule, is much less common.

Bengal rose, methylene blue, and eosin are typical $^1O_2$-generating photosensitizers. However, natural dyes present in food such as curcumin, riboflavin, chlorophylls, and hematoporphyrin are also efficient. In meats, myoglobin and other hemoproteins have been identified as primary photosensitizers of lipid oxidation. Dye additives, for example, erythrosine, may also contribute to photosensitization. One sensitizer molecule may generate 10^3–10^5 $^1O_2$ molecules before becoming inactive (Ryter et al., 2007).

$^1O_2$ can react directly with the high-electron-density double bonds without the formation of alkyl radical forming hydroperoxide, thus initiating lipid peroxidation. It also forms specific products upon reaction with cholesterol (3β-hydroxy-5α-cholest-6-ene-5-hydroperoxide, 3β-hydroxycholest-4-ene-6α-hydroperoxide, and 3β-hydroxycholest-5-ene-7α-hydroperoxide), distinct from free radical induced oxidation products (Jung et al., 1998).

In food, $^1O_2$ reacts with vitamins and other compounds of nutritive value. Vitamin D reacts effectively with $^1O_2$ (reaction rate $k = 2.2 \times 10^7$ M^{-1}s^{-1}), which leads to its photodestruction. Riboflavin is a photosensitizer, but it also reacts with $^1O_2$. Milk exposed to sunlight for 30 min may lose up to 30% of its riboflavin; an 80% loss has been reported in milk stored under light. Light with wavelength of 450 nm (maximum absorption of riboflavin) is the most destructive to this compound. Ascorbic acid is also reactive with singlet oxygen ($k = 3.1 \times 10^8$ M^{-1}s^{-1}); as a result, vitamin C is also easily destroyed by light. Losses of 80%–100% of ascorbic acid have been reported upon 60 min exposure of milk to sunlight (Min and Boff, 2002).

Photochemical deterioration of vegetable oils can be minimized most effectively by removal of natural pigments during the refining process.

9.3.3 LIPOXYGENASES

Lipoxygenase (LOX) catalyzed lipid oxidation differs from the free radical reaction by the formation of hydroperoxides in a certain position of the chain of FA or FA residues in PLs. LOXs use molecular oxygen to catalyze the stereo- and regiospecific oxygenation of PUFA with 1-*cis*, 4-*cis*-pentadiene moieties. LOX can act with more than one methylene carbon on the substrate molecule to yield double oxygenation sites enzymatically (German et al., 1992). The newly formed FA peroxy free radical removes hydrogen from another unsaturated FA molecule to form a conjugated hydroperoxy diene. LOX forms a high-energy (radical) intermediate complex with the substrate; the complex is capable of initiating oxidation of lipids and also other compounds, for example, carotenoids, chlorophyll, tocopherols, thiol compounds, and proteins, which can themselves interact with the enzyme–substrate complex as well (Hammer, 1993; Hultin, 1994). They are responsible for the off-flavor in frozen vegetables (Ganthavorn et al., 1991), lipid oxidation in cereal products, rapeseed, pea, avocado, and for "beany" and bitter flavor. Soybean contains LOX-1 with optimum activity at pH 9 and LOX-2 with optimum at pH 6.5 (there is LOX-3 as well). LOX-1 is specific for linoleic acid and, like other LOXs, catalyzes FFA oxidation. LOX-2 requires no prior lipid hydrolysis and catalyzes oxidation of TAG, carotenoids, and vitamins. Grinding of grains accelerates lipolysis, thus enhancing LOX activity (Frankel, 1998). A change in pH results in reduction in the LOX activity. Activity of the soybean LOX having its optimum at pH 7 was observed to drop to 2% at pH 4; an increase in pH brings about a much slower reduction in activity. Heating to 55°C influenced the soybean activity only slightly (Tedjo et al., 2000). Soybean LOX has been used since 1930 to bleach flour to produce white bread crumb. It is well known that soybean LOX, by oxidizing free lipids, improves the dough-forming properties and baking performance of wheat flour (Hammer, 1993).

Mammalian LOX are categorized according to the positional specificity of oxygen insertion into arachidonic acid. Four isoform positions of arachidonate LOX have been identified: 5-LOX (E.C. 1.13.11.34), 8-LOX, 12-LOX (E.C. 1.13.11.31), and 15-LOX (E.C. 1.13.11.33). The LOX catalyzes oxidation of linoleate (E.C.1.13.11.12) attacking linoleic acid both at positions 9 and 13.

In chicken meat, arachidonate 15-LOX was found to be active during 12-month storage at −20°C (Grossman et al., 1988). In frozen-stored fish, too, LOX contributes to oxidative lipid deterioration. However, LOX in fish is also responsible for the formation of desirable fresh fish flavor- the seaweed flavor (Lindsay, 1994). Some species show a higher activity of 12-LOX, while 15-LOX is more active in other; for this reason, the fresh fish flavor spectrum is species-dependent. The half life of 12- and 15-LOX at 0°C was less than 3h and more than 10h, respectively (German et al., 1992). LOX was observed to be active in cold-stored fish after 48h of storage (Medina et al., 1999). Frozen storage of herring, 3 weeks at −20°C, resulted in the increase in LOXs activity (Kołakowska, 2003), while during prolonged frozen storage a decrease in LOX activity was observed (Samson and Stodolnik, 2001). Although the participation of LOX in the *post-mortem* animal lipid oxidation is acknowledged, the role of LOX in lipid oxidation is much more important in plant than in animal food products.

9.3.4 SECONDARY OXIDATION PRODUCTS

The above-mentioned intermediates and products of lipid peroxidation are subject to further reactions leading to secondary lipid oxidation products. Homolytic β-scission of the C–C bond of alkoxyl radicals produces further degradation products (aldehydes, ketones, acids, alcohols, and short-chain hydrocarbons). One of the main aldehydes generated by lipid peroxidation is 4-hydroxy-2-nonenal, a product of degradation of arachidonic and linoleic acids. PUFA containing three double bonds produce 4-hydroxy-2-hexanal whose concentration in meat and fish products may reach 120 µM. Dialdehydes (malondialdehyde, glyoxal) and acrolein may also be present, malondialdehyde (MA) reaching a concentration of 300 µM in some cases (Kanner, 2007). Food products may also contain protein-bound MA, which is broken down to N-ε-(2-propenal)lysine in the course of digestion. This, and other reactive aldehydes, may be absorbed, react with plasma proteins, and interact with the receptor for advanced glycosylation endproducts (RAGE) and induce secondary generation of ROS. Indeed, consumption of a meal containing lipid breakdown products has been demonstrated to induce oxidative stress and impair vascular function.

Most decomposition products of hydroperoxides are responsible for off-flavor of the oxidized edible oils and other food products. *trans,cis*-2,4-Decadienal was found to be the main compound responsible for the oxidized flavor of oil, followed by *trans,trans*-2,4-decadienal, *trans,cis*-2,4-heptadienal, 1-octen-3-ol, butanal, and hexanal. Hexanal, pentane, and 2,4-decadienal have been suggested and used as indicators to determine the extent of oil oxidation. No single compound is mainly responsible for the oxidized flavor of vegetable oils (Ryter et al., 2007).

Another result of free radical reactions with lipids is the *cis–trans* isomerization leading to the formation of nonbioavailable *trans* isomers of FA and FA residues (Halliwell et al., 1995).

9.3.5 TRANSITION METALS IN LIPID OXIDATION

Transition metal ions have a considerable contribution to lipid peroxidation, especially in food. Firstly, they participate in the Fenton reaction believed to be the main source of the hydroxyl radical in biological systems:

$$H_2O_2 + Fe^{2+} \rightarrow HO^- + {}^\bullet OH + Fe^3$$

This reaction may be perpetuated in the presence of reductants, which may back-reduce ferric ions formed. The hydroxyl radical may initiate the lipid peroxidation chain. In the living organism, compartmentalization of oxidizable components and sequestration of iron contribute to the prevention of undesired ROS formation. In healthy tissue, there are minute amounts of loosely bound iron able to participate in the Fenton reaction and practically no loosely bound copper ions. In food, decompartmentalization and proteolysis stimulate ROS formation and facilitate the Fenton reaction. Decomposition of metalloproteins such as hemoglobin and ferritin in food may be a considerable source of metals able to participate in this reaction. Loss of compartmentalization of cellular components, protein denaturation, and action of reductants may release especially heme and iron from hemoproteins and also copper from copper proteins.

Another contribution of transition metal ions to lipid peroxidation is due to their role in the decomposition of lipid hydroperoxides. These primary products of peroxidation may slowly decompose spontaneously, especially at elevated temperatures, but transition metals ions may accelerate their decomposition resulting in the formation of alkoxyl and peroxyl radicals:

$$ROOH + Fe^{2+} \rightarrow RO^{\bullet} + HO^{-} + Fe^{3+} \text{ (fast reaction)}$$

$$ROOH + Fe^{3+} \rightarrow ROO^{\bullet} + H^{+} + Fe^{2+} \text{ (slow reaction)}$$

The reactivity of ferrous ions in this reaction is higher by about two orders of magnitude, so Fe^{2+} ions are more efficient inducers of lipid peroxidation (Bartosz, 2003).

9.4   ROLE OF ROS IN DISEASES

Oxidative stress has been demonstrated to accompany over a hundred of diseases. In some cases, oxidative stress may be an outcome or side effect of a disease, but in many cases increased ROS production or inadequate antioxidant defense (including insufficient intake of antioxidant vitamins) seems to play a significant role in the development of the pathology.

The causative role of oxidative stress has been demonstrated for *atherosclerosis*, underlying the cardiovascular disease which is currently the main cause of mortality in Western societies. Atherosclerosis is initiated by damage of endothelium mediated by increased production of ROS. Augmented levels of homocysteine (formed in the metabolism of methionine) due to inborn defects of methionine metabolism enzymes or deficiency of cofactors of these enzymes (vitamin B_6 and folate) cause oxidative stress, contributing to the development of atherosclerosis. Injured endothelium stimulates activation of macrophages, which damage smooth muscle cells and activate matrix metalloproteinases in the vessel walls. Oxidized low density lipoprotein (LDL) is taken up by macrophages, which have scavenger receptors for such molecules. Massive uptake of oxidized LDL transforms macrophages into foam cells, which are accumulated in fatty streaks, transformed subsequently into fibrous plaques on vessel walls.

Tissue damage by *ischemia–reperfusion* involves ROS. Ischemia, depriving organ (especially brain or heart) of oxygen, is dangerous by decreasing cellular energy production. Paradoxically, reperfusion (recovery of blood supply) leads to reoxygenation injury mediated by ROS. An important factor in the reoxygenation injury is the increased activity of xanthine oxidase producing $O_2^{\bullet-}$ and H_2O_2. In healthy tissues, xanthine is oxidized mainly by xanthine dehydrogenase reducing NAD^+ not oxygen and producing no ROS. During reperfusion, xanthine dehydrogenase is transformed into xanthine oxidase by limited proteolysis due to Ca^{2+}-activated proteases. Augmented catabolism of adenine nucleotides occurring in the period of hypoxia provides substrates for the enzyme which contributes to the burst of ROS production upon reperfusion. Ischemia–reperfusion contributes to the *transplantation damage* of organs, which are stored and transported under conditions of lowered oxygen supply.

Oxidative stress plays a role in *inflammation* and *chronic inflammatory diseases* including *rheumatoid arthritis, Crohn's disease, ulcerative colitis, asthma*, and *periodontal disease*. It activates macrophages and granulocytes, which participate in the inflammatory response release of ROS (superoxide, hydrogen peroxide, peroxynitrite, hypochlorite) to kill invading pathogens. Excess of so-produced ROS damages own tissues of the host ("innocent bystander effect"). Oxidative stress occurring in *pancreatitis* and *Helicobacter pylori infection*, although not causing these diseases, contributes to their pathological sequelae.

Recruitment and activation of phagocytes in the lung, leading to massive generation of ROS, is the cause of *adult (acute) respiratory distress syndrome* (ARDS).

ROS may be causative factors of carcinogenesis and development of *cancer*. They may participate in all stages of carcinogenesis (initiation, promotion, and progression). Damage of DNA by ROS may initiate cancerogenesis; many carcinogens induce oxidative stress. Chronic inflammation is a carcinogenic factor, apparently by ROS-induced DNA damage. Metabolism of many carcinogens including carcinogenic metals induces oxidative stress.

Oxidative stress induced by light and insufficient antioxidant defense contributes to the development of *cataract* and *age-related macular degeneration* (AMD).

Oxidative stress is blamed for the induction and development of neurodegenerative diseases such as Parkinson's disease, Alzheimer's disease, amyotrophic lateral sclerosis (ALS), and Down's syndrome. The brain is susceptible to oxidative stress due to high rate of oxygen consumption, high abundance of PUFA residues of PLs, easy autoxidation of many neurotransmitters (dopamine, noradrenaline) producing ROS, generation of H_2O_2 by monoamine oxidases that metabolize amine neurotransmitters, and relatively low antioxidant defense. *Parkinson's disease* involves loss of neurons of *substantia nigra*, containing redox-active neuromelanin. Symptoms of oxidative stress are typical for *Alzheimer's disease* and seem to contribute to its development. About 20% of cases of familial ALS (Lou Gehrig's disease) are due to mutations in the gene coding for CuZn-superoxide dismutase. *Down's syndrome* is due to trisomy 21, the chromosome on which the gene coding for CuZnSOD is located. Symptoms of oxidative stress in this disease have been ascribed to increased production of hydrogen peroxide due to the enhanced activity of CuZnSOD.

Oxidative stress has been documented to occur in viral diseases including *AIDS* and *influenza*; antioxidants ameliorated the cause of these diseases in experimental animals.

It has been suggested that oxidative stress may contribute to the origin of type I *diabetes*, the more that β-cells have weak antioxidant defense. Undoubtedly, oxidative stress accompanies diabetes. Increased glucose level leads to enhanced nonenzymatic glycosylation (glycation) of proteins and other amine-containing molecules leading to the inactivation of enzymes. Further transformations of primary glycation products form advanced glycation end produces (AGEs). Glycated proteins and AGEs release ROS. Macrophages and some other cells have RAGE. Activation of RAGE induces ROS production in responding cells. Aldose reductase which reduces glucose to sorbitol consumes NADPH:

$$Glucose + NADPH + H^+ \rightarrow Sorbitol + NADP^+$$

Enhanced consumption of NADPH by aldose reductase due to increased glucose level in diabetes contributes to cataractogenesis and other complications in diabetes.

Nitric oxide is currently thought to be the main factor responsible for the maintenance of low blood pressure. Decrease in the level of NO due to overproduction of superoxide and/or decrease activity of EC-SOD leads to *hypertension.*

Ethanol induces oxidative stress since its metabolism by cytochrome P450 produces free radicals. Moreover, Kupffer cells in the liver are activated after ethanol consumption. *Smoking* involves inhalation of free radicals and other oxidants; oxidative stress can be detected even after consumption of a single cigarette. Consumption of Vitamins C and E is increased in smokers. Oxidants inhaled in the cigarette smoke oxidize methionine in α1-antiproteinase, inactivating this inhibitor of elastase. As a result, elastase released from activated neutrophils is not fully inhibited and may degrade the connective tissue of the lung leading to *emphysema* in chronic smokers.

Interestingly, *emotional stress* was found to be accompanied also by oxidative stress.

According to the free radical theory of *aging* proposed by Harman over 50 years ago, free radicals formed *in vivo* are the cause of aging. The theory has been the subject of many discussions and modifications but is the most popular theory of aging, amenable to experimental testing. Today, not only free radicals but ROS generally are blamed for causing, or contributing to age-related changes in organisms. The process of ageing is often associated with the accumulation of oxidized forms of proteins leading to the formation of deposits of age pigments (lipofuscins). The accumulation of oxidized proteins in living systems may be (1) due to an increase in the steady-state level of ROS and reactive nitrogen species (RNS) and/or to a decrease in the antioxidant capacity of an organism; (2) a decrease in the ability to degrade oxidized proteins due to either a decrease in the protease concentrations and/or to an increase in the levels of protease inhibitors.

Energy restriction, the procedure known to increase the lifespan of almost all organisms studied, including mammals, leads to decreased ROS generation and oxidative modifications of body constituents.

Generation of ROS contributes to the *side effects of* some *drugs* such as the kidney and ear damage by gentamycin, and heart damage by adriamycin and its derivatives used in cancer chemotherapy (Bartosz, 2003; Halliwell and Guteridge, 2007).

The involvement of ROS in the pathogenesis of many diseases and their involvement in aging was a scientific rationale for the use of antioxidant supplements for the prevention of diseases and prolongation of life. The "French paradox," an intriguing phenomenon of low incidence of coronary heart disease in the French population, in spite of diet relatively rich in saturated FA, has been ascribed to Mediterranean diet rich in antioxidants and consumption of high amounts of phenolic antioxidants in the red wine. However, experimental and epidemiological studies concerning antioxidant supplementation generally have not brought clear-cut results. Antioxidant supplementation reduced the risk of various types of cancer, cardiovascular diseases, and cataracts in many but not all studies. In rare cases, antioxidant supplementation increased the mortality (β-carotene given to smokers and workers exposed to asbestos). The effects of antioxidant supplementation on the lifespan of experimental animals are not dramatic (Kirsch and De Groot, 2001; Pihlanto, 2006).

There may be several reasons for this phenomenon. The main reason may concern the dual role of ROS which perform also a signaling role, that is, mediating or augmenting the action of many growth factors. Another aspect may be a downregulation of production of own antioxidants in an attempt to maintain "redox homeostasis." Moreover, compounds known as antioxidants have also other biological effects, which may become disadvantageous when single compounds are given in excess. Vitamins C and E are not toxic even in high doses but some antioxidant supplements may be. Deficiency of selenium, needed for the biosynthesis of selenocysteine present in about 30 selenoproteins including glutathione peroxidases, induces oxidative stress but excess selenium is toxic. In any case, however, deficiency of dietary antioxidants brings oxidative stress and may initiate or accelerate pathologies.

9.5   EFFECTS OF OXIDIZED LIPIDS IN NUTRITION

Western societies age and, at the same time, the Western diet contains large quantities of oxidized lipids because a high proportion of the diet is consumed in a fried, heated, processed, or stored form. It is important to what extent lipid oxidation products contained in a diet may contribute to the *in vivo* destructive activity of ROS. The gastrointestinal tract is constantly exposed to dietary oxidized food compounds; after digestion, a part of them are absorbed into the lymph or directly into the bloodstream. Hydroperoxides are generally thought to be decomposed in the stomach from where they are not transported any farther. On the other hand, the human gastric fluid may be an excellent medium for enhancing the oxidation of lipids and other dietary constituents. The stomach contents of rats fed red meat homogenate showed more than twofold increase in hydroperoxides and MA accumulation (Gorelik et al., 2008). It is possible that, at low doses, FA hydroperoxides are converted to the corresponding hydroxy FA in the mucosal membrane before they are transported to the blood. They can influence endothelial dysfunction, promote thrombosis, and induce atherosclerosis (Riemersma, 2001). The secondary products of lipid autoxidation contain cytotoxic and genotoxic compounds; after digestion, a part of them is absorbed into the lymph or directly into the bloodstream and may cause an increase in oxidative stress and deleterious changes in lipoprotein and platelet metabolism (Kubow, 1992). The aldehydes occur in free form or conjugate with amino acids to be absorbed from the gastrointestinal tract to plasma, muscles, and liver. The absorbable aldehyde adducts with protein from the diet are less toxic than free aldehydes (Kołakowska, 2003). After ingestion of oxidized fats, animals and human have been shown to excrete in urine increased amounts of MA and also lipophilic carbonyl compounds. Oxidized cholesterol in the diet was found to be a source of oxidized lipoproteins in human serum. Some of the dietary advanced lipid oxidation endproducts, which are absorbed from the gut to the circulatory system, seem to act as injurious chemicals that activate an inflammatory response which affects not only the circulatory system but also liver, kidney, lung, and the gut itself. Repeated consumption of oxidized fat in the diet poses a chronic threat to human health (Kanner, 2007). Oxidized oils are absorbed in the intestine, transported as chylomicrones to the liver, and may affect unaltered hepatic cells as well as cause hepatocarcinogenesis. Lipid hydroperoxides of dietary origin may be an important driving force for carcinogenesis in the liver (Rohr-Udilova et al., 2008).

9.6   METHODS FOR DETERMINATION OF LIPID OXIDATION

A review of methods and their critical evaluation can be found in numerous books and papers, the most recent of them including Kuksis et al. (2003), Shahidi and Wanasundara (2002), Kamal-Eldin and Pokorny (2005), Laguerre et al. (2007), as well as the first edition of this book (Kołakowska, 2003). Research of lipid oxidation in foods still relies on conventional chemical methods, used for almost 100 years: the peroxide value (PV) determination was developed by Lee to assay the primary oxidation products, whereas secondary products are most frequently assayed by the 2-thiobarbituric acid test (TBA). The test is used despite the fact that, in addition to MDA, 2-thiobarbituric acid gives a positive reaction also with amino acids, sugars, and other nonlipid food components. Therefore, it is more appropriate to use the term 2-thiobarbituric reactive substances (TBARS). Besides TBA, the concentrations of carbonyl compounds are most often determined, with the anisidine value (AV). The total oxidation is determined by calculating the TOTOX Value $= 2PV + AV$. In simple systems, a convenient and fast method is spectroscopic (UV) measurement of conjugated dienes (at 234 nm) and trienes (at 268 nm), although other compounds are known to exhibit maximum absorption at those wavelengths as well (Kołakowska, 2003).

Attempts are made to replace the tradition methods of lipid oxidation analysis with instrumental ones: GC and HPLC. Various detectors and tandems of detectors and chromatographic techniques are applied. Spectroscopy is more and more often applied to analyze lipid oxidation in food directly, not only as, for example, detectors at the final stage of the analysis. Direct determination of lipid peroxides by LC-MS is one of the most common measurements (Watson et al., 2003). LC-MS or combined techniques such as LC-MS-MS and LC-ESI-MS tandems allow detection of FA hydroperoxides with regioisomers (Watson et al., 2003). Application of GC-MS and GC-ESI-MS renders the analysis to be much more sensitive in determining carbonyl compounds than TBA (Kuksis et al., 2003). However, it requires more complex sample preparation, including derivatization. Mendes et al. (2009) compared, using three fish species, the applicability of the traditional TBA and HPLC separation after TBA or DNPH derivatization to determine lipid oxidation on storage. The performance (accuracy and specificity) of the methods used followed the order of HPLC > MDA-DNPH > HPLC > MDA-TBA > traditional spectrophotometric TBA test. HPLC with simple UV detection was also used by Marcinčák et al. (2004) for the determination of MDA levels in broiler meat. The final assay, however, required preliminary separation involving a solid-phase extraction (SPE) of C18 on the SPE column.

The electron spin resonance (ESR) allows selective detection of free radicals. Due to the very short lifetime (<1 ms) of free radicals, spin trapping or spin scavenging techniques are used. In food lipid oxidation studies, however, ESR is rarely used; it is more frequently applied to determine antioxidant activity against an added source of stable radicals.

High-resolution nuclear magnetic resonance (NMR) spectroscopy allows to follow oil oxidation; correlations were found between the ratio calculated from the ^{1}H NMR spectrum and primary oxidation products (PV and conjugated dienes) (Falch et al., 2004). Medina et al. (2000) used ^{13}C NMR to study changes caused by lipid oxidation in canned fatty fish.

Hydroperoxides are assayed by analysis of x-ray diffraction spectra. Although lipid hydroperoxides are not determined directly, the method provides characteristics of changes in membrane structure caused by hydroperoxide effects on PL chains (Armstrong, 2002). Luminescence-based methods (chemiluminescence (CL) and spectrofluorometry) are enjoying a renewed interest. While earlier studies measured the ultra-weak CL accompanying autoxidation, automated flow injection CL (FICL) has become more popular later on (Bunting and Gray, 2003). A CL detector is used to assay hydroperoxides (HPLC-CL). Fluorescence as an indicator of lipid oxidation extent has been in use for a long time; maximum emission (Em) at a certain excitation wavelength (Ex) is measured, as are the entire emission spectra (Em); in addition, ratios between fluorescence measured at 393/463 nm and that at 327/415 nm are calculated to follow progress in fish rancidity (Aubourg et al., 1998). However, there are too many factors and compounds that interfere with and disturb the direct measurement of lipid peroxide fluorescence; therefore, comparisons with tradition methods (PV, AV) have hardly been published, except for a comparison for thermooxidized lipids (Kołakowska, 2003). In recent years, different fluorescent probes have been used. A probe is usually a nonfluorescent compound that, when oxidized, exhibits increased fluorescence intensities (Chotimarkorn et al., 2006).

Infrared spectroscopy and Fourier transform infrared (FTIR) spectroscopy seem promising. A spectrum collected in the mid-region (4000–800 cm^{-1}) directly from a food product sample and/or from extracted lipids supplies information on the level of oxidation and composition of oxidation products; it allows to carry out a qualitative and quantitative analysis of those products. Most information on oxidation products—carbonyl compounds and *trans* isomers—is obtained from the spectrum wavelengths ranging within about 900–1745 cm^{-1}. The intensity ratio of two bands allows to determine lipid oxidative stability toward hydroperoxides and carbonyls (Guillen and Cabo, 2002; Kołakowska et al., 2006). Spectra collected directly from fish and pork tissues supplied important information on lipid oxidation (mainly on hydroperoxides) in the tissue. On the other hand, the spectrum collected from lipids extracted from a product makes it possible to determine concentration of carbonyl compounds and, to some extent, their composition and *trans* isomers; it is also possible to calculate the susceptibility of lipids in that food to the formation of those oxidation products in pork and fish tissue. Analysis of gradual lipid extraction from tissue subjected to various thermal treatments allowed to demonstrate that, during thermal treatment of rainbow trout, *trans* isomers are formed in the bound lipid fraction, carbonyl products emerging in free lipids. *Trans* isomers in total lipids were below the detection level (Kołakowska et al., 2006). Adhikari et al. (2003) used FTIR to determine hexanal in the headspace gas from the ready-to-eat meal. A more complete picture of lipid oxidation can be obtained by parallel application of FTIR and FT-Raman spectroscopy (Muik et al., 2007).

In recent years, food lipid oxidation analysis was focused on the following topics:

- Analysis of volatile compounds
- Analysis of protein oxidation
- Nondestructive instrumental methods

For some time, much attention has been paid to developing methods of analyzing volatile compounds as indicators of food lipid oxidation. This is particularly pertinent to the analysis of volatile aldehydes that, with their low detection level (10^{-12}), are mostly responsible for the rancid odor profile. GC-olfactometry (GC-O) analysis made it possible to identify some most important short-chain carbonyl compounds, mainly hexanal, pentanal, 2-heptanal, octanal, and nonanal, dominating rancidity profiles of various food types (Angelo, 1996; Venkateshwarlu et al., 2004; Jelen et al., 2007; Varlet et al., 2007). Another example is the application of the SPME fiber (polydimethylosiloxane/divinylbenzene [PDMS/DVB] fiber) in assays—with the SPME-GC-FID technique—of hexanal, 2-nonenal, and 2,4-decadienal as indicators of oxidative microcapsule changes in maltodextrin covers of fish oils (Jonsdottir et al., 2005). Jelen et al. (2007) compared the headspace SPME-MS with the SPME-GC/MS for monitoring rapeseed oil autoxidation. The SPME-GC/MS method allowed detection of 37 volatile compounds, of which 28 were identified. On the other hand, nonvolatile compounds could be identified with SPME-MS, but the method was able to differentiate between low concentrations of hexanal, and correlated with PV and TOTOX. Two-dimensional gas chromatography (2D-GC or GC×GC) is becoming an increasingly popular technique in the analysis of odor compounds. It allows to preserve separation of peaks obtained in the first column, while additionally separating the analytes in the second column.

Volatile compounds are extracted mostly with headspace techniques: static head space (SHS), dynamic headspace (DHS), and DHS/SHS. Analytes are concentrated (and also subjected to derivatization) with solid-phase microextraction (SPME); various SPME fibers have recently been used. Analysis of carbonyl compounds is carried out using GC and HPLC with various detectors. The short-chain aldehydes and hydroxyaldehydes are usually determined by using GC coupled with mass spectrometry (GC-MS) (Watson et al., 2003). In addition to GC-MS, carbonyls are analyzed also with GC-FID, pulsed helium ionization detector (GC-PHID) HPLC-UV, HPLC-MS, and tandems of those techniques. Iglesias and Medina (2008) reported excellent correlations between results of HS-SPME (fiber) coupled with GC-MS and chemical indices (PV, TBARS) in monitoring fish rancidity.

Attempts are made to replace chromatographic methods and the subjective sensory assessment of off-odor rancidity with an electronic nose equipped with a number of sensors to analyze volatile compounds (Shen et al., 2001; Mildner-Szkutlarz et al., 2007, 2008). Pastorelli et al. (2007) reported of a high correlation between electronic nose assays of volatile compounds (hexanal) formed in hazelnuts during storage with results provided by SPME coupled with GC-FID. According to Mildner-Szkutlarz et al. (2008), the applicability of the electronic nose technology to verify sensory and rancidity changes during storage showed promise in quality control of oils.

However, attention has in recent years switched from lipid oxidation products to assays of identical compounds (e.g., carbonyl compounds) in proteins. This has been spurred by destructive effects of ROS on proteins and the resultant *in vivo* effects on the nutritive quality of proteins and other food quality characteristics (Figures 9.1 and 9.2). Lipid and protein oxidation seem to develop simultaneously during food processing and storage, and interact. Protein oxidation is analyzed using methods identical to those applied to lipid oxidation; methods specific to protein analysis are used as well (Kołakowski, 2005). Determination of SPR radicals in proteins, originating

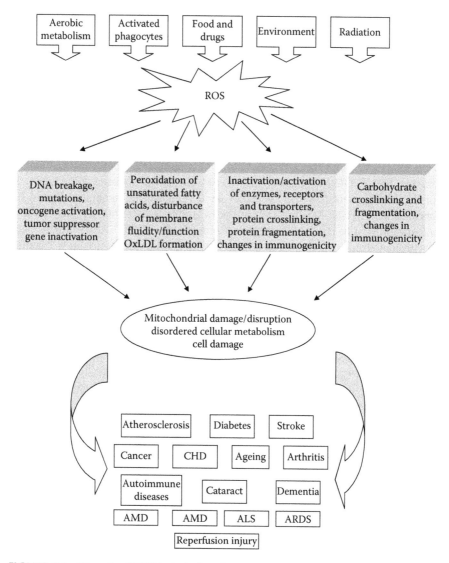

FIGURE 9.2 The role of ROS in inducing diseases. (Adapted from Kołakowska, A., Lipid oxidation in food systems, in *Chemical and Functional Properties of Food Lipids*, Sikorski, Z.E. and Kołakowska, A. (eds.), CRC Press, Boca Raton, FL, 133–166.)

also from lipids, is carried out. Protein carbonyls are most frequently assayed with DNPH, fluorescence spectroscopy, LC-MS, and LC-ESI-MS (Armenteros et al., 2009). In addition, assays used in protein analysis, such as determination of extractability, electrophoresis fractions, including capillary electrophoresis (CE), available lysine, oxidized amino acids are applied as well, as are spectroscopic techniques (FTIR, FT-Raman spectroscopy) which make it possible to simultaneously follow evolution of oxidation products in lipids and proteins (Rodriquez-Casado et al., 2007; Sarkardei and Howell, 2007; Herrero, 2008).

Lipid oxidation in complex foods is affected, in addition to external factors, by the remaining food components and their interactions. It seems that the analysis of diverse systems such as foods will proceed employing, in addition to chromatographic tandems coupled with other, mainly optical techniques, methods that will allow to holistically describe the status of lipid oxidation by direct and noninvasive techniques that require small portions of a sample. Such potential is offered by FTIR, FT-Raman spectroscopy, CL, and front face fluorescence.

9.7 CONCLUDING REMARKS

ROS-induced oxidation is a process important both *in vivo* and in food, contributing to pathologies and aging of the body, and to deterioration of food products. All types of biological molecules are damaged by ROS, but lipid oxidation plays a special role in these processes since the energy barrier to initiate lipid peroxidation is lower than those to initiate oxidation of proteins, carbohydrates, or nucleic acids. Nevertheless, there is an interplay between oxidative reactions of various biomolecules. Free radicals of other molecules may initiate lipid oxidation, while products of lipid oxidation modify proteins and nucleic acids and deplete antioxidants.

ROS and products of lipid oxidation are not only deleterious but may also be elements of cellular signaling pathways, stimulating cell proliferation and other reactions (Kris-Etherton et al., 2002; Bartosz, 2009). Low amounts of lipid peroxidation products in food may be, therefore, not necessarily noxious while, undoubtedly, high levels must be prevented and avoided.

REFERENCES

Adhikari, C., Balasubramaniam, V.M., and Abbot, U.R. 2003. A rapid FTIR method for screening methyl sulfide and hexanal in modified atmosphere meal, ready-to-eat entrees. *Lebensm.-Wiss. U.-Technol.* 36:21–27.

Angelo, A.J.S. 1996. Lipid oxidation in foods. *Crit. Rev. Food Sci. Nutr.* 36:175–224.

Armenteros, M., Heinonen, M., Ollilainen, V., Toldrá, F., and Estévez, M. 2009. Analysis of protein carbonyls in meat products by using the DNPH-method, fluorescence spectroscopy and liquid chromatography–electrospray ionisation–mass spectrometry (LC–ESI–MS). *Meat Sci.* 83, 1:104–112.

Armstrong, D. 2002. *Oxidants and Antioxidants: Ultrastructure and Molecular Biology Protocols.* Humana Press, Totowa, NJ.

Aubourg, S.P., Sotelo, C., and Perez-Martin, R. 1998. Assessment of quality changes in frozen sardine (*Sardina pilchardus*) by fluorescence detection. *J. Am. Oil Chem. Soc.* 75:375–580.

Bartosz, G. 2003. *The Other Face of Oxygen*. Polish Scientific Publishers, Warsaw, Poland.

Bartosz, G. 2009. Reactive oxygen species: Destroyers or messengers? *Biochem. Pharmacol.* 77:1303–1315.

Brannan, R.G., Connolly, B.J., and Decker, E.A. 2001. Peroxynitrite: A potential initiator of lipid oxidation in food. *Trends Food Sci. Technol.* 12:64–173.

Bunting, J.P. and Gray, D.A. 2003. Development of flow injection chemiluminescent assay for the quantification of lipid hydroperoxides. *JAOCS* 80:951–955.

Choe, E. and Min, D.B. 2006a. Chemistry and reactions of reactive oxygen species in foods. *Crit. Rev. Food Sci. Nutr.* 46:1–22.

Choe, E. and Min, D.B. 2006b. Mechanisms and factors for edible oil oxidation. *Compr. Rev. Food Sci. Food Safety* 5:169–186.

Chotimarkorn, C., Ohshima, T., and Ushio, H. 2006. Fluorescent image analysis of lipid hydroperoxides in fish muscle with 3-perylene diphenylphosphine. *Lipids* 41, 3:295–300.

Cullen, P.J., Tiwari, B.K., O'Donnell, C.P., and Muthukumarappan, K. 2009. Modelling approaches to ozone processing of liquid foods. *Trends Food Sci. Technol.* 20:125–136.

Falch, E., Anthonsen, H.W., Axelson, D.E., and Aursand, M. 2004. Correlation between ^{1}H NMR and traditional methods for determining lipid oxidation of ethyl docosahexaenoate. *JAOCS* 81, 12:1105–1110.

Frankel, E.N. 1998. *Lipid Oxidation*. The Oily Press, Dundee, Scotland, Chapter 1.

Ganthavorn, C., Nagel, C.W., and Powers, J.R. 1991. Thermal inactivation of asparagus lipoxidase and peroxidase. *J. Food Sci.* 56, 1:47–49.

German, J.B., Zhang, H., and Berger, R. 1992. Role of lipoxygenases in lipid oxidation in foods. In *Lipid Oxidation in Food*, Angelo, A.J. St. (Ed.), ACS Symposium Series 500. ACS, Washington, DC, pp. 74–92, Chapter 5.

Gómez-López, V.M., Rajkovic, A., Ragaert, A., Smigic, N., and Devlieghere, F. 2009. Chlorine dioxide for minimally processed produce preservation: A review. *Trends Food Sci. Technol.* 20:17–26.

Gorelik, S., Ligumsky, M., Kohen, R., and Kanner, J. 2008. The stomach as a "bioreactor": When red meat meets red wine. *J. Agric. Food Chem.* 56, 13:5002–5007.

Grossman, S., Bergman, M., and Sklan, D. 1988. Lipoxygenase in chicken muscle. *J. Agric. Food Chem.* 36, 1268–1270.

Guillen, M.D. and Cabo, N. 2004. Fourier transform infrared spectra data versus peroxide and anisidine values to determine oxidative stability of edible oils. *Food Chem.* 77:503–510.

Halliwell, B. and Guteridge, J.M.C. 2007. *Free Radicals in Biology and Medicine*. Oxford University Press, Oxford, NY.

Halliwell, B., Murcia, M.A., Chirico, S., and Aruoma, O.I. 1995. Free radicals and antioxidants in food and in vivo: What they do and how they work. *Crit. Rev. Food Sci. Nutr.* 35:7–20.

Hammer, F.E. 1993. Oxidoreductases. In *Enzymes in Food Processing*, Nagodawithana, T. and Reed, G. (Eds.). Academic Press, Inc., San Diego, CA, pp. 221–271.

Herrero, A.M. 2008. Raman spectroscopy a promising technique for quality assessment of meat and fish: A review. *Food Chem.* 107, 4:1642–1651.

Hultin, H.O. 1994. Oxidation of lipids in seafoods. In *Seafoods: Chemistry, Processing Technology and Quality*, Shahidi, F. and Botta, J.R. (Eds.). Blackie Academic & Professional, London, U.K., Chapter 5, pp. 49–74.

Hur, S.J., Park, G.B., and Joo, S.T. 2007. Formation of cholesterol oxidation products (COPs) in animal products. *Food Control* 18:939–947.

Iglesias, J. and Medina, I. 2008. Solid-phase microextraction method for the determination of volatile compounds associated to oxidation of fish muscle. *J. Chromatogr. A* 1192, 1:9–16.

Jeleń, H.H., Mildner-Szkutlarz, S., Jasińska, I., and Wąsowicz, E. 2007. A headspace-SPME-MS method for monitoring rapeseed oil autoxidation. *J. Am. Oil Chem. Soc.* 84:509–517.

Jonsdottir, R., Bragadottir, M., and Arnarson, G.O. 2005. Oxidatively derived volatile compounds in microencapsulated fish oil monitored by solid-phase microextraction (SPME). *J. Food Sci.* 70, 7:C433–C440.

Jung, M.Y., Yoon, S.H., Lee, H.O., and Min, D.B. 1998. Singlet oxygen and ascorbic acid effects on dimethyl disulfide and off-flavor in skim milk exposed to light. *J. Food Sci.* 63:408–412.

Kamal-Eldin, A. and Pokorny, J. (Eds.) 2005. *Analysis of Lipid Oxidation.* AOAC Press, Champaign, IL, pp. 263–280.

Kanner, J. 2007. Dietary advanced lipid oxidation endproducts are risk factors to human health. *Mol. Nutr. Food Res.* 51, 9:1094–1101. Special Issue: Are dietary AGEs/ALEs a health risk.

Kirsch, M. and De Groot, H. 2001. NAD(P)H, a directly operating antioxidant? *FASEB J.* 15: 1569–1574.

Kołakowska, A. 2003. Lipid oxidation in food systems. In *Chemical and Functional Properties of Food Lipids*, Sikorski, Z.E. and Kołakowska, A. (Eds.). CRC, Boca Raton, FL, Chapter 8, pp. 133–166.

Kołakowska, A., Ziobrowski, P., and Wróbel, P. 2006. Determination of lipid changes in chilled and cooked rainbow trout by FT-IR spectroscopy. In *Proceedings of the 37th Science Conference Committee for Food Technology and Chemistry of Polish Academy of Sciences*, Gdynia, Poland, Abstracts, p. 288 (In Polish).

Kołakowski, E. 2005. Analysis of proteins, peptides, and amino acids in foods. In *Methods of Analysis of Food Components and Additives*, Semith, O. (Ed.). CRC Press/Taylor & Francis, Boca Raton, FL, pp. 59–96.

Kris-Etherton, P.M., Hecker, K.D., Bonanome, A., Coval, S.M., Binkoski, A.E., Hilpert, K.F., Griel, A.E., and Etherton, T. 2002. Bioactive compounds in foods: Their role in the prevention of cardiovascular disease and cancer. *Am. J. Med.* 113:71S–88S.

Kubow, S. 1992. Routes of formation and toxic consequences of lipid oxidation products in foods. *Radic. Biol. Med.* 12, 1:63–81.

Kuksis, A., Kamido, H., and Ravandi, A. 2003. Glycerophospholipid core aldehydes: Mechanism of formation, methods of detection, natural occurrence, and biological significance. In *Lipid Oxidation Pathways*, Kamal-Eldin, A. (Ed.). AOCS Press, Champaign, IL, pp. 138–189.

Laguerre, M., Lecomte, J., and Villeneuve, P. 2007. Evaluation of the ability of antioxidants to counteract lipid oxidation: Existing methods, new trends and challenges. *Prog. Lipid Res.* 46, 5:244–282.

Lindsay, R.C. 1994. Flavour of fish. In *Seafoods: Chemistry, Processing Technology and Quality*, Shahidi, F. and Botta, J.R. (Eds.). Blackie Academic & Professional, London, U.K., Chapter 6, pp. 75–84.

Marcinčák, S., Sokol, J., Bystrický, P., Popelka, P., Turek, P., Bhide, M., and Máté, D. 2004. Determination of lipid oxidation level in broiler meat by liquid chromatography. *J. AOAC Int.* 8, 5:1148–1152.

Medina, L.R., Sacchi, R., and Aubourg, S. 2000. Oxidation of fish lipids during thermal stress as studied by ^{13}C nuclear magnetic resonance spectroscopy. *Eur. Food Res. Technol.* 210:176–178.

Medina, I., Saeed, S., and Howell, N., 1999. Enzymatic oxidative activity in sardine (*Sardina pilchardus*) and hessing (*Clupea harengus*) during shilling and correlation with quality. *Eur. Food Res. Technol.* 210:34–38.

Mendes, R., Cardoso, C., and Pestana, C. 2009. Measurement of malondialdehyde in fish: A comparison study between HPLC methods and the traditional spectrophotometric test. *Food Chem.* 112:1038–1045.

Mildner-Szkutlarz, S., Zawirska-Wojtasiak, R., Korczak, J., and Jeleń, H.H. 2007. A comparison of human and electronic nose responses to flavour of various food products of different degree of lipid oxidation. *Pol. J. Food Nutr. Sci.* 57, 2:195–202.

Mildner-Szkutlarz, S., Jeleń, H.H., and Zawirska-Wojtasiak, R. 2008. The use of electronic and human nose for monitoring rapeseed oil autoxidation. *Eur. J. Lipid Sci. Technol.* 110:61–72.

Min, D.B. and Boff, J.M. 2002. Chemistry and reaction of singlet oxygen in foods. *Compr. Rev. Food Sci. Food Safety* 1:58–72.

Morrissey, P.A, Sheehy, P.J.A., Galvin, K., and Kerry, J.P. 1998. Lipid stability in meat and meat products. *Meat Sci.* 49:S73–S86.

Muik, B., Lendl, B., Molina-Diaz, A., Valcarcel, M., and Ayora-Canada, M.J. 2007. Two-dimensional correlation spectroscopy and multivariate curve resolution for the study of lipid oxidation in edible oils monitored by FTIR and FT-Raman spectroscopy. *Anal. Chim. Acta* 593,1, 12:54–47.

Pastorelli, S., Torri, L., Rodriquea, A., Valzacchi, S., Limbo, S., and Simoneau, C. 2007. Solid-phase microextraction (SPME-GC) and sensors as rapid methods for monitoring lipid extraction in nuts. *Food Addit. Contam.* 24, 11:1219–1225.

Pihlanto, A. 2006. Antioxidative peptides derived from milk proteins. *Int. Dairy J.* 16:1306–1314.

Porter, N.A., Caldwell, S.E., and Mills, K.A. 1995. Mechanisms of free radical oxidation of unsaturated lipids. *Lipids* 30:277–290.

Riemersma, R.A. 2001. Oxidized fats in the diet and their putative role in coronary heart disease, Paper HNH-4. In *24th World Congress ISF. Lipids, Fats, and Oils: Reality and Public Perception*, September 16–20, Berlin, Germany.

Rodriquez-Casado, A., Alvarez, I., Toledano A., De Miguel, E., and Carmona, P. 2007. Amphetamine effects on brain protein structure and oxidative stress as revealed by FTIR microspectroscopy. *Biopolymers*, 86, 5–6:437–446.

Rohr-Udilova, N.V., Stolze, K., Sagmeister, S., Nohl, H., Schulte-Hermann, R., and Grasl-Kraupp, B. 2008. Lipid hydroperoxides from processed dietary oils enhance growth of hepatocarcinoma cells. *Mol. Nutr. Food Res.* 52, 3:352–359.

Ryter, S.W., Kim, H.P., Hoetzel, A., Park, J.W., Nakahira, K., Wang, X., and Choi, A.M. 2007. Mechanisms of cell death in oxidative stress. *Antioxid. Redox Signal.* 9:49–89.

Samson, E. and Stodolnik, L. 2001. Effect of freezing and salting on the activity of lipoxygenase of the muscle tissue and roe of Baltic herring. *Acta Ichthyol. Piscat.* 31, 1:97–112.

Sarkardei, S. and Howell, N.K. 2007. The effects of freeze-drying and storage on the FT-Raman spectra of Atlantic mackerel (*Scomber scombrus*) and horse mackerel (*Trachurus trachurus*). *Food Chem.* 103:62–70.

Shahidi, F. and Wanasundara, U.N. 2002. Methods for measuring oxidative rancidity in fats and oils. In *Food Lipids, Chemistry, Nutrition and Biotechnology*, 2nd edn., Akoh, C.C. and Min, D.B. (Eds.). CRC Taylor & Francis, Boca Raton, FL, pp. 465–487.

Shen, N., Moizuddin, S., Wilson, L., Duvick, S., White, P., and Pollak, L. 2001. Relationship of electronic nose analyses and sensory evaluation of vegetable oils during storage. *J. Am. Oil Chem. Soc.* 78:937–940.

Tedjo, W., Eshtiaghi, M.N., and Knorr, D. 2000. Impact of supercritical carbon dioxide and high pressure on lipoxygenase and peroxidase activity. *J. Food Sci.*, 65, 8:1284–1287.

Varlet, V., Prost, C., and Serot, T. 2007. Volatile aldehydes in smoked fish: Analysis methods, occurrence and mechanisms of formation. *Food Chem.* 105:1536–1556.

Venkateshwarlu, G., Let, M.B., Meyer, A.S., and Jacobsen, C. 2004. Modeling the sensory impact of defined combinations of volatile oxidation products on fishy and metallic off-flavors. *J. Agric. Food Chem.* 52, 6:1635–1641.

Watson, D.G., Atsriku, C., and Oliveira, E.J. 2003. Review role of liquid chromatography-mass spectrometry in the analysis of oxidation products and antioxidants in biological systems. *Anal. Chim. Acta* 492:17–47.

10 Antioxidants

Anna Kołakowska and Grzegorz Bartosz

CONTENTS

10.1 INTRODUCTION

In foods, antioxidants have been defined as "substances that in small quantities are able to prevent or greatly retard the oxidation of easily oxidizable materials such as fats" (Chipault, 1962). In biological systems, the definition has been extended to "any substance that, when present at low concentrations compared to those of an oxidizable substrate, significantly delays or prevents oxidation of that substrate" (Halliwell and Gutteridge, 1995). This latter definition covers all oxidizable substrates, i.e., lipids, proteins, DNA, and carbohydrates (Frankel and Meyer, 2000).

Different classifications of antioxidants coexist. Antioxidants can be divided into two major groups: primary, or chain-breaking, antioxidants that react with lipid radicals to produce stable products; and secondary or preventive, antioxidants that retard the process of oxidation by various mechanisms. Antioxidants have also been classified into five types: primary antioxidants, oxygen scavengers, secondary antioxidants, enzymic antioxidants, and chelating agents (Hands, 1996). Pokorny et al. (2001) grouped antioxidants into five classes, depending on the mechanism of their activity: proper antioxidants, hydroperoxide stabilizers, synergists, metalochelators, singlet oxygen quenchers, and hydroperoxide-reducing substances. The occurrence of antioxidants, their characteristics, mechanisms of activity, and applications were discussed elsewhere (Hudson, 1990, Halliwell et al., 1995, Sies, 1997, Decker et al., 2000, Pokorny et al., 2001, Grajek, 2007, Laguerre et al., 2007).

10.2 IN VIVO

10.2.1 Antioxidant Defense System

In living organisms, reactive oxygen species (ROS) are under control of antioxidant proteins and low-molecular-weight antioxidants, keeping their local concentrations at levels optimal for actual functions and signaling purposes. The chemical definitions of antioxidants given above do not seem sufficient in the biological context where antioxidant is perceived as a substance preventing or counteracting effects of uncontrolled oxidation at the cellular or organismal level. A broader biological definition of antioxidants includes several classes of compounds: (1) antioxidant enzymes that decompose ROS, (2) nonenzymatic antioxidant proteins and other macromolecules, (3) low-molecular-weight cofactors of antioxidant enzymes, (4) low-molecular-mass ROS scavengers, (5) tight metal chelators, (6) inhibitors of pro-oxidant enzymes, and (7) inducers of biosynthesis of antioxidant proteins.

This dynamic, spatially distributed, and overlapping network of antioxidants enables efficient control of ROS reactions allowing the performance of their signaling functions and attenuating oxidative stress.

10.2.2 Enzymatic Antioxidants

Most enzymatic antioxidant systems are ubiquitous in the whole living world, but some are confined to certain groups of organisms. The enzymatic system controlling the cellular level of ROS consists of the "enzymatic triad" of superoxide dismutases (SOD), catalases, and glutathione peroxidases, and of other enzymes.

Superoxide dismutases (E.C. 1.15.1.1) catalyze the reaction of dismutation of superoxide radical anion:

$$2O_2^{\bullet-} + 2H^+ \rightarrow H_2O_2 + O_2$$

Mammals have three types of SOD: cytoplasmic CuZnSOD, mitochondrial MnSOD, and extracellular ECSOD, also containing Cu and Zn. CuZnSOD is an unusually stable enzyme, resistant to high temperatures, denaturing agents, and proteolysis, and stable in a broad pH range exceeding 5–10. Bacteria contain FeSOD and MnSOD, some also CuZnSOD, and some NiSOD. Plant cells may have CuZnSOD, MnSOD, and FeSOD.

Catalases (E.C. 1.11.1.6) are heme proteins that catalyze the reaction of dismutation of hydrogen peroxide:

$$2H_2O_2 \rightarrow 2H_2O + O_2$$

Glutathione peroxidases (E.C. 1.11.1.9) containing selenocysteine at the active site reduce H_2O_2 at the expense of oxidation of glutathione (GSH):

$$H_2O_2 + 2GSH \rightarrow 2H_2O + GSSG$$

They may also reduce organic hydroperoxides (LOOH) to corresponding alcohols

$$LOOH + 2GSH \rightarrow LOH + GSSG + H_2O$$

thus preventing reinitiation of lipid peroxidation chain reaction. There are eight human isoenzymes of GSH peroxidases showing different tissue localization.

Glutathione disulfide (GSSG) produced by GSH peroxidase is reduced back by *GSH reductase* (E.C. 1.6.4.2), which oxidizes NADPH:

$$GSSG + NADPH + H^+ \rightarrow 2GSH + NADP^+$$

Excess GSSG may be actively transported out of cells by some of the transporters of the ABCC (MRP) family.

Plants dispose of hydrogen peroxide using the *ascorbate–glutathione system*. H_2O_2 is reduced to water by *ascorbate peroxidase* (E.C. 1.11.1.11), which oxidizes ascorbate to the monodehydroascorbate free radical:

$$Ascorbate\text{-}H + H_2O_2 \rightarrow Monodehydroascorbate^{\bullet} + H_2O$$

Monodehydroascorbate is either reduced by *monodehydroascorbate reductase* (E.C. 1.6.5.4)

$$NADH + H^+ + 2monodehydroascorbate^{\bullet} \rightarrow NAD^+ + 2ascorbate$$

or disproportionates to ascorbate and dehydroascorbate:

$$2monodehydroascorbate^{\bullet} \rightarrow ascorbate + dehydroascorbate$$

Dehydroascorbate is reduced to ascorbate by *dehydroascorbate reductase* (E.C. 1.8.5.1) at the expense of GSH, producing GSSG. Finally, GSSG is reduced by glutathione. Plants are rich in heme-containing *peroxidases* of broad specificity using various electron donors, sometimes named after artificial donors used for colorimetric assays of their activity (e.g., guaiacol peroxidase).

Yeast detoxifies hydrogen peroxide also by oxidation of cytochrome *c* catalyzed by the heme protein *cytochrome c peroxidase* (E.C. 1.11.1.5):

$$H_2O_2 + 2\text{ferrocytochrome } c + 2H^+ \rightarrow 2H_2O + 2\text{ferricytochrome } c$$

Glutathione S-transferases (E.C. 2.5.1.18) catalyze conjugation of electrophile xenobiotics and endogenous compounds, including products of lipid peroxidation, to glutathione, forming less toxic glutathione *S*-conjugates, actively exported by *ABCC transporters*.

Hydrogen peroxide is also removed by *thioredoxin peroxidases* (peroxiredoxins) that catalyze the reaction of oxidation of thiol groups of small proteins, *thioredoxins*:

$$\text{Thioredoxin} + H_2O_2 \rightarrow \text{thioredoxin} + 2H_2O$$

$$\begin{array}{ccc} | \quad | & & | \;\; | \\ HS \quad SH & & S\text{---}S \end{array}$$

Oxidized thioredoxins are reduced by *thioredoxin reductases* at the expense of NADPH:

$$\text{Thioredoxin} + \text{NADPH} + H^+ \rightarrow \text{thioredoxin} + \text{NADP}^+$$

$$\begin{array}{ccc} | \;\; | & & | \quad | \\ S\text{---}S & & HS \quad SH \end{array}$$

The thioredoxin system may also reduce oxidized protein thiol groups although the main role in this case seems to be played by the *glutaredoxin system* in which oxidized *glutaredoxins* are reduced by glutathione and oxidized glutathione by glutathione reductase:

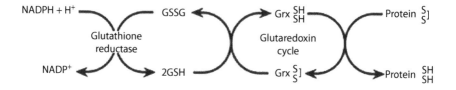

NAD(P)H: quinone oxidoreductase (*DT-diaphorase*; EC 1.6.99.2) is an enzyme catalyzing two-electron reduction of quinones to semiquinones, thus omitting the step of semiquinone-free radical intermediates, which are prone to autoxidation generating superoxide and other ROS. *Epoxide hydrolases* (E.C. 3.3.2.3) convert reactive epoxides into *trans*-dihydrodiols. Both these enzymes play therefore an antioxidant role.

Methionine sulfoxide reductases (E.C. 1.8.4.11) repair methionine residues. *Sulfiredoxins* (E.C. 1.8.98.2) reduce "overoxidized" peroxiredoxins in which thiol groups are oxidized to sulfinic derivatives (believed until recently to be an irreversible step of thiol oxidation). *DNA repair enzymes* are very important in the defense against effects of ROS-induced damage. Proteins (except for repair of oxidized thiol and methionyl groups) are generally not subject to repair, but oxidized proteins are generally better substrates for *proteolytic enzymes*, including proteasomes, which also play a role in the removal of effects of oxidative stress.

10.2.3 Nonenzymatic Antioxidant Proteins and Other Macromolecules

Proteins are the predominant components of cells and extracellular medium in terms of mass fraction and therefore are the main primary target of ROS, especially due to the reactivity of cysteine, methionine, tryptophan, and tyrosine residues. Some of them seem to play an important antioxidant function, protecting other biomolecules. Such a role has been ascribed to *albumin* in blood plasma. *Nuclear proteins*, especially histones, may act as antioxidants with respect to DNA. Antioxidant properties of milk proteins and peptides generated by digestion of these proteins have been studied and suggested to be of potential use in preservation of other food types, especially meat products (Pihlanto, 2006). *Melanins*, dark polymer pigments derived mainly from tyrosine, protect against UV and are also ROS scavengers. Polysaccharides may have also an antioxidant role (e.g., *cell walls* of bacterial cells, fungi, and bacteria, protecting from exogenous ROS).

10.2.4 Nonenzymatic ROS Scavengers

Low-molecular-weight ROS scavengers include glutathione, ascorbic acid, NAD(P)H, and uric acid (hydrophilic antioxidants); tocopherols and tocotrienols; carotenoids; bilirubin; dihydrolipoic acid; and ubiquinol (hydrophobic antioxidants). Low-molecular-weight ROS scavengers may intercept ROS before they react with critical biomolecules (preventive action), inhibit chain of free-radical reactions (interventive action), and repair a biologically important molecule transformed into a free-radical form (repair action) (Kirsch and De Groot, 2001).

Some proteins for which the antioxidant function seems to be important as well as proteins and low-molecular-weight compounds chelating transition metal ions should be also classified here.

10.2.4.1 Hydrophilic ROS Scavengers

The tripeptide *glutathione* (γ-glutamylcysteinylglycine) is the main hydrophilic intracellular antioxidant, present at millimolar concentrations, being a substrate for glutathione peroxidases and glutathione *S*-transferases and a cofactor in the glutaredoxin system, but also acting in a nonenzymatic manner.

Ascorbic acid is present in blood plasma at a concentration of about $50\,\mu M$ and in some cells at higher (even millimolar) concentrations. Upon hydrogen donation to an oxidant, it forms a relatively stable semi-dehydroascorbic radical that may be further oxidized to dehydroascorbic acid. In plant cells, ascorbate is also a substrate for ascorbate peroxidase.

Uric acid is the predominant antioxidant in the blood plasma of human and other primates who lost the ability to metabolize it further; it is present in human blood plasma at concentrations of several hundred μM.

Carnitine, acting as transporter of fatty acids from the cytosol into the mitochondria and peroxisomes for β-oxidation, often used as a nutritional supplement, is also an efficient ROS scavenger (Choe and Min, 2006).

Plants contain a variety (over 8000) of phenolic compounds of antioxidant properties that, however, have limited bioavailability. Flavonoids, the most important polyphenol groups, include flavones, flavonols, flavanones, flavanonols, and anthocyanines. The most ubiquitous flavonoid is quercetin. Soybeans contain high isoflavone levels and are the major dietary source of these compounds (e.g., genistein, daidzein, and glycitein) in humans. Flavonoids inhibit free-radical reactions and complex metal ions (Brannan et al., 2001).

Flavonoids are usually found as glycosides; these derivatives are more hydrophilic than aglycones. However, glycosides are not detected in human blood and urine. Hydrolysis of glycosides is performed by microorganisms in the colon and its efficiency may limit the availability of these compounds. In any case, their antioxidant activity (AAc) may be important for the digestion process to ameliorate the oxidative stress to stomach and intestine.

10.2.4.2 Hydrophobic ROS Scavengers

Four related α-, β-, γ-, and δ-tocopherols and the corresponding four tocotrienols are found in food. Of these, α-tocopherol has been most studied as it has the highest bioavailability and the name of *Vitamin E* has been lately restricted to this compound. It has been claimed that α-tocopherol is the most important lipid-soluble antioxidant, and that it protects cell membranes from oxidation by reacting with lipid radicals produced in the course of peroxidation, thus acting as a chain-breaking antioxidant (Cullen et al., 2009).

Ubiquinone, component of the respiratory chain, is blamed for being a major source of superoxide formation in the mitochondria, due to reaction of its ubisemiquinone-free radical form with oxygen. However, its reduced form (*ubiquinol*) is thought to act as a lipid-soluble antioxidant. A similar role is ascribed to *lipoic acid*.

Bilirubin, the end-product of heme degradation, was demonstrated to be an efficient antioxidant (although its excess is neurotoxic, especially to newborns). It has been postulated that bilirubin, oxidized by ROS to biliverdin and reduced back at the expense of NADPH by *biliverdin reductase*, may be an important antioxidant to protect cell membranes, in an analogous way as glutathione protects the aqueous compartments of the cell (Gómez-López et al., 2009).

Another class of lipid-soluble antioxidants are *carotenoids*. There are two classes of carotenoids: carotenes (tetraterpenoid hydrocarbons without oxygen atoms) and xanthophylls (which contain oxygen in the form of hydroxyl, methoxyl, carboxyl, keto, or epoxy groups). *Lycopene* and β-*carotenes* are typical carotenes, while *lutein* in green leaves and *zeaxanthin* in corn are typical xanthophylls. The double bonds in carotenoids are conjugated, and usually the *all-trans* forms are found in plant tissues. From over 600 naturally occurring carotenoids, more than 400 are regularly consumed in the diet. Carotenoids are most efficient singlet oxygen quenchers in biological systems, their efficiency increasing with the increasing number of double bonds. Carotenoids with seven or less soluble bonds are not effective quenchers. One molecule of β-carotene can quench 250–1000 molecules of singlet oxygen at a rate of 1.3×10^{10} $M^{-1}s^{-1}$. Singlet oxygen quenching by carotenoids is of physical nature, involves no chemical reaction, and generates no oxidizing product (Morrissey et al., 1998).

10.3 METAL ION CHELATORS

As metal ions are important for aggravation of oxidative damage, due to autoxidation and, first of all, participation in the Fenton reaction and decomposition of lipid hydroperoxides, metal-sequestering proteins involved in transport and storage of transition metals (e.g., iron-storage proteins *ferritin* and *lactoferrin*, *haptoglobin* responsible for hemoglobin sequestration, and *hemopexin* responsible for heme sequestration in blood plasma, copper-storage protein *ceruloplasmin*, *transferrin* responsible for iron transport, and *metallothioneins* responsible for binding exogenous metals) are important elements of antioxidant defense. Ceruloplasmin, apart form binding copper, protects also against the prooxidative action of Fe^{2+}, acting as a ferroxidase, i.e., oxidizing Fe^{2+} to Fe^{3+}.

The antioxidant action of the dipeptide *carnosine* β-alanyl-L-histidine) present at high millimolar concentrations in brain and muscle seems to be due mainly to its metal-chelating activity. *Carnitine* and many *flavonoids* are also good metal scavengers.

10.4 OTHER ACTIONS OF ANTIOXIDANTS

Categorization of antioxidants is not absolute since the same compounds may be, e.g., ROS scavengers, cofactors of antioxidant enzymes, and metal chelators. Compounds known as antioxidants may also perform various biological functions indirectly related or not related to their AAc. For example, α-tocopherol inhibits protein kinase C activity and inhibits smooth muscle cell proliferation. Vice versa, classic metabolites (carnitine) and end-products of metabolism (uric acid, bilirubin) may play an antioxidant role.

Interference with ROS signaling and inhibition of pro-oxidant or antioxidant enzymes and induction of their biosynthesis may be an important facet of action of many food components, especially of plant origin.

Under some conditions antioxidants may behave as pro-oxidants. Such situations are not likely to occur *in vivo*, but may take place in food where, for example, antioxidants may reduce Fe^{3+} to Fe^{2+} thus enhancing its prooxidative activity.

10.5 ANTIOXIDANTS IN FOODS

10.5.1 SOURCES OF ANTIOXIDANTS

Antioxidants occurring in foods can be divided into the following:

1. Naturally present in the food in question
2. Formed in food during processing
3. Added on purpose to extend the food's shelf life
   a. Natural
   b. Synthetic
4. Designer foods, i.e., functional foods, genetically modified foods, and nutriceuticals

10.5.2 ANTIOXIDANTS NATURALLY PRESENT IN FOODS

A defense system of antioxidants against ROS is an integral part of plants and species that serve as human diet items and raw materials for food industry. In food, the activity of antioxidant enzymes is limited, although some activities may be relatively well preserved therefore food is more prone to the destructive action of ROS. This action may result in consumption of antioxidants, which is disadvantageous since food is the source of many antioxidants for humans, some of them being vitamins.

Dietary antioxidants have been found to prevent/mitigate the destructive action of ROS (see Chapter 9) in the body (Kanner, 2007). Antioxidant properties that prevent or reduce damage caused by ROS in DNA have been confirmed by *in vivo* studies involving the following bioactive diet components: antioxidant vitamins, Q10 coenzyme, carotenoids, flavonoids, and isoflavones (although there is some controversy regarding daidzein). Among vitamins, the following have antioxidant properties: C, E, A, B_1, and B_2. It has been demonstrated that diet supplementation with vitamin E reduces the susceptibility of the muscle to lipid oxidation. Supplementation with ascorbate did not bring up analogous effect (Morrissey et al., 1998). Although vitamin B9 and vitamin B12 cannot be classified as antioxidants, they play key roles in DNA metabolism and repair (Cemeli et al., 2009).

10.5.2.1 Food of Plant Origin

Phytochemicals, nonnutritive plant chemicals playing various beneficial roles in the body, derived mainly from phenylalanine and tyrosine, are widely distributed in the plant kingdom (over 10,000 phytochemicals are known). The group includes alkaloids (caffeine and theobromine), carotenoids (lycopene), isoflavones (genistein), and phenolic acids (capsaicin, gallic acid, tannic acid). Therefore, raw materials of plant origin are treated as a basic source of dietary antioxidants, for which reason consumption of fruit and vegetables five times a day is recommended. Grains, vegetable oils, nuts, tea, coffee, spices, and soy products are rich antioxidant sources. Plants are sources of vitamins C and E, carotenoids, and are rich in a wide variety of phenolic substances (flavonoids, phenolic acids, simple phenols, hydroxyl cinnamic acid derivatives). Based on total content of phenolic compounds (in catechol equivalents), Kaur and Kapoor (2002) classified 33 vegetables into three groups: vegetables with a high (>200 mg catechol/100 g), average (100–200 mg catechol/100 g), and low (<100 mg catechol/100 g) antioxidant value; for example, broccoli contains 87.5 mg catechol/100 g, while the respective contents in other cruciferous vegetables, Brussels sprouts, white cabbage, and turnip are 105.3, 68.8, 92.5, and 127 mg/100 g. Except for oil plants, nuts, and cereals, plants contain mostly hydrophilic antioxidants.

Researchers have been making attempts to order vegetables and fruits, spices, and other natural-source antioxidants by their AAc.

Vegetables. In terms of AAc, the dominant are the broccoli. Ismail et al. (2004) ordered selected vegetables by their AAc (β-carotene test) in the following series: shallots > spinach > cabbage > kale. In terms of their phenol contents, the series ordering was as follows: spinach > kale > shallots > cabbage. The AAc of red cabbage is much higher than that of green cabbage; red lettuce, too, shows a higher AAc than green lettuce due to the highest anthocyanin content. In terms of the antioxidant capacity [trolox equivalent antioxidant capacity (TEAC) and ferric-reducing antioxidant power (FRAP)], the vegetable series was onion > celeriac > carrot; the antioxidant capacity was found to range from 6 μmol TE/g w.w. in carrot to 25 μmol TE/g w.w. in onion, celeriac showing about 8 μmol TE/g w.w. However, our study (Aranowska A. and Kołakowska A., unpublished data, 2009, Serdyńska, 2009) revealed very wide variations (6–15 μmol TE/g w.w.) in carrot, depending on the batch. The reducing power (FRAP) contributed almost a half (more than 40%) of the total AAc of celeriac, somewhat less in onion, and only several to 23% in carrot. The content of phenols in onion was almost twice that in carrot. Among the onion-like vegetables, garlic shows a higher AAc than onion. Within the *Solanaceae*, the highest AAc is typical of red pepper, followed by tomato.

Fruits are a rich source of vitamin C, carotenoids, and polyphenolic antioxidants. In fruits, grapes contain polyphenolic compounds such as caftaric acid; tartaric acid ester of caffeic acid; flavon-3-ol catechin; anthocyanins; and resveratrol, a stilbene derivative. Berries, including blueberries, strawberries, blackberries, and crowberries, contain large amounts of phenolic compounds such as hydroxylated benzoic acids and cinnamic acid, and flavonoids, including anthocyanins, proanthocyanins, flavonols, and catechins. Citrus fruits contain polyphenols such as hydroxycinnamic acids, including *p*-coumaric, caffeic, and ferulic acids, limonoids, and naringin. Generally, AAc of fruits is almost twice that of vegetables. The highest AAc is typical of berries, particularly black currant. Red currant shows a much lower antioxidant potential, contains less vitamin C and polyphenols (anthocyanins). An AAc similar to that of red currant is shown by strawberries, raspberries, plums, and grapes. In the latter, the highest antioxidant content is shown by the skin (resveratrol) and stones. Total phenolic compounds and anthocyanins in all fruits are concentrated in the skin. Generally, a higher AAc is shown by fruits having a more intense coloration; the highest contents are typical of wild fruits such as blueberries and brambles. Organically grown apples were found to contain more vitamin C, flavonoids, and anthocyanins than apples grown in a conventional way (Rembiałowska et al., 2003). The AAc and phenol profile change during fruit maturation and ripening. Anthocyanins were observed to increase during successive harvest stages, while flavonols and hydroxycinnamic acids decreased from unripe green to ripe blue stage of berry ripening. The blueberry AAc as well as the total phenolic content tended to decrease during ripening (Castrejon et al., 2008).

In nuts, lipophilic antioxidants are dominant. The highest TAC is shown by pecans (179 μmol TE/g), followed by pistachios (80 μmol TE/g) and almonds (17 μmol TE/g); the AAc of cashews is half that of almonds (Wu et al., 2004).

Tea contains large amounts of flavonoids, including catechin, epicatechin, quercetin, epigallocatechin, epicatechin gallate, and epigallocatechin gallate. The maximum AAc in tea is reached after 2 min of steeping, but there are tea-variety-dependent differences. Large amounts of phenolic compounds, which inhibit LDL oxidation, have been reported for wine. Phenolic compounds in wine are *p*-cumaric, cinnamic, caffeic, ferulic, and vanillic acids as well as resveratrol (Porter et al., 1995). The AAc increases as wine ages. Green tea leaves contain up to 36% polyphenols on a dry weight basis, the most abundant compounds being (−) epicatechin, (−) epicatechin gallate, (−) epigallocatechin, (−) epigallocatechin gallate, (+) catechin, and (+) gallocatechin. Catechins and other flavonoids are efficient antioxidants scavenging free radicals and chelating metal ions (Lee et al., 2004).

The AAc-based ordering of vegetables and fruits reported by various authors depends on the method AAc is determined, the matrix, the type of catalyst and solvent used, extraction conditions, and other methodological details. In addition, vegetables and fruits differ in their AAc depending on the species and variety and ripeness; within a species, differences are caused by the site of growing, climatic conditions, year of harvest, and sun exposure. Tomatoes grown in warm-climate countries contain as much as 4–5 times more flavonols than those grown in cooler climates. Ground-grown vegetables (lettuce, tomatoes) contain more polyphenols than those grown in greenhouses. Among greenhouse-grown vegetables, those harvested during the most intense solar radiation are richer in antioxidants (by as much as about 60% in tomatoes) than those picked up in early spring (Czapski, 2007). On the other hand, the vitamin E content in broccoli is higher in shaded than in sunny locations and varies depending on the season of harvest (Kaur et al., 2007). External parts of vegetables and fruits are richer in antioxidants. External leaves of kale contain 150 times more lutein and 200 times more lycopene than the leaves inside. Ripe leaves contain even up to 500 times more lycopene than unripe leaves; however, ripening at higher temperatures (the temperature optimum is 16°C–26°C) results in reduction of lycopene content (transformation to β-carotene) (Czapski, 2007). In genetically modified tomatoes, FRAP and naringenin contents were lower than those in a conventional cultivar (Venneria et al., 2008).

Storage. Storage of vegetables elicits changes in antioxidants, but not necessarily cause losses in the total AAc. Carotenoid losses in stored spinach have been reported, the losses increasing with temperature of storage; after 4–8 days, the original carotenoid content was almost halved. At the same time, the peroxidase activity was observed to increase, but was not significantly affected by storage temperature. The lipoxygenase activity was unaffected by storage time or temperature (Pandrangi and Laborde, 2004). During chilled storage, the AAc of ready-to-eat shredded orange and purple carrots decreased only in the latter, purple carrot containing originally three times as much polyphenols (primarily the less-stable anthocyanins) as orange (Alasalvar et al., 2005). The AAc of carrot and onion assayed (TEAC and FRAP) 5 and 3 months, respectively, after harvest was higher than 1–2 months post harvest. This was, however, the case only with methanol-extracted antioxidants (mainly phenols); water-soluble antioxidant were drastically reduced, and almost totally destroyed in carrot, during storage. Storage of sliced onion for 48 h resulted in an about 20% increase in AAc, mainly due to the parallel increase of reducing properties, but longer storage brought about AAc losses.

Freezing. Freezing of vegetables results in reduced AAc. Freezing of sliced carrot, and 1-month-long storage, brought about a reduction in AAc, FRAP, and phenol content by about 50%. However, with further storage (at −22°C), the AAc gradually increased so that after 6 months the AAc was close to that in the unfrozen carrot, whereas the contents of phenols and FRAP doubled the original level (Aranowska and Kołakowska, 2009, unpublished data). A drastic reduction in polyphenol content was reported from frozen broccoli (Gawlik-Dziki, 2008).

Frozen storage of fruits results in pronounced degradation of polyphenols and anthocyanins as well as in the total AAc and FRAP because of native enzymes, polyphenoloxidases. More than 75% of anthocyanins in frozen cherries were destroyed after 6 months of storage at −23°C. At −70°C, the reduction amounted to as little as 11%–12%. Flavonol glycosides were relatively unchanged; they were to a small degree only affected by polyphenoloxidase (Chaovanalikit and Wrolstad, 2004). Tocols in wholemeal and white flours and bread wheat decreased as a function of time and temperature (−20°C, 5°C, 20°C, 30°C, and 38°C during storage for up to 242 days), following the first-order kinetics. The tocol reduction was faster in white flour than in wholemeal flour (Hidalgo et al., 2009).

Effects of thermal treatment on the AAc in vegetables depend on the vegetable itself, temperature and type of treatment, freshness of the raw material prior to treatment, and method of AAc determination. This is because vegetables differ in the composition of their antioxidants and in thermolability/thermostability of their oxidants. The key role in the total AAc is played by antioxidant interaction, and changed during thermal treatment as well during storage before and after thermal treatment.

Studies, particularly those involving parallel application of a few methods of AAc determination (the results of which seldom are correlated) showed that a mild thermal treatment (about 50°C) of vegetables produced a negligible effect on the AAc or none at all. In broccoli, although the post-harvest treatment (hot air, 48°C/3 h) reduced the phenolic content and antioxidant capacity, they rebound during subsequent (0°C) storage for 3 weeks and were higher than the control (Lemoine et al., 2009).

Cooking was observed to leave the AAc unchanged, to decrease, or to increase it, depending on the type of vegetables and cooking. Cooking increases the AAc in pepper, tomato, spinach, and onion, does not affect the AAc of celeriac, cabbage, shallots, and broccoli, while reducing—by even as much as 50%—the AAc of carrot and red cabbage (Turkmen et al., 2005, 2006, Serdyńska, 2009). Higher thermal losses were recorded during cooking of mixed vegetables than in each of the mix component cooked separately. The cooking method is important for some vegetables only; for example, it is not important for spinach or green beans, but is significant for pepper, peas, and broccoli. The lowest losses are observed during sautéing, microwaving (without water), and baking. In contrast, the highest losses are recorded during steaming (up to 50% in carrot), blanching (also in spinach), boiling, and frying. Vegetables, if fried in antioxidant-rich (vitamin E) oils, may even show a higher AAc, higher than before the treatment. The highest losses occur during boiling. The longer the boiling time, the greater the losses. The magnitude of losses depends on cooking conditions, pH, the presence of other compounds, or cultivar differences; the losses may be as high as 22%–86%. Flavonol losses in tissue during boiling are not generally attributed to a chemical breakdown, but rather to leaching of the compounds into the cooking water; the water in which vegetables were boiled may contain even more than 50% of the total original vegetable AAc. The highest losses occur during canning.

Antioxidant breakdown on thermal treatment concerns mainly thermolabile water-soluble antioxidants, as the amount of phenols in methanol extracts was observed even to increase after cooking; the amount of phenols increases also during refrigerated storage of cooked vegetables. Storage of vegetables prior to cooking tends to increase the losses incurred during thermal treatment, but not in all vegetables: for example, losses during cooking of carrot (in hermetically closed packaging) previously stored for a few months were lower than losses on cooking of fresh carrot.

Thermal treatment of fruits does not reduce their total AAc, although leads to losses of some antioxidants. During canning of cherries, about half the anthocyanins and polyphenols leached from the fruits into the syrup with little total loss (Chaovanalikit and Wrolstad, 2004, Poz.1). Blanching of fruits (blueberries) greatly increased the radical-scavenging activity of the juice, in reaction to a higher recovery of anthocyanin pigments and total cinnamates (Rossi et al., 2003). Rutin, luteolin-7 glucoside, and chlorogenic acid gradually decomposed during heating at 100°C. Even though they rapidly decomposed at 180°C, some decomposition products still retained their radical-scavenging activity. When rutin was heated in the presence of chlorogenic acid, rutin decomposition was totally inhibited at 100°C, but it was reduced at 180°C. Radical scavenging activity is more stable than the content of original polyphenolic compounds in foods during cooking and processing (Murakami et al., 2004).

10.5.2.2 Foods of Animal Origin

Food of animal origin is not treated as a source of antioxidants; consumption of unsaturated fatty acid-rich products such as fish is recommended to be accompanied by consumption of antioxidant-rich vegetables or antioxidant supplements. However, the total AAc (sum of the antioxidants extracted with 5% NaCl, water, and methanol) of beef, pork, poultry, and fish is comparable to or higher than that of vegetables, also when referred to dry weight, except for onion the AAc of which is higher than that in animal products and other vegetables (Figure 10.1). In terms of the total AAc (TEAC), raw materials of animal origin may be ordered to form the following series, beef ≥ chicken meat ≥ pork; the ordering in terms of the reducing power is as follows, fish ≥ pork ≥ beef ≥ chicken meat. The fish studied formed the following average AAc-based series, mackerel > rainbow trout > salmon ≥ cod ≥ herring. There are very high differences between batches of meat, and particularly of fish, even in rainbow trout from one farm. Definitely, the lowest activity (TEAC) was typical of herring

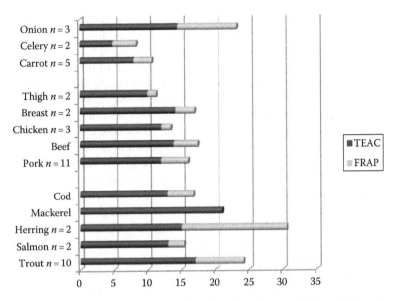

FIGURE 10.1   AAc (μM TE/g. w.w.) TEAC and FRAP (ferric-reducing antioxidant power) of fish muscle tissue, meats, and vegetables, averaged across n batches. (After Plust, D., Serdyńska, A., Kołakowska, A., unpublished data.)

immediately before spawning (6.99 μmol TE/g w.w.), but at the same time it showed the highest reducing power (19. 93 μmol TE/g w.w.). Herring caught less than 2 months earlier showed TEAC of as high as 23.01 μmol TE/g w.w., but a lower reducing power (11 μmol TE/g w.w.). The TEAC + FRAP values were rather similar in all the fish species studied (15–18 μmol TE/g w.w.), those fish (species, batch) showing a higher activity (TEAC) had a higher FRAP and *vice versa*. The lowest between- and within-species variability was shown by phenol concentration that averaged 7.2 mg catechin/g w.w. at the coefficient of variation V = 14.2%. Many circumstances, biotic and abiotic, promote the antioxidant defense response in fish. Factors intrinsic to the fish itself, such as age, phylogenetic position, feeding behavior, and environmental factors such as type of diet available, temperature, dissolved oxygen, toxins present in the water, pathologies, and parasites can either enhance or decrease the antioxidant defenses (Martinez-Alvarez et al., 2005). The defense system remains active also *post mortem*, as suggested by an experiment showing an AAc increase after mildly oxidized herring lipids were added to herring tissue. However, adding strongly oxidized lipids reduced the AAc. Similarly, mild UV irradiation of minced pork enhanced the AAc (Kołakowska et al., 2009). Irradiation of food (up to 30 kGy) was reported to decrease, not affect, or increase the content of phytochemicals, especially antioxidants. The increase was due to stimulation of biosynthesis of phytochemicals, especially in fruits or vegetables (Alothman et al., 2009).

Animal raw materials owe their AAc primarily to compounds extractable with 5% NaCl, i.e., myofibrillar proteins and to water-soluble compounds (sarcoplasmic proteins, products of proteolysis, antioxidant enzymes, vitamins). Contributions of those fractions to the total AAc ranges from about 30% in chicken meat to more than 60% in salmon (the 5% NaCl-extractable fraction) and from about 30% in pork to more than 50% in salmon and poultry meat (water-soluble fraction); contribution of methanol-soluble compounds averages several per cent of the total AAc. Purely lipophilic antioxidants, even in fresh fish, did not contribute to the total tissue AAc.

Refrigerated storage of fish is associated with a drastic reduction in AAc, down to a half of the original value, as early as after 1–2 days of storage; at the same time, a rapid increase in reducing power (FRAP) of the tissue is observed. When the first symptoms of fish spoilage appear, the AAc undergoes a small constant increase due to accumulation of strongly antioxidant peptides. Changes in the AAc during refrigerated storage of fish were found to be negatively correlated with evolution of lipid oxidation. Freezing and storage (−22°C, 1 month) of Baltic herring did not change the AAc in any significant way; however, after 3 months of storage the AAc was found to decrease. The AAc assayed after 6 months amounted to as little as a half of the original value, despite a constant increase of the reducing power and the lack of significant changes in phenol concentrations. Based on results of studies involving fish one may assume that reducing compounds, rather than contributing to the total AAc, provide backup to strong antioxidants when the activity of those decreases (see Chapter 15).

Thermal treatment lowers the AAc, the decrease being more species-specific than dependent on the type of treatment. The most thermostable antioxidants are found in poultry meat and pork, those in beef and fish being much less thermally resistant. Following thermal treatment, the ready-to-eat meats form the following series: poultry meat>pork>beef>fish. Losses in fish and beef average 50%; the highest losses have been recorded in herring, salmon, and rainbow trout. Losses recorded in pork varied, depending on the batch and type of thermal treatment, from a few to 20%, and usually oscillated around 10%. Losses in poultry meat amount to a few percent in breast muscles, whereas thigh muscles showed no losses and even an increase in AAc following thermal treatment. The lowest losses occur during baking and microwaving (without water), the highest losses being incurred during cooking in water when, like in vegetables, antioxidants leach into the water. If, as reported by various authors (particularly with regarded to plant raw materials), AAc losses considered are restricted to the alcohol-extractable fraction, thermal treatment was found (except in some pork batches) to increase the AAc from a few to even 70% (in herring); the highest increase was observed following microwaving and frying (on rapeseed oil). Thermal treatment-induced changes depended also on the freshness of fish: the highest thermal AAc losses were incurred in fresh rainbow trout (Kołakowska et al., 2006a,b). Cold smoking results in doubling of the AAc of pork and fish, whereas hot smoking—similarly to other thermal treatment methods—decreases the summary AAc, but increases the methanol fraction AAc as that was the fraction that mainly scavenged antioxidants from the smoke.

10.5.2.3 Combining Meat and Fish with Vegetables

Combining of minced meat or fish with vegetables almost always produces a synergy of antioxidants. The antioxidant effect of the mix depends on the vegetable, but mostly on the animal component: its type and batch and the original AAc.

TABLE 10.1

Synergistic Effects of Cooked Fish:Vegetable Combination on Antioxidant Capacity (TEAC), Reducing Power (FRAP), and Phenols (FC), Calculated as Percent Ratio between Experimental Data on Cooked 1:1 Fish:Vegetable Combination and Values for the Two Components Cooked Separately

Combination Fish:Vegetable Ratio (1:1)	TEAC	FRAP	Phenols (FC)
Rainbow trout:carrot	12.73	8.36	(−)
Herring:carrot	19.19	−29.57	47.61
Cod:carrot	17.05	47.37	46.91
Salmon:carrot	−48.77	7.29	1.75
Cod:onion	−7.27	−11.55	20.06
Salmon:onion	−0.67	−5.53	−14.41
Rainbow trout:celery	50.40	37.62	(−)

Sources: Data from Serdyńska, A., Antioxidant activity of combination fish-vegetables, PhD thesis, Westpomeranian Technological University, Szczecin, Poland (in Polish), 2009; Aranowska and Kołakowska, unpublished data, 2009.

TEAC, FRAP (μmole TE/g w.w.), FC (mg catechin/g w.w.).

Combining (1:1) carrot, onion, and celeriac with cod, rainbow trout, or salmon was found to increase the AAc (TAC + FRAP) (relative to the sum of AAc's of the products) from about 7% (salmon) to as much as 50%–60% (rainbow trout, cod).

Cooking fish together with vegetables allows to reduce losses produced by thermal processing (in each of the components cooked separately) by about 20% (Table 10.1). Combining vegetables with herring, however, resulted in losses being higher that indicated by the sum of components both raw and during cooking. Also, combining raw pork and onion is associated with losses of AAc, but cooking of the meat with onion allows to diminish thermal losses of antioxidants by about 22%, although there is a simultaneous drastic destruction of lipophilic antioxidants (Serdyńska, 2009). Goerlik et al. (2008) demonstrated that combining meat with red wine reduces, by 75%, absorption of malondialdehyde (MDA), while blending meat (turkey cutlets) with wine prior to cooking totally prevents MDA absorption.

10.5.3 Antioxidants Formed in Food during Processing

Results of combining animal and plant raw materials, usually resulting in positive synergies, provide evidence in favor of using technological processes to increase AAc of foods. It is not known, however, whether the *in vivo* defense system would be enhanced. Even the data evidencing the *in vivo* efficiency of low-molecular antioxidants such as CLA, emerged during processing, are controversial (Martin et al., 2008). However, AAc of some Maillard reaction products has been confirmed, also *in vivo*. Dietary sources of the Maillard reaction products include cookies. At higher temperatures and baking times, replacing sucrose with glucose and adding ammonium bicarbonate promotes the AAc (Morales et al., 2008). The Maillard reaction products are responsible for, a relatively high AAc of soup concentrates. The L-P interaction, a milder form of the Maillard interaction, was elicited by freeze texturization, which stabilizes lipids in minced fish (Kołakowska and Szczygielski, 1994). "Fermentation" technologies involving proteolysis, e.g., in fish processing (fish sauces) by the enzymes already present in fish and/or by different enzymes added at appropriate levels to the fish, produce peptides with strong antioxidant properties (Amarowicz and Shahidi, 1997, Samaranayaka and Li-Chan, 2008). Preparations, hydrolysates, and protein concentrates showing antioxidant properties are manufactured from protein of soybeans, oil plants, milks, and—recently with increasing frequency—fish (capelin, tuna, mackerel, yellowfin sole, Alaska pollack, Atlantic salmon, hoki, conger eel, scad, cod), usually from parts that are usually rejected and not used for food products.

10.5.4 ANTIOXIDANTS ADDED TO FOODS

Antioxidants added to foods are listed in the Codex Allimentarius for Food Additives (Codex Stan 192–1995), adopted in 1995 and revised almost annually (the most recent revision is dated to 2008). The authorization and use of antioxidant as food additives in the European Union are based on the Framework Directive 89/107/EEC (O.J. L40, 11.02.1989, p. 27) and Directive 95/2/EC (O.J.L61,18.03.1995, p. 1). These acts will soon be replaced by regulation (EC) No. 1333/2008 of the European Parliament and of the Council of December 16, 2008 on food additives. The regulation will apply as of January 20, 2010. The regulation will be provided with annexes in the form of Community Lists of Approved Food Additives, which shall be completed by January 20, 2011. Legal underpinning of procedures whereby additives are allowed to be used in foods is contained in Regulation (EC) No. 1331/2008 of the European Parliament and of the Council of December 16, 2008 establishing a common authorization procedure for food additives, food enzymes, and food flavorings (in force from January 2010).

Both the non-obligatory Codex Stan Act and lists of additives allowed in the EU (Table 10.2) and the United States (U.S. Code of Federal Regulations, CFR) contain, in addition to tocopherols, carotenoids, and ascorbic acid and its salt, synthetic antioxidants: butylated hydroxyanisole (BHA), butylated hydroxytoluene (BHT), and tertiary butylohydroquinone (TBHQ) gallates. The maximum level is most frequently 200 mg/kg. Paradoxically, those antioxidants may be added to products listed as rich sources of natural antioxidants, e.g., cocoa and chocolate products, herbs, spices, nuts, smoked products, oats (rolled), dried vegetables, and seaweeds. On one hand, extracts of those sources are used to stabilize food lipids, chocolate being recommended as a medical food (Pucciarelli and Grivetti, 2008), while on the other hand they are enriched with synthetic antioxidants to extend their shelf life or to replace natural protective substances destroyed during processing. Veritably, the

TABLE 10.2

Antioxidants Approved by the European Union

Number	Antioxidant
E 300	Ascorbic acid
E 301	Sodium ascorbate
E 302	Calcium ascorbate
E 304	Fatty acid esters of ascorbic acid
(i)	Ascorbyl palmitate
(ii)	Ascorbyl stearate
E 306	Tocopherol-rich extract
E 307	Alpha-tocopherol
E 308	Gamma-tocopherol
E 309	Delta-tocopherol
E 310	Propyl gallate
E 311	Octyl gallate
E 312	Dodecyl gallate
E 315	Erythorbic acid
E 316	Sodium erythorbate
E 320	Butylated hydroxyanisole (BHA)
E 321	Butylated hydroxytoluene (BHT)

Good Lord must be truly merciful and patient when looking at what we do with the Earth that He has let us manage and where all that is edible, and could lose its quality as a result of oxidation, had been carefully protected by several types of antioxidants. It seems necessary to introduce a HACCP-like system for food processing, with critical control points (CCP) preventing the loss of natural bioactive compounds. At present it is, however, more trendy and lucrative to manufacture designer foods.

An important role in stabilizing food lipids prior to oxidation is played by spices, extracts, and their appropriate mixes used in the food industry. A shortcoming of those natural food additives is the necessity to use much higher, at least 10-fold, amounts than synthetic antioxidants and often undesired effects on sensory properties of the product. In recent years, protein hydrolysates with antioxidant properties have become a topic of great interest for pharmaceutical, health food, as well as food processing/preservation industries (Hogan et al., 2009, Rossini et al., 2009).

10.5.5 Designer Foods with Antioxidants

Designing a food to increase its AAc involves rearing pigs, poultry (for meat and eggs), and farming fish so that modification of the fatty acid profile, i.e., primarily enrichment in n-3 PUFA via feeding, requires an extra antioxidant protection administered via the feed. The antioxidants used include mainly vitamin E and carotenoids, applied as formulae and natural plant sources rich in those antioxidants. The routine practice is to apply carotenoids in salmon cultures to produce the desired meat color. Antioxidants added to feeds reduce fish rancidity flavor and prolong the shelf life, particularly duration of storage, of n-3 PUFA-enriched raw materials and their products.

The dynamically developing functional food market is to a large extent based on antioxidant properties of phytochemicals. Novel functional antioxidant therapeutic diets, based on the veganism, are proposed (Link et al., 2008, Luo et al., 2009). However, functional food, too, requires addition of antioxidants; this is particularly the case of *n*-3 PUFA-enriched food. There is a wide variety of products such as dressings, dairy products, cookies, and desserts, in which enrichment is based on the results of sensory analyses, without regard to possible effects of interactions with matrix components on the nutritive value of the product. The hopes for stability of microencapsulated fish oil used as a food additive failed; microencapsulation requires additional application of antioxidants. Fish oil formulae and the increasing variety of bio-oils, although active components themselves due to the presence of antioxidants in them, require additional stabilization to prolong their shelf life. Although stabilized with antioxidants, microencapsulated fish oil and a variety of concentrated *n*-3 PUFA or *n*-3/*n*-6 PUFA formulae on the market show substantial oxidation, even when their shelf life is to terminate in another 2 years. This is also true with respect to vitamin E and tocopherols, whose ability to inhibit lipid oxidation is reduced.

Research is in progress on engineering of the yeast antioxidant enzyme-enhanced activity to reduce intracellular ROS and to increase yeast tolerance to oxidative stress (Iinoya et al., 2009). An antioxidant gene therapy for myocardial infraction after coronary angioplasty and cerebral ischemic attacks has also been discussed (Wu et al., 2009).

10.6 METHODS FOR DETERMINATION OF ANTIOXIDANT ACTIVITY

10.6.1 ANTIOXIDANT ASSAYS

There are currently between 25 and 100 different methods used to measure antioxidants. These may be classified into two categories. The first category measures the ability of antioxidants in inhibiting oxidation reaction in a model system by monitoring the associated changes using physical, chemical, or instrumental means. The second, involving radical-scavenging assays, includes methods based on hydrogen atom transfer (HAT) or single electron transfer (SET) mechanisms (Figure 10.2) (Frankel and Meyer, 2000, Bartosz, 2003, Kusznierowicz et al., 2006, Bartoszek, 2007, Shahidi and Zhong, 2007, Karadag and Ozcelik, 2009). Other assays are related to lipid peroxidation, and include the TBARS (2-thiobarbituric reactive substances) test, determination of PV (peroxide value) and AV (anisidine value), conjugated dienes and volatile compounds, as well as the instrumental methods such as GC, HS-GC (see Chapter 9) (Laguerre et al., 2007); automated techniques used to test lipid stability, such as Rancimat, are used as well.

10.6.2 HAT-BASED METHODS

The methods usually rely on monitoring the kinetics of reactions involving competition of antioxidants contained in a sample (molecular probe) and an indicator for an oxidant, usually peroxyl radical.

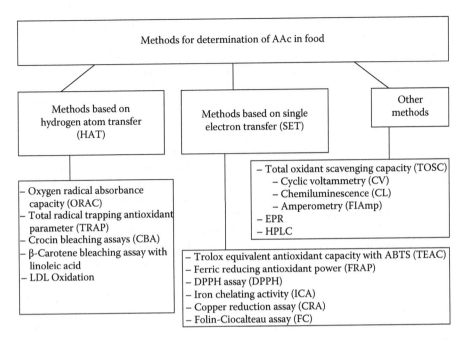

FIGURE 10.2 Methods for determination of AAc in foods. (Adapted from Huang, D. et al., *J. Agric. Food Chem.*, 53, 1841, 2005.)

ORAC (*Oxygen Radical Absorbance Capacity*). In this method, artificially generated peroxyl radicals react with fluorescein. As the reaction progresses, fluorescein is consumed and the fluorescence decreases. Antioxidant capacity is quantified by calculating the net protection area under the time-recorded fluorescence decay curve of fluorescent probe in the presence of an antioxidant. In addition, the ORAC assay estimates the antioxidant capacity of both hydrophilic and lipophilic compounds (Ou et al., 2002, Prior et al., 2003).

TRAP (*Total Radical-Trapping Parameter*). This assay uses peroxyl radicals generated by 2,2′-azobis (2-amidinopropane) dihydrochloride (AAPH) to oxidize antioxidants, and the oxidation is monitored by oxygen absorption (oxygen electrode). The method does not provide a useful estimate of activity of metal chelators and of lipophilic antioxidants such as vitamin E (Frankel and Meyer, 2000).

Chemiluminescence-based TRAP. The principle of this method is that peroxyl radicals produced from AAPH oxidize luminol, leading to formation of luminol radicals that emit light, which can be detected by a luminometer (Prior and Cao, 1999).

CBA (*Crocin Bleaching Assays*) is based on the oxidation (bleaching) of crocin by peroxyl radicals in the absence and presence of antioxidants.

10.6.3 SET MECHANISM-BASED METHODS

Methods based on a SET reaction, such as TEAC, 2,2-diphenyl-1-picrylhydrazyl (DPPH) assay, Folin–Ciocalteau (FC) assay are the most commonly used because of their ease, speed, and sensitivity. The methods involve chromogen compounds of a radical nature to simulate ROS. The presence of antioxidant leads to the disappearance of these radical chromogens, the two most widely used being the ABTS$^{+\cdot}$ and the DPPH$^{\cdot}$ radicals (Arnao, 2000). Both present good stability in certain assay conditions, but also show several important differences in their response to antioxidants and in their manipulation. DPPH$^{\cdot}$ is a free radical that is acquired directly without preparation (ready to dissolve), while ABTS$^{+\cdot}$ must be generated by enzymatic (peroxidase, myoglobin) or chemical (manganese dioxide, potassium persulfate, AAPH) reactions. Another important difference is that ABTS$^{+\cdot}$ can be solubilized in aqueous and in organic media, in which the AAc can be measured due to the hydrophilic and lipophilic nature of the compounds in samples. In contrast, DPPH$^{\cdot}$ can only be dissolved in organic media (especially in alcoholic media) and not in aqueous media, which is an important limitation when interpreting the role of hydrophilic antioxidants. The TEAC assay estimates the ability of an antioxidant to reduce the artificially generated ABTS$^{\cdot+}$ radical cation, expressing the AAc as equivalents of standard antioxidant TROLOX (a vitamin E analogue). Similar to TEAC, the FRAP assay is based on a SET reaction and measures the ability of the antioxidant to reduce a ferric 2,4,6-tripyridyl-s-triazine salt (Fe^{3+}–TPTZ) to a blue-colored ferrous complex (Fe^{2+}–TPTZ) at low pH (Jimenez-Alvarez et al., 2008). An analogous method, the cupric-reducing antioxidant capacity (CUPRAC), measures the copper-reducing activity of antioxidants using a Cu(II)-neocuproine reagent (Kanner, 2007).

10.6.4 OTHER METHODS

According to Calderón et al. (2009) the, LC-MS assay facilitates rapid identification and subsequent determination of relative AAc of individual antioxidant species in complex natural product samples and food products such as cocoa. This assay is based on the comparison of electrospray LC-MS profiles of sample extracts before and after treatment with ROS such as hydrogen peroxide or DPPH. Chen et al. (2003) found high correlations for the AAc (low-molecular-weight fraction of whey) obtained by the ABTS, FRAP, and the flow-injection amperometric procedure. According to Zielińska et al. (2008), cyclic voltammetry (CV), based on electrochemical properties of compounds under the test, is a suitable method for determination of the reducing capacity; they ran parallel assays using TEAC, DPPH, FC, and photochemiluminescence.

Methods that most faithfully reflect AAc in biological systems include ORAC, TRAP, and LDL oxidation (Figure 10.2). The methods used most frequently in food determination of phenols using FC (and the HPLC-based analysis of phenol composition) include FRAP, TEAC, and DPPH. The measurement of AAc, especially of antioxidants that are mixtures, multifunctional, or are acting in complex multiphase systems, cannot be evaluated satisfactorily by a simple antioxidant test without due regard to the many variables influencing the results (Antolovich et al., 2002). Even parallel application of several methods rarely produces similar results. No or very weak correlations were found by Al-Duais et al. (2009) between antioxidant assays TEAC, DPPH, FRAP, ORAC, and total phenolics FC on water and ethanol extracts from *Vitaceae*. Free-radical-scavenging assays such as ORAC and DPPH were not able to consistently predict the ability of compounds to inhibit lipid oxidation in cooked ground beef (Alamed et al., 2009).

Methods used to determine antioxidants in vegetables and fruits cannot be directly applied to complex food and food products of animal origin. To study the latter, methods used in medical research are adapted and preferentially used.

A very important problem is still sample preparation for antioxidant capacity determination. Unfortunately, basically all the widely used methods require preliminary extraction, the conditions of which should be specifically adjusted to each type of complex food. Results of assays are affected by the type of solvent used, sample-to-solvent ratio, extraction conditions, and duration of the reaction with the chromogene used. Most frequent in food research is alcohol extraction; phosphate buffer and water are additionally used to biological materials. In meat (pork, beef, chicken) and fish, from 50% to as much as 80% of the sum of AAc is contributed to antioxidants extracted with 5% NaCl; and then with water, methanol extracted from 20% to 30% of the sum of AAc only (Plust, 2008). There is widespread agreement that antioxidant measurements need to be standardized, but disagreement exists as well on the best method to measure the beneficial compounds, which are thought to reduce the risk of cancer and heart disease as well as fight aging, arthritis, and Alzheimer's disease. Official standard methods could take 2–3 years to develop and must be tested in multiple laboratories to ensure the methods are valid (First International Congress on Antioxidant Methods, held 2004 in Orlando).

10.7   FINAL REMARKS

Although it is theoretically possible to calculate AAc of a product from chemical analysis of activities of individual compounds, the AAc in a complex food is not equal to the sum of the activity of individual, known antioxidants the food contains (Dekker et al., 1999, Alamed et al., 2009). AAc in plant and animal tissues, after harvest or slaughter, reflects the actual antioxidant status. The status is very variable, not only on account of variability in the composition of antioxidants and matrix, but also, and importantly, as a result of changes in interactions between antioxidants themselves and the matrix. This usually results in a synergism or antagonism. Due to the diverse chemical structure, differences in mechanisms of AAc, and genetic background for the activity of antioxidants, changes in antioxidants—elicited by endo- and exogenous factors—may proceed differently, even in identical groups of antioxidants.

Although more and more papers have been recently published on antioxidants of animal origin, and products of protein oxidation are assayed parallel to lipid oxidation, animal antioxidants, except for milk products, are treated as a group requiring backing up by plant antioxidants. However, the AAc of animal tissue is close to, or even higher than, AAc of plant antioxidants, also when referred to dry weight.

There are no data that would allow answering the question if the dominant role of protein in prevention of oxidation in animal raw materials is important only for stabilizing lipids in food or if it can be extended to *in vivo* diet, or—similarly to other large-molecular-configuration antioxidants—if it is very effective *in vitro*, but insignificant *in vivo*.

REFERENCES

Alamed, J., Chaiyasit W., McClements D.J., and Decker E.A. 2009. Relationship between free radical scavenging and antioxidant activity in foods. *Agric. Food Chem.* 57:2969–2976.

Al-Duais, M., Müller, L., Böhm, V., and Jetschke, G. 2009. Antioxidant capacity and total phenolics of *Cyphostemma digitatum* before and after processing: Use of different assays. *Eur. Food Res. Technol.* 228:1438–2377.

Alasalvar C., Al-Farsi, M., Quantick, P.C., Shahidi, F., and Wiktorowicz, R. 2005. Effect of chill storage and modified atmosphere packaging (MAP) on antioxidant activity, anthocyanins, carotenoids, phenolics and sensory quality of ready-to-eat shredded orange and purple carrots. *Food Chem.* 89:69–76.

Alothman, R., Bhat A., and Karim, A.A. 2009. Effects of radiation processing on phytochemicals and antioxidants in plant produce. *Trends Food Sci. Technol.* 20:201–212.

Amarowicz, A.R. and Shahidi, F. 1997. Antioxidant activity of peptide fractions of capelin protein hydrolysates. *Food Chem.* 58(4):355–359.

Antolovich, M., Prenzler, P.D., Patsalides, E., McDonald, S., and Robards, K. 2002. Methods for testing antioxidant activity. *Analyst* 127(1):183–198.

Arnao, M.B. 2000. Some methodological problems in the determination of antioxidant activity using chromogen radicals: A practical case. *Trends Food Sci. Technol.* 11(11):419–421.

Bartosz, G. 2003. Total antioxidant capacity. *Adv. Clin. Chem.* 37:219–292.

Bartoszek, A. 2007. Methods for the determination of the antioxidant activity of foodstuffs, in *Food Antioxidants*, W. Grajek (Ed.), pp. 532–550, WNT, Warszawa, Poland (in Polish).

Brannan, R.G., Connolly, B.J., and Decker, E.A. 2001. Peroxynitrite: A potential initiator of lipid oxidation in food. *Trends Food Sci. Technol.* 12:164–173.

Calderón, A.I., Wright, B.J., Hurst, W.J., and Van Breemen, R.B. 2009. Screening antioxidants using LC-MS: Case study with cocoa. *J. Agric. Food Chem.* 57(13):5693–5699.

Castrejon, A.D.R., Eichholz, I., Rohn, S., Kroh, L.W., and Huyskens-Keil, S. 2008. Phenolic profile and antioxidant activity of highbush blueberry during fruit maturation and ripening. *Food Chem.* 3(1):564–572.

Cemeli, E., Baumgartner, A., Anderson, D. 2009. Antioxidants and the Comet assay. *Mutat. Res./Rev. Mutat. Res.* 681(1):51–67.

Chaovanalikit, A. and Wrolstad, R.E. 2004. Anthocyanins and polyphenolic composition of fresh and processed cherries. *JFS* 69(1):FCT73–FCT83.

Chen, J., Lindmark-Mansson, H., Gorton, L., and Akesson, B. 2003. Antioxidant capacity of bovine milk assayed by spectrophotometric and amperometric methods. *Int. Dairy J.* 13:927–935.

Chipault, J.R. 1962. Antioxidants for use in foods, in *Autooxidation and Antioxidants*, Vol. II, W.O. Lundberg (Ed.), pp. 477–542, Interscience, New York.

Choe, E. and Min, D.B. 2006. Chemistry and reactions of reactive oxygen species in foods. *Crit. Rev. Food Sci. Nutr.* 46:1–22.

Cullen, P.J., Tiwari, B.K., O'Donnell, C.P., and Muthukumarappan, K. 2009. Modelling approaches to ozone processing of liquid foods. *Trends Food Sci. Technol.* 20:125–136.

Czapski, J. 2007. Natural food antioxidants, in *Food Antioxidants*, W. Grajek (Ed.), WNT, Warszawa, Poland, Chapter 3.1.

Dekker, M., Verkerk, R., Van der Sluis, A.A., Khokhar, S., and Jongen, W.M.F. 1999. Analysing the antioxidant activity of food products: Processing and matrix effects. *Toxicol. In Vitro*, 13:797–799.

Decker, E., Faustman, C., and Lopez-Bote, C.J. (Eds.). 2000. *Antioxidants in Muscle Foods. Nutritional Strategies to Improve Quality,* John Wiley and Sons, Inc. Publ., New York.

Frankel, E.N. and Meyer, A.S. 2000. Review. The problems of using one-dimensional methods to evaluate multifunctional food and biological antioxidants. *J. Sci. Food Agric.* 80:1925–1941.

Gawlik-Dziki, U. 2008. Effect of hydrothermal treatment on antioxidant properties of broccoli (*Brassica oleracea var. botrytis italic*) florets. *Food Chem.* 109:393–401.

Goerlik, S., Ligumsky, N., Kohen, R., and Kanner, J. 2008. The stomach as a "bioreactor": When red meat meets red wine. *J. Agric. Food Chem.* 56(13):5002–5007.

Gómez-López, V.M., Rajkovic, A., Ragaert, A., Smigic, N., and Devlieghere, F. 2009. Chlorine dioxide for minimally processed produce preservation: A review. *Trends Food Sci. Technol.* 20:17–26.

Grajek, W. (Ed.). 2007. Przeciwutleniacze wyżywności. Aspekty zdrowotne technologiczne molekularne i analityczne. WNT, Warszawa, Poland (in Polish).

Halliwell, B. and Gutteridge, J.M. 1995. The definition and measurement of antioxidants in biological systems. *Free Radic. Biol. Med.* 18:125–126.

Halliwell, B., Murcia, M.A., Chirico, S., and Auroma, O.I. 1995. Free radicals and antioxidants in food and *in vivo*: What they do and how they work. *Crit. Rev. Food Sci. Nutr.* 35(1/2):7–20.

Hands, E.S. 1996. Antioxidants: Technical and regulatory considerations, in *Bailey's Industrial Oil and Fats Products,* 5th edn., vol. 1, Chapter 13, pp. 523–545, Y.H. Hui (Ed.), John Wiley & Sons, Inc., New York.

Hidalgo, A., Brandolini, A., and Pompei, C. 2009. Kinetics of tocols degradation during the storage of einkorn (*Triticum monococcum* L. ssp. *monococcum*) and breadwheat (*Triticum aestivum* L. ssp. *aestivum*) flours. *Food Chem.* 116(4):821–827.

Hogan, S., Zhang, L., Li, J., Wang, H., and Zhou, K. 2009. Development of antioxidant rich peptides from milk protein by microbial proteases and analysis of their effects on lipid peroxidation in cooked beef. *Food Chem.* 117(3):438–443.

Huang, D., Ou, B., Prior, R.L. 2005. The chemistry behind antioxidant capacity assays. *J. Agric. Food Chem.* 53:1841–1856.

Hudson, B.J.F. (Ed.). 1990. *Food Antioxidants,* Elsevier Applied Science, London, U.K., Chapters 1–7.

Iinoya, K., Kotani, T., Sasano, Y., and Takagi, H. 2009. Engineering of the yeast antioxidant enzyme Mpr1 enhanced activity and stability. *Biotechnol. Bioeng.* 103(2):341–352.

Ismail, A., Marjan, Z.M., and Foong, C.W. 2004. Total antioxidant activity and phenolic content in selected vegetables. *Food Chem.* 87:581–586.

Jimenez-Alvarez, D., Giuffrida, F., Vanrobaeys, F., Golay, P.A., Cotting, C., Lardeau, A., and Keely, B.J. 2008. High-throughput methods to assess lipophilic and hydrophilic antioxidant capacity of food extracts in vitro. *J. Agric. Food Chem.* 56(10):3470–3477.

Kanner, J. 2007. Dietary advanced lipid oxidation endproducts are risk factors to human health. *Mol. Nutr. Food Res.* 51(9):1094–1101, Special Issue: Are Dietary AGEs/ALEs a Health Risk.

Karadag, A. and Ozcelik, B. 2009. Review of methods to determine antioxidant capacities. *Food Anal. Methods* 2:41–60.

Kaur, C. and Kapoor, H.C. 2002. Anti-oxidant activity and total phenolic content of some Asian vegetables. *Int. J. Food Sci. Technol.* 37:153–161.

Kaur, Ch., Kumar, K., Anil, D., and Kapoor, H.C. 2007. Variations in antioxidant activity in broccoli (*Brassica oleracea* L.) cultivars. *J. Food Biochem.* 31:621–638.

Kirsch, M. and De Groot, H. 2001. NAD(P)H, a directly operating antioxidant? *FASEB J.* 15:1569–1574.

Kołakowska, A. and Szczygielski, M. 1994. Stabilization of lipids in minced fish by freeze texturization. *J. Food Sci.* 59(1):88–90.

Kołakowska, A., Domiszewski, Z., Bienkiewicz, G. 2006a. Effect of biological and techno-logical factors on the utility of fish as a source of n-3 PUFA, in *Omega 3 Fatty Acid Research,* M.C. Teale (Ed.), pp. 83–107, Nova Science Publishers, Inc., New York.

Kołakowska, A., Kołakowski, E., and Bienkiewicz, G. 2009. Quality and nutritional value of smoked food. In: *Smoked Food,* E. Kołakowski (ed.), Chap. 10, Wydawnictwo Naukowo-Techniczne, Warsaw, in press, (in Polish).

Kołakowska, A., Kołakowski, E., Domiszewski, Z., Zienkowicz, L., and Bednarczyk, B. 2006b. The effects of freshness on changes of nutritional value of protein and lipids of rainbow trout during cooking, in Report Project No 3 PO6T 060 25. State Committee for Scientific Research (KBN), Warszawa, Poland, pp. 1–146.

Kusznierowicz, B., Wolska, L., Bartoszek, A., and Namieśnik, J. 2006. Methods for in vitro determination of the antioxidant activity of foodstuff samples. Part I. *Bromat. Chem. Toksykol.* 39(3):251–260, Part II, 261–270.

Laguerre, M., Lecomte, J., and Villeneuve, P. 2007. Evaluation of the ability of antioxidants to counteract lipid oxidation: Existing methods, new trends and challenges. *Prog. Lipid Res.* 46(5):244–282.

Lee, J., Koo, N., and Min, D.B. 2004. Reactive oxygen species, aging, and antioxidative nutra-ceuticals. *Comprehens. Rev. Food Sci. Food Safety* 3:21–33.

Lemoine, M.L., Civello, P., Chaves, A., and Martinez, G. 2009. Hot air treatment delays senes-cence and maintains quality of fresh-cut broccoli florets during refrigerated storage. *LWT-Food Sci. Technol.* 42(6):1076–1081.

Link, L.B., Hussaini, N.S., Jacobson, J.S. 2008. Changes in quality of life and immune markers after a stay at a raw vegan institute: A pilot study. *Complement Ther. Med.* 16(3):124–130.

Luo, Y., Chen, G., Li, B., Ji, B., Guo, Y., and Tian, F. 2009. Evaluation of antioxidative and hypolipidemic properties of a novel functional diet formulation of *Auricularia auricula* and hawthorn. *Innov. Food Sci. Emerg. Technol.* 10(2):215–221.

Martin, D., Antequera T., Muriel, E., Andres, A.I., and Ruiz, J. 2008. Oxidative changes of fresh loin from pig, caused by dietary conjugated linoleic acid and monounsaturated fatty acids, during refrigerated storage. *Food Chem.* 111(3):730–737.

Martinez-Alvarez, R.M., Morales, A.E., and Sanz, A. 2005. Antioxidant defenses in fish: Biotic and abiotic factors. *Rev. Fish Biol. Fish.* 15:75–88.

Morales, F.J., Saray, M., Özge, Ç.A., Gema A.-L., and Gökmen, V. 2008. Antioxidant activ-ity of cookies and its relationship with heat-processing contaminants: A risk/benefit approach. *Eur. Food Res. Technol.* 228(3):345–354.

Morrissey, P.A., Sheehy, P.J.A., Galvin, K., and Kerry, J.P. 1998. Lipid stability in meat and meat products. *Meat Sci.* 49:S73–S86.

Murakami, M., Yamaguchi, T., Takamura, H., and Matoba, T. 2004. Effects of thermal treat-ment on radical-scavenging activity of single and Mied polyphenolic compounds. *JFS* 69(1):FCT7–FCT23.

Ou, B., Huang, D., Hampsch-Woodill, M., Flanagan, J.A, and Deemer, E.K. 2002. Analysis of antioxidant activities of common vegetables employing oxygen radical absorbance capacity (ORAC) and ferric reducing antioxidant power (FRAP) assays: A comparative study. *J. Agric. Food Chem.* 50:3122–3128.

Pandrangi, S. and Laborde, L.E. 2004. Retention of folate, carotenoids, and other quality char-acteristics in commercially packaged fresh spinach. *J. Food Sci.* 69(9):7002–7007.

Pihlanto, A. 2006. Antioxidative peptides derived from milk proteins. *Int. Dairy J.* 16:1306–1314.

Plust, D. 2008. Antioxidant capacity of some animals tissues, PhD thesis, Westpomeranian Technological University, Szczecin, Poland (in Polish).

Pokorny, J., Yanishlieva, N., and Gordon, M. (Eds.). 2001. *Antioxidants in Food,* CRC Press LLC, Boca Raton, FL, Chapters 1–15.

Porter, N.A., Caldwell, S.E., and Mills, K.A. 1995. Mechanisms of free radical oxidation of unsaturated lipids. *Lipids* 30:277–290.

Prior, R.L. and Cao, G. 1999. In vivo total antioxidant capacity: Comparison of different analytical methods. *Free Radic. Biol. Med.* 27(11/12):1173–1181.

Prior, R.L., Hoang, H., Gu, L., Wu, X., Bacchiocca, D., Howard, L., Hampsch-Woodil, M., Huang, D., Ou, B., Jacob, R. 2003. Assays for hydrophilic and lipophilic antioxidant capacity (ORAC$_{FL}$) of plasma and other biological and food samples. *J. Agric. Food Chem.* 51:3273–3279.

Pucciarelli, D.L. and Grivetti, L.E. 2008. The medicinal use of chocolate in early North America. *Mol. Nutr. Food Res.* 52(10):1215–1227.

Rembiałowska, E, Adamczyk, M., and Hallmann, E. 2003. Sensory value and selected features of nutritive value in apple from organic and conventional production. *Bromat. Chem. Toksykol.* Suppl.:33–39 (in Polish).

Rossi, M., Giussani, E., Morelli, R., Lo Scalzo, R., Nani, R.C., and Torreggiani, D. 2003. Effect of fruit blanching on phenolics and radical scavenging activity of highbush blueberry juice. *Food Res. Int.*, 36:999–1005.

Rossini, K., Noreña, C.P.Z., Cladera-Olivera, F., and Brandelli, A. 2009. Casein peptides with inhibitory activity on lipid oxidation in beef homogenates and mechanically deboned poultry meat. *LWT Food Sci. Technol.*, 42(4):862–867.

Samaranayaka, A.G.P. and Li-Chan, E.C.Y. 2008. Autolysis-assisted production of fish protein hydrolysates with antioxidant properties from Pacific hake (*Merluccius productus*). *Food Chem.* 107(2):768–776.

Serdyńska, A. 2009. Antioxidant activity of combination fish-vegetables, PhD thesis, Westpomeranian Technological University, Szczecin, Poland (in Polish).

Shahidi, F. and Zhong, Y. 2007. Measurement of antioxidant activity in food and biological systems. *ACS Symp. Ser.* 956:36–66.

Sies, H. (Ed.). 1997. *Antioxidants in Disease Mechanisms and Therapy. Advances in Pharmacology*, Vol. 38, Academic Press, San Diego, CA.

Turkmen, N., Sari, F., and Velioglu, Y.S. 2005. The effect of cooking methods on total phenolics and antioxidant activity of selected green vegetables. *Food Chem.* 93:713–718.

Turkmen, N., Sari, F., Poyrazoglu, E.S., and Velioglu, Y.S. 2006. Effects of prolonged heating on antioxidant activity of honey. *Food Chem.* 95:653–657.

Venneria, E., Fanasca, S., Monastra, G., Finotti, E., Ambra, R., Azzi, E., Durazzo, A., Foddai, M.S., and Maiani, G. 2008. Assessment of the nutritive values of genetically modified wheat, corn and tomato crops. *J. Agric. Food Chem.* 56(19):9206–9214.

Wu, X., Beecher, G.R., Holden, J.M., Haytowitz, D.B., Gebhardt, S.E., and Prior, R.L. 2004. Lipophilic and hydrophilic antioxidant capacities of common foods in the United States. *J. Agric. Food Chem.* 52(12):4026–4037.

Wu, J., Hecker, J.G., and Chiamvimonvat, N. 2009. Antioxidant enzyme gene transfer for ischemic diseases. *Adv. Drug Deliv. Rev.* 61:351–363.

Zielińska, D., Wiczkowski, W., and Piskula, M.K. 2008. Determination of the relative contribution of quercetin and its glucosides to the antioxidant capacity of onion by cyclic voltammetry and spectrophotometric methods. *J. Agric. Food Chem.* 56(10):3524–3531.

11 Dietary Lipids and Coronary Heart Disease

Zdzisław Florian Forycki

CONTENTS

11.1 INTRODUCTION

For almost 50 years, epidemiologic studies have suggested a relation between fat consumption and risk of coronary heart disease (CHD). In the early 1950s, controlled feeding studies demonstrated that fat in food increased serum cholesterol concentration in humans. Direct correlation between dietary lipids, total cholesterol levels, and coronary-related mortality was shown in Seven Countries Study (Keys 1980). Following epidemiological studies found that not only increased serum cholesterol but also dietary lipids consumption seemed to lead to risk of CHD in human population. These discoveries led to the classic diet–heart hypothesis, which postulated a primary role of dietary saturated fat and cholesterol as the cause of atherosclerosis and CHD in humans.

Excessive fats in the diet are considered harmful because they lead to cardiovascular disease. It was the initial hypothesis. About 40 years later, the definition of the best diet is still incomplete, and there is no simple answer to the most frequent question: "What is the healthy diet, what should we be eating to prevent a heart infarction?"

11.2 CORONARY HEART DISEASE AND RISK FACTORS

11.2.1 CORONARY HEART DISEASE

The CHD is a multifactorial, mainly a dietary and age-affected disease. Diet as well as obesity, dyslipoproteinemia, diabetes mellitus, hypertension, and smoking belong to the influential risk factors. A dietary factor is, as opposed to age or genetic predilection, changeable and therefore very attractive as a target for intervention.

CHD refers to the failure of arterial (coronary) circulation to supply sufficient flow (może tak lepiej) of blood to the cardiac muscle. Consequently, oxygen and nutritive agents are not provided to the heart muscle (myocardium). This result is due to a narrowing of coronary arteries in process of atherosclerosis. Fatty, fibrous plaques, rich in inflammatory cells, including calcium deposits (atheromatous plaque), narrow the lumen of the coronary arteries and reduce the volume of blood that can flow through them (Figure 11.1). In such cases, angina pectoris (chest pain) is the usual manifestation. While the symptoms and signs of CHD are noted in the advanced state of disease, most individuals show no evidence of disease for years as it progresses before the first onset of symptoms. Plaque formation; stenosis of coronary artery; and, finally, plaque rupture activating local clotting occludes the lumen of the artery. Limitation of blood flow to the heart causes ischemia, and regional interruption of coronary circulation leads to heart muscle damage and heart muscle death (necrosis of myocardium). Sometimes, the rapid occlusion of coronary artery causes not only myocardial infarction (MI) but also leads to sudden cardiac death (SCD).

Primary aims in many clinical or epidemiological studies for progress of atherosclerosis were: total mortality, cardiovascular mortality, sudden cardiovascular death, nonfatal myocardial infarction, stroke, angina pectoris, heart failure, angioplasty, or coronary artery by-pass grafting. Especially, incidence of non-fatal MI and SCD has been useful for estimation of risk factors influence on CHD.

The term "incidence" of CHD refers to the annual diagnosis rate or the number of new cases diagnosed each year. The term "prevalence" usually refers to the estimated population of people who are managing CHD at any given time; 16.8 million American adults have CHD (American Heart Association-AHA 2004).

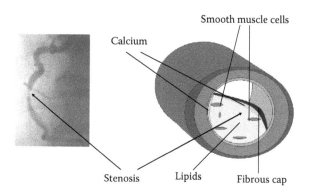

FIGURE 11.1 Coronary angiogram and coronary stenosis (narrowing).

11.2.2 RISK FACTORS

Numerous clinical and epidemiological studies have identified several factors that increase the risk of CHD. Major risk factors are those that research has shown to significantly increase the risk of CHD. AHA identified also factors associated with increased risk and contributing factors of cardiovascular disease, but their significance and prevalence have not yet been precisely determined.

The major risk factors cannot be changed. The race is one of them. African–Americans have higher risk of CHD than Caucasians. Heart disease risk is also higher among Mexican Americans and American–Indians, but lower among Japanese. Heart disease can run in the family. Men have a greater risk of MI than women do. However, after menopause, the women's death rate from coronary disease increases.

The strongest risk factor is increasing age. Over 83% of people who die of CHD are 65 or older. The greater the age, the greater is the chance of developing CHD. The incidence rises from 3 per 1000 men at ages 35–44 to 74 per 1000 at ages 85–94. For women, comparable rates occur 10 years later in life. The gap narrows with advancing age.

One of the most important modifiable risk factors for CHD is cigarette smoking. Other important risk factors as high blood pressure, obesity, or hypercholesterolemia can only be reduced, but smoking could be nearly totally eliminated. Passive smoking "secondhand smoke" coming from another person, increases the risk of heart and blood vessel disease too. A cigarette smoker has two to three times the risk of having a heart attack than a nonsmoker. Quitting smoking is associated with a substantial reduction in risk of all-cause mortality among patients with CHD.

High blood pressure is another major, influential risk factor for heart disease. It is a condition where the pressure of the blood in the arteries is too high, defined as systolic pressure of 140 mmHg or greater and/or diastolic pressure of 90 mmHg or greater. The heart, brain, kidneys, and arteries are affected. There are often no symptoms to signal high blood pressure. Lowering blood pressure by changes in lifestyle or by medication can lower the risk of CHD.

Obesity and high blood cholesterol, together with glucose intolerance and high blood pressure, contribute importantly to the development of multiple risk factors; this complex of multiple metabolic risk factors is called the metabolic syndrome. People who have excess body fat, especially if a lot of it is at the waist, are more likely to develop heart disease.

Diabetes mellitus seriously increases the risk of developing cardiovascular disease. The risk is greater if blood sugar is not well controlled. At least 65% of people with diabetes die of some form of heart or blood vessel disease.

An inactive lifestyle is an underestimated risk factor for CHD. Regular physical activity helps prevent heart disease independently from other risk factor. The more vigorous is the activity, the greater the benefits. However, even moderate-intensity activities help if done regularly and for long term. Physical activity has an influence on other risk factors, particularly on diabetes and obesity, and also helps lower blood pressure in some people.

Stress had been shown to be a significant contributing factor. Under stress, the body can produce hormones that encourage the production of low density lipoproteins (LDL). Stress usually immediately provokes the acute symptoms of CHD; together with other contributing factors as diet and lack of exercise, it belongs to the lifestyle risk factors complex.

11.3 THE ROLE OF LIPIDS IN CORONARY HEART DISEASE: CLINICAL AND EPIDEMIOLOGICAL STUDIES

11.3.1 LIPIDS RICH IN SATURATED FATTY ACIDS

For prevention of CHD, one of the main questions is the optimal quantity and quality of consumed fat. Diet modification, the best proportion of saturated fatty acids (SFA) to polyenoic fatty acids (PUFA) in the diet is the next question. SFA, PUFA, n-3 FA, alfa-linolenic acid (ALA), and trans-fatty acids (TFA) in the diet are permanently on the topics in the discussion about healthy diet. On the other hand, products such as fish oil, eggs, sunflower, soybean, olive oil, walnut, bovine milk, and others were used in the studies title.

Total dietary fat is usually composed of many types of lipids and contains cholesterol, and saturated and unsaturated FA. Since the middle of the previous century, low-fat diets have been intensively recommended as a way to reduce the cholesterol level and the development of atherosclerosis (McGill 1979). The populations, particularly men, with low intakes of saturated and total fat tend to be at lower risk. Saturated fat increases the concentration of LDL in the blood—high LDL is recognized as one of the most important risk factors. The basis for recommendation to decrease the intake of SFA is the evidence that dietary saturated fats generally increase blood cholesterol concentration. The state of the art was the Seven Countries Study (Keys 1980). Coronary death rates across 16 populations in 7 countries were strongly correlated with intake of SFA as percentage of total food energy. The study was an opinion creator for years. The diet–heart idea was born (Gordon 1988).

Although both clinical and epidemiological evidence indicates that diets rich in SFA have negative influence on cardiovascular health, there is still the question of what is the most healthful mixture of the different dietary fats. In other words, how low should SFA in the diet go? Public health recommendation for the U.S. population in 1977 were to reduce the total fat intake to as low as 30% and SFA to 10% of total energy (Krauss et al. 2000).

The results of prospective epidemiological investigation of dietary total fat and CHD for primary prevention have been inconsistent (Ascherio et al. 1996, Boniface and Tefft 2002, Hu and Walter 2002). What is the reason? The randomized studies regarded mostly isocaloric low-fat but high-carbohydrate diet. This constellation reduces high-density lipoprotein (HDL) cholesterol level and increases the level of triacylglycerols (Sacks and Katan 2002). Both changes are independent risk factors of CHD. In this way, consumption of a low fat diet (20% of energy from fat) was subsequently shown to induce atherogenic dyslipidemia (German and Dillard 2004). There are epidemiological suggestions that SFA are harmful when consumed in a mixed diet. This does not necessarily mean that SFA are harmful when consumed in a carbohydrate-restricted diet. The recommendation "as low as possible," appears as to be not profitable. It was recently pointed out that the recommendation for reducing the fat content in the diet below 30% of energy is not supported by experimental evidence, and the advice to decrease total fat intake has failed to have any effect on the prevalence of the metabolic syndrome and CHD. However, some former randomized studies showed correlation (Pietinen et al. 1997, Hu et al. 1999).

The total fat intake and its association with CHD was examined in Women's Health Initiative (Howard et al. 2006). The 8.2% fat reduction had no influence on CHD morbidity. On the other hand, the Nurses' Health Study with 20 years follow up shows a 5% SFA substitution with PUFA and a 42% CHD risk reduction (Oh et al. 2005). This study (78,778 women) is an example concerning dietary trends in the last 20 years in the United States; total fat intake decreased by 26% and intake of saturated fat by 40% from 1980 to 1998. Similar trends are observed in other countries. The studies suggest a relation between specific types of fat and risk of CHD but no evident relation to total fat. However, intakes of longer-chain SFA were associated with a small increase in risk (Hu et al. 1999, Xu et al. 2006). In other words, higher consumption of high-fat dairy products and red meat, the main sources of SFA in the diet, was associated with greater risk of CHD.

The SFA have a bad reputation and have been long recognized as precursors for development of CHD, but what are the details? The lauric, myristic, and palmitic acids definitely raise plasma cholesterol concentrations. On the other hand, stearic acid and medium chain FA seemed to be neutral in this regard. Excess of stearic acid is converted to the monounsaturated oleic acid, not known to be hypercholesterolemic. This property was used by chocolate lovers, because the fat in chocolate is high in stearic acid. However, finally (Hu et al. 1999) with 80,082 women during 14 years showed small, directly proportional relation between dietary stearic acid and CHD.

One of the SFA that has been associated with an increased risk of CHD is myristic acid. Together with lauric acid, it is most strongly related to the average serum cholesterol concentration in the blood. Palmitic acid is dietary relevant because it is present at 22% in bovine milk lipids. The comparison between the effects of palmitic acid in normolipemic humans and effects of dietary laurate and myristate showed that palmitic acid lowers the serum cholesterol. Except a negative effect of dietary SFA on the cardiovascular system, they have pleiotropic influences on human health. They have multifocal, regulatory functions as antiviral and antibacterial agents. The antitumor activity, immunomodulative, and homeostatic functions were experimentally confirmed.

Actually, research on how specific SFA contribute to CHD and on the role played by each specific SFA is not sufficient to make global recommendations for everybody to reduce SFA in his diet. The persons differ in their response to dietary fat. The Nutrition Committee of AHA emphasized that more clinical studies with cardiovascular endpoints are needed to evaluate the effects of fat quantity. There is a lack of knowledge of how low SFA intake can be tolerated without the risk of potentially deleterious health outcomes. The question about an appropriate amount remains.

11.3.2 Lipids Rich in Monoenoic Fatty Acids

Oleic acid in olive oil is probably the most consumed monoenoic fatty acid (MUFA). Epidemiological studies suggest that higher proportion of MUFA in the diet is linked with reduction in the risk of CHD (Hu et al. 2002). The results of other, earlier studies were neutral or showed an inverse relation between food consumption rich in MUFA and risk of CHD. To achieve a possible benefit, MUFA is to replace a similar amount of saturated fat and not increase the total energy value of the daily diet (Kris-Etherton 1999).

However, evaluation of single nutrients is limited. It concerns not only MUFA and has to be postulated generally. Food components of the diet engage in interactions and are complex. It is a little easier to assess food groups in relation to the occurrence of a disease. Therefore, contemporary trials focus the attention from the evaluation of single nutrients to the dietary pattern as a whole. From this point of view, MUFA can be well integrated in the Mediterranean diet (MedD) (see below).

11.3.3 Lipids Rich in PUFA

In humans, *n*-3 FA and linoleic acid (LA) are essential. It means they cannot be synthesized from saturated or monoenoic FA and must be supplied with the diet. Nutritive lack of them, seldom in case of inadequate parenteral nutrition, leads to a deficiency syndrome, sensory neuropathy, and visual symptoms.

ALA is an intermediate-chain *n*-3 PUFA that is often overshadowed by the more famous long-chain representatives of the *n*-3 family, namely eicosapentaenoic (EPA) and docosahexaenoic (DHA) acids. Recently, significant advance was made in respect to setting recommendation for *n*-3 FA for the prevention of CHD (Harris et al. 2009). Numerous epidemiological observations and interventional studies for prevention have shown a beneficial association. EPA and DHA are marine–derived and ALA is mainly a plant-derived FA. Due to the different origin, both have to be discussed separately.

ALA is one of the components of the MedD. After ingestion (flaxseed, walnuts, and canola oil), ALA is in some percentage converted into EPA and DHA. The benefit is not only expected to be in this way. ALA has probably direct antiarrhythmic properties independent from EPA and DHA (Albert et al. 2005, Mozaffarian et al. 2005a). In an animal model, the risk of cardiac arrhythmia is reduced, particularly ventricular fibrillation is lower. ALA, probably more than marine derived oils, increases the electrical threshold of ventricular fibrillation. The reduction was observed in animals fed red meat supplemented with fish oil and in those with canola oil. The antiarrhythmic effect was greater in the canola oil than in the fish oil group. Moreover, favorable effect of ALA on the cholesterol level would predict positive influence on CHD risk.

Further benefits are associated with an antithrombosis influence, an important function in the pathogenesis of CHD. Inverse correlation with ALA consumption showed blood pressure (Miura et al. 2008). ALA may exert favorable influence on coronary endothelial functions, particularly as an anti-inflammatory effect.

The benefit of ALA on the risk of fatal CHD has been observed in most human studies (Ascherio et al. 1996, de Lorgeril et al. 1999, Djousse et al. 2001), but single datum could not confirm it (Wang et al. 2006). By replacing butter by oil and margarine rich in ALA, platelet aggregation was significantly diminished, and suggestion about antithrombotic properties of FA and in consequence the inhibitory effect on the development of CHD was confirmed.

The main clinical dietary effort with ALA was focused on the reduction of SCD (Albert et al. 2005). SCD is in *ca*. 90% of the cases dependent on ventricular arrhythmia. The prospective data published by Albert shows that increasing dietary intake of ALA may reduce the risk of SCD but not other types of fatal CHD. The ALA and EPA + DPA have a common pathway. One would expect to observe a weaker association between ALA intake and SCD among those who consumed more of marine-derived *n*-3 acids, but the benefit of marine-derived *n*-3 was not affected by low or high consumption of ALA. An interaction between PUFAs is possible (Mozaffarian et al. 2005a). In contrast, the data of Albert suggested that increased ALA intake was associated with a lower risk of SCD, even among women with a higher intake of EPA + DHA. The additive effect of the PUFA complex is probable but needs confirmation.

Not only SCD showed a strong inverse association with a diet rich in ALA. Non fatal acute MI in some studies was also reduced (Laaksonen et al. 2005). However, these trials were not specifically designed to test the effects of ALA.

Recently, Campos et al. (2008) postulated the cardiovascular protection of ALA in Costa Rica. Greater ALA level, assessed by gas chromatography in adipose tissue and a validated food frequency questionnaire, was associated with lower risk of myocardial infarction. Consumption of ALA was protective, particularly evident among subjects with low intakes. The relationship between ALA and MI was nonlinear; the risk did not decrease with intakes >1.79 g/day. It is meaningful, in this study in Costa Rican population, that fish or DPA and DHA did not modify the observed association.

Health Professionals Follow-up Study showed heart infarct reduction associated with a 1% increase in linolenic intake (expressed as % of total energy). In the former, Multiple Risk Factor Intervention Trial, the highest ALA intake determined the lowest CHD mortality (Dolecek 1992). However, in this study, the investigation of the relation between ALA intake and CHD was not the main focus. In the Nurses' Health Study, Oh et al. (2005) found a 54% lower risk of fatal CHD among women who consumed ALA in the form of oil and vinegar salad dressing frequently (5 and more times a week) compared with those who consumed ALA very rarely (less than once a month). As final statement: the beneficial effect of ALA in CHD is not questionable (Harris 2005).

The protective effects of marine *n*-3 FA on a cardiovascular system are complex. The fish oil may have anti-arrhythmic effects, and reduces triacylglycerol levels and platelet aggregation. Moreover, *n*-3 FA supplementation improves endothelial-dependent vasomotor function.

Interesting and the most promising nutritional intervention is fish consumption (heart-healthy diet) (Kris-Etherton 2002). The observation of low rate of CHD in populations with very high intake of fish, such as Eskimos, Alaskan native Americans, and Japanese, suggests that fish oil may be protective against atherosclerosis. Japanese had 2-fold higher levels of marine-derived *n*-3 FA than whites and

Japanese Americans in the United States. It may contribute to lower the burden of atherosclerosis in Japanese (Sekikawa et al. 2008). Genetic determinants are rather, in this context, unlikely. Prospective cohort studies have found an inverse association between fish consumption and risk of cardiovascular morbidity and mortality in diverse populations.

The Western Electric Study (Shekelle et al. 1981) and Seven Countries Study showed that men who consumed 30 g or more of fish per day had a 40% or 50% lower CHD mortality than men who rarely ate fish. The fish consumption is probably more protective against fatal CHD than nonfatal MI (Ka et al. 2004, Albert et al. 2005). A similar relation was observed among women (Nurses' Health Study). During 16 years of follow-up, an association between higher consumption of fish and n-3 FA and risk of CHD was revealed in 84,688 healthy women, particularly CHD deaths (Hu et al. 2002).

PUFA seems more important than fat quantity in the reduction of cardiovascular mortality. Bucher et al. (2002) published a meta-analysis of randomized controlled trials and association between n-3 PUFA and risk of CHD. A diet rich in PUFA as well as dietary supplementation with PUFA reduced overall mortality, mortality due to MI, and SCD in patients with CHD. The Health Professionals Follow-up Study seems to be less spectacular. There was a non-significant trend for a reduction in risk for fatal CHD with increasing fish consumption (Ascherio et al. 1996).

More recently, Hooper's et al. (2006) systematic review about the risk and benefits of n-3 FA for mortality, cardiovascular disease, and cancer showed no clear effect. 41 cohort studies were analyzed and the results were inconclusive. Composite end points, mixing of primary and secondary prevention, heterogeneity, and exclusion of 108 potential cohort that have no n-3 assessment are probably responsible for the confusion. These uncertainties reflect difficulties related both to the imprecision and variability in the assessment of fish-oil content in the diet and many variables that may influence the development of CHD. Despite these inconclusive results, the majority of the evidence support positive effect on risk of CHD and leave no doubt about the beneficial effects of PUFA in fish diet in primary prevention. In conclusion, AHA and other institutions recommend consumption of fatty fish at least two times per week (Gebauer et al. 2006).

The secondary prevention study with 2033 MI survivors (Burr et al. 1989) showed that subjects who received advice for fish consumption had a significant reduction in all-cause mortality of 29% after 2 years. Meta-analysis (Mozaffarian et al. 2005) with 30 clinical trials has shown that fish-oil consumption reduces heart rate, thus providing evidence that marine PUFA affects cardiac electrophysiology. Lower heart rate is known as predictor of lower risk to cardiovascular death. Epidemiological and clinical studies have shown that intake of EPA and DHA from fish reduces cardiac mortality and sudden death. The problem is the availability of fish and other marine products worldwide. The intake of long-chain n-3 "marine" FA is low in many countries.

ALA, more available and consumed as soybean, canola, or flaxseed oil, could be a viable cardioprotective alternative. ALA is present in the MedD, however, the greatest exponent is olive oil with MUFA as a prime component.

The MedD is plant-based with natural vegetable oils (olive oil) as the basis and foods such as fruits, vegetables, nuts, and grains. This diet is characterized by a low intake of total and saturated fats and increased intake of plant oils (MUFA and plant-PUFA), marine PUFA, and is not intended primarily to reduce cholesterol intake (eggs are not excluded).

One of the first to give information about the beneficial influence of the MedD was Seven Countries Study (Keys 1980). The lowest CHD rates were in Crete, the highest in Finland, and both regions had the same amount of total fat intake. High fat intake in Crete (about 40%) was mainly olive oil (MUFA), in Finland it was dairy fat (SFA). The type of fat is important, suggested Keys. Meta-analysis (Sofi et al. 2008) proved a relation between adherence to the MedD and health status. The cumulative analysis among eight cohorts (514,816 subjects) showed significantly reduced risk of cardiovascular and global mortality in population with greater adherence to the MedD.

The last published contribution to this topic is Nurses' Health Study (Fung et al. 2009) with 74,886 women. Fung computed an alternate MedD score from self-reported data collection, which showed significantly lower cardiovascular mortality (29% reduced risk of MI) among women in the top quintile of a MedD score.

In secondary prevention, the Lyon Diet Heart Study (de Lorgeril et al. 1999) showed reduced recurrence of CHD in the group of MedD compared with a prudent Western-type diet. The study suggests an impressive protective effect of the MedD, and the authors indicate the prominent role of ALA in this dietary pattern.

11.3.4 Trans-Fatty Acids

Previous (Keys 1980) and recent evidence (Mozaffarian et al. 2006) indicates that TFA increase the risk of CHD. In comparison with an equal amount of energy from SFA or cis PUFA, the consumption of TFA raised the levels of LDL and total cholesterol and reduced the levels of HDL cholesterol. Because TFA increase LDL cholesterol to levels similar to those induced by SFA and also decrease HDL cholesterol, the net effect of TFA on the ratio of LDL to HDL cholesterol is approximately double that of SFA. Moreover, TFA increases the blood level of triacylglycerols and lipoproteins, which are also powerful predictors of CHD.

Parallel to changes in lipids profile, inflammation is promoted by consumption of TFA. Increased activity of the tumor necrosis factor (TNF), interleukin-6, and C-reactive protein CRP are observed. In addition, Clifton et al. (2004) showed a positive association between levels of TFA in adipose tissue and risk of nonfatal MI. These pathophysiological effects of TFA conform to epidemiological studies. The Seven Countries Study not only confirmed that SFA was strongly correlated with the risk of death from CHD but also showed that consumption of TFA was correlated with the risk of death from CHD (Kromhout et al. 1995). TFA consumption is recognized as a significant risk of CHD more than any other macronutrient (Clarke and Lewington 2006). In a meta-analysis of three prospective cohort studies involving 139,161 subjects, a 2% increased intake of TFA was associated with a nearly 23% increase in the incidence of CHD (Ascherio et al. 1996, Pietinen et al. 1997, Oh et al. 2005).

The results are convincing. Experimental studies and dietary trials, particularly prospective studies, create a basis of evidence about the harmfulness of TFA from partially hydrogenated oils in diet. Gerberding (2009) notes that, gram for gram, TFA are more potent than SFA in increasing the risk for heart disease, and if exposure is eliminated, approximately 30,000 to 100,000 U.S. deaths related to heart disease could be prevented each year.

The question about SCD and nutrition with TFA is not yet finally cleared. Lemaitre et al. (2006) showed that TFA in the diet might affect the risk of SCD as well. Roberts et al. (1995a) investigated subjects after SCD and found no evidence of a relation between trans-isomers of oleic and linoleic acids in adipose tissue and SCD. In conclusion, metabolic and epidemiological studies indicate an adverse effect of TFA on the risk of CHD. Moreover, many scientists agree that the amount of TFA should be labeled on foods. This action is actually in process in the United States and Europe.

Small amounts of TFA of natural origin are present in dairy products of ruminants. Prospective studies evaluating the relation between the intake of TFA from ruminants and risk of CHD showed absence of high risk (Mozaffarian et al. 2006). The lower contents of TFA in products from ruminants are probably an explanation for the non-adverse effect. There is perhaps a link to the old Paracelsus sentence: "All things are poison and nothing is without poison, only the dose permits something not to be poisonous."

11.4 ENDOGENOUS AND DIETARY CHOLESTEROL

Elevated cholesterol in blood is an important component of metabolic syndromes and main risk factor of CHD. Cholesterol, a substrate in humans for many hormones, is abundant in youth and in adulthood, and has several evolutionary advantages. On the contrary, in older subjects, this excess has a detrimental effect on the metabolic state and creates a cardiovascular risk. The relation between fat intake and cholesterol concentration in serum has been recognized to have a link to CHD for nearly 50 years (Keys and Parlin 1966). The effect of diet on CHD can be mediated through multiple biochemical pathways, particularly by high blood cholesterol level (hypercholesterolemia). Many epidemiological studies found that increased total cholesterol or LDL level are recognized risks for developing atherosclerosis and incidence of CHD. This relation is significant; the higher blood cholesterol, the higher incidence of fatal SCD or nonfatal heart infarction.

It has a link to the statement about the influence of diet on serum cholesterol concentration. Most reviews of the lipids–heart problem deal with change of serum cholesterol concentration by alteration of more than one dietary component and the resulting effect (or lack of effect) on CHD incidence. Dietary intakes of SFA, cholesterol, and total fat were each associated with highly significant increases in blood total cholesterol (Clarke et al. 1997). Serum cholesterol concentration is not a linear function of dietary cholesterol intake, and individual predisposition determines the serum cholesterol response to changes in dietary lipids. Foods rich in SFA generally contain substantial amounts of dietary cholesterol and therefore it is difficult to differentiate the influence of both risk factors on the development of CHD. In this context, cholesterol has a special position. Apart from dietary intake, the human liver produces *ca.* 1 g cholesterol daily. Cholesterol synthesis in humans is genetically regulated. Both processes, cholesterol synthesis and cholesterol absorption in the gut, interfere permanently, and the normal HDL- as well as LDL-cholesterol levels stay in balance. Demonstration of the adverse effects of levels of LDL much in excess of the normal range is afforded by observation in familial hypercholesterolemia (FHC). Homozygous individuals develop atherosclerosis in childhood and die early. Such cases are rare and the much more common heterozygous form with lesser degrees of hypercholesterolemia is observed. Levels of LDL-cholesterol typical of industrial population are very much lower than of FHC heterozygote but above physiological levels and sufficiently high for development of CHD. In individuals with oversized cholesterol synthesis, low cholesterol diet may be insufficient for balancing LDL–HDL in the normal range. Only 10%–20% reduction of cholesterol serum level may be expected in low cholesterol diet.

One of the dietary recommendations for CHD prevention is to limit egg consumption (one egg contains *ca.* 200 mg cholesterol) (Eckel 2008). Eggs are in the prudent Western-type diet a major source of dietary cholesterol. Several epidemiological studies found no relation between egg consumption and risk of CHD, however higher egg consumption elevates the ratio of total to HDL cholesterol and the concentration of total and LDL cholesterol. The recommendation to limit cholesterol intake, particularly in individuals with CHD risk, is still valid (Djousse and Gaziano 2008).

Some investigators have extrapolated the reduction in CHD that might be expected from reduced levels of serum cholesterol concentration. The reduction may be realized in two ways: lower intake of SFA, cholesterol, and FTA, or cholesterol-lowering drug intervention. In prevention trials using diet, it was found that in general, the greater the cholesterol reduction, the greater the decrease in CHD risk. The 10% reduction of cholesterol corresponded with a 15% reduction in risk. In corresponding analysis of randomized trials of drug cholesterol-lowering therapy, more expressed relations were reported.

Is every depletion of cholesterol concentration in serum due to intervention beneficial and induce CHD reduction? Generally, dietary modifications are safe, with limited efficiency, and not expensive. Drug therapy with statins appears worldwide as salutary, is more effective, and relatively safe but cost intensive. Contemporary efforts for improving the cholesterol-lowering are disappointing. Recently tested Torcetrapib (transfer of cholesterol from HDL to LDL) showed an adverse imbalance of mortality and cardiovascular events compared with placebo. Another example was early fibrates used for hyperlipidemia. They seemed to increase mortality from other causes. The benefit from this reduction in cholesterol concentration in blood was outweighed by adverse effects of the drug themselves.

In conclusion, decreasing the concentration of blood cholesterol with no other dietary or medical interventions gives no guarantee for clinical improvement of CHD patients.

11.5 THE CLINICIAN'S VIEW ON THE EFFECTIVENESS OF DIETARY SUPPLEMENTS

The fatty fish diet was partly substituted and applied in the form of EPA/DHA concentrate as a supplement. In this way, EPA/DHA may be exactly dosaged, even in high range for use in intervention clinical trials. The advantage of a high dosage of both FA, difficult to achieve with standard fish diets, was needed for more effect and better differentiation with placebo in antiarrhythmic studies. However, supplementation is easy for application, but is "nutrition" the proper expression in such a case? Are EPA/DHA "dietary" lipids or kind of medication? Where is the limit for using the term "treatment"? Nutrition as the pill in the pocket? This difficulty is not finally cleared.

An interventional study has evaluated whether fish-oil supplementation reduces coronary mortality among 11,324 MI patients (GISSI 1999). The n-3 FA supplementation resulted in a 10%–15% reduction in the main endpoints (death, nonfatal MI, and stroke). An important observation was the decrease in cardiovascular death, especially SCD. The GISSI study provides support for a prevention as well as therapeutic role of fish oil in the "treatment" of MI patients. Furthermore, the GISSI trial supports the hypothesis about suggested antiarrhythmic action of EPA and DHA.

Another advantage of supplementation (capsules and fish oil) should be mentioned. As supplement they are free, not contaminated ingredients. Over the years, frequently, methylmercury and others chemical pollutants were found in certain fish. This fact may diminish the health benefit and mask the salutary effects of PUFA in a fish "heart-healthy diet" (Weaver et al. 2008).

On the other hand, the PUFA supplementation is probably not enough to compensate the benefits of the "healthy fish diet." The "fish diet" means usually a replacement of SFA (red meat and milk products) with more frequent fish consumption. Related to risk of CHD, synergistic effects of two beneficial factors are expected; reduction of SFA (see above) and positive influence of PUFA. Contrary to fish diet, supplementation may suggest a prudent Western diet and probably lack of the benefits related to SFA reduction.

The consideration may be useful to explain the conflicting data on n-3 supplementation in cardiac arrhythmia. For years, many trials suggested that n-3 PUFA provide cardiovascular protection and prevent arrhythmias (Siebert et al. 1993, GISSI 1999). Number of randomized trials has failed to show a protective effect on PUFAs against arrhythmias (London et al. 2007). Lately published trials and meta-analysis conducted on 1148 patients with implanted defibrillator did not support this opinion and showed no effect of n-3 supplementation on cardiac arrhythmia in patients (Brouwer et al. 2006). However, a synthesis of 12 studies with 32,779 patients (León et al. 2008) treated with supplements showed significant reduction in death from cardiac causes but no effect on arrhythmias. The small meta-analysis (Jenkins et al. 2008) showed that fish-oil supplementation failed to protect against ventricular arrhythmia in implantable cardiodefibrillator patients.

Hooper's review showed that n-3 fats do not have any clear effect on total mortality and combined cardiovascular events (reviewed were 44 trials but 36 with capsules and 6 with fish oil and only 2 with oily fish). In conclusion, the advantage of n-3 FA supplementation alone is probably less relevant than a fatty fish diet. The last word comes from the OMEGA study (Senges 2009). Supplements, EPA 380 mg and DHA 1 g/day, have no additional benefits in patients who have suffered an acute MI. The patients had optimal medical care and the cardiac event rate became very low. The n-3 FA supplements have no potential to improve the cardiac status. Probably, optimal MI care eliminates the benefits of n-3 FA intake.

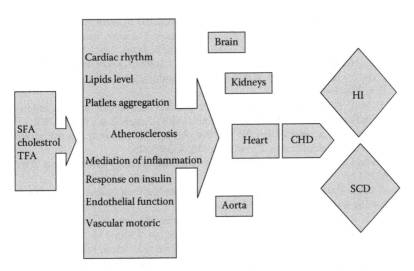

FIGURE 11.2 Dietary lipids (SFA, TFA, cholesterol) and possible ways to coronary heart disease (CHD) and endpoint: heart infarction (HI) and sudden cardiac death (SCD).

11.6 FINAL REMARKS

Lipids (SFA, cholesterol, and TFA) have more than other dietary components an impact on human health, particularly on the incidence of cardiovascular disease (Figure 11.2). The limitation of total fat in the diet brought a relative benefit only. In the last recommendation, FDA limited the amount of lipids in diet to 30% of the total diet energy per day. The qualitative modification of dietary lipids recommendation appears essential. Replacement of SFA with carbohydrates brought hardly any beneficial effect. Only replacing SFA with MUFA and PUFA reduced significantly the risk of CHD. Recent diet recommendation has been made for ALA, PUFA, EPA-DHA supplements, and fish consumption for the prevention of CHD.

Antiarrhythmic action of EPA and DHA supplement in humans was not consistent in clinical trials. They are effective for CHD prevention but not for "treatment" of arrhythmias.

Hydrogenated TFA of nonruminant origin should be eliminated or minimalized because even a low amount shows a detrimental effect on the cardiovascular system. TFA from ruminant are not recognized as harmful, but a final opinion cannot be actually defined. Contrary to TFA and SFA, the MUFA and PUFA in the diet are effective in reducing CHD incidence and are favored particularly in the MedD.

REFERENCES

Albert, C. M. et al. 2005. Dietary alpha-linolenic acid intake and risk of sudden cardiac death and coronary heart disease. *Circulation* 112, 21, 3232–3238.

Ascherio, A. et al. 1996. Dietary fat and risk of coronary heart disease in men: Cohort follow up study in the United States. *BMJ* 313, 7049, 84–90.

Boniface, D. R. and Tefft, M. E. 2002. Dietary fats and 16-year coronary heart disease mortality in a cohort of men and women in Great Britain. *Eur J C Nutr* 56, 8, 786–792.

Brouwer, I. A. et al. 2006. Effect of fish oil on ventricular tachyarrhythmia and death in patients with implantable cardioverter defibrillators: The study on omega-3 fatty acids and ventricular arrhythmia (SOFA) randomized trial. *JAMA* 295, 22, 2613–2619.

Bucher, H. C., Hengstler, P., Schindler, C., and Meier, G. 2002. *N*-3 polyunsaturated fatty acids in coronary heart disease: A meta-analysis of randomized controlled trials. *Am J Med* 112, 4, 298–304.

Burr, M. L. et al. 1989. Effects of changes in fat, fish, and fibre intakes on death and myocardial reinfarction: Diet and reinfarction trial (DART). *Lancet* 2, 8666, 757–761.

Campos, H., Baylin, A., and Willett, W. C. 2008. Alpha-Linolenic acid and risk of nonfatal acute myocardial infarction. *Circulation* 118, 4, 339–345.

Clarke, R. and Lewington, S. 2006. Trans fatty acids and coronary heart disease. *BMJ* 333, 7561, 214.

Clarke, R. et al. 1997. Dietary lipids and blood cholesterol: Quantitative meta-analysis of metabolic ward studies. *BMJ* 314, 7074, 112.

Clifton, P. M., Keogh, J. B., and Noakes M. 2004. Trans fatty acids in adipose tissue and the food supply are associated with myocardial infarction. *J Nutr* 134, 4, 874–879.

de Lorgeril, M. et al. 1999. Mediterranean diet, traditional risk factors, and the rate of cardiovascular complications after myocardial infarction: Final report of the Lyon Diet Heart Study. *Circulation* 99, 6, 779–785.

Djousse, L. and Gaziano, J. M. 2008. Egg consumption and risk of heart failure in the physicians' health study. *Circulation* 117, 4, 512–516.

Djousse, L. et al. 2001. Relation between dietary linolenic acid and coronary artery disease in the National Heart, Lung, and Blood Institute family heart study. *Am J Clin Nutr* 74, 5, 612–619.

Dolecek, T. A. 1992 Epidemiological evidence of relationships between dietary polyunsaturated fatty acids and mortality in the multiple risk factor intervention trial. *Proc Soc Exp Biol Med* 2001, 177–182.

Eckel, R. H. 2008. Egg consumption in relation to cardiovascular disease and mortality: The story gets more complex. *Am J Clin Nutr* 87, 4, 799–800.

Fung, T. T. et al. 2009. Mediterranean diet and incidence of and mortality from coronary heart disease and stroke in women. *Circulation* 119, 8, 1093–1100.

Gebauer, S. K., Psota, T. L., Harris, W. H., and Kris-Etherton, P. M. 2006. *n*-3 Fatty acid dietary recommendations and food sources to achieve essentiality and cardiovascular benefits. *Am J Clin Nutr* 83, 6, 1526–1535.

Gerberding, J. L. 2009 Safer fats for healthier hearts: The case for eliminating dietary artificial trans fat intake. *Ann Intern Med* 151, 2, 137–138.

German, J. B. and Dillard, C. J. 2004. Saturated fats: What dietary intake? *Am J Clin Nutr* 80, 3, 550–559.

Gordon, T. 1988. The diet-heart idea: Outline of a history. *Am J Epidemiol* 127, 220–225.

Gruppo Italiano per lo Studio della Sopravvivenza nell'Infarto miocardico (GISSI). 1999. Dietary supplementation with *n*-3 polyunsaturated fatty acids and vitamin E after myocardial infarction: Results from the GISSI-Prevenzione trial. *Lancet* 354, 447–455.

Harris, W. S. 2005. Alpha-linolenic acid: A gift from the land? *Circulation* 111, 22, 2872–2874.

Harris, W. S. et al. 2009. AHA Science Advisory. Omega-6 fatty acids and risk for cardiovascular disease. *Circulation* 119, 0.

Hooper, L. et al. 2006. Risks and benefits of omega 3 fats for mortality, cardiovascular disease, and cancer: Systematic review. *BMJ* 332, 7544, 752–760.

Howard, B. V. et al. 2006. Low-fat dietary pattern and risk of cardiovascular disease: The women's health initiative randomized controlled dietary modification trial. *JAMA* 295, 6, 655–666.

Hu, F. B. and Walter, C. W. 2002. Optimal diets for prevention of coronary heart disease. *JAMA* 288, 20, 2569–2578.

Hu, F. B. et al. 1999. Dietary saturated fats and their food sources in relation to the risk of coronary heart disease in women. *Am J Clin Nutr* 70, 6, 1001–1008.

Hu, F. B. et al. 2002. Fish and omega-3 fatty acid intake and risk of coronary heart disease in women. *JAMA* 287, 14, 1815–1821.

Jenkins, D. J. A. et al. 2008. Fish-oil supplementation in patients with implantable cardioverter defibrillators: A meta-analysis. *CMAJ (Canadian Medical Association Journal)* 178, 2, 157–164.

Ka, H. et al. 2004. Accumulated evidence on fish consumption and coronary heart disease mortality: A meta-analysis of cohort studies. *Circulation* 109, 22, 2705–2711.

Keys, A. 1980. *Seven Countries: A Multivariate Analysis of Death and Coronary Heart Disease*. Cambridge, MA: Harvard University Press.

Keys, A. and Parlin, W. 1966. Serum response to changes in dietary lipids. *Am J Clin Nutr* 19, 175–181.

Krauss, R. M. et al. 2000. AHA Dietary guidelines revision 2000: A statement for healthcare professionals from the Nutrition Committee of the American Heart Association. *Circulation* 102, 2284.

Kris-Etherton, P. M. 1999. Monounsaturated fatty acids and risk of cardiovascular disease. *Circulation* 100, 11, 1253–1258.

Kris-Etherton, P. M. 2002. Fish consumption, fish oil, omega-3 fatty acids, and cardiovascular disease. *Circulation* 106, 2747.

Kromhout, D. et al. 1995. Dietary saturated and trans fatty acids and cholesterol and 25-year mortality from coronary heart disease: The seven countries study. *Prev Med* 24, 3, 308–315.

Laaksonen, D. E. et al. 2005. Prediction of cardiovascular mortality in middle-aged men by dietary and serum linoleic and polyunsaturated fatty acids. *Arch Intern Med* 165, 2, 193–199.

Lemaitre, R. N. et al. 2006. Trans-fatty acids and sudden cardiac death. *Atheroscler Suppl* 7, 2, 13–15.

León, H. et al. 2008. Effect of fish oil on arrhythmias and mortality: Systematic review. *BMJ* 337, a2931.

London, B. et al. 2007. Omega-3 fatty acids and cardiac arrhythmias: Prior studies and recommendations for future research: A Report from the National Heart, Lung, and Blood Institute and Office of Dietary Supplements Omega-3 Fatty Acids and Their Role in Cardiac Arrhythmogenesis Workshop. *Circulation* 116, 10, e320–e335.

McGill, H. C. 1979. The relationship of dietary cholesterol to serum cholesterol concentration and to atherosclerosis in man. *Am J Clin Nutr* 32, 2664–2702.

Miura, K. et al. 2008. Relationship of dietary linoleic acid to blood pressure. *Hypertension* 52, 2, 408–414.

Mozaffarian, D. et al. 2005a. Interplay between different polyunsaturated fatty acids and risk of coronary heart disease in men. *Circulation* 111, 2, 157–164.

Mozaffarian, D. et al. 2005b. Effect of fish oil on heart rate in humans: A meta-analysis of randomized controlled trials. *Circulation* 112, 13, 1945–1952.

Mozaffarian, D. et al. 2006. Trans fatty acids and cardiovascular disease. *NEJM* 354, 15, 1601–1613.

Oh, K. et al. 2005. Dietary fat intake and risk of coronary heart disease in women: 20 Years of follow-up of the Nurses' Health Study. *Am J Epidemiol* 161, 7, 672–679.

Pietinen, P. et al. 1997. Intake of fatty acids and risk of coronary heart disease in a cohort of Finnish men: The alpha-tocopherol, beta-carotene cancer prevention study. *Am J Epidemiol* 145, 10, 876–887.

Roberts, T. L. et al. 1995a. trans-Isomers of oleic and linoleic acids in adipose tissue and sudden cardiac death. *Lancet* 345, 8945, 278–282.

Sacks, F. M. and Katan, M. 2002. Randomized clinical trials on the effects of dietary fat and carbohydrate on plasma lipoproteins and cardiovascular disease. *Am J Med* December 30, 113, Suppl 9B, 13S–24S.

Sekikawa, A. J. et al. 2008. Marine-derived *n*-3 fatty acids and atherosclerosis in Japanese, Japanese-American, and white men: A cross-sectional study. *J Am Coll Cardiol* 52, 6, 417–424.

Senges, J. 2009. Randomised trial of omega-3 fatty acids on top of modern therapy after acute MI. The OMEGA – Trial. *58th Annual Scientific Session of the American College of Cardiology*. March, Orlando, FL.

Shekelle, R. B. et al. 1981. Diet, serum cholesterol, and death from coronary heart disease. The Western Electric study. *NEJM*, 304, 65–70.

Siebert, B. D., McLennan, P. L., Woodhouse, J. A., and Charnock, J. S. 1993. Cardiac arrhythmia in rats in response to dietary *n*-3 fatty acids from red meat, fish oil and canola oil. *Nutr Res* 13, 1407–1418.

Sofi, F. et al. 2008. Adherence to Mediterranean diet and health status: Meta-analysis. *BMJ* 337, a1344.

Wang, C. et al. 2006. *n*-3 Fatty acids from fish or fish-oil supplements, but not alpha-linolenic acid, benefit cardiovascular disease outcomes in primary- and secondary-prevention studies: A systematic review. *Am J Clin Nutr* 84, 1, 5–17.

Weaver, K. L. et al. 2008. The content of favorable and unfavorable polyunsaturated fatty acids found in commonly eaten fish. *J Am Diet Assoc* 108, 7, 1178–1185.

Xu, J. et al. 2006. Dietary fat intake and risk of coronary heart disease: The strong heart study. *Am J Clin Nutr* 84, 4, 894–902.

12 Role of Lipids in Children's Health and Development

Grażyna Sikorska-Wiśniewska

CONTENTS

12.1 THE ROLE OF FAT IN CHILDREN'S DIET

12.1.1 LIPIDS AS A SOURCE OF ENERGY FOR GROWTH AND DEVELOPMENT

The intensively growing organisms of neonates, infants, and children have great energy requirements. They vary considerably according to age, gender, body size, genetically conditioned rate of growth, and physical activity (Table 12.1).

Lipids, being the main source of concentrated energy in foods, deliver about 40% of energy consumed by an infant and baby. This amount of fat is contained in the human milk and in most infant formulas. Fats are the main energy source in food, since in the process of being broken down in the organism they produce more than twice the energy provided by proteins and carbohydrates. Fatty acids (FAs) are used as a source of energy by the majority of human tissues except erythrocytes and central nervous system cells, which utilize glucose for their nutritional needs. Improper fat intake in children may contribute to disturbances in somatic, biological, and psychomotor development. It means, that there is a risk that the baby may not achieve the expected weight and height, its maturation may be delayed (dentition, changing body proportions, sexual maturation), and also may not attain proper achievements in gross motor and fine motor skills, emotional and social development, and speech.

12.1.2 Dietary Fats as a Source of Bodybuilding Materials

TABLE 12.1

Daily Energy Demands According to Age and Gender

Age	kcal/kg
Preterm babies	75
Term babies	100–200
Infants	95–115
Preschool children	90–95
7–10 years children	75
Girls 11–14 years	55
Boys 11–14 years	65
Girls 15–18 years	40
Boys 15–18 years	60

Source: Lissauer, T. and Clayden, G., *Illustrated Textbook of Paediatrics*, 3rd rev. edn., Elsevier Health Sciences, London, U.K., 2007. With permission.

Fatty tissue is an indispensable component of the human organism. Its amount, similar to muscles and bone tissue, depends on age and sex. It determines the body's structure and appearance. Fat constitutes approximately 10% of body mass. In obese children its amount may reach 25%–30% of total body mass. Lipids serve as an energy reserve. They build a subcutaneous tissue, serve as the constituent of body cavities, and surround interior organs such as heart, liver, spleen, intestines, and kidneys. The adipose tissue occurs in mammals in two forms. Brown adipose is especially abundant in newborns; it makes up about 5% of the body mass, and diminishes with age. It is located mainly on the back and along the upper half of the spine. Its primary function is to generate body heat; its presence protects immature infants against hypothermia, which is the major risk of death in preterm neonates. The brown fat serves also as an amortization layer, preserving deeply located organs against harmful mechanic factors. Many researchers suggest also, that brown fat plays an important role in regulating body weight and that higher level of brown fat prevents obesity. White adipose tissue is located under the skin and deeply in the abdominal cavity. Its crucial role is to store energy to be used in case of increase of needs which exceed the actual supply with the diet.

FA residues present in food lipids serve as building materials in the body. They are necessary to form cell membranes and organelles, regulating the permeability of these structures to nutritional factors. Cholesterol is used as a precursor of steroid hormones and in bile acid synthesis.

12.1.3 THE ROLE OF LIPIDS IN THE ABSORPTION OF FAT-SOLUBLE VITAMINS AND CAROTENOIDS

A, D, E, K vitamins and carotenoids are all nonpolar lipids with extremely low solubility in aqueous media. They are absorbed from the small intestine, along with dietary fat. The bile salts and micelle formation play a crucial role in the absorption of nonpolar lipids. In cases of improper fat absorption resulting from various diseases, e.g., prolonged intestinal infections, Crohn's disease, ulcerative colitis, cystic fibrosis, and celiac disease, the affected child may lack these vitamins and present clinical symptoms of their deficiency. Malabsorption of vitamin D in adult celiac patients seems to be proportional to the degree of steatorrhea (Thompson et al., 1966). Severe malabsorption of fat-soluble vitamins in children with pancreatic insufficiency, biliary obstruction, and chronic liver diseases is an indication for a prolonged vitamins supplementation.

12.1.4 THE SIGNIFICANCE OF POLYENOIC FA IN CHILDREN'S HEALTH AND DEVELOPMENT

12.1.4.1 Role in Immune and Inflammatory Process

Linoleic and α-linolenic acids, which must be delivered with food are important precursors in further metabolic processes contributing to the formation of long-chain polyenoic FAs (LC-PUFAs). Arachidonic acid (AA) and docosahexaenoic acid (DHA) have the highest biological activity. They are the precursors of eicosanoids such as prostaglandins, prostacyclins, leukotrienes, and thromboxanes. Eicosanoids, derived from PUFAs play a crucial role in the inflammatory and immune processes. Their physiological significance is of high interest, but still not enough understood. Many studies have documented that eicosapentaenoic acid (EPA)–derived eicosanoids are less potent than those of AA origin; those (thromboxane A2 and leukotriene B4) are severe pro-inflammatory agents. Their crucial role in the pathogenesis of autoimmune diseases is inducing leukocyte chemotaxis and adherence, and vasoconstriction. Many data suggest that the imbalance between pro- and anti-inflammatory agents, which are the consequence of immune cells activation, plays an important role in the pathogenesis of autoimmune diseases: inflammatory bowel diseases, bronchial asthma, rheumatoid arthritis, and psoriasis. It is now well documented that adequate supplementation of LC n-3 PUFA may help to ameliorate the clinical symptoms of these chronic diseases and allow a reduction of the dose of anti-inflammatory medications in affected patients (Prescott and Cadler, 2004).

12.1.4.2 Influence on Human Homeostasis and Cardiovascular System

PUFAs alter platelet aggregation and processes, which influence the coagulation and fibrinolysis. The n-3 LC-PUFAs prolong the template bleeding time, impair platelet aggregation and thromboxane formation, and exert a beneficial effect on erythrocyte flexibility as they are incorporated into cellular membranes, thus exerting a favorable effect on thrombosis and preventing thrombotic consequences in patients with coronary heart disease (Knapp, 1997). The favorable effect of DHA and EPA on hemostasis is also achieved via endothelium by decreasing inflammatory reaction as a result of free radicals production in granulocytes.

TABLE 12.2

Potential Protective Mechanisms of n-3 PUFAs on the Cardiovascular System

Reducing triacylglycerols plasma concentration

Antithrombogenic

Reducing heart susceptibility to ventricular arrhythmia

Retarding growth of atherogenic plaques

Anti-inflammatory by lowering pro-inflammatory eicosanoids and cytokines

Mild hypotensive

Promoting nitric oxide–induced endothelial relaxation

Decreasing of smooth muscle proliferation

The mechanisms responsible for the effects of n-3 PUFAs are complex and still not well documented. One of the most beneficial is the ability to reduce fasting and postprandial triacylglycerols in plasma (Schmidt et al., 1990).

It is proved, that sufficient consumption of PUFAs decreases the concentration of the low-density lipoprotein (LDL) fraction in serum and induces an increase in the high-density lipoprotein (HDL) level, thus diminishing the risk of hypertension, arrhythmias, sudden heart death, and other diseases connected with heart vessels factors. The anti-arrhythmic activity of n-3 PUFAs has two potential mechanisms: cardiocentric and neurocentric. The first mechanism is based on increased resistance of cardiomyocytes to pro-arrhythmic factors, while the second is based on the modulation of the autonomic system (Leaf, 2003). The recommended sea food consumption includes two fish meals weekly (Hu et al., 2002). The new evidences of protective influence on cardiovascular system of human body are still emerging (Table 12.2).

12.1.4.3 Proper Functioning of the Brain and Retina

The brain and retina are rich in DHA and AA, which are rapidly built into the cell membranes determining their fluidity during the third trimester of pregnancy and after delivery. They have a crucial importance in myelinization and formation of synapses. PUFAs also control the physiological functions of the brain, being involved in synthesis and functioning of neurotransmitters, and in the molecules of the immune system. There are two main life periods, when PUFAs are especially important in brain functioning: childhood and aging.

In the first years of life, children's brains grow intensively and are particularly vulnerable to FA deficiency. The special needs for LC-PUFAs in newborns have clinical importance; it is proved now that preterm infants who are breast-fed or fed with formulas enriched with n-3 PUFAs reach better neuronal development measured by electroretinogram recordings, visual evoked potentials, and performance in psychometric tests than those who are not supplemented. In breast-fed infants, whose mothers were supplemented with DHA, a positive correlation has been found between the levels of DHA in breast milk and neurodevelopmental scores in infancy. After 1 year of age such correlation is not obvious. It may be explained by the fact, that FA incorporating into the brain cells membranes is a dynamic system and thus the functional effects of special diet may be transient.

The loss of DHA from the brain has been associated with disturbances of nervous system function in experimental animals as well as in human infants fed with vegetable oil–based formulas (Moriguchi et al., 2001). PUFA deficiency during the postnatal period may contribute to myelinization disturbances. A major delay in this process may be accompanied by impaired learning as well as motor, vision, and auditory abnormalities (Stockard et al., 2000). Appropriate PUFA consumption may probably also play an important role in prevention and management of neurodegenerative diseases associated with the aging process.

12.2 CLINICAL IMPORTANCE OF PUFAs IN PEDIATRIC DISEASES

Fish oil, as a rich source of n-3 FA has been suggested for the treatment of inflammatory bowel diseases because of its anti-inflammatory effect mediated by a lowering of the generation of Leukotriene B4 (LTB4). It is well documented that the concentration of this powerful pro-inflammatory cytokine is increased in inflamed intestinal mucosa. It has been also proved that n-3 PUFAs can diminish the production of Interleukine 1β, tumor necrosis factor α (TNF α), reduce cell-mediated free radicals production, and can also increase production of important anti-inflammatory Leukotriene B5. The supplementation with n-3 PUFAs contained fish oil of patients with inflammatory bowel diseases allows in some cases the reduction of steroid's dose, which has a crucial importance especially in children. Prolonged high-dose steroid therapy has many severe side effects including growth retardation, Cushing-type obesity, and osteoporosis.

In patients receiving n-3 PUFAs and nonsteroidal anti-inflammatory drugs due to rheumatoid arthritis, the dose of the latter can also be decreased. The reduction of edema and rigidity of the affected joints is probably correlated with decrease of LTB4 and TNFα in the patient's blood.

The lack of n-6 PUFAs in children's diet may contribute to chronic skin diseases, desquamation, thickening, and decreased skin pigmentation. δ-6 desaturase deficiency, the enzyme responsible for gamma-linolenic acid production in metabolic changes of LA contributes to skin dryness and is also common in atopic dermatitis. Disturbances of epidermis keratosis in case of dryness due to PUFA deficiency are important factors in pathogenesis of psoriasis, simple acne, and fish skin disease. Lack of PUFAs in puberty contributes to characteristic acne changes: sebum becomes less liquid thus doweling the orifice of sebaceous glands and creating the comedones, and later inflammatory process. Improper pH reaction may lead to bacterial and fungal colonization of the skin. Data concerning clinical effects of PUFAs in skin diseases are inconsistent. It has been reported that fish oil supplementation improved scaling, itching, and subjective severity of atopic dermatitis (Bjorneboe et al., 1987). Other studies revealed no effect of fish oil on the clinical course of the disease and positive results after intravenous administration of n-3 FA (Berth-Jones and Graham-Brown, 1993, Mayser et al., 2002). EPA supplementation modulates eicosanoid biosynthesis and is also a promising tool to prevent the negative effects of UV irradiation on human health (Pupe et al., 2002).

Few studies in children proved that fish oil supplementation was responsible for the formation of less aggravating leukotrienes, reduced asthmatic symptoms, improved the parameters of spirometric tests, and enabled medication reduce. It is believed now that eating fair amounts of fish starting early in life may decrease the risk of the later development of bronchial asthma. As the data on these subjects are still scanty, these observations highlight the importance of further studies on PUFA therapy in children with bronchial asthma.

Depletion of n-3 PUFAs is considered as an important factor in pathogenesis of attention-deficit hyperactivity disorder (ADHD), specific difficulties with reading and spelling (dyslexia), and developmental coordination disorder (DCD). Several clinical trials have evaluated the impact of n-3 PUFAs on children with ADHD, revealing significant improvement in symptoms (Belanger et al., 2009). A growing number of studies suggest that other mental illnesses, particularly mood and attentional disorders, are associated with reduced dietary supply of n-3 PUFAs, but data are still insufficient to recommend these acids in a mono- or adjunctive therapy in any mental illness.

12.3 RECOMMENDATIONS OF FAT CONSUMPTION IN INFANTS AND CHILDREN

12.3.1 LIPID COMPOSITION IN HUMAN MILK AND ARTIFICIAL FORMULAS

Lipids in human milk originate from mother's diet, mobilization of her fat stores, and FA synthesis by the mammary gland. The total lipid content ranges between 3.2 and 4.5 g/100 cm³; it delivers 40%–55% of the total energy of milk. The lipid amount may vary due to ethnic differences as well as to timing of lactation, the time of the day, mother's health, and state of nutrition. Triacylglycerols account for about 98% of total milk fat; cholesterol, diacylglycerols, phospholipids, and free FA constitute, respectively, 0.5%, 0.01%, 0.8%, and 0.08% (Koletzko et al., 2001).

The composition of mother's milk fat is highly dependent on her diet. High consumption of sea foods increases the content of EPA and DHA, while vegetarian diet enriches milk in LA. Hence, the quality of maternal diet in pregnancy and during lactation is of special importance for breast-fed baby. Saturated FA are the predominant lipid milk fraction (about 45%), whereas monoenoic and polyenoic ones constitute respectively about 39% and 14% of total fat. PUFAs make the rest of FA. They include mainly LA, but also small amounts of the other FAs in the n-6 and n-3 series (Table 12.3).

Particularly, high amount of LC-PUFAs in colostrum may play a significant role in proper development of neonate, mainly due to intensive deposition of these acids in the brain and retina, which starts in the last trimester of pregnancy. Data received from cord blood and maternal blood revealed higher amounts of LC-PUFAs in the cord blood, which is the evidence for preferential and selective LC-PUFAs transport from the mother to fetus (Berghaus et al., 1998). The mean supply of PUFAs in infants who are exclusively breast-fed, based on the average milk composition, is about 100 mg/kg body weight per day (Koletzko, 1992).

According to the guidelines of European Society for Pediatric Gastroenterology, Hepatology and Nutrition (ESPGAN) Committee on Nutrition (Com. Directive, 1996),

TABLE 12.3
Content of FA in Human Milk in Europe

FA	Mean Value (% of Total FA)	Range (% of Total FA)
Saturated	45.2	39.0–51.3
Monoenoic	38.8	34.2–44.9
Polyenoic	13.6	8.5–19.6
LA	11.0	6.9–16.4
ALA	0.9	0.7–1.3
Arachidonic acid	0.5	0.2–1.2
DHA	0.3	0.1–0.6
n-6 LC-PUFAs, total	1.2	0.4–2.2
n-3 LC-PUFAs, total	0.6	0.3–1.8
LA/ALA	12.1	8.6–16.9
n-6/n-3	2.7	0.3–3.7

Source: Stolarczyk, A., Pediatria Współczesna. Gastroen-terologia, *Hepatologia i Żywienie Dziecka*, 1(2/3), 155, 1999. With permission.

the total fat content in starting formulas should range from 4.4 to 6.6 g/418 kJ and 3.3 to 6.5 g/418 kJ in follow-up formulas. The fat used in milk production is usually a blend of several vegetable oils, mainly sunflower, coconut, soybean, safflower, and palm oil. These oils do not contain cholesterol; therefore, formula-fed infants up to 6 months receive less cholesterol than breast-fed babies due to its high concentration in human milk (30 mg/100 cm^3). As the beneficial effects of cholesterol supplementation into milk formulas have not been established, it is not routinely added.

Due to the very important role of PUFAs in infants' diet, there is much interest not only in the lipid amount, but also its composition, in infant formulas (Table 12.4). Regarding human milk as a standard, the ESPGAN Committee has recommended the contents of LA and ALA in infant formulas. LA should supply 4.5%–10.8% of total energy. Due to the competitive antagonism of the n-3 and n-6 acids, the proportion of the concentrations of LA:ALA has been set as 5:1 to 15:1 (Aggett et al., 1991).

It is proved that plasma PUFA levels in formula-fed babies can be achieved with PUFA supplementation into the formulas without adverse effects of this enrichment (Koletzko et al., 2003).

12.3.2 PROPER FAT INTAKE IN INFANCY, IN SCHOOL CHILDREN, AND ADOLESCENTS

The requirements for fat intake is usually given for total fat, n-6 PUFAs, n-3 PUFAs, cholesterol, and saturated FA. In childhood, the total food intake is recommended as 30%–35% of total energy requirement (Prentice et al., 2004).

TABLE 12.4

The Contents of FA in Infant Formulas

FA	Contents in Formulas (% of Total Fat)
Lauric	Maximum 15
Mirystic	Maximum 15
LA	300–1200 mg/100 kcal
ALA	Minimum 50 mg/100 kcal
LA/ALA	5:1 to 15:1
trans-FA	Maximum 1
Erucic	Maximum 1
LC-PUFA (C20–22)	
Total n-3	Maximum 4
Total n-6	Maximum 2
Arachidonic	Maximum 1
EPA	<DHA

Source: Com. Directive 96/4/EC, *Off. J. Eur. Commun.*, No. L49/13. With permission.

Note: LA, linoleic acid 18:2 (n-6); AA, arachidonic 20:4 (n-6); ALA, α-linolenic 18:3 (n-3).

The recommended fat intake in infants is based on adequate intake (AI), which reflects mean lipid supplies in breast-fed babies and constitutes 31 g/day (Food and Nutrition Board, 2005). As the mean energy content of mature human milk is 2717 kJ/dm^3, fat delivers about 50%–55% of the total energy to breast-fed infants. The proportion of total energy delivered with fat subsequently decreases in the next 6 months due to a wider variety of foods in the baby's diet. In the second half of the first year, fat delivers about 40% of total energy.

After infancy, the daily recommendation for fat intake ranges from 3.1 to 3.3 g/kg of body weight, which constitutes approximately 32% of the total energy requirement. During next years of life, total fat consumption decreases to 30% of total energy intake (Table 12.5).

12.3.3 Special Needs for Premature Babies

Premature infants and babies with intrauterine growth retardation present insufficient activity of elongase and saturase, the enzymes indispensable for PUFAs production. They also have limited essential FA stores. As their needs for PUFAs are especially high, enrichment of the formulas for them with PUFAs (1% of n-6 and 0.5% for n-3 of total FA) have been considered to improve substrate supply and is beneficial for early growth and development of the baby. A close dose-response relationship between the

TABLE 12.5
Dietary Reference Intake of Lipids

	Range (Percentage of Energy)		
	Children 1–3 Years	Children 4–8 Years	Children 9–13 Years
Total fat	30–40	25–35	25–35
n-6 PUFA	5–10	5–10	5–10
n-3 PUFA	0.6–1.2	0.6–1.2	0.6–1.2

Source: Food and Nutrition Board, Institute of Medicine of the National Academies, *Dietary Reference Intakes for Energy, Carbohydrate. Fiber, Fat, Fatty Acids, Cholesterol, Protein, and Amino Acids*, National Academies Press, Washington, DC, 2002/2005. http://www.nap.edu. With permission.

DHA formula supplementation and visual acuity in preterm babies has been found (Koletzko et al., 2003). PUFAs for milk supplementation are received from fish oil, fractionated yolk lipids, as well as from grape, and black currant seeds. The high degree of unsaturation makes n-3 and n-6 PUFAs susceptible to oxidation leading to oxidative stress. Thus enrichment of PUFAs requires substantial amounts of antioxidants to preserve their structure and physiological activity. Sufficient availability of vitamin E is of great value for the response to oxidative stress, which may contribute to many disorders characteristic for low birth weight infants: necrotizing enterocolitis, retinopathy, bronchopulmonary disease, intracranial hemorrhage, and infection. European recommendation advises to supplement formulas with vitamin E in amount of 0.5 mg α- tocopherol/g linoleic acid and 0.75 mg α-tocopherol/g γ-linolenic acid.

12.4 UNDESIRABLE EFFECTS OF IMPROPER FAT CONSUMPTION ON THE CHILD'S HEALTH

12.4.1 Obesity and Metabolic Syndrome as a Consequence of Junk Food Consumption

Obesity, recognized in many countries as a social disease, has a multifactor pathogenesis, in which behavioral and environmental influences are of major importance. Elevated intake of fat has been recognized as a crucial dietary factor of children obesity. However, the important factor is not the fat amount, but the total energy of the diet supplied as well by saccharides and proteins, the energy expenditure related to the children's physical activities, and the organism's genetic predisposition (Butte et al., 2006).

Obesity is associated with significant health problems: hypertension, endocrine disturbances (impaired glucose tolerance, type 2 diabetes mellitus, menstrual irregularity, hyperandrogenemia), orthopedic complications, nonalcoholic fatty liver disease, and mental disorders (low self-esteem, depression). Overweight in adolescence may contribute to the metabolic syndrome that increases in adults the risk of atherosclerosis (ischemic heart disease, cerebral vessel disease, peripheral blood vessel disease) and diabetes. The metabolic syndrome is diagnosed in case of central obesity, high blood pressure, impaired fasting glucose, and dyslipidemia. Recent studies have shown that the prevalence of this syndrome was over 38% among a group of nearly 450 obese children in the United States (Weiss et al., 2004).

Fast food, usually eaten in large portion size, includes high content of refined carbohydrates and *trans*-FA, as well as low content of fiber and micronutrients. Studies on a big group of over 6000 children in the United States have indicated that children's energy consumption was 187 kcal/day higher on days when fast food was consumed, compared with those without fast food (Bowman et al., 2004). Moreover, some individuals may be especially prone to adverse effects of fast food. In a prospective analysis of young adults, the authors proved that individuals with the highest consumption of fast food over a 15 year period had gained an extra 4.5 kg of bodyweight and had a twofold greater increase in insulin resistance (Pereira et al., 2005). On the contrary, consumption of marine lipids may have an anti-obesity effect due to the role of EPA and DHA in lipid metabolism (Toyoshima et al., 2004). For adipose tissue reduction, replacing foods high in saturated FA and *trans*-FA with foods rich in plant-based fat sources may be effective.

12.4.2 LOW-FAT DIETS

Fat is an irreplaceable component of a well-balanced diet. Fat, compared with other food components has the advantage of carrying more energy in a smaller volume, which is of vital importance for children who have limited intake capacity and extraordinarily high energy needs. Consuming fat-reduced diet by children is controversial; very low fat diet (supplying less than 20% of total energy from fat) may have an undesirable effect on child's health.

Young children eating fat-restricted diet appear to grow normally, but they are prone to consume insufficient amounts of macro- and micronutrients, especially Ca, Zn, Mg, P, vitamin E, B_{12}, thiamin, niacin, and riboflavin (Olson, 2000). Fat-soluble vitamins' deficiencies are the result in their diminished absorption from the intestines. Low-fat diets due to deficits of PUFAs could lead to decreased immunological functioning, skin problems, and unsatisfactory visual abilities, and may be detrimental rather than beneficial. A well-balanced diet should have sufficient amount of cholesterol necessary for the production of many hormones.

There is also no evidence that restriction in fat intake in childhood will prevent atherosclerosis in adulthood. There are no long-term prospective studies linking hypercholesterolemia in children with coronary artery disease in adults. There is also no strict correlation between fatty streaks in childhood and prevalence of atheromatous plaques in adulthood; they do not become significant until much beyond puberty. The effect of fat-restricted diet and low cholesterol consumption on serum lipoproteins and serum lipids, are of lesser magnitude in children than in adults. These diets have also been ineffective in reducing serum cholesterol levels over prolonged periods.

It is, however, proven that moderate fat restriction, 25%–30% of total energy, is safe and assures normal child's growth.

REFERENCES

Aggett, P. et al. 1991. Committee Report: Comment on the content and composition of lipids in infant formulas ESPGAN Committee on Nutrition. *Acta Paediatr Scand* 80:887–896.

Belanger, S.A. et al. 2009. Omega-3 fatty acid treatment of children with attention-deficit hyperactivity disorder: A randomized, double-blind, placebo-controlled study. *Pediatr Child Health* 14(2):89–98.

Bergahaus, T., Demmelmair, H., and Koletzko, B. 1998. Fatty acid composition of lipid classes in maternal and cord plasma at birth. *Eur J Pediatr* 157:163–168.

Berth-Jones, J. and Graham-Brown, R.A. 1993. Placebo-controlled trial of fatty acid supplementation in atopic dermatitis. *Lancet* 341:1557–1560.

Bjorneboe, A. et al. 1987. Effect of dietary supplementation with eicosapentoic acid in the treatment of atopic dermatitis. *Br J Dermatol* 117:463–469.

Bowman, B.A. et al. 2004. Effects of fast food consumption on energy intake and diet quality among children in a national household survey. *Pediatrics* 113:112–118.

Butte, N.F. et al. 2006. Genetics of childhood obesity. In *Handbook of Pediatric Obesity*, M.I. Goran and M.S. Sothern (eds.), pp. 79–96. Boca Raton, FL: CRC Press.

Commission Directive 96/4/EC, of February 16, 1996, amending Directive 91/321/EEC on infant formulae and follow-up formulae. *Official Journal of the European Communities*, No. L49/13.

Food and Nutrition Board. Institute of Medicine of the National Academies. 2002/2005. *Dietary Reference Intakes for Energy, Carbohydrate. Fiber, Fat, Fatty Acids, Cholesterol, Protein, and Amino Acids.* Washington, DC: National Academies Press. http://www.nap.edu

Hu, F.B., Bonner, L., and Willet, W.C. 2002. Fish and omega-3 fatty acid intake and risk of coronary heart disease in women. *JAMA* 287:1815–1821.

Knapp, H.R. 1997. Dietary fatty acids in human thrombosis and hemostasis. *Am J Clin Nutr* 65(Suppl. 5):1687–1698.

Koletzko, B. 1992. Fats for brains. *Eur J Clin Nutr* 46 (Suppl. 1):51–62.

Koletzko, B. et al. 2001. Physiological aspects of human milk lipids. *Early Hum Dev* 65 (Suppl.):3–18.

Koletzko, B. et al. 2003. Fatty acid profiles, antioxidant status, and growth of preterm infants fed diets or with long-chain polyunsaturated fatty acids. A randomizes clinical trial. *Eur J Nutr* 42(5):243–253.

Leaf, A. 2003. Clinical prevention of sudden cardiac death by n-polyunsaturated fatty acid in mechanism of prevention of arrhythmias by n-3 fish oils. *Circulation* 107:2646–2656.

Lissauer, T. and Clayden, G. 2007. *Illustrated Textbook of Paediatrics*, 3rd rev. edn. London, U.K.: Elsevier Health Sciences.

Mayser, P. et al. 2002. A double-blind, randomized, placebo-controlled trial of n-3 versus n-6 fatty acid-based lipid infusion in atopic dermatitis. *J Parenter Enteral Nutr* 6(3):151–158.

Moriguchi, T. et al. 2001. Reversal of docosahexaenoic acid deficiency in the rat brain, retina, liver, and serum. *J Lipid Res* 42:419–427.

Olson, R.E. 2000. Is it wise to restrict fat in the diets of children? *J Am Diet Assoc* 100:28–32.

Pereira, M.A. et al. 2005. Fast food habits, weight gain, and insulin resistance (the CARDIA study) 15-year prospective analysis. *Lancet* 365:36–42.

Prentice, A. et al. 2004. Energy and nutrient dietary reference values for children in Europe: Methodological approaches and current nutritional recommendations. *Br J Nutr* 92(Suppl. 2):83–146.

Prescott, S.L. and Cadler, P.C. 2004. N-3 polyunsaturated fatty acids and allergic disease. *Curr Opin Clin Nutr Metab Care* 7:123–129.

Pupe, A. et al. 2002. Eicosapentaenoic acid, a n-3 polyunsaturated fatty acid differentially modulates TNF-α, IL-1α, IL-6 and PGE$_2$ expression in UVB-irradiated human keratinocytes. *J Investig Dermatol* 118:692–698.

Schmidt, E.B. et al. 1990. The effect of n-3 polyunsaturated fatty acids on lipids, platelet function, coagulation, fibrinolysis and monocyte chemotaxis in patients with hypertension. *Clin Chim Acta* 189:25–32.

Stockard, J.E. et al. 2000. Effect of docosahexaenoic acid content of maternal diet on auditory brainstem conduction times in rat pups. *Dev Neurosci* 22:494–499.

Stolarczyk, A. 1999. Fats in infant formulas and special formulas. *Pediatria Współczesna. Gastroenterologia, Hepatologia i Żywienie Dziecka* 1(2/3):155–160.

Thompson, G.R., Lewis, B., and Booth, C.C. 1966. Absorption of vitamin D3-3H in control subjects and patients with intestinal malabsorption. *J Clin Invest* 45: 94–102.

Toyoshima, K. et al. 2004. Separation of sardine oil without heating from surimi waste and its effect on lipid metabolism in rats. *J Agric Food Chem* 52:2372–2375.

Weiss, R. et al. 2004. Obesity and the metabolic syndrome in children and adolescents. *N Engl Med* 350(23):2362–2374.

13 Lipids and the Human Vision

Paweł Lipowski

CONTENTS

13.1 INTRODUCTION

Vision is a very complicated process dependent on the transparent media of the eye as the cornea, the lens, and the vitreous humor, providing a clear image on the retina. The change in the shape of the lens in the accommodation process assures clear images at different distances. The 120 million rods and 6 million cones of the human retina together with three layers of nuclei perform billions of calculations per second before the signal reaches the optic nerve. Then the cerebral cortex processes the information in the vision centers. The three different pigments in the retinal cones constitute the main source of color vision. The eye adapts to light and dark over an intensity range of 1–100,000 (Newell, 1992). The location of human eyeballs in the orbits in front of the head and the decussation of axons from nasal part of each retina permit a retinal correspondence and stereoscopic vision. The vision process requires energy to provide its metabolic needs by several metabolic pathways.

13.2 THE METABOLIC ASPECTS OF VISION

Glucose is the main source of energy in the cornea, the lens, and especially in the retina and the optic nerve, which are part of the central nervous system. Glucose is

239

first phosphorylated to glucose-6-phosphate, so it may be utilized by the cells. The reaction is catalyzed by the enzyme hexokinase. In the retina, glucose is converted into glucose-6-phosphate in the retinal pigment epithelium (RPE) and diffuses to the photoreceptor cells. The mitochondria of the photoreceptor cells convert the glucose-6-phosphate to two molecules of glyceraldehyde, which is converted to pyruvic acid (Berman, 1991). The retina requires more oxygen than any other tissue. The pyruvic acid is then metabolized to carbon dioxide and water, and this reaction produces adenosine triphosphate (ATP). The ATP provides energy for axonal transport, outer segment renewal of photoreceptors, and biosynthesis of cell membranes, where the lipids play a very important role. All metabolites from the outer segments of photoreceptors pass through the RPE cells tightly adherent at their apices, which create the blood–retinal barrier. The bases of the RPE cells have receptors for retinol-binding protein that transports vitamin A in the plasma (Uchida et al., 2005). The outer segments of rods and cones are removed diurnally by the phagocytic process of the RPE cells. For this reason, the outer segments must be constantly synthesized by inner segments of the photoreceptors. The rods and cones are the light-sensitive elements of the sensory retina. The rods function at low levels of illumination (scotopic vision), and the cones are active at medium and high levels of illumination and in color perception (photopic vision). The sensory retina is divided into central part (macula), which contains the fovea centralis that functions in photopic vision, and into periphery that functions in scotopic vision.

Retinol (vitamin A) is utilized in the retina in its aldehyde form (retinal). When the 11 position is *cis*, it is able to bind the protein opsin. The absorption of light causes an isomerization of 11 *cis* retinal to all *trans* retinal. The change in the shape of retinal is followed by chemical reactions that disassociate the photopigment to provide free opsin and all *trans* retinal. It is called the change of the rhodopsin, as it is activated by absorption of a photon of light that initiates the visual impulse.

13.3 LIPIDS AND THE VISUAL SYSTEM

13.3.1 THE LENS

The concentration of lipids in the lens is low, below that in other tissues of the eye. The lenticular lipids are composed chiefly of cholesterol and phospholipids (PL). Cholesterol is the most often examined lipid of the lens because of the frequent appearance of cholesterol crystals in cataract or even in transparent lens (Rae, 1994).

Our understanding of the lens metabolism is still evolving. The lens like other tissues requires energy. The principal sources within the cell are ATP and nicotinamide adenine dinucleotide phosphate (NADPH). NADPH is the source of reducing power in the biosynthesis of essential cellular components such as fatty acids and glutathione. The pathways synthesizing ATP and NADPH also produce the cell membrane components (PL, sphingolipids, and cholesterol) (Berman, 1991). PL play a role in the regulation of lens internal calcium. Ca^{2+} can be sequestered at binding sites in the plasma membranes and cytosol (PL, crystallines, and calmodulin-binding proteins). High external Ca^{2+} levels in the aqueous and vitreous humor have an important role in the control of membrane permeability to Na^+ and K^+ (Duncan et al., 1994).

The fully differentiated lens cell (fibril) has anterior and posterior processes. They are composed of a protein matrix surrounded by a lipid bilayer membrane that allows the cell to preserve the shape and permeability of the cell membrane.

13.3.2 THE CORNEA

The cornea is composed mainly of stroma with squamous epithelium on its outer surface and single layer of endothelial cells on its inner surface. The anterior epithelium is bathed with relatively stagnant layer of tears called the precorneal tear film. The tear film contains three layers: a thin anterior lipid layer derived from the palpebral meibomian glands, sebaceous glands of Zeis, and sweet glands of Moll; a thick middle aqueous layer; and a thin mucin layer. The lipid layer retards the evaporation of tears and assures a smooth and regular anterior optical surface. The anterior lipid layer composition depends on the meibomian gland secretion that includes wax esters—35%, cholesterol esters—30%, PL—16%, triacylglycerols—4%, free fatty acids—2%, and free sterols—2% (Newell, 1992).

The corneal epithelium is readily permeable to lipid-soluble substances because the cell membranes are composed of a lipoprotein. The histochemical investigation of the basal membrane of the corneal epithelium revealed the presence of a lipid layer (McCarthy et al., 1994). To pass through the stroma and endothelium, the compounds must be water soluble. To penetrate the normal cornea, topical medications must be water and lipid soluble.

Structural constituents of the cornea indicate the complete difference in the nature of epithelium and stroma. The lipid content of the cornea is low, although it is higher in the epithelium. The difference in cellularity is reflected in the difference of constituents such as nucleic acids, ATP, and lipids. The corneal epithelium is the site of most metabolic activity, as it contains 15–20 times more cells than the stroma and endothelium.

13.3.3 THE RETINA

The retina is anatomically a part of the brain, so the lipid concentration in the retina is expected to be high. The separated rod outer segments have a higher content of PL than the whole retina. The concentration of PL in the retina is one of the highest in any tissue or tissue fraction. The outer segment membranes of photoreceptors contain an unusually high proportion of polyenoic FA in their PL and low cholesterol content. Usually, about 50% and sometimes even 90% of the fatty acids in the photoreceptors are polyenoic (Robison et al., 1982). The high degree of unsaturation is critical for providing high fluidity and other characteristics of the microenvironment required for the normal function of rhodopsin and the stability of the membrane structure. The major PUFA found in the retina are docosahexaenoic acid (DHA) and arachidonic acid (AA). They are both primarily detected in neural and vascular cell membrane PL. Eicosapentaenoic acid (EPA), the precursor of DHA, is found in the retinal vascular endothelium.

13.4 THE ROLE OF LIPIDS IN THE PATHOPHYSIOLOGY OF THE VISUAL SYSTEM

13.4.1 THE RETINA

Vitamin A, vitamin E, and PL containing PUFA play important roles in both the structure and function of retina, especially with regard to the photoreceptor cells. Deficiency in dietary vitamin A leads to night blindness, loss of photoreceptors outer segments, irreversible loss of photoreceptors nuclei, and to complete and permanent blindness (Panozzo et al., 1998). The photosensitive portions (outer segments of rods) consist of densely arranged membranes (discs). The main protein in these membrane discs is opsin, which, combined with 11 *cis*-retinaldehyde, constitutes the visual pigment rhodopsin. Rhodopsin is an essential component of the disc membranes. The absence of sufficient vitamin A to form pigment compromises the integrity of the membrane structure. Vitamin E deficiency also results in the disruption of the outer segment disc membranes and the eventual loss of photoreceptor nuclei. The loss of membrane integrity probably results from increased autoxidation of membrane lipids in the absence of the antioxidant activity of vitamin E (Robison et al., 1980).

The outer segment membranes contain PUFA, which make them susceptible to damage by autoxidation. The importance of antioxidants in the maintenance of photoreceptor membranes and the contribution of oxidized membrane products to the formation of lipofuscin (aging pigment) in the RPE is a well-known fact (Bazan et al., 1990). Vitamin A as well as aging, vitamin E, and PUFA are implicated in determining the rates of formation of lipofuscin granules in the RPE. The main components of lipofuscin are lipids (5%), proteins (30%), and a small amount of saccharides. Melanin is detected sometimes as an additional component. The lipofuscin is autofluorescent, which is connected with lipid peroxidation (Feeney, 1978).

A major difference between retinas from young and old individuals is the greater lipofuscin content in old retinas. What is interesting is that the lipofuscin is not located in the neuron cell bodies or any other part of the neural retina, as might be predicted from aging studies of the spinal cord and brain, but in the RPE, which is situated between the choriocapillaris and the photoreceptor cells. The RPE performs functions essential for photoreceptor cell function and survival, including participation in the retinoid visual cycle. Impairment of RPE activity appears to be involved in a number of inherited retinal degenerative disorders as well as in age-related macular degeneration (AMD), one of the most important causes of serious visual impairment in developed countries. The buildup of the lysosomal storage bodies (lipofuscin) has been implicated in AMD (Algvere and Seregard, 2002).

The retinoids in the RPE–retina complex play a major role in RPE lipofuscin formation. The absence of retinal (all *trans* and 11 *cis*) in mice without a functional Rpe 65 gene drastically reduced the formation of lipofuscin (Katz and Redmond, 2001). Other molecular constituents (e.g., lipids and proteins) of the photoreceptor outer segments in addition to the retinoid-derived compounds are involved in the formation of RPE lipofuscin. The very interesting fact, that rats with normal retinas had accumulated more lipofuscin than rats with the gene for retinal dystrophy (rdy), suggests that photoreceptor cells play a significant role in lipofuscin deposition in RPE (Robison et al., 1982).

The lipids like vitamin A are bound in the membrane of the outer segments of photoreceptors and are sometimes processed or stored in the RPE. The RPE is a post-mitotic tissue, which ingests, processes, and transports extraordinary amounts of unsaturated lipids and vitamin A, both of which are susceptible to oxidation. All of these reactions are done in high oxygen flux. At the same time, the RPE exhibits lipofuscin accumulations that are influenced by increased age, antioxidant deficiencies, and vitamin A levels. Lack of dietary vitamin A results in a striking decrease in lipofuscin formation (Katz et al., 1996). Better understanding of the mechanisms and effects of lipofuscin deposition in the RPE may aid in the diagnosis and treatment of retinal diseases. Retinitis pigmentosa, ceroid lipofuscinosis, fundus flavimaculatus, and Stargardt disease exhibit ocular changes, which involve a degeneration of photoreceptor cells and an accumulation of lipofuscin in RPE.

The role of lipids in angiogenesis is becoming more clear. The deposits of insoluble material accumulating between RPE and Bruch's membrane are called druses. They are an integral element of the diagnosis of AMD. Small hard druses contain PL; larger in diameter soft druses contain neutral lipids. The accumulation of neutral lipids in Bruch's membrane is followed by the formation of drusenoid (elevation of RPE), which creates the impermeable barrier between the RPE and choroids, and is one of the factors initiating the inflammatory reaction and neovascularization.

Ocular neovascularization is the cause of blindness in all age groups; retinopathy of prematurity in children, diabetic retinopathy in adults, and AMD in the elderly. Dietary sources of n-3 PUFA and n-6 PUFA as well as PUFA released as free fatty acids contribute to many substrates. These can be converted to bioactive intermediaries such as eicosanoids from AA, neuroprotectins from DHA, D series resolvins from DHA, and E series resolvins from EPA. Knowledge of lipid mediators and epidemiologic data linking PUFA and neovascular AMD indicate that EPA, DHA, and AA may function in vivo to regulate retinal vaso-obliteration and neovascularization.

The n-3 and n-6 PUFA influence vascular growth and pathology. EPA, DHA, and their bioactive products reduce at physiological levels pathologic neovascularization through enhanced vessel regrowth after vascular loss and injury. Macrophages or microglia are important components of retinal vascular growth and repair. Increased dietary n-6 PUFA (AA) increases in the retina microglial production of TNF alfa, which is suppressed with elevated n-3 PUFA levels (DHA, EPA) (Connor et al., 2007).

Lipids are the components of retinal deposits called hard or fatty exudates. They follow localized areas of retinal edema and consist of fats and lipid-filled macrophages in the outer plexiform layer of the retina. Several patterns of retinal deposits distribution are observed: a ring in circinate retinopathy, a foveal star in stellate retinopathy, and scattered irregular deposits. A circinate pattern may be seen in diabetes mellitus, a stellate pattern in vascular hypertension, and scattered deposits in Coats retinal telangiectasis. In Coats disease, there is a massive outpouring of lipids, mainly cholesterol, into and beneath the retina.

13.4.2 THE LENS

The most important role of glutathione in the lens is protection against oxidative damage. Reactive oxygen species (ROS) and free radicals have the potential to damage lipids, proteins, saccharides, and nucleic acids. ROS have two origins in the tissues—cell metabolism and photochemical reactions. The continuous entry of optical radiation into the lens, in particular the preferential absorption of shorter wavelengths (295–400 nm), makes lens tissue susceptible to photochemical reactions. ROS can also enter the lens from the surrounding humors. They have a great capacity to damage the lens by peroxidizing membrane lipids, which results in the formation of malondialdehyde. It can form cross-links between membrane lipids and proteins. This reaction would rapidly result in lens damage. Protection against damage induced by ROS in the lens is achieved in a number of pathways. One of them results in the formation of H_2O_2. The glutathione system is thought to provide the most protection against H_2O_2. This system also protects against the lipid free radicals chain reaction by the neutralization of lipid peroxidases (Berman, 1991).

As the lens ages, many morphological changes occur to the epithelial cells, fibers, and capsule. Lens fibers show a loss or partial degeneration of a number of plasma membranes and cytoskeletal proteins. By 80 years of age, the expression of these cytoskeletal proteins is restricted to the epithelial cells. The cholesterol-to-PL ratio of fiber cell plasma membranes increases throughout the life, and membrane fluidity decreases. These changes are greatest in the nucleus of the lens and are partially responsible for the increase in the nuclear sclerosis (Checchin et al., 2006). The deeper cortical fibers and nucleus are not prone to the ruptures because the high cholesterol content of their membranes makes them more resistant to damage.

Changes to the cellular junctions and alterations in cation permeability occur as the lens ages. Thus, the Na^+/K^+ permeability ratio increases, which results in a greater sodium content of the lens. The change in the levels of these two ions correlates with an increase in the optical density of the lens. This change in ion permeability with increased fiber age is thought to occur due to a decrease in membrane fluidity as a result of the age-related increase in the cholesterol-to-PL ratio. It is thought that the Ca^{2+} ATP-ase may be inhibited by the decrease in membrane fluidity (Duncan et al., 1994). The metabolic activity of the lens, as well as the activity of many glycolytic and oxidative enzymes, decreases with age. This is attributed to decreasing activities in the cortex and nucleus. The activity of many enzymes involved in the metabolism of glucose also decreases with age. Although overall metabolic activity decreases, the lens still maintains the capacity to synthesize proteins, fatty acids, and cholesterol at substantial rates. Decreased metabolic activity does not serve as a limiting factor for the production of new lens fibers (Rae, 1994). A reduction in the activity of many antioxidants occurs with increasing age. Glutathione peroxidase, which is involved in the breakdown of lipid peroxides, increases from birth until 15 years of age and then slowly decreases throughout adulthood.

A deficiency of the enzyme α-galactosidase results in accumulation of the glycolipid—ceramide trihexoside, which is detected in Fabry's disease. It is an X-linked lysosomal storage disorder, where the mild cataract occurs. The opacities in the lens are thought to result from the incorporation of abnormal glycolipid into cell membranes.

13.4.3 THE CORNEA

The pathological lipid accumulation in the cornea is one of the main symptoms of many corneal dystrophies and degenerations, resulting very often in visual acuity deterioration.

Deposition in the corneal periphery of a gray or yellow band opacity is called corneal arcus. The arcus is almost always bilateral and is associated with aging. The deposits of arcus are composed of extracellular steroid esters of lipoproteins, most of a low density. Limbal capillaries are the source of lipid material; its central flow is limited by a functional barrier to the flow of large molecules in the cornea, which locates the lipid deposits in the periphery of the cornea (limbus). Corneal arcus is associated with increased plasma cholesterol and low density lipoprotein cholesterol (LDL). Young patients with corneal arcus also have an increased risk for type IIa dyslipoproteinemia. Patients with arcus juvenilis have relative risk of mortality from coronary heart disease and cardiovascular disease (Moss et al., 2000).

Lipid keratopathy may be central, diffuse, or peripheral and occurs as a primary or more often as a secondary form. White or yellow stromal deposits are separated by a narrow clear zone from corneal stromal neovascularization. Lipid deposits follow corneal edema as in corneal hydrops, herpes zoster keratitis (Shapiro and Farkas, 1977). Histopathologically, the material consists of intra- and extracellular lipids. Primary lipid keratopathy has attributes of corneal dystrophy. It is usually central, often with cholesterol crystals, and may severely decrease visual acuity.

Terrien's marginal corneal degeneration is characterized by marginal corneal ectasia and occurs at any age. At the beginning, the peripheral corneal haze, usually superiorly, is observed. Then it gradually vascularizes superficially and is followed by corneal thinning. The lipid deposits along the central edge of changed cornea had been detected.

The granular corneal dystrophy (Groenouw type I) is an autosomal dominant disorder with rod-like deposits and microfibrils, present in keratocytes and in epithelial cells. The material is thought to be PL.

The Schnyder crystalline dystrophy is an autosomal dominant disorder that begins with central subepithelial corneal crystals accumulated in a ring pattern. Systemic hypercholesterolemia is frequent in affected family members. Histopathological investigations revealed oil-red O-positive lipid material throughout the stroma. Cholesterol clefts had been detected in the anterior stroma. The material appears to be PL and esterified and unesterified cholesterol (Weiss, 1996).

In the Fleck corneal dystrophy, an autosomal dominant disease, small white and gray opacities are observed in the stroma. Membrane-bound vacuoles filled with electron-dense material had been found in keratocytes. Electron microscopy shows the presence of lipid and acid mucopolysaccharide.

A number of systemic metabolic disorders of genetic origin affect the anterior segment of the eye. These disorders are usually autosomal recessive, and a single enzyme deficiency accounts for the clinical manifestations. The disorders are subdivided according to the biochemical group, in which the abnormality is found for sphingolipidoses, dyslipoproteinemias, and mucolipidoses (Barchiesi et al., 1991).

13.5 FINAL REMARKS

Lipids are very important components of all living cells. Especially in the retina, they are a building material of cell and neuron membranes, where the amount of PUFA is very high. Consumption of EPA and DHA results in better uptake of these fatty acids into the cell membrane PL layers. EPA is a precursor to less active eicosanoids and, in this respect, is considered as anti-inflammatory. The minimum dietary intake of EPA and DHA is 3 g/day in patients with fish oil supplementation in a number of inflammatory conditions, also of the retina (Kremer, 2000). Intake of dietary long chain n-3 fatty acids and fish has been suggested to prevent AMD, the leading cause of vision loss in old people. They protect against oxygenic, inflammatory, and age-related retinal damage, which are very important in the development of AMD. Consumption of fish and foods rich in n-3 fatty acids is connected with lower risk of AMD. Fish intake at least twice a week was associated with reduced risk for early and late form of AMD (Smith et al., 2000). There are numerous studies related to n-3 fatty acids and pregnancy. Maternal DHA levels during pregnancy have an influence on correct neural development. The fact that DHA is concentrated in the brain and retina gives us the conviction that the DHA is essential in visual development. Human milk DHA content is a result of mothers' dietary intake. The highest levels of DHA in human milk are observed in Japanese and Chinese females because of their dietary customs. The formula of human milk DHA consumption levels should contain 0.3%–0.4% DHA of total fatty acids. Such supplementation of DHA leads to improved visual acuity in children (Malcolm et al., 2003). There are two main roles of DHA in the retina: the structural role, which is responsible of the correct development of the retina, and the protective role. DHA as a protective factor elongates the photoreceptors' life and has anti-apoptotic activity. It also protects the retina against the ischemic diseases and decreases the lipofuscin accumulation in RPE. The results of many prospective cohort studies, such as the Blue Mountains Eye study, the Health Professional Follow-Up study, and Cho et al. study, were consistent with protective effect of n-3 fatty acids against AMD (Cho et al., 2001). They also revealed that saturated fatty acids increase the risk of AMD. Consumption of fish and foods rich in n-3 fatty acids is a kind of wide prophylaxis of especially degenerative and dystrophic retinal diseases. Dietary supplements available on the market are purified and concentrated to make the sum EPA + DHA about 60% of the total PUFA content. Such products should be also purified to decrease the heavy metals and other toxin levels. When the first symptoms of AMD are observed, to the special diet, rich in n-3 fatty acids, the supplements of n-3 fatty acids with lutein, vitamin E, and zeaxanthin as an adjunct therapy should be introduced.

REFERENCES

Algvere P.V. and Seregard S. 2002. Age-related maculopathy: Pathogenic features and new treatment modalities. *Acta Ophthalmol. Scand.*, 80, 136–143.

Barchiesi B.J., Eckel R.H., and Ellis P.P. 1991. The cornea and disorders of lipid metabolism. *Surv. Ophthalmol.*, 36, 1–22.

Bazan H.E., Bazan N.G., Feeney-Burns L. et al. 1990. Lipids in human lipofuscin-enriched subcellular fractions of two age populations. Comparison with rod outer segments and neural retina. *Invest. Ophthalmol. Vis. Sci.*, 31, 1433–1443.

Berman E.R. 1991. Lens. In: *Biochemistry of the Eye*, Blakemore C. (ed.). New York: Plenum Press, pp. 201–290.

Checchin D., Sennlaub F., Levavasseur E., Leduc M., and Chemtob S. 2006. Potential role of microglia in retinal blood vessel formation. *Invest. Ophthalmol. Vis. Sci.*, 47, 3595–3602.

Cho E., Hung S., Willet W.C., Spiegelman D., Rimm E.B., Seddon J.M. et al. 2001. Prospective study of dietary fat and the risk of age-related macular degeneration. *Am. J. Clin. Nutr.*, 73, 209–218.

Connor K.M., SanGiovanni J.P., Lofquist C., Aderman C.M., Chen J., and Higuchi A. 2007. Increased dietary intake of n-3-polyunsaturated fatty acids reduces pathological retinal angiogenesis. *Nat. Med.*, 13, 868–873.

Duncan G., Williams M.R., and Riach R.A. 1994. Calcium cell signaling and cataract. *Prog. Retin. Eye Res.*, 13, 623–652.

Feeney L. 1978. Lipofuscin and melanin of the human retinal pigment epithelium: Fluorescence, enzyme cytochemical and ultrastructural studies. *Invest. Ophthalmol. Vis. Sci.*, 17, 583–600.

Katz M.L. and Redmond T.M. 2001. Effect of Rpe 65 knockout on accumulation of lipofuscin fluorophores in the retinal pigment epithelium. *Invest. Ophthalmol. Vis. Sci.*, 42, 3023–3030.

Katz M.L., Gao C., and Rice L.M. 1996. Formation of lipofuscin-like fluorophores by reaction of retinal with photoreceptor outer segments and liposomes. *Mech. Ageing Dev.*, 92, 159–174.

Kremer J.M. 2000. N-3 fatty acid supplements in rheumatoid arthritis. *Am. J. Clin. Nutr.*, 71, 349–351.

Malcolm C.A., Hamilton R., McCulloch D.L., Montgomery C., and Weaver L.T. 2003. Scotopic electroretinogram in term infants born of mothers supplemented with docosahexaenoic acid during pregnancy. *Invest. Ophthalmol. Vis. Sci.*, 44, 3685–3691.

McCarthy M., Innis S., Dubord P., and White V. 1994. Panstromal Schnyder corneal dystrophy: A clinical pathologic report with quantitative analysis of corneal lipid composition. *Ophthalmology*, 101, 895–901.

Moss S.E., Klein R., and Klein B.E. 2000. Arcus senilis and mortality in a population with diabetes. *Am. J. Ophthalmol.*, 129, 676–678.

Newell F.W. 1992. *Ophthalmology: Principles and Concepts*, 7th edn. St. Louis, MO: Mosby-Year Book, Inc., pp. 71–98.

Panozzo G., Babighian S., and Bonora A. 1998. Association of xerophthalmia, flecked retina, and pseudotumor cerebri caused by hypovitaminosis A. *Am. J. Ophthalmol.*, 125, 708–710.

Rae J. 1994. Physiology of the lens. In: *Principles and Practice of Ophthalmology: Basic Sciences*, Albert D.M. and Jakobiec F.A. (eds.). Philadelphia, PA: WB Saunders, pp. 123–146.

Robison W.G. Jr., Kuwabara T., and Bieri J.G. 1980. Deficiencies of vitamins E and A in the rat. Retinal damage and lipofuscin accumulation. *Invest. Ophthalmol. Vis. Sci.*, 19, 1030–1037.

Robison W.G. Jr., Kubuwara T., and Bieri J.G. 1982. The roles of vitamin E and unsaturated fatty acids in the visual process. *Retina*, 2, 263–281.

Shapiro L.A. and Farkas T.G. 1977. Lipid keratopathy following corneal hydrops. *Arch. Ophthalmol.*, 95, 456–458.

Smith W., Mitchell P., and Leeder S.R. 2000. Dietary fat and fish intake and age-related maculopathy. *Arch. Ophthalmol.*, 118, 401–404.

Uchida H., Hayashi H., Kuroki M., Uno K., Hamada H., Yamashita Y., Tombran-Tink J., Kuroki M., and Oshima K. 2005. Vitamin A up-regulates the expression of thrombospondin-1 and pigment epithelium-derived factor in retinal pigment epithelial cells. *Exp. Eye Res.*, 80, 23–30.

Weiss J.S. 1996. Schnyder crystalline dystrophy sine crystals; recommendation for a revision of nomenclature. *Ophthalmology*, 103, 465–473.

14 Plant Lipids and Oils

Jan Pokorný and Štefan Schmidt

CONTENTS

14.1 INTRODUCTION

Plant lipids belong to two categories: (1) structural lipids necessary for the metabolism of cells, all plant tissues, and the whole organism; and (2) storage lipids, which serve as a source of energy for germinating plant seeds. With a few exceptions, only the latter lipids are used in the food industry. Four large research areas are the center of interest: (1) new breeding technologies, (2) interesterification of triacylglycerols (TGA), (3) *trans*-acid free products, and (4) products with low energy content.

Structural lipids have their own significance in nutrition. They are present in all plant tissues and, therefore, they are consumed in food. They affect the sensory value of foods. They influence the oxidation of foods on storage and cooking of meals.

14.2 SOURCES OF PLANT LIPIDS

14.2.1 STRUCTURAL LIPIDS

Structural lipid sources are almost ubiquitous in all plants. Small amounts of lipids, usually 0.5%–2.0% (dry weight), are present in all cells, where they have an essential role in plant metabolism, mainly as constituents in intracellular membranes. Only a part is bound in glycerol esters. The fatty acids (FA) are mainly bound in phospholipids, glycolipids, gangliosides, and mucolipids or in lipoproteins. Contrary to storage lipids, chemical structures and composition of FA are rather similar in all plant families. Plants of some species contain an unusual FA in the storage seed oil or a few. In such cases, the particular FA is either absent or present only in traces bound in structural lipids. Lipids of algae are not included here because their FA composition is more similar to that of marine invertebrates than to lipids of terrestrial plants. Some microbial lipids are suitable as a substitute of seed lipids and are used as feeds.

Structural lipids are often bound to proteins in the form of relatively polar, water-soluble lipoproteins. The ratio of lipids and proteins may vary considerably in lipoproteins, but they are still water soluble and able to transport nutrients. Therefore, structural lipids are not completely extracted with nonpolar organic solvents, unless the lipid–protein bonds of lipoproteins have been destroyed, either by heating (best, with steam) or by addition of alcohols. Some lipids are also bound to amylose and other high molecular weight carbohydrates (Chapter 22). Bonds between the protein and structural lipids are essential, and nearly no covalent bonds are present, excepting some S–S bonds. Physical operations are then sufficient for their cleavage.

14.2.2 PLANT WAXES

The composition of lipids on the surface of leaves, stems, and fruits is quite different from that of structural intercellular lipids. Their role is the protection of sensitive plant tissues against the loss or intake of water, gases, and biologically important volatiles such as terpenes. These lipids on the surface are mainly waxes.

Waxes are esters of FA with monofunctional alcohols or bifunctional alcohols, mostly aliphatic. The FA composition is completely different from that of structural lipids of the particular plant. They contain both esterified and free saturated long-chain FA, sometimes branched or hydroxyacids. Both the FA and alcohols may contain a keto and one or a few hydroxyl groups.

Some plant waxes are of commercial importance, such as carnauba or candellila wax. They are solid at ambient temperature in countries with a temperate climate, but sometimes plastic or even liquid in tropical countries. An exception is liquid jojoba wax, which is liquid even in Central Europe, but it belongs to storage lipids.

Thin film of waxes on the surface of apples and other fruits from the temperate zone are solid or semisolid pastes, consisting of nonvolatile terpenes, ceryl cerotate, ceryl palmitate, and other esters of similar composition. As waxes are expensive, they are sometimes supplemented by hydrocarbons or similar compounds in the industry.

In the wax on lettuce leaves, higher alcohols prevail, with only small amounts of free FA (Bakker et al., 1998). Other components, such as alkenes, ketones, esters, and secondary alcohols, were also detected in other vegetables (e.g., in kale or rutabaga).

14.2.3 STORAGE LIPIDS IN SEEDS

New progress concerning sources of plant lipids was reviewed by Gunstone (2006), where he classifies major and minor sources of oils.

Seeds must always contain energy reserves, which are consumed by the germ during germination. The energy source can be starch (or other polysaccharides, such as inulin) or lipids (chiefly oil in form of lipoproteins). Both energy sources occur only rarely in the same plant seed, a typical example are soybeans. However, even the starchy seeds have high lipid content in the germ, e.g., in wheat, corn germs (Table 14.1).

Lipids liquid in the country of production are called oils, semisolid lipids used to called butter, such as cocoa butter or shea butter. This term is now avoided, because of the confusion with milk butter, and is replaced by the term fat. Solid lipids are called fats, and all plant butters are solid in the temperate zone. Coconut or other lipids of plants are liquid in the producing country, but in Europe they are semisolid, but still they are called oils. Storage lipids consist mainly of TAGs with a specific steric configuration. Only rarely (jojoba oil) they consist of waxes.

The lipid content in seeds may be higher than 30%, in several cases 40%–50%, only rarely it is lower, e.g., in soybeans (Table 14.2). The seeds may be used directly as food, dried or roasted, mainly nuts (such as peanuts, almonds, and walnuts). In most cases, they are processed in the industry to obtain plant oils. During ripening, TAGs are synthesized from FA and glycerol via monoacylglycerol (MAG) and

TABLE 14.1
The Main Fatty Acids in Cereal Lipids (% Total Fatty Acids)

Fatty acid	Palmitic	Stearic	Oleic	Linoleic	Linolenic
Corn germ	11–17	2–3	24–42	39–60	<1
Rice bran	16–28	2–4	42–48	15–36	<1
Oat oil	14–23	1–3	29–53	24–48	1–3
Rye germ	9–21	0–2	7–35	48–72	3–8
Wheat germ	12–19	<3	14–23	50–56	3–7

Source: Adapted from AOCS.

TABLE 14.2
Lipid Content of Food Materials (% Dry Weight)

Lipid Source	Lipid Content
Copra (*Cocos nucifera*)	63–70
Palm kernels (*Elaeis guineensis*)	40–52
Cocoa beans (*Theobroma cacao*)	54–58
Shea beans (*Butyrospermum parkii*)	45–55
Cottonseed (*Gossypium hirsutum*)	22–24
Hazelnuts (*Corylus avellana*)	60–68
Almonds (*Prunus amygdalus*)	60–65
Sesame seeds (*Sesamum indicum*)	50–55
Sunflower seeds (*Helianthus annuus*)	42–63
Flaxseed (*Linum usitatissimum*)	38–44
Hempseed (*Cannabis sativa*)	30–35
Poppyseed (*Papaver somniferum*)	40–51
Walnuts (*Juglans regia*)	56–59
Peanuts (*Arachis hypogaea*)	45–50
Soybeans (*Soja hispida*)	18–22
Rapeseeds (*Brassica napus*)	40–48
Palm fruits (*Elaeis guineensis*)	30–40
Olive fruits (*Olea europaea*)	12–50
Avocado fruits (*Persea americana*)	40–80
Corn germ (*Zea mays*)	15–20
Wheat germs (*Triticum aestivum*)	7–12
Rice bran (*Oryza sativa*)	8–16
Wheat flours	0.8–1.6
Wheat breads	0.9–1.7
Chocolate	32–40
Beans, bean meal	1.6–1.9
Peas, pea meal	1.5–1.7
Mushrooms	3.1–3.4
Fruits (different)	1.0–2.8
Vegetables (different)	1.3–4.0
Potatoes (peeled)	0.7–0.9
Sugar	0

diacylglycerol (DAG). The latter are present in small amounts in most plant oils produced in the industry, particularly in case of crops in unfavorable climatic conditions (rains), when seeds are cropped either unripe or already starting germination. The opposite process proceeds during seed germination (Figure 14.1), when TAGs are hydrolyzed by lipases to DAGs, and further to MAGs and free FA.

In addition to seeds, seed pericarp is sometimes rich in oil, such as in olives, avocado, or palm pericarp. Oils produced from these sources, all are used in the industry, have a rather different FA composition from those of the seeds of the same plant.

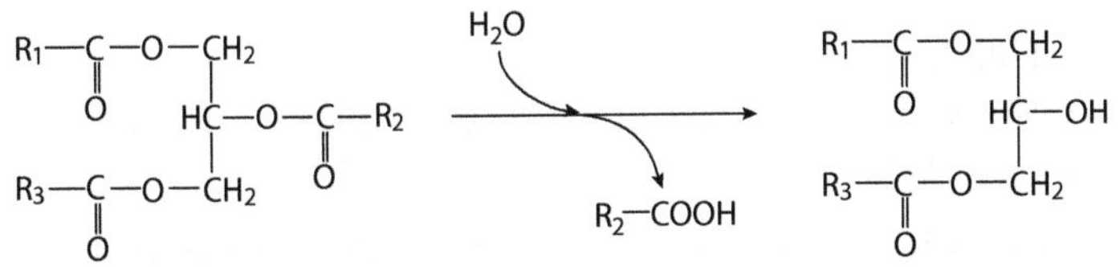

FIGURE 14.1   Enzymic hydrolysis of triacylglycerols during storage.

Lipids are distributed in seed cells in tiny droplets (about 1 µm in diameter) emulsified in protein containing medium as lipoproteins (Chapter 22). Small amounts of lipids are bound to protein as lipoproteins to complexes with starch. In the emulsion, lipids are primarily in liquid form, at least at ambient temperature during the ripening of fruits. Under these conditions, the melting point is generally lower than 10°C, and after processing, they form viscous liquids called oils after olive oil. If some precipitate appears at +5°C, it is removed in the industry by filtration.

For edible purposes, TAGs have the greatest nutritional importance of oil components. Therefore, the processing conditions are selected in such a way as to obtain TAGs in high yield, high purity, and low price (Section 14.5.5).

14.3   EFFECT OF PLANT SPECIES ON LIPID COMPOSITION AND QUALITY

Plant lipids consist of neutral lipids (Table 14.2), mostly TAGs, small amounts of phospholipids and glycolipids. From the standpoint of utilization in human nutrition, the TAGs are the most important components. Many different TAGs exist in every plant lipid fraction, differing in the distribution of their FA. Three FA are bound to a molecule of glycerol, but the three FA are rarely identical in plant lipid TAGs, or only as minor components. If the FA bound in the 1- and 3-positions are different, the carbon atom at the 2-position becomes asymmetric so that the number of isomers thus increases. Therefore, it is better to use the *sn* system (Chapter 1) in giving the exact number of TAGs. Because of large molecular weight, the difference between the two isomers is small.

The distribution of acyls among different TAGs is not random among different TAGs because of the stereospecificity of plant lipases. The *sn*-1 position bound acyls are preferably occupied by saturated FA, while the position *sn*-2 mainly by polyunsaturated acyls. The amount of 1-palmitoyl-2,3-dioleoyl glycerol in olive oil is approximately 20–50 times greater than that of the isomeric 2-palmitoyl-1,3-dioleolyl glycerol (Boskou, 1996).

Of course, there is no strict selectivity (as in life generally), which increases the number of isomers present. An example is shown in Table 14.3. In some lipids (for example in cocoa butter) the number of different major TAGs is reduced to just a few, which results in certain specific properties of this fat. Transesterification results in an increased melting point by 20°C only by random rearrangement of the TAGs structure.

The reactivity and the nutritional value of a particular TAG depend on stereospecific isomerism. If a particular TAG composition is undesirable to any of the

TABLE 14.3

The Main Fatty Acids of Palm Seed Lipids (% Total Fatty Acids)

Fatty Acid	Coconut Oil	Palm Kernel Oil	Babassu Oil
Octanoic	4.6–9.4	2.1–4.7	2.6–7.7
Decanoic	5.5–7.8	2.6–4.5	1.2–7.6
Lauric	45.1–50.3	43.6–53.2	40.0–55.0
Myristic	16.8–20.6	15.3–17.2	11.0–27.0
Palmitic	7.7–10.2	7.1–10.0	5.3–11.0
Stearic	2.3–3.5	1.3–3.0	1.8–7.4
Oleic	5.4–8.1	11.9–19.3	9.0–20.0
Linoleic	1.0–2.1	1.4–3.3	1.3–6.6

Source: Adapted from AOCS, *Official Methods and Recommended Practices of the AOCS*, 5th edn., AOCS Press, Champaign, IL, 1997.

above mentioned purposes, it is possible to randomize the product, most often by treatment with alkaline catalysts. In a randomized fat or oil, the FA distribution among TAG species is quite random, not selective as before. The randomization is particularly important in solid fats, because the melting interval and the texture are changed. The melting point of cocoa butter increases from 34.8°C to 54.2°C due to randomization.

The specific distribution of FA in a molecule also occurs in polar lipids, such as DAGs or phospholipids. However, it does not affect the properties of the respective food products, fats, and oils in very high degree as they are only trace or minor components when compared with TAGs.

In a mixture of TAGs, such as edible oils, cooking fats, or margarine, the reactivities of individual TAG species differ significantly because the more saturated species are less reactive than TAGs containing polyenoic FA. For example, dilinoleoylmonooleoylglycerol is oxidized more easily in a mixture than monolinoleoyldioleoylglycerol (Pánek et al., 1995). Therefore, the TAG composition changes a little during food processing or storage.

14.4 EFFECT OF PLANT SPECIES ON THE FATTY ACID COMPOSITION

14.4.1 FATTY ACID COMPOSITION OF TRADITIONAL EDIBLE PLANT OIL

The FA composition of plant lipids differs considerably from that of lipids of other origin, even when the most common FA (e.g., palmitic and oleic acids) are present in all types of lipids. While structural plant lipids have similar FA composition in all plant species, storage lipids are somewhat different. Their composition is often specific for some plant families or even for a small group of related plant species.

Seed and fruit lipids of different plant species have been formerly determined according to their texture. Nowadays, they are distinguished according to their FA composition (Hilditch and Williams, 1964). They can be classified into groups according to similar FA compositions of their oils:

Group 1: A group that, typically, has a high content of lauric acid and medium-chain FA of low unsaturation (Table 14.3). Oils from palm seeds belong to this group, such as coconut and palm kernel oil. These oils begin to solidify below 25°C and are very stable under storage or during frying.

Group 2: A group of seed oils of some tropical trees, such as cocoa or shea butter, and also palm oil, which is produced from the pericarp, not from the seed. These oils are characterized (Table 14.3) by their high content of palmitic acid and moderately lower content of stearic acid. They are solid at ambient temperature in temperate climates (Table 14.4).

Group 3: A group of high-oleic acid, low-linoleic acid, such as olive oil from the pericarp of olive fruit. Seed oils, such as almond, hazelnut, or avocado oils (Werman and Neeman, 1987), also belong to this group (Table 14.5).

Group 4: A group with high-linoleic acid content, but containing only traces of linolenic acid. Examples include sunflower, safflower, sesame, peanut, and cottonseed oils.

Group 5: A group containing linolenic acid as the most unsaturated FA. Examples include soybean, rapeseed, hempseed, linseed, and perilla oils.

Germs oil belong either to the Group 4 or Group 5, but the linolenic acid content is very low (Table 14.1). The canola sprouts from different locations had 27%–33% saturated FA and the ratio of 18:2/18:3 differed between 1.00 and 2.09 (Yoshida et al., 2007).

All unsaturated FA in the previously mentioned groups have mainly the *cis*-configuration of double bonds, but also contain small amounts of *trans*-double bonds (less than 1%), i.e., less than in milk butter. In refined or virgin oils, the content ≥1% is a sign of improper technology, but concentrations up to 1% are common in oils processed by refining (see Section 14.6). They are formed most often in course of deodorization. They may be present already in crude oils because they are formed

TABLE 14.4
Main Fatty Acids of Seed Butters

Fatty Acid	Illipe Butter	Cocoa Butter	Shea Butter	Borneo Tallow
Palmitic	23	25–27	4–8	18–21
Stearic	23	33–37	36–41	39–43
Oleic	34	34–36	45–50	34–37
Linoleic	14	3–4	4–8	<1

Source: Adapted from AOCS, *Official Methods and Recommended Practices of the AOCS*, 5th edn., AOCS Press, Champaign, IL, 1997.

TABLE 14.5
The Main Fatty Acids of Liquid Edible Oils (% Total Fatty Acids)

Oil Source	Palmitic	Stearic	Oleic	Linoleic	Linolenic
Almond	4–13	0–10	43–60	20–34	0
Avocado	9–18	<1	56–74	10–17	0–2
Cashew nut	4–17	2–11	61–80	19–22	<1
Coffee bean	35–42	7–11	8–10	36–43	<1
Cottonseed	21–26	2–3	15–22	46–58	<1
Hazelnut	5–7	1–3	72–84	6–22	<1
Hempseed	6–12	1–2	11–16	45–66	15–30
Linseed	7	4	20	17	52
Olive	8–20	1–5	55–83	4–21	<1
Palm oil	40–46	4–7	36–41	9–12	<1
Palm olein	37–43	4–5	40–44	10–13	<1
Palm stearin	48–74	4–6	15–36	3–10	<1
Peanut	8–14	2–4	36–67	14–43	Trace
Pecan nut	5–11	1–6	49–69	19–40	0–3
Perilla	6–7	1–2	13–15	14–17	44–64
Poppyseed	7–11	1–4	16–30	62–73	Trace
Rapeseed (canola)	3–6	1–3	52–67	16–25	6–14
Safflower	5–8	2–3	8–21	68–73	Trace
Sesame	8–10	5–6	36–48	41–48	<1
Soybean	10–13	3–5	18–25	50–57	5–10
Sunflower	6–4	2–7	13–40	40–74	<1
Walnut	7–8	2	17–19	56–60	1–14

Source: Adapted from AOCS, 1987.

there by enzyme-catalyzed oxidation of polyunsaturated lipids during storage of seeds in elevators.

Plant desaturases are not entirely specific; therefore, oleic (9-octadecenoic) acid is accompanied by small amounts of 11-octadecenoic acid. Linoleic (9,12-octadecadienoic) acid is accompanied by small amounts of 9,15-octadecadienoic acid. By modern breeding techniques and genetic manipulation, modified oil may now belong to another group than indicated above. For more details, see later in this chapter.

14.4.2 OILS CONTAINING UNUSUAL OR SPECIFIC FATTY ACIDS

The lipids of some plant species contain special FA not found in other plant families (see Table 14.6). They contain the double bond in a different position, like in parsley oil (Diedrich and Henschel, 1991a), *trans-* or conjugated double bonds, and acetylenic bonds (Diedrich and Henschel, 1991b). Some are suitable for human nutrition, at least in small amounts (*Umbelliferae* and *Cruciferae* plants). Others, like tung oil, are consumed only in subtropical or tropical countries in Asia. Still other oils

TABLE 14.6
Examples of Unusual Fatty Acids in Plant Lipids

Source (English Name)	Source (Latin Name)	Trivial Name of Fatty Acid	Systematic Name of Fatty Acid
Parsley and related plants	*Umbelliferae*		
	Brassicaceae	Petroselinic acid	6-Octadecenoic
Mustard oil and related	*Ulmus americana*	Erucic acid	13-Docosenoic
	Pinus pinea	Capric acid	Decanoic
Elm	*Limnanthes douglasii*		5,11,14-Eicosatrienoic
Pine nut	*Pinus pinea*		5,13-Dosadienoic
Evening primrose	*Oenothera*	γ-Linolenic	6,9,12-Octadecatrienoic
Tung oil	*Calendula officinalis*		*cis,cis,trans*-9,11,13-Octadecatrienoic
Marigold	*Parinarium laurinum*	Isanic	*trans,trans,cis*-8,10,12-Octadecatrienoic
			9,11,13,15-Octadeca-tetraenoic acid
			Octadeca-17-en-9,11-diynoic
Boleko	*Ongokea gore engler*	Malvalic, sterculic	See Figure 14.3
Sterculia	*Sterculia foetida*	Vernolic	See Figure 14.3
		Chaulmoogric	See Figure 14.3
Chaulmoogric oil	*Vernonia antihelmintica*	Licanic	4-Keto-9,11,13-octadecatrienoic
Oiticica	*Licania rigida*	Ricinoleic acid	12-Hydroxy-9-octadecenoic
Castor oil	*Ricinus communis*		

Source: Adapted from AOCS, *Official Methods and Recommended Practices of the AOCS*, 5th edn, AOCS Press, Champaign, IL, 1997.

are used for nonedible purposes. Some examples of unsaturated FA are given in Figure 14.2. Included are also those FA, which normally exist only in animal lipids, but they can be found in genetically manipulated plants, too.

Some FA are substituted with a keto, epoxy, of hydroxy functional groups (Table 14.6 and Figure 14.3). They may be used for nonedible purposes (such as castor oil or oiticica oil). Chaulmoogra oil with some related FA is used for pharmaceutical purposes.

14.4.3 Oils with Modified Fatty Acid Composition

Some traditional oils are not suitable for nutrition or are less suitable for specific technological uses (such as deep frying). Therefore, their FA composition has been

$$H_3C—(CH_2)_{10}—CH=CH—(CH_2)_4—COOH \qquad A$$

$$H_3C—(CH_2)_7—CH=CH—(CH_2)_{11}—COOH \qquad B$$

$$H_3C—(CH_2)_3—(CH=CH)_2—(CH_2)_7—COOH \qquad C$$

$$H_3C—CH_2—(CH=CH)_4—(CH_2)_7—COOH \qquad D$$

FIGURE 14.2 Fatty acids with different double bond number and positions. A, petroselinic acid; B, erucic acid; C, conjugated linoleic acid; D, conjugated octadecatetraenoic acid.

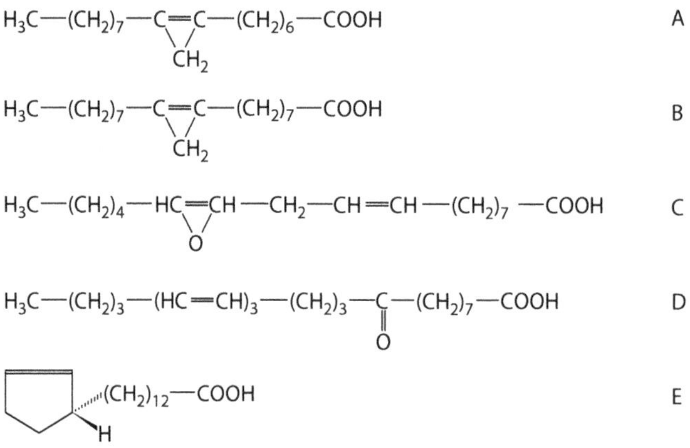

FIGURE 14.3 Unusual chain-substituted fatty acids. A, malvalic acid; B, sterculic acid; C, vernolic acid; D, licanic acid; E, chaulmoogric acid.

modified (Table 14.7), usually by classical or modified breeding methods, in recent years also by genetic manipulation (e.g., to introduce highly polyunsaturated FA (Figure 14.4), common otherwise only in fish, or for uses of stable frying oil.

The best example is rapeseed oil (either *Brassica napus* or *Brassica campestris*). The traditional rapeseed oil contained erucic acid (13-docosenoic acid) as its major FA. Its consumption in higher doses was found unsuitable from the point of view

TABLE 14.7
The Main Fatty Acid Groups of Modified Oils (% Total Fatty Acids)

Oil		Saturated	Monoenoic	Dienoic	Trienoic
Safflower oil	Traditional	7–10	8–12	68–83	Trace
	Modified	6–8	74–80	13–18	Trace
Sunflower oil	Traditional	7–14	13–41	40–74	Trace
	Modified	6–9	70–87	3–20	Trace
Linseed oil	Traditional	11	20	17	52
	Modified	4	22	72	2
Peanut oil	Traditional	18	44	37	Trace
	Modified	13	81	6	Trace

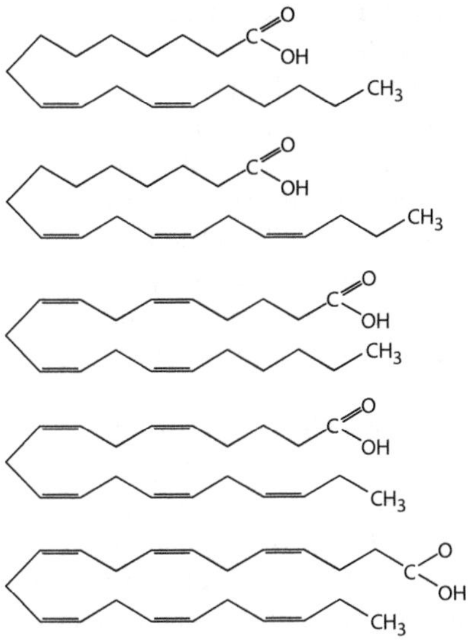

FIGURE 14.4 Chemical structures of polyenoic fatty acids. A, linoleic acid; B, linolenic acid; C, arachidonic acid; D, eicosapentaenoic acid (EPA); E, docosahexaenoic acid (DHA).

of health risks. After breeding for many years, rape plant varieties now prevail in Europe (*B. napus*, winter rape) and North America (*B. campestris*, summer rape). They do not contain now more than 1% erucic acid. As the non-erucic acid variety was developed in Canada, it is called canola oil.

14.5 CHANGES DURING RIPENING AND POSTHARVEST STORAGE

14.5.1 CHANGES DURING RIPENING OF PLANT SEEDS

The lipid composition of unripened plant seeds is similar to that of structural lipids in other tissues of the respective plant. The lipid content in seeds is rather low (up to 3%–5%, dry weight). During ripening, lipids are gradually synthesized and the composition of these new lipids corresponds to that of storage lipids. DAG are rapidly formed as intermediates during the TGA synthesis, e.g., in rapeseed (Perry and Harwood, 1993). Free FA are incorporated into the DAG molecule, and their synthesis ceases when the TAG synthesis ceases (May and Hume, 1995). Some free FA usually remain in seeds. At the same time, the content of polar lipids decreases because newly synthesized lipids contain mostly TAGs, and thus dilute the original polar lipids.

On the contrary, the content of phospholipids and galactolipids decreased even during the ripening of tomatoes, in which TAGs do not accumulate substantially (Whitaker, 1994). In this connection, it is interesting that with increasing age in human subjects, the content of linoleic acid in blood plasma decreases and that of eicosapentaenoic acid (EPA) and docosahexaenoic acid (DHA) rises (DeGroot et al., 2009).

The increase of lipid content is fast during the last 2–4 weeks prior to full ripeness. The FA composition at that time is equal to the composition of fully ripe seed. In high-oleic sunflower oil, the oil first develops similarly to that of a traditional cultivar, but oleic acid is preferentially formed during the final stages of ripening. It is important as oleic acid has a peculiar structure, which is crucial for the total structure of some lipids (Figure 14.5).

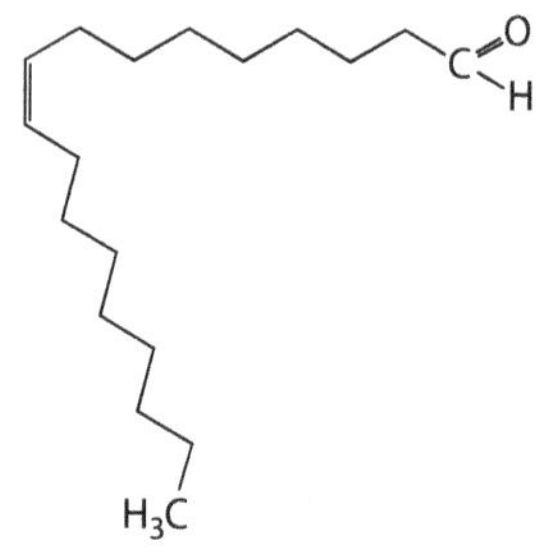

FIGURE 14.5 Chemical structure of oleic acid.

Not only the FA composition, but also their distribution among the TAG molecules changes. For example, the content of palmitic acid in 1,3-positions increased, but that of oleic acid decreased during ripening of palm fruit (George and Arumughan, 1994). At the same time, the enzymic activities of the respective enzymes increase. Tocopherols are formed at the same time as unsaturated FA, so that the stability against oxidation reaches the optimum at the stage of full ripeness.

Changes in plant tissues other than the seeds are much smaller prior to harvest ripeness, except for the pericarp in olives and palm fruit, where the changes are similar to those occurring in seeds.

14.5.2 Postharvest Changes of Plant Lipids

After the harvest of seeds, the lipids are relatively stable if the seeds are dry. In the rainy season during the harvest, it is necessary to dry seeds within a short time; otherwise, postharvest changes are very fast, because lipases and lipoxygenases are activated by the presence of water, and the seed quality deteriorates. Molds growing on the surface of moist seeds contribute to the deterioration of seed lipids. TAGs are hydrolyzed to DAG. It is preferable to store the harvested seeds intact; they should be free of any broken seeds. Damaged seeds are rapidly spoiled (e.g., poppyseeds or peanuts). Nuts are much more stable if they remain unshelled.

Under optimum storage conditions, seeds rich in oil can be stored until the next harvest without significant changes, even in case of relatively polyunsaturated lipids, such as in poppyseeds, hempseeds, and walnuts. The most suitable moisture content depends on the particular seed and varies between approximately 8%–10% dry weight, in most cases. Very dry seeds may be less stable.

Hydrolases have low activity in air-dried seeds, but still, the content of free FA in the extract slowly increases if the storage time increases by several months. Phospholipids are partially converted into lysophospholipids and phosphatidic acid. Hydrolytic changes catalyzed by enzymes are very fast in cereal bran and germ lipids. For example, the free FA content nearly doubles within two days in moist rice germ (Dong and Chung, 1999), but it did not change significantly if the material has been dried immediately. Therefore, germs and brans should be rapidly processed by steaming. Similarly, rapid changes may be observed in olive oil. If the free FA content reaches a value greater than approximately 10 mg/kg, bitter substances are formed, and the oil is refined only with difficulty and with great losses. Plant tissues other than the seed, such as lettuce or onion leaves, have higher water content and should be rapidly blanched to deactivate enzymes.

Another group of postharvest processes are enzyme-catalyzed oxidation reactions, such as the degradation by lipoxygenases. Hydroperoxides are slowly formed even in relatively dry seeds, but are again decomposed by hydroperoxide lyases and other hydroperoxide-cleaving enzymes. Volatile decomposition products cause rancidity, which is particularly intensive in lipids containing bound linolenic acid. Hay-like off-flavors are produced in green parts of the plants as they are rich in linolenic acid, the oxidation products of which give rise to such off-flavors. If the oxidation reactions have proceeded to a low degree only, they may be considered as desirable. For example, the typical flavors of cucumbers and virgin olive oil are due to lipid oxidation products. Even if oxidoreductases are inactivated by heating or if seeds are dry, the oxidation may take place via other mechanisms (such as autooxidation). For details, see Chapter 9.

Contrary to the lipids in animal products, plant lipids always contain natural antioxidants, mostly tocopherols (Hall, 2001; Yanishlieva and Heinonen, 2001). They control the oxidation that occurs during storage, but they are slowly consumed by inhibition reactions with free radicals. Therefore, the resistance of plant lipids to oxidation slowly decreases on storage, especially in presence of light.

14.6 EFFECT OF PROCESSING ON THE FUNCTIONAL PROPERTIES AND NUTRITIONAL VALUE OF PLANT LIPIDS

14.6.1 CHANGES DURING THE PROCESSING OF OILSEEDS OR OIL-BEARING FRUIT PERICARP

Oilseeds are crushed, lipid–protein bonds are destroyed by steaming, and the resulting material is expeller pressed. Cakes obtained in this operation should contain 10% oil, but sometimes even 25% oil, to make further extraction economic. They are extracted with a solvent, primarily hexane. Other solvents are possible, too, but hydrocarbon solvents are cheaper than diethyl ether and similar organic solvents. The extracted meal still contains 2%–3% lipids, but the residual lipids are utilized as feed, so that the low content of oil is not objectionable. Crude oils obtained by expeller pressing and solvent extraction are usually mixed and refined together. The first refining step is the removal of phospholipids. This step is called degumming, even when no gums are present there.

Degummed oils are further refined either by alkali refining or by physical refining (Figure 14.6). In the alkali refining process, free FA are first neutralized with sodium hydroxide or sodium carbonate and removed as a water solution. The residues of alkali are removed by washing, and the deacidified oil is dried. Both the chlorophylls and carotenoids are removed by bleaching with bleaching earth and filtration. The resulting oil still contains some volatile oxidation products and other compounds, imparting off-flavors to bleached oil. They are removed in the process of deodorization, consisting of the application of superheated steam under reduced pressure and

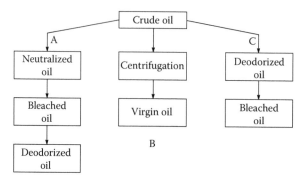

FIGURE 14.6 Procedures of refining of crude degummed oils. A, classical alkali refining; B, virgin oil production; C, physical refining.

at temperatures sometimes exceeding 200°C. The lower the temperature, the longer time of deodorization is required. The time depends on the equipment too.

In the physical refining process (Čmolík and Pokorný, 2000), phospholipids are efficiently eliminated, and the degummed oil is thoroughly bleached and deodorized—distilled off with other volatiles during the deodorization step. This technological process is possible only in case that the free FA content is low.

During the refining, the sensory value of the oil is substantially improved, but the nutritional value is impaired by partial removal of carotenoids, tocopherols, and phytosterols, and by moderate transformation of *cis*-unsaturated acids to *trans*-unsaturated acids; linolenic acid is especially sensitive (Čmolík and Pokorný, 2000). The industry hesitates to buy oils with more than 1.0% *trans*-unsaturated FA. Refined oils are nearly flavorless, tasteless, and colorless, and possess good stability against autooxidation; they remain without adverse changes for more than one year in a refrigerator.

Olive fruits are processed in another way (DiGiovacchino, 1996). Because the pericarp contains water, they should be processed by pressing very rapidly, without any preliminary heating. Such oils are called virgin olive oils. They have a typical characteristic flavor and greenish color. Only oils of lower quality are refined and cannot be declared as virgin oils, but only as olive oil.

It is possible to produce virgin oils from other sources; however, their flavor is usually not sufficiently acceptable and equal to virgin olive oils. Sunflower, hazelnut, and almond oils are best among them. Some consumers do not like to purchase refined oils in fear of the negative effect of technological treatment. They are willing to buy virgin oils as they are healthier.

Palm oils should also be processed quickly; they are rarely sold as virgin oils, and are primarily refined. Because the nice red or orange color of crude oils disappears, the carotenoid pigments are often returned back to refined oils. Consumers like red color as they believe the oil has not been refined.

Solid fats can be produced by an alternative method. High melting or fully hydrogenated oils are mixed with oils and interesterified by catalytic support of sodium methoxide or metallic sodium. After this ester interchange between solid TAGs and liquid oil occurs, the resulting fat is plastic, but does not contain any *trans*-FA.

Low-energy lipids are formed by ester interchange of high-melting and, thus, less digestible TAGs with medium chain-length FA, which are not deposited in the adipose tissue. They have substantially lower energy content than normal fats, as three oxygen atoms in the molecule are the same in normal oils and in interesterified fats, but the number of carbon atoms in the latter is substantially smaller.

The TAG composition can be also modified by interesterification, catalyzed by stereospecific lipases, mostly of microbial origin (Xu, 2000). Cocoa butter substitutes are produced from palm olein by introducing stearoyl residues into the molecules. The product should not be labeled as pure cocoa butter.

14.6.2   Effect of Blanching and Germination

Lipids in fruits and vegetables would be rapidly destroyed by the enzyme activity and by microorganisms. Therefore, they are often treated with steam to inactivate the enzymes and kill the microorganisms. The heating should be rapid; otherwise, both hydrolytic and oxidative lipid degradation would take place to a significant degree.

In some technological processes, seeds are germinated as a special step, e.g., in pulses, in order to remove the flatulence factors. They are soaked in water to initiate enzymatic processes. When the germination reaches the appropriate degree, the processes are interrupted by rapid drying. Malting of barley and other cereal grains are typical examples. Lipids are rapidly hydrolyzed during the germination, and liberated free FA are consumed as an energy source, and therefore, the lipid content decreases (Patterson, 1998). Malted cereals are used for brewing or in the production of breakfast cereals.

14.6.3   Changes of Lipids during Culinary Food Preparation

Changes occur with the addition of water to a food material (e.g., during soaking or dough making). During boiling, lipid degradation is slow because enzymes are soon destroyed from TAGs, and oxygen in water is rapidly consumed or removed with steam. If boiling takes a long time, however, TAG hydrolysis might take place, and the broth becomes turbid because of dispersed liberated FA, and partially esterified glycerol esters.

During extrusion cooking, no perceptible changes occur because the process is very fast and the temperature sufficiently high for enzyme inactivation in less than 1 min. The content of extractable lipids may be affected by lipid–starch interactions (Chapter 22). The maximum temperature is usually too low to initiate autooxidation reactions, but partial destruction of natural antioxidants does take place.

During baking, lipids are minimally affected, except in the surface layers. Moderate oxidation may occur, and the oxidation products reacting with proteins influence the texture. Roasting proceeds at high temperatures, often at more than 200°C, but the process takes only up to 20 min. Nevertheless, oxidation reactions may be pronounced, especially in the case of coffee or peanuts and coffee substitutes. Microwave heating causes oxidative changes if the heating time is sufficiently long and the temperature attains high values. As the rate of heating is higher in a microwave oven than in a conventional oven, the lipid damage is only moderate.

The most severe degradation takes place during deep fat frying in oils (see Chapter 21), especially in frying oil and in the surface layers of fried food. Fortunately, plant foods are usually fried in a batter so that the inner layers are protected against degradation processes, and several food components inhibit oxidation reactions (Pokorný, 1999).

14.7 ROLE AND CHANGES OF PLANT LIPIDS IN PROCESSED FOODS

Plant lipids increase the nutritional value of food; they contain more essential FA than animal fats. They also contain tocopherols and tocotrienols, which are the most important sources of vitamin E. There exists a disadvantage that plant lipids (similarly to animal fats) considerably increase the content of available energy. Therefore, the consumption of fried foods should be restricted because they are rich in fat.

Lipids affect the functional properties of food; for example, they help to retain carbon dioxide in fermenting dough, thus increasing the final volume of bakery products. Polar lipids are useful in improving the cohesiveness of food components. They contribute to the pasting properties of cereal dough, stabilize the dispersion, and also affect the surface properties. Phospholipids and MAGs are useful in this respect as they increase the surface activity.

The main importance of lipids is their influence on sensory properties. They affect the texture and increase the viscosity of the morsel by mixing with saliva during chewing. High viscosity is appreciated by most consumers as the preference is inherited. Their interaction products with amino acids, proteins, and carbohydrates (Chapter 22) have favorable effects on the color of food surfaces. Coating food with a layer of oil produces a glossy aspect and improves the appearance of the food product. Oil is frequently applied on ready-to-eat meals.

The most desirable influence of lipids is their effect on the odor and flavor of food products. Plant lipids, being more unsaturated than animal lipids, produce different flavor notes as a result of culinary operations. Flavors originating at roasting or frying temperatures are particularly appreciated.

Fat migration takes place during storage of emulsions, and either lipid and water droplets gradually separate. In chocolate snacks, the solid-to-liquid ratio of TAGs continually changes due to small variations in the storage temperature. The result is the separation of white solid TAGs on the surface, called blooming. In tropical countries, cocoa butter is partially hydrogenated to prevent blooming.

During the storage of foods, volatile compounds produced by lipid oxidation cause rancidity, especially if the lipids contain linolenic acid. Dark flours become easily rancid on storage. Crackers and other durable bakery products should be stored in an inert gas atmosphere or in vacuum or be protected by antioxidants. Roasted peanuts or similar food products may change their agreeable flavor if stored in air. Therefore, they are stored either in nitrogen or under reduced atmospheric pressure. Fried products are most sensitive to oxidative rancidification, especially fried products with weak flavor, such as fried bread, French fries, or potato chips. Dry and deep-frozen products are generally rather sensitive to oxidation because of easier access of air into the inner layers of the food product.

14.8 MODIFICATION OF PLANT LIPIDS

14.8.1 CHEMICAL AND ENZYMIC INTERESTERIFICATION OF LIPIDS FOR USE IN FOODS

There are two important methods for interesterification of lipids: (1) chemical interesterification and (2) enzymic interesterification. Transesterification is the change within the single source of lipids, interesterification means changes between two or several different TAGs or with addition of free FA, ester interchange means either the same as interesterification or addition of new TAGs to the reaction mixture.

The new state and trends in interesterification were reviewed from both aspects (Xu et al., 2006). The chemical interesterification consists in rearranging the distribution of FA in TAGs, without affecting the FA profile. The most widely used chemical catalysts are sodium/potassium alloys or sodium alcoholate. A hydrogen atom is abstracted from the TAG moiety by the action of the catalyst. The enolate anion is thus formed, which abstracts a proton from a partial glycerol ester, and the resulting alcoholate attacks the carbonyl group. This first step is then repeated many times (Dijkstra et al., 2005). The chemical interesterification is now widely used for transforming TAGs in the respective mixture of methyl esters for gas chromatography.

The ester–ester interchange may take place between two or several TAGs and also between methyl or ethyl esters. The products from enzymic interesterification are different from the products of chemical interesterification due to the regional specificity of lipases used. Enzymic acidolysis or alcoholysis are also possible. The enzymic interesterification may proceed in a batch or a packed bed reactor. The latter procedure is more advantageous as the ratio of enzyme to lipid is higher and the course of reaction is better regulated. It may be used for continuous processes. The use of Lipozyme TL IM™ for catalysis of the interesterification gives good results. The reduction of water gives no problem. The stability of the products is higher than in the case of chemical interesterification. For preparation of structured lipids, the acidolysis of 1,3-selective lipase is the first step, if the 2-position should be occupied by a specific FA. The acyl migration during the reaction leads to nonspecificity of the product. The acyl migration can be reduced by temperature programming of the reaction (Yang et al., 2005).

14.8.2 FRACTIONATION OF LIPIDS FOR USE IN FOOD

The interest in fractionation techniques was initiated by elimination of hydrogenated fats from human nutrition. In case of fractionation, lipid material is first crystallized, and the liquid phase is then separated from the solids. Fractional distillation is convenient for the separation of FA or their methyl esters according to their different volatilities corresponding to different molecular weights. The short part devices are necessary to prevent degeneration of TAGs during the reaction. The distillate has a lower solid fat content than the residual fraction left after distillation. The supercritical extraction is based on the different solubility of lipid fractions. The reaction is carried out at high pressure of carbon dioxide (100–400 bar) at the temperature of $30°C–60°C$. The techniques have many applications.

The membrane separation (Parmentier et al., 2003) is performed at a temperature near above the solidification point. These micellar structures improve the filterability. Different conformations of the lipid structures have different lipophilicity between membranes and TGA components. The method is in state of experiments, but the results are promising.

Fractional crystallization is based on the ability of fats to produce crystals. In the wet fractionation, crystals are wetted by detergent, and separated in water medium by centrifugation. In the solvent procedure, lipids are dissolved in acetone or hexane, the dilute solution is cooled, and the crystals are then separated by filtration. The exact procedure depends on the task and lipid composition. Dry fractionation is the simplest and cheapest technique. No additional substances are necessary. The oil is kept refrigerated under controlled conditions. The temperature program depends on oil to be separated and on the requirements for the properties of the fractions.

Winterization at a temperature sufficiently low to crystallize the TAGs of high melting point is a procedure for removing solid fraction from liquid oil, and prevents its cloudiness on storage or the formation of solid precipitates.

14.8.3 HYDROGENATION OF LIPIDS FOR USE IN FOOD

Refined oils can be hydrogenated by gaseous hydrogen in the presence of a nickel catalyst. Double bonds of polyunsaturated FA are converted to monounsaturated FA. The hydrogenation can proceed further with production of saturated FA. The advantage is that hydrogenated oils are solid. They were used for the production of margarines and cooking fats. The disadvantage is that they contain 10%–40% trans-unsaturated acids, which are not desirable for human nutrition in such high amounts, higher than in milk butter (2%). The structure of FA is very different from that of the *cis*-isomers (Figure 14.7). The *trans*-FA has a similar straight chain as saturated FA, while the *cis*-isomer has the chain curbed. Therefore, hydrogenated oils are now used rather for the production of soaps and other nonedible products.

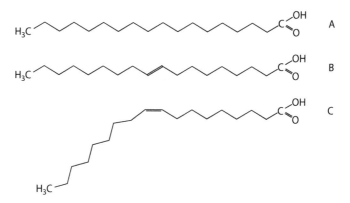

FIGURE 14.7 Effect of *trans*-fatty acid formation on the chemical structure. A, stearic acid; B, elaidic acid (*trans*-octadecenoic acid); C, oleic acid.

Oil is heated with a suspension of nickel/nickel oxide catalyst. It is often mixed by circulating hydrogen in order to prevent sedimentation of the catalyst. The temperature and the pressure during the hydrogenation depend on the industrial equipment. The content of monoenoic and saturated FA can be regulated, but the problem is the formation of *trans*-FA, the content of which in margarines can vary from 12% to 100% in different countries. The lowest values were reported from the United States and Canada, and also in the European Union.

In addition to margarines and cooking fats (shortenings), another application is the use as frying fats and oils for long storage of fried foods.

Experiments were reported of hydrogenated oils with low *trans*-FA content, e.g., decreasing the temperature or using platinum metals (e.g., palladium) as catalysts. A more perspective way is the production of zero-*trans* lipids by interesterification or fractionation. Fractionated palm oils (palm stearin and palm olein) are substitutes of solid plastic fats, with great success. High-stearic acid soybean oil is solid at room temperature, and good to replace hydrogenated fats, when it is combined with palm stearin. Another variant is the interesterification of liquid soybean oil with completely (100%) hydrogenated soybean oils, where the *trans*-FA were also hydrogenated.

The zero-*trans* margarines are now common in North America and European Union. In spite of moderately higher price, consumers and marketing specialists succeeded to introduce them.

14.8.4 New Oils Produced by Breeding or Genetic Manipulation

The progress of the modification of plant oils was recently reviewed (Murphy, 2006), and it is discussed in Chapter 19 of this book. Since about 30 years, experts and consumers have been asking for a plant oil with specific properties. Introduction of oilseed plants from other countries is only a partial solution. The efficient breeding techniques were necessary, and canola oil was the first success of modern breeding. High-oleic and low-linoleic acid sunflower oils were produced on a large scale. A cultivar containing 80% oleic acid and 10% linoleic acid was met with formal difficulties due to patent protection, so that a compromise was developed, called NuSun (it contains about 65% oleic acid, 25% linoleic acid, and 10% saturated FA), which is now in the process of developing in order to be introduced with success.

In presence of *Aeromonas hydrophila* isolates, vegetable oils are converted: rapeseed oil into 7,10,13-hexadecatrienoic acid and 5,8-hexadecadienoicacid, in case of linseed oil also 9,12,15-tetradecadienoic acid; the corresponding alcohols are also formed (Nagao et al., 2009).

The rape plant was found very flexible and easily influenced by breeding. The case of Canola has been already discussed. One cultivar was breeded to contain reduced linolenic acid content so that it would have better resistance against rancidification. Rapeseed oil with high stearic acid content is solid at room temperature. It can be used as a substitute of hydrogenated oils in margarine or cooking fat production. Another cultivar has high lauric acid content, and its oil can be used as a substitute of coconut oil in soap making. Another cultivar has, on the contrary, very high erucic acid content and can be used for technical purposes (e.g., cleavage

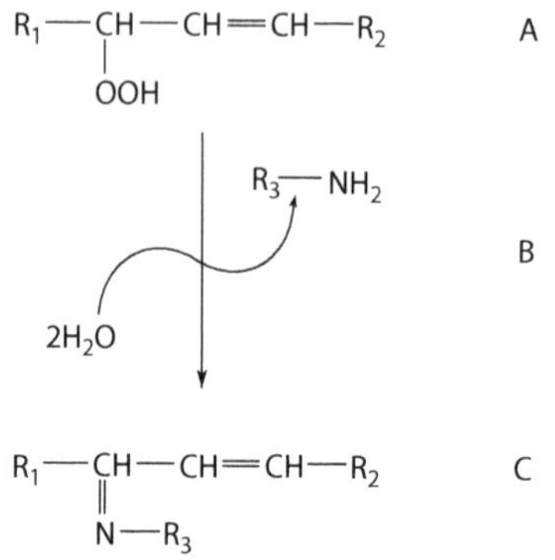

$$R_1-\underset{\underset{OOH}{|}}{CH}-CH=CH-R_2 \qquad A$$

$$R_3-NH_2$$

B

$$2H_2O$$

$$R_1-\underset{\underset{N-R_3}{\|}}{CH}-CH=CH-R_2 \qquad C$$

FIGURE 14.8 Reaction of oxidized lipids with protein. A, lipid hydroperoxide; B, lysine bound in protein; C, unsaturated imine.

to 13-aminotridecanoic acid, the monomer for nylon 13). A cultivar was announced that contains both EPA or DHA. Such an oil can replace fish oil for pharmaceutical purposes.

Of course, rapeseed oil is not the only modified oil (Krawczyk, 1999). Polyunsaturated FA cause low resistance against oxidation (e.g., sunflower seed oil), especially under frying conditions, but also on storage or microwave heating. The lipid hydroperoxides react with proteins, forming very reactive imines (Figure 14.8). Therefore, low-linoleic and high-oleic are now available, in which the polyunsaturated FA content has been reduced substantially or at least partially (Table 14.7). Reduced linoleic acid-sunflower seed oil is available for many years, and high-oleic sesame, safflower, or peanut oils are available. Even special cultivars of linseed oil are available, containing only trace amounts of linolenic acid and a low percentage of linoleic acid, so that it can be used as edible oil. All these modified oils are suitable for prolonged storage, deep frying, and microwave heating. The yield and availability of some of the more recently introduced oils are still not guaranteed. Sufficient experience is also lacking in seed and oil processing factories. In spite of these difficulties, it is expected that in the next several years, they would be as well established as low-erucic acid rapeseed oil is now.

Genetic manipulations are modern techniques, but they are not much used in the fat and oil production, except for herbicide tolerance and insect resistance. The technique was adapted in South America and in China. The application of genetic manipulations to change the FA would still require intensive research.

14.8.5 Reduced and Zero-Energy Lipids in Foods

Many consumers eat too much and their physical work is low, so that the excess energy is deposited as adipose fat. Nearly 50% of same populations is either overweight or obese. The medical research went to the conclusion that obesity is a serious disease, which should be cured. Therefore, consumers try to reduce their energy intake, as it is easier than to start physical exercise. Among several other methods, reducing the content of available energy in lipids is of interest. Low-fat foods were the first attempt, e.g., taking whey instead of milk, low-fat dressings, and mayonnaise.

Other more recent procedure is to use nutritionally unavailable lipid substitutes, like olestra (sucrose esterified with FA). Propoxylated glycerols have properties near those of normal oil, but they are not digested. FA partially esterified with polysaccharides or carbohydrate FA esters are another possibility. Alkyl glycoside FA esters, consisting of 1–50 alkoxy groups, can be used as frying oils. Caprenin, salatrim, and other products are slowly and incompletely digested, so that they can be counted among lipids with reduced available-energy content. Olibra, a product from palm oil and oat oil, prolongs the feeling of satiety and reduces efforts to eat. Long-chain *n*-3 polyunsaturated FA have also an anti obesity effect (Buckley and Howe, 2009). All reduced or zero-energy lipids were used in baked goods, table spreads, or frying oils.

14.8.6 PRODUCTION OF SPECIALTY LIPID PRODUCTS

Many consumers are looking for some plant oil products for special uses, which are needed only in low quantities; they are usually produced by small companies (Bhattacharyya, 2006). They were reviewed by Krawczyk (1999), but many new products have been developed since then.

Because of the great number of such products, it is only possible to give a few examples. A very important application is the production of chocolate and confectionary fats, cocoa butter replacers or filling fats (Norberg, 2006), special frying oils, various pharmaceutical preparation (Bhattacharyya, 2006), natural antioxidants, and products for cosmetic industry (Gunstone, 2006).

REFERENCES

AOCS. 1997. *Official Methods and Recommended Practices of the AOCS*, 5th edn., AOCS Press, Champaign, IL.

Artz, W. E. 2006. Reduced and zero calorie lipids in foods, in *Modifying Lipids for Use in Fats*, Gunstone, F. D. (Ed.), Woodhead Publishing, Cambridge, U.K., pp. 444–461.

Bakker, M. I., Bass, W. J., Sijm, D. T. H., and Kollofel, C. 1998. Leaf wax of *Lactuca sativa* and *Plantago major*, *Phytochemistry*, 47, 1489–1493.

Barnes, P. J. 1987. Wheat grain lipids and their role in the bread making process, in *Recent Advances in Chemistry and Technology of Fats and Oils*, Hamilton, R. J. and Bhati, A. (Eds.), Elsevier Applied Science, London, U.K., pp. 78–108.

Bhardvaj, U. L. and Hamama, A. A. 2009. Cultivar and growing location effect on oil, and fatty acids in canola sprouts, *Hortic. Sci.*, 44, 1628–1631.

Bhattacharyya, K. 2006. Speciality oils and their applications in foods, in *Modifying Lipids for Use in Fats*, Gunstone, F. D. (Ed.), Woodhead Publishing, Cambridge, U.K., pp. 539–566.

Boskou, D. 1996. Olive oil composition, in *Olive Oil Chemistry and Technology*, Boskou, D. (Ed.), AOCS Press, Champaign, IL, pp. 52–83.

Buckley, J. D. and Howe, P. R. G. 2009. Antiobesity effects of long chain ∞-3 polyunsaturated fatty acids, *Obes. Rev.*, 10, 648–659.

Chong, C. N., Hoh, Y. M., and Wang, C. W. 1992. Fractionation procedure for obtaining cocoa-butter-like fat from enzymatically interesterified palm olein, *JAOCS*, 69, 137–140.

Čmolík, J. and Pokorný, J. 2000. Physical refining of edible oils, *Eur. J. Lipid Sci. Technol.*, 102, 472–486.

DeGroot, R. H. M., Van Boxtel, M. P. J., Schiepers, O. J. G., Hornstra, G., and Joles, J. 2009. Age dependence of plasma phospholipids fatty acid levels: Potential role of linoleic acid in the age associated increase in DHA and EPA concentration, *Br. J. Nutr.*, 102, 1058–1064.

Diedrich, H. and Henschel, K. P. 1991a. The natural occurrence of unusual fatty acids, 2. Even numbered fatty acids with unusual positions of the double bonds, *Nahrung*, 35, 85–95.

Diedrich, M. and Henschel, K. P. 1991b. The natural occurrence of unusual fatty acids, 3. Acetylenic fatty acids, *Nahrung*, 35, 193–202.

Dijkstra, A. L., Töke, E. R., Kolenits, P., Recseg, K., Kövári, K., and Poppe, L. 2005. The base catalyzed, low temperature interesterification mechanism revisited, *Eur. J. Lipid Sci. Technol.*, 107, 912–921.

DiGiovacchino, L. 1996. Olive harvesting and olive oil extraction, in *Olive Oil Chemistry and Technology*, Boskou, D. (Ed.), AOCS Press, Champaign, IL, pp. 12–51.

Dong, H. S. and Chung, C. K. 1999. Chemical composition of the rice germ from rice milling and its stability during storage, *Korean J. Food Sci. Technol.*, 30, 241–243.

Flöter, E. and Van Duijn, G. 2006. Trans-free fats for use in Foods, in *Modifying Lipids for Use in Foods*, Gunstone, F. D. (Ed.), Woodhead Publishing, Cambridge, U.K., pp. 429–443.

George, H. S. and Arumughan, C. 1994. Positional distribution of fatty acids in the triacylglycerols of developing palm fruit, *JAOCS*, 70, 1255–1258.

Gertz, C. 2006. Development in frying oils, in *Modifying Lipids for Use in Fats*, Gunstone, F. D. (Ed.), Woodhead Publishing, Cambridge, U.K., pp. 517–538.

Gibon, V. 2006. Fractionation of lipids for use in foods, in *Modifying Lipids for Use in Foods*, Gunstone, F. D. (Ed.), Woodhead Publishing, Cambridge, U.K., pp. 201–233.

Gunstone, F. 2006. Vegetable sources of lipids, in *Modifying Lipids for Use in Food*, Gunstone, F. D. (Ed.), Woodhead Publishing, Cambridge, U.K., pp. 11–27.

Hall, C. III. 2001. Sources of natural antioxidants: Oilseeds, nuts, cereals, legumes, animal products and microbial sources, in *Antioxidants in Food*, Pokorný, J., Yanishlieva, N., and Gordon, M. (Eds.), CRC Press, Boca Raton, FL, pp. 158–209.

Haumann, B. F. 1994. Modified oil may be key to sunflower future, *INFORM*, 5, 1118–1210.

Krawczyk, T. 1999. Edible speciality oils, *INFORM*, 10, 552–561.

List, G. R. 2006. Hydrogenation of lipids for use in food, in *Modifying Lipids for Use in Foods*, Gunstone, F. D. (Ed.), Woodhead Publishing, Cambridge, U.K., pp. 173–200.

May, W. E. and Hume, D. J. 1995. Free fatty acids contents in developing seed of three summer rape cultivars in Ontario, *Can. J. Plant Sci.*, 75, 111–116.

Murphy, D. J. 2006. Plant breeding to change lipid composition for use in food, in *Modifying Lipids for Use in Foods*, Gunstone, F. D. (Ed.), Woodhead Publishing, Cambridge, U.K., pp. 273–305.

Nagao, T., Watanabe, Y., Hiraoka, K., Kishimoto, N., Fujita, T., and Shimada, Y. 2009. Microbial conversion of vegetable oil to rare unsaturated fatty acids and fatty alcohols by an *Aeromonas hydrophila* isolate, *JAOCS*, 86, 1189–1192.

Norberg, S. 2006. Chocolate and confectionery fats, in *Modifying Lipids for Use in Fats*, Gunstone, F. D. (Ed.), Woodhead Publishing, Cambridge, U.K., pp. 488–516.

Pánek, J., Pokorný, J., and Réblová, Z. 1995. Reaction rates of oxidation of triacylglycerol species in edible oils under storage conditions, in *Oils, Fats, Lipids*, Vol. 5, Castenmiller, W. A. M. (Ed.), Barnes, Bridgewater, NJ, pp. 291–293.

Parkányiová, L., Trojáková, L., Réblová, Z., Zainuddin, A., Nguyen, H. T. T., Sakurai, H., Miyahara, M., and Pokorný, J. 2000. Resistance of high-oleic acid peanut oil against autooxidation under storage and deep frying conditions, *Czech J. Food Sci.*, 18S, 125–126.

Parmentier, M., Fanni, J., and Linder, M. 1995. Membrane technology in lipotechnique, in *Lipides et Corps Gras Alimentaiers*, Graille, J. (Ed.), Lavoisier, Paris, France, pp. 107–143.

Patterson, D. M. 1998. Malting oats: Effects of chemical composition of hull-less and hulled genotypes, *Cereal Chem.*, 75, 230–234.

Perry, H. J. and Harwood, J. L. 1993. Changes in the lipid content of developing seeds *Brassica napus*, *Phytochemistry*, 32, 1411–1416.

Pokorný, J. 1999. Changes of nutrients at frying temperature, in *Frying of Food*, Boskou, D. and Elmadfa, I. (Eds.), Technomic Publishing Co. Inc., Lancaster, PA/Basel, Switzerland, pp. 69–103.

Rossel, J. B., King, B., and Downes, M. J. 1985. Composition of oil, *JAOCS*, 62, 221–230.

Steegala, S. M., Willett, W. C., and Mozafforon, D. 2009. Consumption and health effects of *trans* fatty acids: Review, *JAOC Int.*, 12, 1250–1257.

Werman, B. D. 1994. Lipid changes in mature-green tomatoes during ripening, during chilling, and after rewarming subsequent to chilling, *J. Am. Hortic. Sci.*, 119, 994–999.

Werman, M. J. and Neeman, I. 1987. Avocado oil production and chemical characteristics, *JAOCS*, 64, 229–232.

Wolf, R. B., Kleiman, R., and England, R. E. 1983. New sources of gamma-linolenic acid, *JAOCS*, 60, 1858–1860.

Xu, X. 2000. Production of specific-structural triacylglycerols by lipase-catalyzed reactions: A review, *Eur. J. Lipid Sci. Technol.*, 102, 287–303.

Xu, X., Guo, Z., Zhang, A., Vikbjerg, A. F., and Dumstrup, M. L. 2006. Chemical and enzymatic interesterification of lipids for use in food, in *Modifying Lipids for Use in Foods*, Gunstone, F. D. (Ed.), Woodhead Publishing, Cambridge, U.K., pp. 234–272.

Yang, T. K., Fruekilde, M. B., and Xu, X. 2005. Suppression of acyl migration in enzymic production of structural lipids through temperature programming, *Food Chem.*, 92, 101–107.

Yanishlieva, N. and Heinonen, I. M. 2001. Sources of natural antioxidants: Vegetables, fruits, herbs, spices and teas, in *Antioxidants in Food*, Pokorný, J., Yanishlieva, N., and Gordon, M. (Eds.), CRC Press, Boca Raton, FL, pp. 210–265.

15 Fish Lipids

Anna Kołakowska

CONTENTS

15.1 INTRODUCTION

And they had a few small fishes: and he blessed, and commended to set them also before them. So they did eat, and were filled and they took 'up of the broken meat that was left seven baskets. And they that had eaten were about four thousand; and he sent them away. (Mark 8. 7–9.)

It takes usually long before we, as individuals and as the humanity, realize that what God has communicated to us in the Holy Scriptures serves our welfare. We should have realized long before that the fish are a particular food, as 2000 years ago it was the fish and bread that were multiplied by Jesus to feed the hungry. The fish are a good, low-energy source of easily digestible proteins; vitamins A, D, and B; calcium; phosphorus; and rare microelements (iodine, selenium, and fluorine), but it is lipids that make the fish a particularly healthy food. Fish lipids contain the long chain n-3 polyunsaturated fatty acids (LC n-3 PUFA): eicosapentaenoic acid 20:5(n-3) (EPA) and docosahexaenoic acid 22:6(n-3) (DHA), which are physiologically and psychologically beneficial by directly affecting human health, but also by affecting gene activity. They are also necessary for the human fetus to develop. Crawford et al. (1999) are even of the opinion that reduced contribution of those fatty acids in the diet could have been involved in slowing down the civilizational progress of peoples migrating inland. Thus, when talking about functional properties of fish lipids, we mean—like in other food lipids discussed in this book—the role of lipids in controlling and shaping desirable and undesirable food properties as well as the *functionality* as scientifically documented beneficial health effect. In addition to the health claim in terms of heart disease, there is a pool of evidence indicating reduced risk of other diseases (immune deficiencies, allergies, and tumors) as well as beneficial effects on mental health (e.g., stress reduction, therapeutic effects in depressions).

Fishery products mean all seawater or freshwater animals (except for live bivalve molluscs, live echinoderms, live tunicates and live marine gastropods, and all mammals, reptiles and frogs) whether wild or farmed and including all edible forms, parts and products of such animals (EC 853/2004). Humans consume also marine plants – the algae.

More than 1000 fish species are eaten by humans; approximately 350 of those species can be regarded as commercially valuable.

According to the 2006 data, captured fish and aquaculture supplied the world with about 110 million tonnes of food fish. Of that total, aquaculture accounted for 47%. In addition to its signature freshwater fish, it is more and more frequently that aquaculture supplies increasing number of marine fish species and most of the shellfish present in the market. The growth rate of worldwide aquaculture has been sustained and rapid, averaging about 8% per annum for over 30 years, while the take from wild fisheries has been essentially flat for the last decade. More than 20% of the total world fisheries is accounted for by non-food uses, but some of those is recycled as feed for the farmed fish.

Edible parts of fish include: fillets (mainly the muscle tissue called the fish meat, with or without skin), carcasses (e.g., canned), and by-products of some fish species, mainly gonads (roe) and livers (of lean fish). Some traditional local cultures utilize also other parts, e.g., fins. The meat recycled from post-filleting remains (separated from the backbone) may be used as food as well, although post-filleting rests, viscera, heads, etc., are generally processed directly into feeds. Very often, such remains of marine fish are processed into fishmeal and fish oil, and serve also as a raw material with which to derive enzymes, collagen, protein preparations, and other biotechnological products. The major aquatic food resource is, however, fish meat, which, depending on fish species, size, and recovery technology, accounts for about 35%–75% of fish weight. The meats content of lipids, their composition and properties are not only species-dependent, but depend also on effects of various biological factors. Lipid content and nutritional value in fish products labeled by a particular fish species name is a derivative of a summary effect of various biological and technological factors, and—although established based on the available nutritional tables—may in reality deviate strongly from the expected standards.

15.2   LIPID CONTENT

Fish meat contains from 60% to 82% water and from 15% to 23% (w. w.) protein; lipid content can vary over a wide range (0.3%–45% w. w.). There is a strong negative correlation between lipid and water contents; lipid and protein contents correlate positively in juveniles, but negatively in older and large fish (Tobin et al. 2006).

Depending on lipid content, fish can be divided into four basic groups:

- Lean: <2% fat (cod, haddock, hake, Alaska pollock, and blue whiting)
- Medium-fat: 2%–7% (sole, flatfish, tuna, roach, bream, wild salmon, and rainbow trout)
- Fat: 7%–15% (herring, sprat, mackerel, horse mackerel, salmon, and carp)
- Highly fat: >15% (eel, capelin, catfish, and carp)

Shellfish, which are mostly crustaceans (prawns, crabs, lobster, and crayfish) and molluscs (bivalves and squids), usually contain as little as 1%–2% lipids.

Although different fish species are categorized as above, muscle lipid contents are affected by numerous biological factors. This is particularly evident in temperate fatty species due to their migratory patterns and highly seasonal feeding and reproductive cycles; moreover, muscle lipid content tends to vary widely in some farmed species. Variations in lipid content in herring and mackerel are so wide that the species, depending on when captured, could be assigned to any of the groups shown above. For instance, depending on the season, lipid content can vary from 4% to more than 30% and from 2% to 25% in mackerel and herring, respectively. The lowest lipid contents are recorded in spawning fish, which usually do not feed at that time; they begin to feed intensively right after spawning is over and accumulate lipids in muscles and the mesenteric tissue. Meat lipid contents of non-spawning and spawning Pacific herring were reported to amount to almost 11% and 2.23%, respectively (Huynh et al. 2007). In anchovy, lipid content was observed to vary from 9.00% to 15.3%. Muscle lipid contents in lean fish vary as well, but the variations are more pronounced in the liver, which accumulates stored lipids. In hake, the muscle lipid content varied over the year from 1.1% to 3.4%, the mean annual content amounting to 1.5% ± 1.1% (Mendez and Gonzalez 1997). The flesh of temperate latitude species generally contains more lipid than that of the leaner tropical species.

Freshwater fish species living in natural habitats of rivers and lakes of temperate latitudes usually belong to the medium-fat group (roach, bream, tench, and whitefish). However, the same species living in warm climates are leaner. For instance, zander and tench from Poland were found to contain three times as much lipids as the same species from Turkey (Kołakowska et al. 2000, Guler et al. 2007).

Lipid content and composition in freshwater fish, too, undergo seasonal variations, although the changes are less pronounced than in the marine species. Lipid content in zander from Turkey was observed to change from 0.58% in autumn to 1.26% in winter (Guler et al. 2007). The lacustrine carp, too, exhibited seasonal changes in lipid content, the changes being significantly influenced by the feeding period and season (Guler et al. 2008).

In farmed fish, the factors of key importance for lipid content include genetics and diet. Farmed fish contain more lipids than their wild conspecifics. More and more fish species, both wild and farmed, are transferred to different climatic conditions. When adapting to a changed climate, a specie may use different strategies of storage lipid accumulation and composition, as demonstrated in eelpout (Brodte et al. 2008). Introduction to a different environment results also in changes in populations of wild fish.

Lipids in fish are located in the subcutaneous tissue, belly flap, muscle, skin, head, liver, gonads, and the mesenteric tissue. In many species, lipid contents in offal are higher than in fillets with skin (Kołakowska et al. 2006a). Lean fish deposit lipids in the liver; in cod, the liver accounts for about 4%–9% of body weight and may contain as much as 70% lipids, in contrast to the muscle content of less than 1%. Cod roe contains from 0.3%–1.5% lipids.

Lipid content in the liver and perivisceral tissue increases with fish age and size, as the water content decreases. Muscle lipids differ primarily between white and red muscles; the latter may contain even 10 times as much lipid as white muscles. Red muscles are particularly well developed in migratory pelagic species (tuna, herring, mackerel, sprat, anchovy, and sardine); in clupeids, red muscles account for 20% of all muscles, while in lean, less mobile fish, the proportion is only about 10% (Kołakowska et al. 2003). Lipid distribution within the fillet was observed to vary depending on fish size; generally, the ventral and the frontal-lateral portions had a higher concentration of lipids and less water (Palmeri et al. 2007). However, the proportions are also species-dependent; for instance, the ventral muscle of farmed sea bass was reported to contain three times more lipids than the dorsal muscle, while the differences were much less pronounced in rainbow trout (Testi et al. 2006).

15.3 LIPID COMPOSITION: LIPID CLASSES

The main constituents of fish lipids are triacylglycerols (TAG), phospholipids (PL), sterols, and wax esters; in addition, fish lipids contain also minor quantities of metabolic products of these, as well as small amounts of unusual lipids such as glycerol ethers (liver of sharks and other elasmobranchs), hydrocarbons, glycolipids, and sulfolipids. PL as structural lipids of the cell and organelle membranes in the fish muscle tissue occur in a relatively constant low concentrations of about 0.3–0.5 g/100 g tissue (the maximum content may reach 1%). Fish sterol contents are relatively stable and range within 40–60 mg/100 g of edible fish muscle (Krzynowek et al. 1990, Ackman 1994), 250–650 mg/100 g of roe (Sikorski et al. 1990), and 480–1150 mg/100 g oil in liver of cod, herring, menhaden, and salmon (Kinsella 1987, Kennish et al. 1992). In finfish, 95% of the total sterol is accounted for by cholesterol. Shellfish usually contain two to three times more total sterols than finfish, so a high consumption of shellfish is considered as a risk of coronary heart disease (Matheson et al. 2009). The highest sterol contents (120–160 mg/100 g) were recorded in prawns, squid, octopus, and scallops (Nichols et al. 2002).

TAG are the most quantitatively variable lipid component. They constitute the storage lipids in almost all commercial fish species. The remaining components such as free fatty acids (FFA), diacylglycerols (DAG), and monoacylglycerols (MAG) occur in fish, directly post mortem, in small quantities. FFA contribute about 2%–7% to lipids in fresh fish, but the contribution in lean fish is much higher because the FFA there are derived mainly from PL, phospholipases being usually more active than lipases. The composition of fish lipids depends on the lipid content. The proportion of TAG in lipids increases, and that of PL and sterols decrease, with increasing lipid content. Thus, lipids of lean fish consist mainly of PL, which may account even for more than 80% of total lipids (e.g., in cod); PL in herring, depending on the capture season and lipid content, contribute from 12% to almost 40% of total lipids. In invertebrates, PL are the main lipid component as well. At a mean shellfish lipid content of 0.9% ± 0.5%, PL contributed 76.9 ± 14.7; in crustaceans, at a mean lipid content of 0.8% ± 0.4%, as much as 85.7% ± 7.7% were contributed by PL (Nichols et al. 2002). Antarctic krill accumulates PL as storage lipids, therefore even at the lipid content of about 3.8%, PL are the major component. The cholesterol content in lipids (as calculated on the w. w. basis) is about 2–3 times higher in shellfish than in finfish with a similar total lipid content. The high sterol contents of 120–160 mg/100 g were recorded in shellfish such as prawns, squid, octopus, and scallops. In molluscs such as oysters and abalone, cholesterol is—like in finfish—usually the main sterol, but a complex suite of other sterols is present as well (Kołakowska et al. 2003).

A typical fish polar lipid fraction contains about 60% phosphatidylcholine (PC), 20% phosphatidylethanolamine (PE), and several percent phosphatidylserine and sphingomyelin, while the rest is made up by other minor PL. In fish, the PC:PE ratio is generally 2–3:1. Various cod tissues were found to differ in their PL composition. The PC:PE ratio of white and dark muscles of cod were 3.5:1 and 2.7:1, respectively (Lie and Lambertsen 1991). The ovaries and testes of skipjack tuna differed in their PC:PE ratio. In the ovaries, the PC (plus lyso-PC): PE was 3.47, in contrast to the testes ratio of 1.38. The testes contained also substantial amount of phosphatidylserine (Hirasuka et al. 2004).

Compared to vertebrates, invertebrate PL contains less PC. In edible bivalves (clams and mussels), the PC:PE ratio was 1, most PL occurring in the plasmalogen form (Hanuš et al. 2009). On the other hand, in Antarctic krill, the PC content is three times that of PE; the former serves also as a storage lipid because, as the krill fasted, the PC contents were observed to decrease (Kołakowska 1985, Kołakowski and Kołakowska 1993). A low PC content is also typical of marine plants. Shellfish contain also glycolipids; e.g., glycolipids in edible clams and mussels accounted for 7%–24% of total lipids (Hanus et al. 2009).

In some deepwater fishes, such as oilfish, escolar, orange roughy, species of oreos, and myctophids, lipids are dominated by wax esters; more than 90% of the total oil is indigestible wax esters. They are present in different muscle parts such as dorsal, ventral, subcutaneous, and periosteum. In oil fish, wax esters were identified additionally in various internal organs, including liver, gall bladder, and testes. These fishes are substituted as cod steaks (Ling et al. 2008). The livers of certain sharks may be rich in glycerol ethers and hydrocarbons such as squalene, which are also thought to aid in buoyancy.

15.4 FATTY ACID COMPOSITION

15.4.1 FATTY ACID GROUPS

The main groups of fatty acids (FA) may be saturated (SFA), monounsaturated (MUFA), or PUFA. The proportion of each FA group in the total lipid from fish tissues differs and depends mostly on the total lipid content (and hence lipid class composition) and can be affected by numerous biological factors. The mean contribution of individual FA classes amounts to about 30% SFA, from about 20% to more than 40% MUFA, and from about 20% to more than 40% PUFA. Extreme deviations from those ranges occur usually in farmed fish. SFA contributed less than 20% FA in sturgeon and rainbow trout; MUFA accounted for less than 20% in perch and zander, but contributed as much as 60% FA in carp, bream, and grass carp. Very low PUFA contents (8% FA) were reported from lipids of grass carp. High contributions of PUFA, even in excess of 50% FA, are typical of lipids of marine lean fish such as cod and in freshwater predatory species (Kołakowska et al. 2003, Bienkiewicz et al. 2008). Generally, the proportion of FA groups in lean and low-fat fish (<4%) forms a series of PUFA>MUFA>SFA; the series of MUFA>PUFA>SFA is observed in more fatty fish (>4%), but sardine, belonging to the medium-fat group, showed a series of PUFA>SFA>MUFA (Huynh and Kitts 2009). Fatty fish such as herring, sprat, and mackerel showed MUFA>SFA>PUFA. Mean annual contents in herring amounted to: SFA, 32.11% (coefficient of variation, $V = 11.96\%$); MUFA, 40.28 ($V = 25.5\%$); and PUFA, 27.78 ($V = 25.31\%$). In winter months, contributions of different FA groups were mostly uniform (about 30%). The high coefficients of variation emerged on account of spring–summer, as an effect of gonad development/spawning and changes in food accessibility and feeding intensity. The lowest coefficients of variation were typical of SFA, while the contributions of MUFA and PUFA are more variable, the two latter groups being inversely correlated both in the lean and fatty fish (Nichols et al. 2002, Kołakowska et al. 2003). A similar relationship was revealed also within a fillet: MUFA in the farmed murray cod tended to increase exponentially ($R^2 = 0.86$) and PUFA to decrease exponentially ($R^2 = 0.86$), while SFA remained unchanged ($R^2 = 0.01$) with increasing lipid levels (Palmeri et al. 2007).

TABLE 15.1
Atheriosclerosis Index–Ratio SFA + MUFA/PUFA of Fish Lipids

Seawater		Freshwater		Farmed	
Cod	0.88	Zander	1.74	Trout	2.79±0.66
Coalfish	1.02	Tench	2.33	Salmon	1.81
Pollock	2.61	Roach	2.33	Catfish	12; 8.6
Hake	1.15	Bream	5.53	Butterfish	32
Redfish	1.88	Eel	4.76	Nile perch	3.54
Mackerel	2.43	Perch	1.08	Tilapia	2.41
Herring	2.6	Grass carp	6.04	Sturgeon	3.29
Flounder	3.66	Crucian	2.87		

Source: Data from Bienkiewicz, G. et al. *Fish Ind. Mag.*, 3(63), 58, 2008; Bienkiewicz, G. and Domiszewski, Z., *Fish Ind. Mag.*, 2(62), 45, 2008; Kołakowska, A. et al., *Acta Ichth. Piscat.*, 30(2), 59, 2000.

In a study on Antarctic, temperate Australian, and tropical Australian fish, Dunstan et al. (1999) showed that, with increasing latitude, the proportion of total MUFA and SFA decreased and that of PUFA remained fairly constant.

The (SFA+MUFA):PUFA ratio was found to correlate with atherosclerosis progression and increased risk of cardiac events (Ma et al. 1997). The ratio is species-dependent, but intraspecific differences occur as well. Generally, it is at its lowest in marine fish, followed by freshwater species, the highest ratio being shown by farmed fish (Table 15.1). The ratio of approximately 4 being reported from tilapia and catfish, farmed trout and salmon show favorable ratios <2, (Weaver et al. 2008). The (SFA + MUFA):PUFA ratios in various batches of rainbow trout from a single fish farmer were found to range from 1.73–3.73 (Kołakowska et al. 2006d).

Compared to the lean fish of a similar lipid content, shellfish showed somewhat different proportions of various fatty acid groups. Invertebrate lipids, particularly those of shellfish, contained less MUFA and more PUFA than finfish (Nichols et al. 2002).

15.4.2 THE FATTY ACID PROFILE

The FA profile of a fish is its fingerprint revealing its origin, hatchery, or diet. All the fish show, however, some common features of individual FA groups: SFA are dominated by palmitic acid (16:0), followed by 14:0 and 18:0 acids, the 15:0, 17:0, and 20:0 occurring in contents lower than 1% FA; in MUFA, the dominant oleic acid (18:1) > (16:1), some species showing 20:1 and 22:1 as well as 17:1, 24:1, the latter two not exceeding 1% FA; PUFA are dominated by long-chain n-3 PUFA (22:6)/(20:5), (DHA/EPA), small amounts of docosapentaenoic acid 22:5 (DPA), (18:3), and (20:4); n-6 PUFA linoleic acid (18:2), and 20:4 in small amounts (about 1%).

The combined contribution of palmitic acid (16:0) and myristic acid (14:0) accounts for about 90% of saturated acids in the Baltic herring lipids. There are also a few percent of stearic acid (18:0), acid which is more abundant in the warmer water fish. Also present are 15:0, 17:0, and 20:0 acids, occurring at about 1% or less and generally considered to be of a limited nutritive value.

The MUFA are dominated by oleic acid [18:1 (n-9)], followed by the three times less abundant palmitooleic acid [16:1 (9n-9)]. Lipids of species such as mackerel and herring contain high amounts of gadoleic [20:1 (n-9)] and cetoleic [22:1 (n-11)] acids. This is considered to be due to a diet of wax ester-rich (and hence long-chain MUFA-rich) copepods. During some periods, 20:1 (n-9) and especially 22:1 (n-11) acids dominate the Baltic herring MUFA. The 22:1 (n-11) acid content can be used as a biomarker of the Baltic herring, a fish that forms numerous distinct populations migrating within the Baltic Sea. There is a close ($r=0.96$) inverse relationship between the 18:1 and 22:1 acids over the annual cycle of herring, which could be a result of a shift in the diet and/or an endogenous synthesis of 18:1 (n-9) by shortening the chain of dietary 22:1 (n-9) (Kołakowska et al. 2003).

The PUFA composition is the most characteristic trait of fish lipids. In aquatic animals, PUFA are usually LC, with (n-3) configuration. Quantitatively, they are mostly EPA and DHA. Marine fish usually contain more DHA than EPA: the DHA/EPA ratio in herring and sprat was observed to range from 1.5 to about 20. In most carnivorous fish and invertebrates, DHA is usually more (up to 2–3 times) abundant than EPA. Generally, these predatory fish contain more DHA than do non-predatory ones, while EPA is more abundant in herbivorous/omnivorous fish than in carnivorous species. A predominance of EPA over DHA is typical of phytophagous fish.

The proportion of EPA is higher in shellfish (crustaceans, cephalopods, bivalves, and gastropods) than in finfish (bony fish, sharks, and rays) (Dunstan et al. 1999). An example is furnished by the blue crab with its lipid content of 0.37%–0.65% and n-3 PUFA accounting for 40% of FA (EPA content about twice as high as that of DHA) (Naczk et al. 2004).

Fish flesh may also contain small amounts (1%–3% of FA) of DPA. Exceptionally high amounts of DPA (13%–14%) were recorded in the abalone in which DPA >> DHA (Dunstan et al. 1999). The mammalian depot fats contain almost equivalent amounts of EPA, DHA, and DPA. The 18:3(n-3), 18:4(n-3), and 20:4(n-3) acids occur in small (a few percent at most) amounts in fish lipids as well (these acids may be more abundant in freshwater fish). The flatfish lipids were found to contain the very long-chain 24:6(n-3) fatty acid, contributing as much as 6%–9% of total FA in the flesh, 3% in the liver and 6% in the viscera (Ota et al. 1994). This FA is an intermediate in biosynthesis of 22:66 DHA from 22:5 DPA (Tomita and Ando 2009).

The (n-6) PUFA family may be represented in fish by the following acids: 16:2(n-6), 18:2(n-6), 20:3(n-6), 20:4(n-6), 22:2(n-6), 22:4(n-6), and 22:5(n-6). Some (n-6) PUFA such as 18:2(n-6) and 22:5(n-6) may contribute a few percent; however, most are present in marine fish in trace amounts only. Arachidonic acid 20:4 (n-6) (AA) accounts usually for about 0.3% FA, although higher contents (in excess of 1%) were reported from mackerel icefish and Alaska pollock (Bienkiewicz et al. 2008). Particularly high concentrations of n-6 PUFA, including AA, were reported from farmed tilapia and catfish (Weaver et al. 2008) and also from shellfish (Cherif et al. 2008).

FA consumed by all fish are of algal origin or are modified from algal FA that may contain odd-numbered PUFA, such as 21–29 carbon atoms, and even longer chain lengths (as much as 30 carbon atoms), which may also occur in fish oils. The contribution of those acids usually does not exceed 1% FA. The most widespread FA is the 21:5(n-3), 21:4(n-6) being less common (Řezanka and Sigler 2009).

Unusual FA found in some fish are furan FA and nonmethylene-interrupted dienoic acids (NMI) (Ota et al. 1994). NMI were also present in abalone at 3%–6% of FA, in oysters at 4.5%, and at around 20% in the dusky shark. Saito (2008) found novel n-4 and n-7 NMI PUFA (20:3n-4, 21:3n-4, 20:2n-7, 21:2n-7) in cold-seep clam. Such unusual FA in clam are assimilated by the symbiotic bacteria.

15.4.3   (N-3)/(N-6) RATIO

Reduction of the ratio in the western diet during recent decades is considered to be the cause of more and more common pro-inflammatory diseases, heart condition, depression, and other pathologies. The (n-6)/(n-3) ratio in nutritional recommendations has been, in recent years, steadily reduced, to almost 1:1. It is also suggested that, because of differences in effects and activity, nutritional recommendations concerned the LC (n-3)/(n-6) ratio.

In fish lipids, the (n-3)/(n-6) ratio is higher than 1. The recommendations are aimed not only at bringing the ratio to the most appropriate one, but also for the fish to supplement the dietary LC n-3 PUFA deficiency and to counterbalance the excess of n-6 PUFA typical of the western diet. In marine fish, the n-3 acids occur at higher proportions and the n-6 at lower than in the freshwater fish, allowing differentiation between freshwater and marine species based on the ratio of these two types of PUFA. The in marine fish is substantially, by a factor as high as 10, higher than that of the freshwater fish. The n-3/n-6 ratio in marine fish ranges from less than 10 (6 in grenadier, 9.9 in Patagonian grenadier) to 30 in cod (Bienkiewicz et al. 2008) and increases with latitude (Dunstan et al. 1999). In a single species, it varies throughout the year; for instance, in the Baltic herring lipids, the ratio averages 5, but varies from 3 to 12, with a decrease during spawning (Kołakowska 2003). In females of common sole from Turkey, the n-3/n-6 ratio was observed to range from 1.45 to 3.84 during the year; the n-6 series was utilized as a source of energy more than the n-3 one (Gokce et al. 2004). The n-3/n-6 ratio in freshwater fish varies from about 2 in bream, roach, perch, and eel (1.93) to 5 in zander. The ratio is lower in farmed than in wild fish; for example, the n-3/n-6 ratios in the wild and farmed salmon fillets were 10.68 and 1.47–1.5, respectively. Lipids of the wild salmon contained almost twice as much LC n-3 PUFA than the farmed salmon (Domiszewski et al. 2008). Although the farmed rainbow trout showed an average ratio of 2.52, it was less than 1 in some species, e.g., in tilapia (0.28) and pangasius (0.19) (Bienkiewicz and Domiszewski 2008, Bienkiewicz et al. 2008). The LC n-3/n-6 ratio in fish is usually higher than all (n-3)/(n-6) PUFA, because fish contain low amounts of LC n-6 PUFA.

15.4.4 Distribution of Fatty Acids in Lipid Structure

The FA distribution among the three *sn*-positions of the glycerol backbone is nonrandom. Long-chain PUFA tend to be preferentially esterified at the 2-position in fish and invertebrate triacyl-*sn*-glycerol and in PL SFA and/or MUFA tend to be preferentially esterified at the *sn*-1 and/or *sn*-3 positions of TAG. Almost 70% of DHA are located in *sn*-2 position of TAG in fish; more than 70% of DHA being located there in salmon lipids and cod liver, while almost 40% of EPA are found in the 2-position (Aursand et al. 1995). The distribution of DHA, DPA, and EPA was found to be affected by total amounts of DHA and 20:1 and 22:1 in triacyl-*sn*-glycerol (Ando 1992). The 24:6n-3 acid, too, in TAG of the flathead flounder is preferentially located in the *sn*-2 position (Tomita and Ando 2009). On the other hand, LC PUFA in the subcutaneous fat of certain marine animals are located mainly in the *sn*-1,3 position of TAG. In PL, too, PUFA preferred the *sn*-2 position (Kołakowska et al. 2003, Simonetti et al. 2008). In tuna PL, the *sn*-2 position contained elevated levels of PUFA (DHA, EPA, and AA) and 18:1(n-9), whereas elevated levels of SFA (16:0 and 18:0) and of 18:1(n-7) were located in the *sn*-1 position (Aubourg et al. 1996).

15.5 BIOLOGICAL FACTORS AFFECTING n-3 POLYENOIC FATTY ACIDS IN FISH

15.5.1 Lipid Content

The n-3 PUFA contents in fish depend on, i.a., species, geographic area, season, and diet (Kołakowska et al. 2003, 2006c,d). The key factor for than-3 PUFA/LC n-3 PUFA in fish lipids and in muscle is the lipid content. With increasing lipid content in fish muscle, the LC n-3 PUFA decreases exponentially, a still more substantial reduction being observed in DHA (Figure 15.1). Within the lean fish group itself, correlation between lipid content and n-3 PUFA is weak. Thus, the relatively highest amounts of n-3 PUFA are found in lipids of lean fish and shellfish. Nevertheless, despite a few exceptions (eels and some farmed fish), the richest source of n-3 PUFA (and of LC n-3 PUFA) are the fatty fish, followed by medium-fat and lean fish. To obtain the

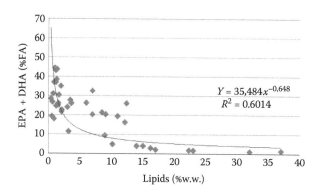

FIGURE 15.1 Relationship between EPA + DHA [% FA] and content of lipids in fish muscle tissue [% w.w.].

recommended daily allowance of 1 g EPA + DHA, it is sufficient to consume, e.g., 50 g (half of a fillet) herring or still less, 30 g mackerel, but as much as 0.5 kg cod.

At similar lipid content, n-3 PUFA contents in cold-water fish are higher than those in the tropical fish. Marine fish are n-3 PUFA-richer than freshwater species within the same climate zone. The rainbow trout kept in seawater showed higher n-3/n-6 and DHA/EPA ratios than their conspecifics kept in fresh water and fed identical diet (Haliloglu et al. 2004).

15.5.2 SEASONAL VARIATIONS

The fish developmental cycle (lipid metabolism) and diet (availability of, competition for, and composition of food) are most important for the n-3 PUFA content and composition. Changes in n-3 PUFA throughout the year are a net result of interactions between the two factors and are most pronounced in the high-fat pelagic fish.

Seasonal variations in muscle lipids of herring caught in different years from the same area reflected a fairly constant annual pattern, associated with the biological cycle (gonad maturation and spawning, both being lipid-consuming processes) and with feeding (Figure 15.2), the coefficient of variation of EPA contents being more variable than that of DHA; most variable, however, were the 18:3 and 18:4 n-3 acids. This demonstrates a significant effect of food: the herring that did not feed immediately prior to spawning contained trace amounts of the 18:3 acid only. Sprat, a species similar to herring in its lipid content and composition, but feeding throughout the year, showed less extensive annual variability in n-3 PUFA, and particularly in

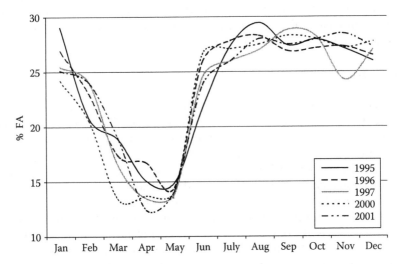

FIGURE 15.2 Seasonal variations in n-3 PUFA of Baltic herring; data collected over several years. (Adapted from Kołakowska, A. et al., Effects of biological and technological factors on the utility of fish as a source of n-3 PUFA, in *Omega 3 Fatty Acid Research*, M.C. Teale (ed.), Nova Science Publishers, Inc., New York, pp. 83–108, 2005. With permission.)

the 18:3 acid. Seasonal changes in LC n-3 PUFA show a pattern that is similar to that of changes in total lipid contents in muscles (Ackman 1994, Kołakowska et al. 2003). Changes in lipid contents were much more extensive, no straightforward relationship between the variables tested being revealed (Kołakowska et al. 2006d). Differences between herrings caught at different times may be still greater: the edible portion (fillet with skin) may contain more than 3.5 (autumn) or less than 0.5 (spawning herring) g n-3 PUFA/100 g w. w.

15.5.3 Gonad Maturity

Gonad maturity may be of significance, also in farmed fish, for differences in the n-3 PUFA content. The gonad maturity stage proved to have significantly influenced the male rainbow trout lipid content and n-3 PUFA, no such strong effect being observed in DHA. Females and males with gonads at stage IV did not differ significantly, but a significant difference was observed at a later (V) maturity stage.

15.5.4 Fishing Ground

Differences in FA composition depend also on the fishing ground. The herrings at an identical maturity stage, caught on the same day from the Baltic fishing grounds about 400 km apart, were found to differ in their n-3 PUFA contents. Freshwater fish species living in natural habitats of temperate rivers and lakes are mostly medium-fat fish (roach, bream, perch, and whitefish). The same species in warm climate belong to the lean fish group, whose lipids are n-3 PUFA richer; however, a portion of a fish (zander and tench) caught in Turkey supplies much less LC n-3 PUFA than a portion of its conspecific caught in colder waters of Poland. Celik et al. (2005) demonstrated significant differences in n-3 PUFA between the zander from two different lakes (situated under different climatic conditions).

15.5.5 Diet

One of the main factors affecting the lipid composition of fish is diet. Fish diet does not affect the FA composition of PL as much as it does that of TAG. TAG reflect the lipid composition of the diet. However, fish can adjust their FA composition according to environmental pressure or physiological demands. Marine planktons contain more long-chain (n-3) PUFA than freshwater planktons. This is regarded as the major reason for the basic difference in the composition of FA between marine and freshwater fish. In contrast to marine fish, most freshwater fish (e.g., carps, tilapias, eels, and the anadromous salmonids) consume substantial amounts of 18:3(n-3) and 18:2(n-6), derived from terrestrial and aquatic plants and insects, and obtain substantially less DHA with their natural diet. Therefore, most of the freshwater fish and some marine species have active elongase and desaturase systems allowing a rapid conversion to DHA and AA from shorter chain (n-3) and (n-6) FA, respectively (Sargent 1995). As much as 70% of 18:3(n-3) in rainbow trout diet are converted to DHA, while the predacious pike is not capable of converting 18:3(n-3) into long-chain (n-3) PUFA and must obtain its DHA from the diet (Shulman and Love

1999). As opposed to freshwater fish, marine species cannot convert C18 to C20 (n-6) PUFA; thus, AA is also an essential FA in marine fish (Sargent et al. 1999). The FA profile of the total muscle lipids of farmed fish may be modified through diet variation. It reflects the feed FA, but the degree of similarity is species-dependent; the profile may differ within a species as well. For example, the rainbow trouts from the same culture, fed identical feed, were found to differ in their FA content and in their n-3 PUFA composition (Kołakowska et al. 2006c,d). According to Jobling et al. (2008), it should be possible to manipulate the FA compositions also of cod tissues in desires directions by introducing mixtures of different feed oils at various points in the production cycle.

15.5.6 BETWEEN-INDIVIDUAL VARIABILITY

Differences between-individuals in the cultured brook trout (3 females and 2 males at gonad maturity stage III) were not statistically significant. Within a cultured trout batch, differences between males and females in terms of their total n-3 PUFA proved nonsignificant. On the other hand, there was a significant difference in the abdominal muscle DHA content, higher contents being recorded in males; dorsal muscles showed no sex-dependent differences in this respect (Kołakowska et al. 2006c,d).

15.6 SUSCEPTIBILITY TO OXIDATION

Compared with other food lipids, fish lipids are regarded as most susceptible to oxidation. This is because of their high degree of unsaturation and low content of lipophilous antioxidants compared to unsaturated vegetable oils. The α-tocopherol (vitamin E) content in fish muscle lipids ranged from 1 to 75 mg/100 g of oil (Kinsella 1987). In herrings, the vitamin E content was found to vary from trace amounts to 66 mg/100 g of lipids, the seasonal variability being recorded as well (Seidler 1997). The fish muscle pro-oxidative system, both enzymatic and non-enzymatic, is more active than that in homoiotherms. The role of lipoxygenases (LOX) in fish lipid oxidation is less pronounced than in vegetable oil oxidation; the most active LOX are in the gill and skin. Fish gill tissue contains two separate LOX, 12-LOX and 15-LOX, each active toward PUFA (AA, EPA, and DHA), but exhibiting different hydroperoxide addition sites. LOX from sardine skin was more active toward α-linolenic acid [18:3(n-3)] than linoleic acid [18:2(n-6)] (Mohri et al. 1990). The skin contains carotenoids and the melanin radical trapping mechanism, which significantly affects fish lipid oxidation. In addition, the skin provides a physical boundary protecting subcutaneous lipids from light and oxygen. The average carotenoid content in fish muscle varied between 0.043–0.283 μg/g w. w., but was much higher in the whitefish: 0.776 μg/g w. w. in muscles, 1.45 μg/g in fins, and 2 μg/g w. w. in farmed rainbow trout. More carotenoids are contained in the meat of farmed salmon and other salmonids whose feed is carotenoid-enriched to enhance the muscle color. On the other hand, crustaceans such as Antarctic krill have exceptionally high carotenoid contents (from about 500 to 2947 μg/g). The major carotenoid of both fish and shellfish is astaxanthin.

Compared to other food lipids, fish lipids isolated from the tissue are very sensitive to photo-oxidation, thermo-oxidation being less pronounced. Seasonal changes in the susceptibility of Baltic herring muscle lipid to UV-catalyzed oxidation are larger than changes in lipid content and FA composition. This is also true for sprat, bream, and Antarctic krill. Most susceptible to oxidation were lipids of spawning fish. Lipid oxidation in fish muscle only partially depends on lipid susceptibility to oxidation. The most important role is played by the antioxidant activity (AAc) of tissue. It is not lower than that of many vegetables, regarded as a source of dietary antioxidants. The AAc of fish tissue is in 80% due to myofibrillar proteins; less important are water-soluble antioxidant, lipophilous antioxidants being least important (Chapters 9 and 10). AAc of fish was observed to vary between batches of fish, most probably as an effect of feeding-related factors (pro- and antioxidant contents in feed) and oxidative stress.

15.7 EFFECTS OF PROCESSING ON FISH LIPIDS

15.7.1 STORAGE

15.7.1.1 Chilling

To preserve quality of wild and farmed fish prior to delivery to the market, they have to be chilled, the former immediately after capture and the latter directly after killing. The most common method is to use ice; recently, slurry ice and ozone-slurry have been gaining in popularity. Onboard chilling frequently involves using sea water. Occasionally, the fish are chilled by merely storing them at a reduced temperature, preferably 2°C.

The major process occurring in lipids of chilled fish, depending on the storage time, is lipolysis. The content of FFA, released from both TGA and PL (earlier) increases as storage continues, the FFA content–storage duration relationship being significant ($p < 0.05$). At the same time, the contribution of DAG to total lipids increases. It is only at an advanced stage of autolysis that the MAG content increases as well. FFA formation occurs during the first stage of the chilling process (up to A/B quality classes), triggered by the activity of endogenous enzyme; later, only microbial activity plays a role as well. Digestive lipolytic enzymes and active phospholipases (PLA$_2$) are contained in fish intestines, mainly in their pyloric caeca. Digestive lipolytic enzymes isolated from fish (cod, salmon, and rainbow trout) intestines preferentially hydrolyze LC (n-3) PUFA (Halldorsson et al. 2001). The rate of lipolytic changes is faster in whole than in gutted fish and fillets. In terms of tissue lipase and PLA activities, lipolysis proceeds more rapidly in muscles of feeding and pre-spawning fish than in spawning fish. At the final stage of the rainbow trout shelf life (14 days), the FFA content was observed to increase by 18% (8% total lipids). At that time, 25% TAG became hydrolyzed in whole and slightly less in gutted rainbow trout stored in ice, whereas most PL hydrolyzed as early as after 7 days of storage (Kołakowska et al. 2006b). An extreme example is furnished by Antarctic krill, whose high absolute phospholipid content is accompanied by a high PLA activity. After 72 h of cold storage of Antarctic krill, hydrolysis affected 20% of the PL; in particular, PC was reduced to half of its initial amount. Ohshima (1984) recorded changes in the PL structure, and re-esterification of free FA proceeded during cold storage of cod.

Oxidative changes in lipids in cold-stored fish are secondary to autolytic, and later on microbiological, changes that are decisive for consumer safety and sensory properties of chilled fish. Obviously, oxidation is involved in quality changes of refrigerated fish. However, lipid oxidation indicators such as PV and AV are not significantly correlated with the time of storage. The desirable smell ("of the sea", "of marine plants") of fresh marine fish is imparted by volatile compounds during LOX-catalyzed oxidation of LC n-3 PUFA (Lindsay 1994). The subsequent progress in lipid oxidation depends on numerous biological and technological factors, fishing/killing methods, initial processing (whole or gutted fish, carcasses, and fillets), chilling method (ice, sea water, slurry, ozone slurry, etc.), temperature, and storage conditions. Those factors could be compressed to internal and external. The internal (endogenous) factors include lipid content, FA profile, and tissue AAc. The external (exogenous) factors include temperature and oxygen availability. Most important are the lipid content, AAc, and temperature. The catalytic effects of oxygen in whole fish are limited by placing ice on the fish surface and by the skin and melanin it contains as a barrier. On the other hand, as initial processing progresses (carcasses, fillets, and skinned fillets), lipid oxidation is intensified and, importantly, proceeds differently. Lipid oxidation in refrigerated fish is more problematic in fatty fish and those with a high proportion of dark muscle due to faster oxidation, color change, and enhancement of rancid flavor (Sohn et al. 2005). In addition to volatile compounds produced by lipid oxidation, the rancid odor profile is influenced also by products of interaction between carbonyl and nitrogen compounds. Oxidative stability of lipids in chilled fish depends on the AAc, which decreases during storage. The most pronounced changes occur during the initial 3–4 days of storage. In iced rainbow trout, the AAc was reduced—depending on fish batch (all batches from one farmer)—by 20%–70%, more drastic reduction being observed in whole than in gutted fish. The AAc continued decreasing until day 7 of storage (A/B class), i.e., until the first signs of spoilage appeared. At the same time, however, the tissues exhibited a gradual, slow increase of their reducing activity. The initially rapid AAc reduction was accompanied by increasing lipid oxidation. A similar relationship had been earlier reported by Undeland et al. (1999) in herring, the increase in oxidation products being linearly correlated with reduction in antioxidants; the most distinct reduction in α-tocopherol occurred on day 3 of storage. Antioxidants in herring decreased in the following order: α-tocopherol > ascorbic acid > glutathione peroxidase (Undeland et al. 1999). Lipid oxidation in rainbow trout was, however, very low. During the shelf life, i.e., 14–18 days of storage in ice, the Totox value—despite some changes—did not exceed its original value. The rainbow trout batches of a very high initial AAc, despite a distinct drop during the first days of storage, lipid oxidation was not accelerated. Most probably, different antioxidant groups were replacing one another, producing lipid stability in refrigerated rainbow trout, despite their high degree of unsaturation. Herring, stored under identical conditions, with shelf life as short as 7 days, showed a fivefold increase in Totox. Average PV and AV for 3 batches of herring and sprat during 7 days were 12.25 μEqO/kg lipid and AV = 8.00, respectively. During an identical length of time, lipid oxidation in 3 batches of rainbow trout could be summarized as PV = 2.36 µEqO/kg lipids, and AV = 4.34 (Zienkowicz and Kołakowska, 2009). The difference in the lipid oxidation rate between chilled rainbow trout and herring may be sought in differences between their AAc. The AAc of the rainbow trout and herring batches analyzed was 17.06 and 6.99 µM TE/g m. m., respectively (Plust 2008). The fatty acid profile does not seem to be decisive for lipid oxidation rate in refrigerated fish. The rainbow trout and herring batches mentioned above showed a similar degree of unsaturation and similar PUFA contents; besides, the lipids themselves, subjected to UV-catalyzed oxidation, were more sensitive to oxidation in rainbow trout than in herring. Judging by the duration of induction period, the herring lipids are better protected by lipophilous antioxidants.

Losses of LC n-3 PUFA during cold storage usually amount to several percent at most, but in the case of DHA, they may be as high as 20% or even 30%, e.g., in bream. The losses often occur at the transition from class A to class B rather than taking place at the stage of advanced spoilage.

To sum up, changes in lipids of chilled fish, particularly lipids oxidation, are not a major quality problem; sensory effects are perceivable mainly in fatty fish and affect the odor profile. On the other hand, advanced changes in chilled fish lipids pose a threat by accelerated oxidation during processing and preparation for consumption of the fish in which such changes have taken place.

15.7.1.2 Frozen Storage

The most important processes, typically reducing the quality of frozen fish on storage, include lipid oxidation–rancidity and myofibrillar protein denaturation, which result in texture changes (to dry and fibrous) and drip. The two processes interact (Sikorski and Kołakowska 1990). All the remaining quality defects reported from frozen fish are predominantly a result of reduced pre-freezing quality and/or technological errors in freezing and storage conditions. While protein denaturation does not directly affect food safety, advanced lipid oxidation poses a health hazard and significantly reduces nutritive quality of a food (see Chapter 9).

Lipolysis. Storage of frozen fish is accompanied by accumulation of FFA, the accumulation being usually linearly correlated $p < 0.05$) with duration of storage. For instance, in herring fillets stored at −15°C, FFA contributed 1% and 10% to total lipids after 16 days and 8.5 months, respectively. The first to be released were C 16:0, C18:1n-9, and C 22:6n-3; after 8.5 months, FFA were dominated, in addition to the above mentioned acids, by C16:1n-7 and C 20:5 (Ingemansson et al. 1992). FFA originate from TAG and PL; the contents of PC and PE decrease, whereas lyso-PC and lyso-PE increase. Lipolytic changes were found to be more important in lean fish such as gadoids than in fatty fish, because PL breakdown during frozen storage is faster than TAG lipolysis, and FFA participate in protein denaturation in frozen fish (Sikorski and Kołakowska 1994). In mackerel, caught in May when it is at its leanest, FFA increased very rapidly, up to 12 g/100 g lipids, but only during the initial 3 months of storage at −20°C; on the other hand, lipolysis in fat mackerel was very slow during the entire 12-month storage, the final FFA content being as low as 3.5 g/100 g (Aubourg et al. 2005). The rate of FFA formation decreased with decreasing temperature of storage and was higher for the PL than for neutral lipids. PL of

Antarctic krill are very active at low temperatures, probably as an adaption to the krill habitat where enzymes have to operate at low temperatures. After 6 months of storage, approximately 70% of PL were hydrolyzed and the FFA content increased by a factor of 6–20; FFA accounted for more than 50% of total lipids. Generally, lipolysis is faster in whole and minced fish than in carcasses and fillets with skin (as reported from herring, mackerel, rainbow trout, and horse mackerel). However, lipolysis in skinned herring fillets was even somewhat faster than in whole and minced fish. Most probably, once the skin (and the subcutaneous fat) is removed, the PL contribution to total lipids "increased." Simeonidou et al. (1997), too, found higher FFA contents in frozen fillets than in whole hake. FFA are first released from *sn*-1,3 TAG. Because LC, and particularly DHA, are located primarily at the *sn*-2 position, DHA is somewhat better protected from oxidation than other unsaturated FA.

Lipid oxidation. The rate of lipid oxidation-induced changes in frozen fish depends on a number of biological and technological factors: species, lipid content, fishing season, extent of dressing, freshness, and conditions of freezing and storage.

The extent of lipid oxidation in frozen lean fish is usually high due to their high PUFA content and because the membrane lipids are the first to be oxidized (Hultin 1994). However, if lean fish is stored frozen at an appropriate temperature, lipid oxidation may not be perceivable until after 1 year of storage, due to the low contribution of lipids to muscle rancidity. This effect is described as "cardboardy" and is related to the presence of *cis*-4-heptenal (Hardy et al. 1979). In shellfish (prawns), regardless of the absence of direct sensory signals of rancidity, lipid oxidation is evidenced by, e.g., a 70% decrease in DHA and EPA content after 6 months at −18°C (see references in Sikorski and Kołakowska 1990). Lipid oxidation during frozen storage of fat and medium-fat fish is an important quality problem, mainly due to effects on sensory properties such as rancid off-odor (appearing earlier than off-flavor), orange-brown discoloration, and texture changes. During a several-month-long storage at an proper temperature, the hydroperoxide content increases several times, the increase being dependent on species, dressing, lipid content, and pre-freezing history. Carbonyl compounds increase later on, after a certain induction period, usually unnoticeable in hyperoxide changes. After a few months (about 6 months in herring) of storage, a periodic decrease is observed, particularly in aldehydes, due to their interaction with nitrogenous substances (see Chapter 22), which is often manifest in frozen fish as a periodic disappearance of intensive rancid off-odor and a change in rancid off-odor profile. Next, the amount of all the oxidation products increases rapidly. Activity of the enzymatic and non-enzymatic pro-oxidative system in muscle was observed to increase slightly during the first 3 weeks of frozen storage (−18°C) of herring and sprat, compared to that in unfrozen fish, particularly sprat (Kołakowska 2003). Lipid oxidation in Baltic herring during storage at −22°C was closely correlated ($p < 0.05$) with changes in the total AAc (Figure 15.3). The highest significant correlation coefficients were revealed in relationships between the summary AAc and PV, particularly that of hydroperoxide as calculated from tissue w. w. With time of frozen storage, PV and AV increased, while AAc decreased. On the other hand, the reducing power (FRAP) and phenol contents increased with increasing lipid oxidation during frozen storage of herring.

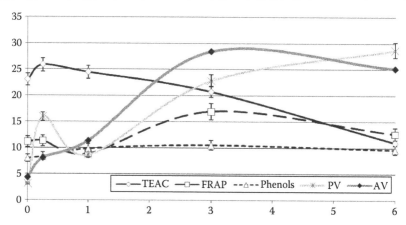

FIGURE 15.3 Changes in total antioxidant capacity (TEAC), reducing power (FRAP), phenols (F-C), and lipid oxidation (PV and AV) during storage of whole Baltic herring at −18°C. (From Aranowska, A. and Kołakowska, A., unpublished data.)

During storage at −27°C, lipid oxidation rate in different batches of mackerel, herring, Baltic herring, and horse mackerel was inversely proportional to the lipid content of fish muscle. However, rancidity was sensorily perceivable first in those batches showing the highest lipid content, although the rancid off-odor profiles were not identical. In batches of herring and mackerel of low lipid content, the fastest oxidation rate was recorded in the batches harvested during the spawning season. On the other hand, the fish captured when feeding actively after spring spawning showed higher concentrations of natural antioxidants in the flesh during frozen storage; lipid oxidation in muscle was markedly reduced (see references in Kołakowska et al. 2003).

The rate of lipid oxidation during frozen storage of Baltic herring increased in relation to *the extent of dressing*: whole fish < fillets with skin < minced (without skin) < carcasses < skinless fillets (Kołakowska et al. 2003). Oxidation rate of dark muscles was faster than that of light muscles, the reverse being the case in lipolysis (Undeland et al. 1998, Aubourg et al. 1999). The presence of skin protects lipids, particularly the subcutaneous deposits, from oxidation. Skin provides mechanical protection and the presence of melatonin has a protective effect; skin also contains an abundance of pro-oxidants. The pro-oxidative enzymatic and non-enzymatic activity in the muscle with skin was 50% higher than in skinless muscle. Minced meat without skin does not become rancid any faster. As found in minced cod and bream, lipid oxidation and (n-3) PUFA losses can be reduced by using directional freezing instead of conventional technology; this is most probably an effect of weak lipid–protein binding that stabilizes lipids. Most likely, such interactions are involved in inhibition of oxidation processes in frozen minces, when appropriate mixing technology is applied prior to freezing (Kołakowska and Szczygielski, 1994).

Washing during fat fish surimi production increased the (n-3) PUFA contribution to lipids remaining in the mince. However, after two rinses of minced herring, its (n-3) PUFA content decreased, most likely as a result of oxidation and interaction with protein. LC n-3 PUFA are preferentially involved in lipid-protein (L-P) complexes. During a half-year storage, both free and bound lipids of washed minces were more oxidized than those in unwashed minces.

Lipid oxidation is faster at higher *storage temperatures*, but the relationship, as a function of a_w, is not linear. Auto oxidation occurs most rapidly at very low a_w levels, higher levels producing diminution in rates until the a_w of 0.4 is obtained. Further increases in a_w tend to increase and than once again to decrease oxidation rates (Labuza 1972). During storage of blue whiting fillets at −30°C, a significant increase in FFA and oxidation (PV, TBA, and conjugated dienes) with time of storage was observed, but lipid oxidation products did not correlate with storage time at −10°C. At both temperatures of storage, the fluorescence ratio —an indicator of L-P interactions was observed to increase (Aubourg 1999). When studying temperature effects (−20°C, −30°C, −80°C) on oxidative changes in lipids and protein of rainbow trout, Baron et al. (2007) concluded that, at −20°C, lipid and protein oxidation seemed to develop simultaneously. At −30°C and −80°C, no lipid or protein oxidation occurred during 13 months of storage; it was only after prolonged (8 months) storage at −30°C that the content of volatile carbonyl compounds increased; the presence of those compounds was ascribed to reactions of oxidized lipids with protein.

Fish freshness at the moment of freezing is important in determining the type and extent of changes in lipids once the fish is frozen. Cold storage of krill for several hours and cold storage of whole herring and herring mince for 1–2 days prior to freezing, inhibited lipid oxidation during frozen storage of minces. Lipid oxidation and hydrolysis proceeded at a faster rate in minces and whole fish frozen immediately after capture and mincing, and also in those stored for a prolonged time (6–7 days) prior to freezing. Optimal timing of freezing of herring mince and whole fish with respect to inhibition of oxidation varied with duration of pre-mincing storage, fish species, and batch (fishing season) and is most probably related to changes in AAc and antioxidants profile.

Changes in n-3 PUFA contents during frozen storage are species-dependent; in a single species, the changes differ in relation to the catching season, similar to other alterations in lipids, particularly their oxidation. Therefore, the losses of n-3 PUFA in, e.g., 3 batches of herring and 3 batches of sprat frozen-stored at −18°C ranged from a few to more than 20%, depending on a batch, the DHA losses reaching 27%. The herring and sprat, caught in the same area on the same day of July and kept frozen under identical conditions for 5 months, showed 3% and 21% losses of EPA+DHA, respectively; a reverse pattern of losses (23% and 5%) was revealed in herring and sprat caught in November. The differences were caused mainly by DHA lability, EPA losses being less important.

Depending on the product and the freezing technique, the freezing process itself induces changes in the form of L-P bonding and most often releases some of the lipids. Frozen storage results in further changes in L-P interactions that affect lipid extractability and PUFA stability. Freezing of Baltic herring mince resulted in 20% more EPA+DHA being extracted, compared to the amount extracted prior to freezing. During 6 months of mince storage at −18°C, contents of those acids varied within 10%, the variability being identical to that in the amount of extracted lipids (Kołakowska et al. 2006c).

15.7.2 COOKING

15.7.2.1 General Changes of Lipids

Changes in lipids during heating of fish depend on temperature, duration of heating, thermal treatment technique, the medium in which the fish is being heated, and the fish susceptibility to thermal degradation. Muscle susceptibility to thermal effects may be influenced by species, season of catching, and pre-treatment history (time and conditions of storage).

Changes in fish lipids heated at a given temperature take different courses, depending on whether isolated or muscle-contained lipids are being subjected to thermal treatment. Heating of isolated fish lipids results in breakdown of hydroperoxides and carbonyls, the breakdown severity increasing with heating temperature. As heating continues, carbonyls gradually accumulate; the higher the cooking temperature, the earlier carbonyl accumulation begins. Heating to above 100°C causes heating time-dependent losses of n-3 PUFA, losses of DHA being higher than those of EPA and total n-3 PUFA. After deodorization of fish oils at 220°C and 250°C, Fournier et al. (2006) found geometrical isomers of EPA and DHA, polymers, and cyclic FA monomers, but only minor changes were observed in the EPA and DHA *trans* isomer content after deodorization at 180°C.

Cooking of fish causes disruption of cellular integrity, dislodgement of iron from heme, an about 50% reduction of AAc and reducing power of muscle tissue, inactivation of enzymes: lipases, phospholipases, LOX, and others associated with lipid oxidation. Lipid-tissue bonds are regrouped, lipids being partly released from the naturally occurring L-P complexes, including covalent bonds; new products of L-P interactions emerge. Some of the latter, formed particularly at higher temperatures, are antioxidative products of the Maillard reaction.

Cooking induces hydrolysis of PL and TAG, and decreases/increases FFA, hydroperoxides, and, to a lesser extent, carbonyls, and decreases conjugated dienes (Kołakowska et al. 2003, Ericson 2002, Rodrigues et al. 2008, Weber et al. 2008). As lipids, even in fresh fish, contain some amount of FFA and lipids oxidation products that break down during heating and participate in L-P interactions, their contents in cooked fish is usually lower than before cooking. A significant ($p < 0.05$) relationship was revealed between PV of uncooked and cooked fish. The lipid oxidation level, at which break-down begins on cooking, depends on fish species and thermal treatment applied, but it could be pinpointed at about PV = 6 (5–8) mgO/100 g lipids (3.75 µEq active oxygen/kg lipids) and AV > 4 in the Baltic herring (11 batches) and farmed rainbow trout (10 batches) subjected to boiling and frying, the AV of raw and cooked fish being poorly correlated. However, at a low oxidation level, PV and AV increase on cooking. Most important in affecting oxidative changes taking place during cooking is the AAc of raw fish tissue, specifically the FRAP of methanol-extractable antioxidants (i.e., phenols, amino acids, peptides and others). The relationship, observed under identical cooking conditions, was found to differ between rainbow trout, cod, salmon, and herring. In salmon (and rainbow trout), most important effects ($r = 0.98$, $p < 0.001$) those antioxidants exerted on the hydroperoxide content in tissue, slightly lower effect ($R = 0.94$) being observed on PV in lipids; in herring, AV was most affected ($r = 0.81$, $p < 0.05$). Correlation coefficients were lower in cod; the amount of lipid oxidation products in cod was more dependent on the total AAc, particularly AV. Cooking fish together with vegetables allows to reduce AAc losses produced by thermal processing (in each of the components cooked separately) by about 20% (see Chapter 10).

Lipids extracted from cooked herring were three times as sensitive to photooxidation than those extracted from raw fish, which could have been related to changes in lipid and FA profiles and, most probably, with reduced—due to cooking—content of lipophilous antioxidants.

Lipids extracted from cooked fresh fish contain more LC n-3 PUFA, and the n-3/n-6 ratio is higher than that before cooking; there is significantly LC n-3 PUFA in fish flesh, even when calculated on a dry weight basis. Such results were obtained for virtually all fish species we examined (rainbow trout, salmon, herring, sprat, mackerel, and others) (Kołakowska et al. 2003, 2006c,d); similar results were earlier reported from vendace by Agren and Hanninen (1993). Numerous authors (Sinclair et al. 1992, Sahin and Sumnu 2001, Al-Saghir et al. 2004, Gladyshev et al. 2006, 2007, Weber et al. 2008, and others) found no losses of n-3 PUFA and LC-n-3 PUFA during heat treatment of various fish species. The lack of losses and higher contents of LC n-3 PUFA in cooked products than in raw fish (calculated in flesh w. w.) result from the loss of water and some amount of LC n-3 PUFA-poorer lipids on cooking. The lack of losses and/or increase (as %) of FA may be also caused by a better extractability of lipids from cooked than from raw fish and also by a more effective saponification of cooked tissue if FA are assayed without lipid extraction (Domiszewski 2000). Moreover, DHA is preferentially located in the sn-2 position of TAG, which ensures protection from oxidation. Aubourg et al. (1990) found the content of DHA in the sn-2 TAG position to increases due to cooking of albacore. This, in part, explains the "increase" in the n-3 PUFA content in cooked fish lipids, compared to the raw material.

15.7.2.2 Cooking Methods

Effects of a cooking method on lipids in fish depend on fish species, freshness, antioxidant activity, and extent of lipid oxidation. Figure 15.4 illustrates a comparison of different cooking methods (to 70°C inside the fillet) of very fresh rainbow trout at a low lipid oxidation level. The largest changes were induced by microwave cooking with water (Mw) and by boiling (C). All the methods, except for baking (B) and frying (F), resulted in a several percent increase in the lipid DHA content. On the other hand, microwaving (but only with water) of the Baltic herring resulted in significant losses of DHA and EPA; both microwaving techniques resulted in doubling of the hydroperoxide content (Kołakowska et al. 2003). The literature contains evidence on both an increase and a decrease of lipid oxidation products during cooking with various methods, depending on the raw material used [(see references in Kołakowska et al. (2003), Weber et al. (2008), Plust (2008)]. All the authors who compared different cooking techniques found the largest lipid changes to be caused by frying and grilling; in all the techniques, except for frying, there were no losses in LC n-3 PUFA.

All the cooking methods result in L-P interactions; the strongest binding of lipids, however, occurs during Mw, as indicated by the series Mw>C>M>B>F. This is also the case in interactions involving covalent bonds formed in the order Mw>M>C. The presence of water on microwaving results in an increase, by 16% and 66%, in weakly

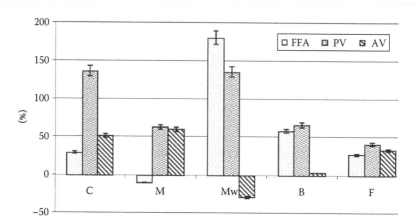

FIGURE 15.4 Changes in lipid oxidation (PV: peroxide value, AV: anisidine value) and free fatty acid (FFA) in muscle tissue fresh rainbow trout, caused by cooking: C, cooking in boiling water; M, microwaving without water; Mw, microwaving with water; B, baked; F, fried. (Adapted from Kołakowska, A. et al., Report Project No. 3 PO6T 060 25 State Committee for Scientific Research, 2006d.)

and strongly bonding lipids, respectively. The interactions involve LC n-3 PUFA; the contribution of EPA and DHA in the strongly bonding lipid fraction increases in microwaved herring, but decreases in boiled herring. Baking and frying tend to result in breakup of the existing strong bonds. Such strongly bonding lipids account for, at most, a few percent of total lipids of rainbow trout, weakly bonding lipids making up from several to 30%. The contribution of free lipids in herring is higher; higher is also the share of strongly binding lipids. Selective extraction combined with Fourier Transform-Infrared Spectroscopy (FT-IR) makes it possible to demonstrate *trans* isomers to be formed on cooking, mainly in bonded lipids, primarily during microwaving and baking. On the other hand, carbonyl compounds accumulate in the free lipid fraction.

15.7.2.3 Frying

Due to absorption of frying fat, frying results in a drastic alteration of fish FA composition. Depending on a frying fat (sunflower oil, rapeseed oil, and hydrogenated frying fat), lipid content of fried herring doubled or tripled and was inversely correlated with unsaturation of the frying fat used, while lipid oxidation was proportional to the frying fat unsaturation. At the same time, frying causes some LC n-3 PUFA to be transferred from fish to the frying fat; losses incurred during deep-frying were lower than those caused by pan-frying. Generally, lean fish absorb more frying fat than fatty fish, and the less fresh fish absorb more frying fat than the fresh fish. During frying on rainbow trout in rapeseed oil, some of the oil absorbed is bonded, also with covalent bonds. Breading reduced the amount of frying fat absorbed: the total fat content increased approximately 10 times in the non-breaded fillets of the black pomfret and less than twice in the breaded fillets (Yazdan et al. 2009). Frying without oil, on the other hand, does not produce such large changes in FA profiles, despite the high temperature used. During salmon pan-frying without oil (the pan surface temperature of 190°C), LC n-3 PUFA decreased nonsignificantly (EPA decreased from 4.9 to 4.1 and DHA from 4.6% to 3.9% FA), but the losses were significant when converted to FA content (Kitson et al. 2009). On the other hand, when pan-frying salmon without oil, Al-Saghir (2004) found EPA to increase from 4.3 to 4.4, and DHA from 6.7% to 6.8% FA. At the same time, the sum of cholesterol oxidation product increased from 0.9 µg/g in raw fish to 9.9 µg/g in the fish pan-fried without oil (Al-Saghir et al. 2004). Frying reduces, by 50% on the average, the lipid oxidation level in fish fried in rapeseed oil. As lipids extracted from fried fish consist mostly of the frying fat (from about 50% to about 80% of lipids extracted from fried herring and rainbow trout, and still more in lean fish), the FA profile in lipids of fried fish is decided upon by the FA profile of the frying fat. Frying fats are much poorer in LC n-3PUFA. For example, the DHA content of 12.29% FA in herring before frying was observed to decrease to 3.03% FA after frying in rapeseed oil, the reduction in EPA was still more substantial (from 5.5% to 1.19% FA). On the other hand, frying of rainbow trout, which absorbs less fat, the reductions were lower: from 12.70% to 7.23% FA in the case of DHA. Due to the loss of water, the LC n-3 PUFA content—converted to 100 g w. w. fried product—is not lower than that in 100 g raw fish but its calorific value increases dramatically. The possible losses of LC n-3 PUFA on frying are more an effect of leaching with the frying oil than of thermal oxidation.

15.7.2.4 Effects of Fish Freshness on Lipid Changes during Cooking

Fish freshness affects changes in lipid lipolysis, oxidation, and FA composition during cooking. TAG lipolysis was slight only during cooking of fresh rainbow trout, but was intensive in the fish bordering on spoilage (14–18 days), and was stronger in oven-baked than in boiled rainbow trout. On the other hand, most PL hydrolyzed during cooking still in a good quality rainbow trout. As lipid oxidation products accumulate, the increase in their content slows down and breakdown occurs: hydroperoxides are the first to break down, followed by carbonyls (AV). The AV of boiled rainbow trout decreased as time of storage of the fish increased ($p < 0.05$, $r = 0.95$). At the same time, AV of lipids leached to the water during boiling increased. Therefore, regardless of small exceptions (such as spawning herring), there is no correlation between lipid oxidation in cooked fish and extent of lipid oxidation in the raw material and the time of storage before treatment, particularly the high temperature-technique such as baking and frying. Another possible reason is that L-P interactions are at their most intensive when the fish being high-temperature treated (baked and fried) border their shelf life.

Frying was found to level off the quality of the raw material and the extent of oxidation; when assessed with a scale of 1–6, the sensory quality of fish was "improved" by about 1 point (Zienkowicz 2009). Inasmuch as the mean Totox over the entire period of 14 or 18 days of storage was in raw rainbow trout about 20, and from about 50 to 70 in herring stored for 7 days, fried fish—regardless of species and batch—showed a similar level of about 20 all the time.

Precooking storage was found to significantly increase losses in LC n-3 PUFA (Figure 15.5). However, when reporting the FA content in g/mg per 100 g w. w. or portion (necessary for providing the guideline daily amount (GDA) information on food labels), the freshness of the raw product "is not important." The EPA and DHA

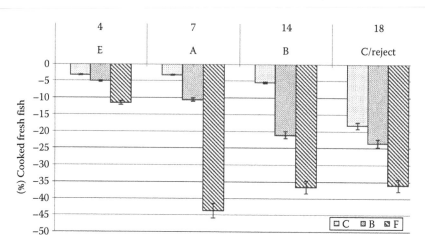

FIGURE 15.5 Effects of rainbow trout freshness on losses of EPA + DHA during cooking: C, cooking in boiling water; B, baked; F, fried; 4, 7, 14, 18 days, fish stored before cooking; freshness class E, Extra; A, B, C, fish reject. (Adapted from Kołakowska, A. et al., Report Project No. 3 PO6T 060 25 State Committee for Scientific Research, 2006d.)

contents may even be higher in cooked product of less fresh fish, because as raw fish storage time increases, losses of fish weight on cooking increase as well.

Effects of frozen storage are visible in lipid oxidation of cooked fish. Although lipid oxidation after cooking (in tightly closed containers, in boiling water bath) of herring during 6-month storage at −23°C were lower than in the raw fish, but was closely dependent ($p<0.05$) on PV and AV of the frozen fish. The strongest relationship was found between the amount of carbonyl compounds in the cooked fish and in the raw material ($r=0.94$). Lipid stability (PV) in herring during cooking depended on the raw fish AAc ($p<0.05$, $R=-0.97$). Most important were myofibrillar proteins (the fraction extracted with 5% NaCl; $R=-0.96$), followed by the water-extractable fraction ($R=-0.88$). On the other hand, lipophilous antioxidants, phenols, and the reducing power of tissue were of minor importance.

Time of catching affected changes in herring FA on thermal treatment. While the autumn herring muscle heated for 30 min at 160°C showed no significant changes in n-3 PUFA, 20% losses were incurred in the spawning herring muscle n-3 PUFA after heating under identical conditions (Kołakowska et al. 2006c).

15.7.2.5 Storage of Precooked Fish

Lipid oxidation in precooked fish proceeded faster than in the fish stored raw for the same period of time at a similar temperature of cold storage. For instance, raw herring mice showed a slower lipid oxidation and smaller changes in PUFA during 4 days at 5°C than minces heated for 30 min at 60°C, 100°C, and 160°C. Jittrepotch et al. (2006) cooked minced sardine (100°C/15 min) and stored the mince for 14 days at 2°C; they found PUFA to decrease by about 30%, the losses occurring as early as after a few days of storage. As early as after 24 h of storage of rainbow trout fillets after heat treatment, lipid oxidation was observed to increase and the PUFA content decreased in the following order of thermal treatment methods: C > Mw > B ≥ F > M.

In boiled fillets, PV doubled, and DHA decreased by 28%; the changes were minimal only in the fillets microwaved (without water) (Kołakowska et al. 2006d). The amount of hydroperoxides accumulated in fried rainbow trout and herring during 3–4 days of storage was twice the amount of hydroperoxides accumulated at the same time by raw fish. Lipid oxidation in the pre-fried stored fish was almost three times as high as than in the fish assayed immediately after frying from raw fish stored for an identical period of time, which shows that a better practice is to store raw fish prior to frying than to fry them first and then store (Zienkowicz 2009). Storage of fried fish leads also to losses in LC n-3 PUFA in lipids. The largest losses were incurred by fried fish stored for 1 day in ice; after 11 days (the shelf life of fried fish), the EPA+DHA content decreased by almost 50%, a 37% decrease (relative to the proportion of those acids in fish lipids immediately after frying) being recorded after 4 days.

Freezing and reheating the pre-fried, pre-baked, and pre-grilled sardine fillets was found to significantly ($p < 0.001$) decrease n-3 PUFA; the oleic acid content in the FA profile increasing and the contents of the n-3PUFA decreasing more after reheating in microwave oven than in conventional oven (García-Arias 2003). FA and cholesterol were slightly affected by warm holding (65°C, 3 h) of fried (in sunflower oil) sardines, mackerel, and salmon (Candela et al. 1998).

15.7.3 CANNED FISH

Changes in lipids during canning of fish are affected by

- The type of raw material
- Fish history prior to processing/freshness
- Preliminary thermal treatment
- Type of liquid coating
- Conditions of sterilization
- Time and temperature of canned product storage

Canning involves mainly fatty fish (mackerel, herring, sardine, sprat, anchovy, and salmon), but also tuna, lean marine fish, and all the common freshwater fish species. Canning is also applied to species of low commercial value, difficult to process, containing abundant bones that, in addition to backbones left in canned fish, are softened on sterilization and become an additional source of minerals. Sterilization conditions ensure microbial safety; the diversity of additives applied mask deficiencies of the raw material; therefore, in no fish product, the danger of using low quality/freshness product is as apparent as in canning. Manufacturers have to cold- or frozen-store the raw materials before they are canned. Storage of chilled coho salmon for 0.5 and 9 days before canning (cooking and sterilization) did not cause any differences in rancid odor of the canned product. However, FFA, PV, TBA, and fluorescence ration in canned salmon were observed to increase. Fluorescence and browning increased both in fish muscle and in the coating oil of the canned product analyzed after 3 months (Rodrigues et al. 2009). A few day-long cold storage of fish, even if does not lead to substantial lipid oxidation and losses of n-3 PUFA, significantly weakens AAc of the tissue and changes its susceptibility to thermal oxidation during heat treatment (usually involving two procedures). Frozen fish, after several months of storage, usually show advanced lipid oxidation. Canning of mackerel (manufacture of mackerel in juice), frozen-stored for 4 months, resulted in an increased lipid oxidation level and a reduction in (n-3) PUFA, including 13% and 15% reductions in EPA and DHA, respectively (Kołakowska et al. 2003).

Changes taking place during initial thermal treatment depend on fish species, biological factors, fish freshness, and type of cooking. The cooking procedures applies usually include steaming, drying, smoking, but also frying. Those procedures are associated with about 20% loss of water, breakdown of oxidation products (mainly hydroperoxides), and their participation in L-P interactions. Precooking was reported to cause no n-3 PUFA losses, but increased their content in solids of tuna (Garcia-Arias et al. 1994) and sardine (Ruiz-Roso et al. 1998) via losses of some of the water and, in the fish canned without precooking, also by the loss of some n-3 PUFA-poorer lipids.

Thermal drip is usually removed. Leaving the drip, when canning fresh fish and those fish frozen-stored for a short time, is beneficial for the canned product quality by inhibiting lipid oxidation. On the other hand, the longer the storage of frozen fish (herring, mackerel, and horse mackerel), the better quality, and the longer shelf life of canned product is achieved by removing the thermal drip. The n-3 PUFA content in canned mackerel in juice (manufactured from whole fish stored for 3 months at −18°C) was observed to decrease, relative to that in the raw material, by 11% in cans with drip and by almost 15% in cans with thermal drip removed.

Changes taking place during sterilization depend on the type of liquid coating and on conditions of sterilization itself. Prolonged sterilization, particularly at high temperatures, may lead to substantial losses of LC n-3 PUFA. Sterilization of canned Baltic herring at 126°C and $F_0 = 20$ resulted in several-fold higher % losses of DHA than at $F_0 = 8$; extension of sterilization time to $F_0 = 55$ resulted in only trace amounts of DHA (0.22% FA) remaining (Bykowski et al. 2003). Sterilization of canned pink salmon fillet at 121.1°C for 10, 30, and 60 min resulted in a slight increase in PV and TBA within the first 10 min of heating, followed by a significant reduction as heating progressed and no loss of PUFA (Kong et al. 2008). PUFA losses during sterilization of minced bigeye tuna and halibut at 124°C were recorded in the FFA fraction only (Koizumi et al. 1986). Differences in sterilization conditions ($F_0 = 10$ and 7) did not, however, affect lipid oxidation in canned tuna in oil during 6 years of canned product storage (Aubourg 1998).

The type of liquid coating significantly affects the LC n-3 PUFA content in the canned fish. Lipid composition in canned fish in oil changes after sterilization. The lipid content in fish increases and, depending on the oil used, its FA profile changes. At the same time, the LC n-3 PUFA of the fish canned in oil are leached from the solids during sterilization and with time of storage (Aubourg et al. 1990). In consequence, the vegetable oil in canned fish stored for a long time (even during the period of warranty, exceeding a year) may be LC n-3 PUFA-richer than lipids of the fish solids (Kołakowska et al. 2006c). Thus, the lipid composition of canned fish, particularly those with a lower lipid content, tended to be similar to that of the oil used for coating. However, due to the whole fat enrichment, some canned products, e.g., canned tuna (g/100g wet fish), showed only a slight loss of EPA, a 25%–37% loss of DHA, and an increase in the contents of FA from vegetable oil (oleic, linoleic, and linolenic), compared to steamed tuna (Garcia-Arias 1994). Caponio et al. (2003) compared canned tuna in different coating oils (olive, virgin olive, soybean, sunflower, and corn). The coatings were found to contain FA typical of the fish; in addition, each coating, except for virgin olive oil, showed the presence of *trans* isomers. Canned products may also contain substantial amounts of cholesterol oxdation products (COP) (Chapter 7). Contents of COP in the canned herring we studied slightly exceeded 40 µg/g fish lipid, but in canned tuna analyzed by Zunin et al. (2001) ranged within 40–350 µg/g lipids.

Muscles of canned fish in tomato sauce, most probably due to the antioxidant properties of tomato and the lack of extraction by the coating, contain more EPA and DHA than canned fish in oil (Tarley et al. 2004, Kołakowska et al. 2006c).

Lipid oxidation in a canned fish product during storage depends on temperature of storage and on absorption of metal ions from the can, but is generally slow and takes a number of months. Lipid oxidation measured (PV and TBA) in canned tuna in brine stored for 12 months at room temperature showed gradual increases, with storage time, PL, TAG, FFA and sterols, an opposite trend being shown. Concentrations of total PUFA and total n-3 and n-6 PUFA significantly decreased, stepwise, with increased storage time; DHA and DPA started to decrease significantly after six months, EPA and AA decreased after 3 months of storage. The n-3/n-6 ratio decreased significantly from 3.8 at month 0 to 3.2 at month 9 (Siriamornpun et al. 2008).

Oxidative changes in canned fish in oil are substantially affected by the oil added to the coating. The fish in the can contains more oxidation products than the coating oil. Addition of α-tocopherol to the latter efficiently inhibited lipid oxidation in fish and in oil for up to a year after canning of herring. After 2 years, the amount of DHA in herring and in the coating oil was by 31% and 26% higher, respectively, in α-tocopherol enriched products than in those without the additive. In an extreme example, canned mackerel stored for 17 years could not be consumed due to sensory changes, but it still contained about half of the initial EPA content and about 1/3–1/4 of the initial DHA content. Compared to fish juice, the coating oil protects fish lipids during prolonged storage.

Canned fish available on the market are regarded as a good source of LC (n-3) PUFA (Ota et al. 1990, Kołakowska et al. 2000, 2003, 2006c, Sirot et al. 2008, Usydus et al. 2008, Gladyshev et al. 2009). Their LC-(n-3) PUFA contents depend primarily on the type of raw material, and specifically on the lipid content thereof. For example, the EPA+DHA contents in canned herring (the entire can content) amounted to slightly more than 1.5–2.0g/100g; canned sprat and mackerel showed usually slightly more than 2.0 and about 2.0–5g/100g, respectively. Canned fish with tomatoes always contained more LC n-3 PUFA than canned fish with oil only (Bienkiewicz et al. 2008). On the other hand, canned tuna in its own fluid was found to contain as little as 0.18g/100g. Canned cod liver is a particularly rich source of LC n-3 PUFA. The EPA+DHA contents in solids of canned fish of different manufacturers varied from 9 to 18g/100g, the contents in oil varying from 23 to 28g/100g (Kołakowska et al. 2002).

However, canned fish may contain an abundance of oxidation products, including COP. Some samples of the coating oil of canned sardines, available on Italian and Swiss markets, were found to contain *trans*-FA in amounts exceeding the EU-proposed tolerance level (Cavallaro et al. 1996).

15.7.4 SMOKING

Factors involved in smoking, mainly drying, salting, heating, and impregnation with smoke constituents may have some pro-oxidative as well as antioxidative effects on lipids in fish muscle (Sikorski and Kołakowska 2001, Gómez-Guillén 2007).

The classic hot smoking of fish involves a combined action of smoke and temperature (internal temperature 72°C, air temperature about 130°C). Effects of smoking on fish lipids are similar to those of cooking because the smoke impregnation stage is preceded by cooking proper. Hot smoking of fresh fish, or of fish stored frozen for a short time, resulted in increases in hydroperoxides, carbonyls, FFA, conjugated FA, and fluorescence in fish lipids. In FA, regrouping occurs only between PUFA and MUFA, but no increase of SFA and no reduction of (n-3) PUFA takes place. Effects of smoking on the herring and sprat FA profiles did not differ from those exerted by the heat treatment alone (Kołakowska et al. 2006c, Stołyhwo et al. 2006). On the other hand, small (about 10%) losses of DHA and EPA on smoking were found in PL of sardine (Beltran and Moral 1991). When lipids in frozen mackerel were already considerably oxidized (PV = 79 mgO/100 g lipids), smoking resulted in disintegration of hydroperoxides, AV decreased, but the amount of conjugated dienes increased, while (n-3) PUFA decreased by almost 20% (Kołakowska et al. 1998).

Practically, during smoking of fatty fish (for the sake of sensory aspects, mostly fatty fish are being smoked), losses in LC n-3 PUFA amount to as a few percent at most, and frequently the content of those FA is higher than in lipids before smoking, particularly when calculated on fish w. w. Smoking induces an about 20% loss of water; in addition, some n-3 PUFA-poorer lipids are lost, particularly during smoking of partly dressed fatty fish. On the other hand, PUFA losses during hot smoking of lean fish are substantial, and are more pronounced in n-3 than in n-6 PUFA. For example, smoking of red fish resulted in more than 19% losses of n-3 PUFA; the loss of n-6 PUFA and DHA amounted to 14 and as much as 33%, respectively. Moreover, smoking may induce formation of oxysterols; smoked cod roe was found to contain more oxysterols than the fresh raw roe (Pickova and Dutta 2003).

Cold smoking involves combination of salting (6%–10% NaCl in the product) and prolonged exposure to smoke (30°C). Compared with the raw fish (frozen mackerel), lipids extracted from cold-smoked mackerel contained by 22% less EPA+DHA, the loss being incurred by DHA only.

Smoking, particularly hot smoking, induces losses of natural antioxidants (Espe et al. 2002, Sampels et al. 2004). In cold-smoked (at 20°C or 30°C) Atlantic salmon, ascorbic acid was observed to decrease by about 80%, compared to the original level; tocopherol proved more resistant to smoking (Espe et al. 2002). Although hot smoking of mackerel resulted in doubling of phenols in the tissue, the total AAc decreased by 30% (Kołakowska et al. unpublished data, 2009).

Storage of cold- or hot-smoked fish is associated with increase in lipid oxidation products. After 3 days of dark storage at 4°C of hot-smoked mackerel, PV was observed to increase by 40%, AV increased by a factor of 8, but no reduction in LC (n-3) PUFA was recorded. After 8 days of storage at 4°C, cold smoked mackerel showed a PV increase by 60% and a few percent reduction in DHA (Kołakowska et al. 2003). Smoked chub stored at 4°C for as long as 4 weeks showed only a few percent reduction in LC (n-3) PUFA (Wang et al. 1990). After 3 months of refrigerated storage of smoked tuna, its 22:6/C16:0 ratio decreased by 15%–20% (Zotos et al. 2001). Although smoked chub mackerel stored for 30 days at 2°C showed increasing lipid oxidation, the increase was low, compared to that in non-smoked samples stored at identical conditions (Goulas and Kontaminas 2005). Although antioxidant effects of smoking are in practice not particularly convincing due to results of cooking involved in hot smoking, and of salting involved in cold smoking, and basically seem to boil down to rancidity masking by the smoke, smoking does exert a successful antioxidant effect on fish lipids. This is demonstrated by the efficiency of smoking, as revealed by FT-IR spectroscopy, by a high stability of lipids in UV-irradiated tissues of hot-smoked mackerel, compared to lipids in raw UV-irradiated tissue of mackerel.

To sum up, hot smoking does not inhibit lipid oxidation; changes, particularly in LC n-3 PUFA, are even more extensive than those induced by cooking. However, smoking increases lipid stability during storage of smoked products, compared to both raw and precooked fish. The increase is due to antioxidants migrating from the smoke into the tissue rather than due to any direct effect on lipids.

15.7.5 Salting

Salting involves removal of some amount of water from the tissue as a result of it being penetrated by NaCl, whereupon a solution inhibiting microbial activity and stimulating endogenous enzymes is formed (Kołakowski 2005). Salted fish, particularly herring, are traditional products in Central and Eastern Europe as well as in Scandinavia. Salted fish, ready for consumption depending on type of salting (weak, medium, and strong), contain from about 6% to 14% NaCl w. w. Traditional technologies of salted-dried fish and fermented products have also emerged in adaptation to climatic conditions.

Sodium chloride affects both hydrolysis and oxidation of lipids contained in fish. Lipid hydrolysis occurs at lower NaCl concentrations, high (20%) concentrations inhibiting activity of lipases. FFA accumulate a few days after salting, unless the accumulation has already occurred in the raw material as a result of prior cold or frozen storage, particularly in the case of, e.g., herring caught during post-spawning feeding season when lipases are at their most active stage. During salting, FA are released from TAG, but primarily from PL; after 4 days of salting the Baltic herring, PL contained only traces of DHA. Salting is associated with increased fish lipid oxidation, particularly intensive in salted-dried fish. NaCl catalyzes the oxidation process, the effect increasing with the strength of NaCl solution. Most probably the mechanism involves enhancement of oxidation effect of iron and copper ions, which takes place particularly in dark muscles and depends also on the degree to which the brine is contaminated with metal ions. Lipid oxidation in salted fish proceeds mainly via autoxidation, but LOX are involved as well. LOX activity is enhanced by the presence of peroxides; LOX activity is inhibited with increasing peroxide concentration. Results of lipid oxidation, visible in about 14–20 days after NaCl was applied to herring (20% NaCl brine, 10°C), are noticeable as a significant damage to lipids (Kołakowska et al. 2003). At the same time, losses of antioxidant vitamins (tocopherol and ascorbic acid) were observed in salted salmon (Espe et al. 2001); however, ripening resulted also in an increase in contents of peptides and amino acids, most of which show AAc. Such activity is shown also by sodium benzoate, an additive allowed in salted products. Therefore, changes in the content and composition of lipid oxidation products are primary effects observed in lipids during storage of salted products (Kołakowska et al. 2003, 2006c, Aro et al. 2005). Such pattern of oxidative changes is produced also by interactions between oxidation products, particularly interactions of carbonyl compounds and products of proteolysis. The interactions result in pigmentation (browning and yellowing) and changes in the profile of volatile rancid odor compounds.

Lipids of salted Baltic and open ocean herring are poorer in n-3 PUFA, by a few to several percent on the average, compared to the unsalted fish; this is particularly true with respect to PL, but also TAG. The PUFA losses taking place during salting of sardine amounted to about 30%; a reduced contribution of PUFA was accompanied by increased shares of MUFA and SFA. The Indian mackerel DHA content dropped after salting by 23%, compared to the raw material (Piggot and Tucker 1990).

The extent of oxidative changes in salted fish is evidenced by about fivefold increase in the amount of conjugated dienes and the presence of *trans* isomers and their 40% increase after 1.5 month storage of salted herring manufactured from frozen fish. Despite the progress of lipid oxidation in salted fish, herring, mackerel, and sardine, and evident losses of LC n-3 PUFA, a 100 g portion of, e.g., salted herring, did not show a lower amount of those acids, compared to a 100 g portion of the raw material, the contents being occasionally even higher in the salted fish due to increased lipid content produced by mass exchange between the fish and the brine. Salted fish products (salted matjes herring) contain from 1.69 to 2.67 g LC n-3 PUFA, i.e., much more than (twice as much as) the recommended dose of those acids (ISRAFAL), but owing to the load of FA oxidation products and high NaCl content (a 100 g portion exceeds the daily NaCl allowance), they are not a recommended source of LC-n-3 PUFA. This is particularly true with respect to dried salted products in which contents of oxidation products, especially those of cholesterol oxidation as well as losses of n-3 PUFA and antioxidants, are much higher (Ohshima et al. 1993, Davis et al. 1995, Guillen and Ruiz 2004). On the other hand, Aro et al. (2005) contend that weakly salted (sugar salted, spice-salted) herrings are good and stable sources of LC n-3PUFA and vitamin D.

15.7.6 Marinading

Marinading is applied to pelagic fatty fish, in which catepsin activity at reduced pH results in ripening after a few days in marinading batch (acetic acid solution and NaCl). The shelf life of the fish marinaded cold, without any preservatives added and stored cold, is 14 days. In addition to the classic marinaded products, there are numerous other types of marinades. The shelf life of marinades is extended by as much as several months by addition of ascorbic acid and preservatives allowed in EU (benzoate, sorbic acids and their salts, and nitrates) that also exert antioxidant effects. The marinades manufactured commercially from open ocean herring contained—despite considerable technological differences—a similar amount (about 3 g) of n-3 PUFA in 100 g marinades fish and showed a low lipid oxidation level. The oil-containing marinade was found to contain substantial amounts of n-3 PUFA also in the oil, which evidenced elution of the acids from the fish by the marinading fluid, for that reason, the product showed the lowest n-3 PUFA content in fish; the lipid oxidation level was also 2–3 times higher than in oil-free marinades.

Major changes occurring in lipids on fish marinade production involve lipid hydrolysis, whereas—despite the presence of NaCl—lipid oxidation is inhibited. The lipid oxidation product content during the initial days of marinading is usually lower than in the raw material. Although it may increase for some time during storage, depending on the type of a marinade, it is generally low. During 22 days of marinading the Baltic herring, PV (in three different batches) was observed to reach the maximum value of 12 mg/100 g, i.e., was equal to the value found in fresh raw herring. Because of the oxidative stability of lipids, losses of EPA and DHA relative to the content in raw materials are as low as a few percent, although EPA and DHA (as %FA) tend to decrease during marinading and storage (Kołakowska et al. 2003, 2006c, Tomczyk 2008).

The time of capture was found to affect changes during marinading and marinade storage. Compared to the raw materials used in three experiments, mean losses of EPA + DHA per 100 g of marinaded product occurred in one Baltic herring batch only, the batch consisting of the herring caught in April during spawning; a 22% loss was recorded. At that time, herrings are most susceptible to lipid oxidation, their susceptibility to lipid hydrolysis being at the lowest. On the other hand, the amount of EPA + DHA in the marinaded herring caught in March was by 18% higher than in the raw material, while mean EPA + DHA contents in the remaining batch did not differ significantly between the marinaded and raw fish. After a prolonged storage, the marinades showed even higher n-3 PUFA content, compared to freshly manufactured marinades, which could have been related to mass exchange between the curing liquid and the fish and to elution of lipids during ripening and storage. The fish weight losses were from several percent after 60 days (open ocean herring fillets) (Tomczyk 2008) to even as high as 20% in the Baltic herring.

Thus, there is no reason (notwithstanding the presence of acetic acid and NaCl) to conclude that marinading can reduce the value of herring as a source of n-3 PUFA, particularly that marinading is not accompanied by any increase in the amount of oxidation products.

15.8 DIETARY SUPPLEMENTS AND FOOD ENRICHMENT WITH FISH OIL

In most western countries, food consumption is insufficient to cover the daily recommended (ISSAFAL) dose of LC n-3 PUFA. This is because of reduced landings, particularly those of sensorily attractive species, and high prices of food products.

Numerous people are not fond of fish. People who eat fish regularly are few, but their fish consumption is high. However, the fish eaten most often are not the recommended fatty fish, but—at least in the United States—lean fish and shellfish (prawns). Those people who do not eat fish are a target of increasing wide spectrum of commercially available dietary supplements and foods enriched with fish oils. Supplements are available as encapsulated or liquid cod liver oil, fish muscle oil, and oil manufactured from whole fish and offal, even or reduced freshness (Aidos et al. 2002, Kołakowska et al. 2006a). Fish oils are subjected to high temperatures during manufacture, refining, and deodorizing. Fish oil capsules from cold oceanic water fish, e.g., from mackerel, contain more EPA than DHA, although the DHA content in the original fish is twice that of EPA. This may evidence incomplete extraction during oil production, but also oxidative destruction of the more unsaturated DHA. Although the shelf life of those products, due to addition of antioxidants declared by manufacturers, is about 2 years, capsules are typically characterized by rancid flavor, elevated values of oxidation indicators (PV, AV) and their wide variations during the declared shelf life. Microencapsulation does not ensure oxidative stability of fish oil either (Kolanowski et al. 2006, Velasco et al. 2006). The fragmentary data on consumption of dietary n-3 PUFA supplements demonstrate their negligible role in the actual supply of LC n-3 PUFA. Special DHA-rich algal supplements are advertised for consumption by vegans (Arterburn et al. 2008). The trend toward functional foods and the business interest in such products bring attention toward fish oil addition to various food products available on the market. Fish oils in the form of dried microcapsules make it possible to enrich also low-fat foods. Studies on n-3 PUFA enrichment are focused on changes in sensory characteristics and oxidative stability of a product; on the other hand, effects of such supplements on nutritive value of the remaining components of the matrix are virtually unknown. Another tendency involves increasing the -3 PUFA content in meat, milk, eggs, and—most successfully—in farmed fish via food and via genetic modification of plants and animals. Dietary modifications involving feed enrichments with n-3 PUFA and/or fish oils are increasingly frequently accompanied by feed enrichment in natural antioxidants that increase stability of the product, also during storage and processing.

15.9 FINAL REMARKS

Due to the presence of LC (n-3) PUFA, fish lipids play an important nutritive and therapeutic role. It is already known that LC (n-3) PUFA cannot be substituted by α-linolenic acid, and that EPA and DHA serve different functions in the human body. The importance of fish as a source of those valuable PUFA depends on the latter's content in the raw material and its treatment along the entire route *from the farm (waterbody) to the plate*. The treatment applied should prevent LC n-3 PUFA losses and protect fish lipids from oxidation. The concentration of lipids and their n-3 PUFA content in fish are affected by a number of biological factors, such as species, region from which the fish are harvested (and in which they are farmed), life cycle stage (age, gonad development stage), season of capture, size, sex, and diet; the latter is decisive for lipid profile of farmed fish. Effects of biological factors is so important that some fish batches can be successfully used to demonstrate significant interspecific differences, but within each species there will be batches more similar in their lipid content and profile to another species than to other conspecific batches. The dynamic development of aquaculture, including mariculture, will most probably suppress the significance of biological differences and will render fish processing problems to those related to, e.g., poultry processing. So far, however, even rainbow trout batches obtained from the same farmer differ in their lipid content, profile, and susceptibility to oxidation. Disregard to those factors, particularly the season of capture, when labeling fish products (general nutritive value tables are then used) may lead to even 100% errors in LC n-3 PUFA declared for a given food product type.

Although lipids in lean fish—consisting mostly of PL—are EPA- and DHA-richer than lipids of fatty fish in which PL account for several percent only, a portion of a fatty and medium-fat fish (except for extreme cases of farmed fish) supplies more LC n-3 PUFA than a portion of lean fish does. Effects of technological factors on the n-3 PUFA content are generally lower than could be expected. Although fish storage, chilled storage (but particularly frozen storage), and fish processing are associated with damage of fish lipids, their destructive effects on LC n-3 PUFA are much lower than commonly thought, considering their high unsaturation. On the one hand, the reason may be sought in concentration of the FA in question in the less readily available PL, including TAG and PL, particularly DHA at the *sn*-2 position. Decisive for protection during thermal treatments is the high AAc of fish tissue, more important than tocopherol contents in the lipophilous fraction. Losses of LC n-3 PUFA on storage or thermal treatment are estimated at about 20%, which seems low, close to the error induced by extraction from raw or cooked tissue. Although the highest thermal stress (increase of lipid oxidation, decrease of AAc) takes place during processing of very fresh fish, thermal treatment of fish that have been chilled or frozen-stored is accompanied by break-down of lipid oxidation products accumulated earlier, their participation in L-P interactions and profound changes in bonded lipid fraction (*trans* isomers) results in higher LC n-3 PUFA losses. Because, however, thermal treatment is accompanied by loss of water and some n-3 PUFA-poorer lipids (thermal drip is higher in low-freshness fish), a portion of, e.g., 100 g product for all practical reasons does not contain less, and even may be richer in, LC n-3 PUFA than a 100 g portion of raw fish. This practical aspect is, however, not sufficient to forego a possibility of gaining better knowledge on using, and creating in farmed fish, an antioxidative potential in fish tissue so that the LC n-3 PUFA content will be transferred unchanged to the final product. On the other hand, enriching a whole spectrum of foods in n-3 PUFA without gaining knowledge on their interactions with the remaining components of the matrix and effects of added, already substantially oxidized, fish oil on nutritive value of the matrix component seems controversial.

REFERENCES

Ackman, R.G. 1994. Seafood lipids, in: *Seafoods: Chemistry, Processing Technology and Quality*, F. Shahidi and J.R. Botta (Eds.), pp. 34–48, Chapman & Hall, London, U.K.

Agren, J.J. and Hanninen, O. 1993. Effects of cooking on the fatty acids of three freshwater fish species, *Food Chem.*, 46, 4:377–382.

Aidos I., van der Padt, A., Luten, J.B., and Boom, R.M. 2002. Seasonal changes in crude and lipid. Composition of herring fillets, by products, and respective produced oils, *J. Agric. Food Chem.*, 50:4589–4599.

Al-Saghir, S., Thurner, K., Wagner, K.-H., Frisch, G., Luf, W., Razzazi-Fazeli, E., and Elmadfa, I. 2004. Effects of different cooking procedures on lipid quality and cholesterol oxidation of farmed salmon fish (*Salmo salar*), *J. Agric. Food Chem.*, 52, 16:5290–5296.

Ando, Y., Nishimura, N., and Takagi, T. 1992. Stereospecific analysis of fish oil triacyl-*sn*-glycerols. *JAOCS*, 69, 5:417–424.

Aro, T.L., Larmo, P.S., Backman, C.H., Kallio, H.P., and Tahvonen, R.L. 2005. Fatty acids and fat-soluble vitamins in salted herring (*Clupea harengus*) products, *J. Agric. Food Chem.*, 53:1482–1488.

Arterburn, L.M., Oken, H., Bailey, A., Hall, E., Hamersley, J., Kuratko, C.N., and Hoffman, J.P. 2008. Algal-oil capsules and cooked salmon: Nutritionally equivalent sources of docosahexaenoic acid, *J. Am. Diet. Assoc.*, 108:1204–1209.

Aubourg, S.P. 1998. Lipid changes during long-term storage of canned tuna, *Z. Lebensm Unters Forsch A*, 206:33–37.

Aubourg, S.P., Rey-Mansilla, M., and Sotelo, C.G. 1999. Differential lipid damage in various muscle zones of frozen hake (*Merluccius merluccius*). *Z.-Lebensm. Unters. Forsch.*, 208:189–193.

Aubourg, S.P., Rodríguez, A., and Gallardo, J.M. 2005. Rancidity development during frozen storage of mackerel (*Scomber scombrus*): Effect of catching season and commercial presentation. *Eur. J. Lipid Sci. Technol.*, 107, 5:316–323.

Aubourg, S.P., Sotelo, G.C., and Gallardo, M.J. 1990. Changes in flesh lipids and fill oils of Albacore (*Thunnus alalunga*) during canning and storage, *J. Agric. Food Chem.*, 38:809–812.

Aubourg, S.P., Medina, I., and Perez-Martin, R. 1996. Polyunsaturated fatty acids in tuna phospholipids: Distribution in the *sn*-2 location and changes during cooking, *J. Agric. Food Chem.*, 44:585–589.

Aursand, M., Jorgensen, L., and Grasdalen, H. 1995. Positional distribution of ω3 fatty Acids in marine lipid triacylglycerols by high-resolution ^{13}C nuclear magnetic resonance spectroscopy, *JAOCS*, 72, 3:293–2297.

Baron, C.P., Kjærsgård, I.V.H., Jessen, F., and Jacobsen, C. 2007. Protein and lipid oxidation during frozen storage of rainbow trout, *J. Agric. Food Chem.*, 55:8118–8125.

Beltran, A. and Moral, A. 1991. Changes in fatty acid composition of fresh and frozen sardine (*Sardina pilchardus* W.) during smoking, *Food Chem.*, 42:99–109.

Bienkiewicz, G. and Domiszewski, Z. 2008. Content of long chain unsaturated omega 3 fatty acids (LC) n-3 PUFA in selected fish species, *Fish Ind. Mag.*, 2, 62:45–46 (in Polish).

Bienkiewicz, G, Domiszewski, Z., and Kuszyński, T. 2008. Fresh water fish as a source essential unsaturated fatty acids, *Fish Ind. Mag.*, 3, 63:58–59 (in Polish).

Brodte, E., Graeve, M., Jacob, U., Knust, R., and Portner, H.O. 2008. Temperature-dependent lipid levels and components in polar and temperate eelpout (*Zoarcidae*), *Fish Physiol. Biochem.*, 34, 3:261–274.

Bykowski, P., Kołodziejski, W., and Pawlikowski, B. 2003. Optimization of heat sterilization of tinned fish considering the food safety and quality, *Żywienie Człowieka i Metab.*, 30, 3(4):906–911 (in Polish).

Candela, M., Astiasarán, I., and Bello, J. 1998. Deep-fat frying modifies high-fat fish lipid fraction, *J. Agric. Food Chem.*, 46, 7:2793–2796.

Caponio, F., Gomes, T., and Summo, C. 2003. Quality assessment of edible vegetable oils used as liquid medium in canned tuna, *Eur. Food Res. Technol.*, 216:104–108.

Cavallaro, A., Bizzorero, N., Carnelli, L., and Renon, P. 1996. Fatty acid composition and trans unsaturation of the covering oil of canned sardines, *Industrie Alimentari*, 35, 350:801–805, 812. *Abstract* AN: 97-03-R0027.

Celik, M., Diler, A., and Kucukgulmez, A. 2005. A comparison of the proximate composition and fatty acid profiles of zander (*Sander lucioperca*) from two different regions and climate conditions, *Food Chem.*, 92:637–641.

Cherif, S., Frigha, F., Gargouri, Y., and Miled, N. 2008. Fatty acid composition of green crab (*Cacinus mediterraneus*) from the Tunesian mediterranean coasts, *Food Chem.*, 111, 4:930–933.

Crawford, M.A., Bloom, M., Broadhurst, C.L., Schmidt, W.F., Cunnane, S.C., Galli, C., Gehbremeskel, K., Liseisen, F., Lloyd-Smith, J., and Parkington, J. 1999. Evidence for unique function of docosahexaenoic acid during the evolution of the modern hominid brain, *Lipids*, 34:S39–S47.

Davis, L., Smith, G., and Hole, M. 1995. Lipid oxidation in salted-dried fish. *J. Sci. Food Agric.*, 67, 4:493–499.

Domiszewski, Z. 2000. The comparison of fatty acids composition fish lipids with the use of direct and indirect saponification of a sample. *Science Conference on Chromatography Methods*. Szczyrk, Poland, June 7–9, p. 123 (in Polish).

Domiszewski, Z., Bienkiewicz, G., and Kołakowska, A. 2008. Fatty acid composition in wild and farmed salmon: Fresh and cold- and hot-smoked, in: *Oil, Fats and Lipids in the 3rd Millennium Challenges, Achievements and Perspectives, 6th Euro Fed Lipid Congress*, Athens, Greece, HUNU, p. 79.

Dunstan, G.A, Olley, J., and Ratkowsky, D.A. 1999. Major environmental and biological factors influencing the fatty acid composition of seafood from Indo-Pacific to Antarctic waters, *Recent Res. Dev. Lipid Res.*, 3:63–86.

Ericson, M.C. 2002. Lipid oxidation of muscle foods, in: *Food Lipids: Chemistry, Nutrition, and Biotechnology*, C.C.A. Akoh and D.M. Min (Eds.), pp. 365–412, CRC Taylor & Francis, Boca Raton, FL.

Espe, M., Nortvedt, R., Lie, O., and Hafsteinsson, H. 2001. Atlantic salmon (*Salmo salar*, L.) as raw material for smoking industry. I: Effect of different salting methods on the oxidation of lipids. *Food Chem.*, 75, 4:411–416.

Espe, M., Nortvedt, R., Lie, O., and Hafsteinsson, H. 2002. Atlantic salmon (*Salmo salar* L.) as raw material for the smoking industry. II: Effect of different salting methods on losses of nutrients and on the oxidation of lipids, *Food Chem.*, 77:41–46.

Fournier, V., Destaillats, F., Juaneda, P., Dionisi, F., Lambelet, P., Sebedio, J.-L., and Bordeaux, O. 2006. Thermal degradation of long-chain polyunsaturated fatty acids during deodorization of fish oil, *Eur. J. Lipid Sci. Technol.*, 108, 1:33–42.

García-Arias, M.T., Álvarez Pontes, E., García-Linares, M.C., García-Fernández, M.C., and Sánchez-Muniz, F.J. 2003. Cooking-freezing-reheating (CFR) of sardine (*Sardina pilchardus*) fillets. Effect of different cooking and reheating procedures on the proximate and fatty acid compositions. *Food Chem.*, 83, 3:349–356.

García-Arias, M.T., Sánchez-Muniz, F.J., Castrillón, A.M., and Pilar Navarro, M. 1994. White tuna canning, total fat, and fatty acid changes during processing and storage, *J. Food Compos. Anal.*, 7, 1–2:119–130.

Gladyshev, M.I., Sushchik, N.N., and Makhutova, O.N. 2009. Content of essential polyunsaturated fatty acids in three canned fish species. *Int. J. Food Sci. Nutr.*, 60, 3: 224–230.

Gladyshev, M.I., Sushchik, N.N., Gubanenko, G.A., Demirchieva, S.M., and Kalachova, G.S. 2006. Effect of way of cooking on content of essential polyunsaturated fatty acids in muscle tissue of humpback salmon (*Oncorhynchus gorbuscha*), *Food Chem.*, 96:446–451.

Gladyshev, M.I., Sushchik, N.N., Gubanenko, G.A., Demirchieva, S.M., and Kalachova, G.S. 2007. Effect of boiling and frying on content of essential polyunsaturated fatty acids in muscle tissue of four fish species, *Food Chem.*, 101:1694–1700.

Gokce, M.A., Tasbozan, O., Celik, M., and Tabakoglu, S.S. 2004. Seasonal variations in proximate and fatty acid compositions of female common sole (*Solea solea*). Food Chem., 88:419–423.

Gómez-Guillén, M.C. and Montero, M.P. 2007. Polyphenol uses in seafood conservation. *Am. J. Food Technol.*, 2, 7:593–601.

Goulas, A.E. and Kontaminas, M.G. 2005. Effect of salting and smoking-method on the keeping quality of chub mackerel (*Scomber japonicus*): Biochemical and sensory attributes, Food Chem., 93:511–520.

Guillen, M.D. and Ruiz, A. 2004. Study of the oxidative stability of salted and unsalted salmon fillets by ^{1}H nuclear magnetic resonance, *Food Chem.*, 86:297–304.

Guler, G.O., Aktumsek, A., Citil, O.B., Arslan, A., and Torlak, E. 2007. Seasonal variations on total fatty acid composition of fillets of zander (*Sander lucioperca*) in Beysehir Lake (Turkey), *Food Chem.*, 103:1241–1246.

Guler, G.O., Kitzanir, B., Aktumsek, A., Citill, O.B., and Ozparlak, H. 2008. Determination of the seasonal changes on total fatty acid composition and ω3/ω6 ratios of carp (*Cyprinus carpio* L.) muscle lipids in Beysehir Lake (Turkey), *Food Chem.*, 108, 2:689–694.

Haliloglu, H.I., Bayir, A., Sirkecioglu, A.N., Aras, N.M., and Atamanalp, M. 2004. Comparison of fatty acid composition in some tissues of rainbow trout (*Oncorhynchus mykiss*) living in seawater and freshwater, *Food Chem.*, 86, 1:55–59.

Halldorsson, A., Kristinsson, B., and Haraldsson, G.G., 2001. Lipase selectivity toward fish oil fatty acids. Lipidforum. *Proceedings of the 21st Nordic Lipid Symposium*, Bergen, Norway, June 5th–8th, 2001.

Hanuš, L.O., Levitsky, D.O., Shkrob, I., and Dembitsky, V.M. 2009. Plasmalogens, fatty acids and alkyl glyceryl ethers of marine and freshwater clams and mussels, *Food Chem.*, 116, 2:491–498.

Hardy, R., McGill, A.S., and Gunstone, F.D. 1979. Lipid and autoxidative changes in cold stored cod (*Gadus morhua*). *J. Sci. Food Agric.*, 30:999–1006.

Hirasuka, S., Kitagawa, T., Matsue, Y., Hashidume, M., and Wada, S. 2004. Lipid class and fatty acid composition of phospholipids from the gonads of skipjack tuna, *Fisheries Sci.*, 0, 5:903–909.

Hultin, H.O. 1994. Oxidation of lipids in seafoods, in: *Seafoods: Chemistry, Processing Technology and Quality*, F. Shahidi and J.R. Botta (Eds.), pp. 49–74, Blackie Academic & Professional, London, U.K.

Huynh, M.D. and Kitts, D. 2009. Evaluating nutritional quality of pacific fish species from fatty acid signatures, *Food Chem.*, 114, 3:912–918.

Huynh, M.D., Kitts, D.D., Hu, Ch., and Trites, A.W. 2007. Comparison of fatty acid profiles of spawning and non-spawning Pacific herring, *Clupea harengus pallasi*, *Comp. Biochem. Physiol.*, Part B, 14:504–511.

Ingemansson, T., Kaufmann P., and Pettersson, A. 1992. Lipid deterioration in frozen storage in light and dark muscle of frozen rainbow trout (*Oncorhynchus mykiss*), fillets in relation to water temperature and astaxanthin content, in: *Quality Assurance in the Fish Industry*. H.H. Huss (Ed.), pp. 29–38, Elsevier Science Publishers, Amsterdam, the Netherlands.

Jittrepotch, N., Ushio, H., and Ohshima, T. 2006. Effects of EDTA and a combined use of nitrite and ascorbate on lipid oxidation in cooked Japanese sardine (*Sardinops melanostictus*) during refrigerated storage, *Food Chem.*, 99:70–82.

Jobling, M., Leknes, O., Saether, B.-S., and Bendiksen, E.A. 2008. Lipid and fatty acid dynamics in Atlantic cod, *Gadus morhua*, tissues: Influence of dietary lipid concentrations and feed oil sources. *Aquaculture*, 28, 1–4:87–94.

Kennish, J.M., Chambers, K.A., Whipple, W.J., Sharp-Dahl, J.I., and Rice, S.D. 1992. Differences in lipid, fatty acid composition and cholesterol levels among tissues and among stocks for pen-reared Chinook salmon (*Oncorchynchus tshawytsccha*) fed on a commercial diet, in: *Seafood Science and Technology*, E.G. Bligh (Ed.), pp. 46–57, Fishing News Books, Oxford, U.K.

Kinsella, J.E. 1987. Potential sources of fish oil: Fatty fish in U.S. waters, in: *Seafood and Fish Oils in Human Health and Diseases*, Marcel Dekker, Inc., New York, pp. 239–300.

Kitson, A.P., Patterson, A.C., Izadi, H., and Stark, K.D. 2009. Pan—Frying salmon in an eicosapentaenoic acid (EPA) and docosahexaenoic acid (DHA) enriched margarine prevents EPA and DHA loss, *Food Chem.*, 114, 3:927–932.

Koizumi, C., Takada M., Ohshima, T., and Wada, S. 1986. Changes in the composition of lipids in fish meats on thermal processing at high temperature, *Bull. Jpn. Soc. Sci. Fisheries*, 52, 6:1095–1102.

Kołakowska, A. 2003. Lipid oxidation, in food systems, in: *Chemical and Functional Properties of Food Lipids*. Z.E. Sikorski and A. Kołakowska (Eds.), Chap. 8, pp. 133–166, CRC Press, Boca Raton, FL.

Kołakowska, A. 1985. Changes of krill lipid content and composition, in annual cycle, in *Proceedings of the XII Polar Symposium*, Szczecin, Poland, pp. 167–173 (in Polish).

Kołakowska, A. and Szczygielski, M. 1994. Stabilization of lipids in minced fish by freeze texturization, *J. Food Sci.*, 59, 1:88–90.

Kołakowska, A., Macur, I., Pankiewicz, A., and Szczygielski, M. 1998. Smoked mackerel as a source of n-3 polyunsaturated fatty acids, in *Proceedings of the Evening Primose and Other Oils Containing n-3 or n-6 Fatty Acids in Prevention and Treatment*, A. Stołyhwo (Ed.), pp. 71–78, Sulejow, Poland.

Kołakowska, A., Szczygielski, M., Bienkiewicz, G., and Zienkowicz, L. 2000. Some of fish species as a source of n-3 polyunsaturated fatty acids, *Acta Ichth. Piscat.*, 30, 2:59–70.

Kołakowska, A., Stypko, K., Domiszewski, Z., Bienkiewicz, G., Perkowska, A., and Witczak, A. 2002. Canned cod liver as a source of n-3 polyunsaturated fatty acids, with a reference to contamination, *Nahrung/Food*, 46, 1:40–45.

Kołakowska, A., Olley, J., and Dustan, G.A. 2003. Fish lipids, in: *Chemical and Functional Properties of Food Lipids*, Z.E. Sikorski and A. Kołakowska (Eds.), Chapter 12, pp. 221–264, CRC Press, Boca Raton, FL.

Kołakowska, A., Domiszewski, Z., Kozłowski, D., and Gajowniczek, M. 2006a. Effects of rainbow trout freshness on n-3 polyunsaturated fatty acids in fish offal, *Eur. J. Lipid Sci. Technol.*, 108:723–729.

Kołakowska, A., Zienkowicz, L., Domiszewski, Z., and Bienkiewicz, G. 2006b. Lipid changes and sensory quality of whole-and gutted rainbow trout during storage in ice, *Acta Ichth. Piscat*, 36, 1:39–47.

Kołakowska, A., Domiszewski, Z., and Bienkiewicz, G. 2006c. Effect of biological and technological factors on the utility of fish as a source of n-3 PUFA, in: *Omega 3 Fatty Acid Research*, M.C. Teale (Ed.), pp. 83–107, Nova Science Publishers, Inc., New York.

Kołakowska, A., Kołakowski, E., Domiszewski, Z., Zienkowicz, L., and Bednarczyk, B. 2006d. The effects of freshness on changes of nutritional value of protein and lipids of rainbow trout during cooking, in: Report Project No 3 PO6T 060 25 *State Committee for Scientific Research* (KBN Poland), pp. 1–146.

Kołakowski, E. 2005. Enzymatic modification of proteins, in: *Enzymatic Modification of Food Components*, E. Kołakowski, W. Bednarski, and S. Bielecki (Eds.), pp. 30–99, Wyd. AR w Szczecin, Poland (in Polish).

Kołakowski, E. and Kołakowska, A. 1993. Krill composition and autolytic activity, in: *The Maritime Antarctic Coastal Ecosystem of Admirality Bay*, S. Rakusa-Suszczewski (Ed.), pp. 72–79, Department of Antarctic Biology Polish Academy of Sciences, Warsaw, Poland.

Kolanowski, W., Ziołkowski, M., Weizbrodt, J., kunz, B., and Laufenberg, G. 2006. Microencapsulation of Fish oil by spray drying-impact on oxidative stability. Part 1, *Eur. Food Res. Technol.*, 222:336–342.

Kong, F., Oliveira, A., Tang, J., Rasco, B., and Crapo, C. 2008. Salt effect on heat-induced physical and chemical changes of salmon fillet (*O. gorbuscha*) during thermal processing, *Food Chem.*, 106, 3:957–966.

Krzynowek, J., Murphy, J., Pariser, E.R., and Clifton, A.B. 1990. Six Northwest Atlantic finfish species as a potential fish oil source, *J. Food Sci.*, 55:1743–1744.

Labuza, T.P., McNally, L., Gallagher, D., Hawkes, J., and Hurado, F. 1972. Stability of intermediate moisture foods. 1. Lipid oxidation, *J. Food Sci.*, 37:154–159.

Lie, O. and Lambertsen, G. 1991. Fatty acid composition of glycerophospholipids in seven tissues of cod (*Gadus morhua*), determined by combined high-performance liquid chromatography and gas chromatography, *J. Chromatogr.*, 565:119–129.

Ling, K.H., Cheung, C.W., Cheng, S.W., Cheng, L., Li, S.-L., Nichols, P.D., Ward, R.D., Graham, A., and But, P.P.-H. 2008. Rapid detection of oilfish and escolar in fish steaks: A tool to prevent keriorrhea episodes, *Food Chem.*, 110, 2:538–546.

Ma, J., Folsom, A.R., Lewis, L., and Eckfeldt, J.H. 1997. Relation of plasma phospholipid and cholesterol ester fatty acid composition to carotid artery intima-media thickness: The Atherosclerosis Risk in Communities (ARIC) Study. *Am. J. Clin. Nutr.*, 65:551–559.

Matheson, E.M., Mainous III, A.G., Hill, E.G., and Carnemolla, M.A. 2009. Shellfish consumption and risk of coronary heart disease, *J. Am. Diet. Assoc.*, 1098:1422–1426.

Mendez, E. and Gonzalez, R.M. 1997. Seasonal changes in the chemical composition of fillets of the Southwest Atlantic hake (*Merlucius hubbsi*), *Food Chem.*, 59, 2:213–217.

Mohri, S., Cho, S.Y., Endo, Y., and Fujimoto, K. 1990. Lipoxygenase activity in sardine skin. *Agric. Biol. Chem.*, 54:1889–1991.

Naczk, M., Williams, J., Brennan, K., Liyanapathirana, Ch., and Shahidi, F. 2004. Compositional characteristics of green crab (*Carcinus maenas*), *Food Chem.*, 88, 3:429–434.

Nichols, P., Mooney, B., Virtue, P., and Elliott, N. 2002. Nutritional value of Australian Fish: Oil, fatty acid and cholesterol composition of edible species. Final Report, CSIRO Marine Research, Hobart, Australia, Fisheries Research and Development Corporation Project 1995/122. ISBN 0643061789.

Ohshima, T., Wada, S., and Koizumi, C. 1984. Enzymatic hydrolysis of phospholipids in cod flesh during storage in ice. *Bull. Jpn. Soc. Sci. Fish.*, 50, 1:107–114.

Ohshima, T., Nan, L., and Koizumi, C. 1993. Oxidative decomposition of cholesterol in Fish products, *J. Am. Oil Chem. Soc.*, 70, 6:595–600.

Ota, T., Sasaki, S., and Abe, T. 1990. Fatty acid compositions of the lipids obtained from commercial salmon products, *Nippon Suisan Gakk.*, 56, 2:323–327.

Ota, T., Chihara, Y., Itabashi, Y., and Takagio, T. 1994. Occurrence of all-*cis* −6,9,12,15,18,21-tetracosahexaenoic acid in flatfish lipids, *Fisheries Sci.*, 60, 2:171–175.

Palmeri, G., Turchini, G.M., and De Silva, S.S. 2007. Lipid characterization and distribution in the fillet of the farmed Australian native fish, Murray cod (*Maccullochella peelii peelii*), *Food Chem.*, 102:796–807.

Pickova, J. and Dutta, P.C. 2003. Cholesterol oxidation in some processed fish products, *JAOCS*, 80, 10:993–996.

Pigott, G.M. and Tucker, B.W. 1990. *Seafood Effects of Technology on Nutrition*, Marcel Dekker, Inc., New York/Basel, Switzerland.

Plust, D. 2008. Antioxidant capacity of some kind of meat, PhD thesis, Westpomeranian Technological University, Szczecin, Poland (in Polish).

Řezanka, T. and Sigler, K. 2009. Odd-numbered very-long-chain fatty acids from the microbial, animal and plant kingdoms, *Prog. Lipid Res.*, 48, 3–4:206–238.

Rodríguez, A., Carriles, N., Cruz, J.M., and Aubourg, S.P. 2008. Changes in the flesh of cooked farmed salmon (*Oncorhynchus kisutch*) with previous storage in slurry ice (−1.5°C), *LWT: Food Sci. Technol.*, 41, 9:1726–1732.

Rodriguez, A., Carriles, N., Gallardo, J.M., and Aubourg, S.P. 2009. Chemical changes during farmed coho salmon (*Oncorhynchus kisutch*) canning: Effect of preliminary chilled storage, *Food Chem.*, 112:362–368.

Ruiz-Roso, B., Cuesta, I., Perez, M., Borrego, E., Perez-Olleroz, L., and Varela, G. 1998. Lipid composition and palatability of canned sardines, influence of the canning process and storage in olive oil for five years, *J. Sci. Food Agric.*, 77:244–250.

Sahin, S. and Sumnu, G. 2001. Effects of microwave cooking on fish quality, *Int. J. Food Prop.*, 4, 3:501–512.

Saito, H. 2008. Unusual novel n-4 polyunsaturated fatty acids in cold-seep mussels (*Bathymodiolus japonicus* and *Bathymodiolus platifrons*), originating from symbiotic methanotrophic bacteria. *J. Chromatogr. A.*, 1200, 2:242–254.

Sampels, S., Pickova, J., and Wiklund, E. 2004. Fatty acids, antioxidants and oxidation stability of processed reindeer meat. *Meat Sci.*, 67, 3:523–532.

Sargent, J.R. 1995. (*n-3*) Polyunsaturated fatty acids and farmed fish, in: *Fish Oil Technology, Nutrition and Marketing*, R.J. Hamilton and R.D. Rice (Eds.), pp. 67–94, PJ Barnes & Associates, High Wycombe, U.K.

Sargent, J., McEvoy, L., Estevez, A., Bell, G., Bell, M., Henderson, J., and Tocher, D. 1999. Lipid nutrition of marine fish during early development: Current status and future directions, *Aquaculture*, 179:217–229.

Seidler, T. 1997. α-Tocopherol content in muscle tissue of Baltic herring, caught at different season, in *Proceedings of the 28th Symposium*, vol. 303, Gdańsk, Poland (in Polish).

Shulman, G.E. and Love, R.M. 1999. *The Biochemical Ecology of Marine fishes. Advances in Marine Biology*, Vol. 36, A.J. Southward, P.A. Tyler, and C.M. Young (Eds.), pp. 1–351, Series, Academic Press, San Diego, CA.

Sikorski, Z.E. and Kołakowska, A. 1990. Freezing of marine foods. In: *Seafood: Resources, Nutritional Composition, and Preservation*, Z.E. Sikorski (Ed.), Chap. 7., pp. 111–124, CRC Press, Boca Raton, FL.

Sikorski, Z.E. and Kołakowska, A. 1994. Changes in proteins in frozen stored fish, in: *Seafood Proteins*, Z.E. Sikorski, B.S. Pan, and F. Shahidi, pp. 99–112, Chapter 8, Chapman & Hall, New York/London, U.K.

Sikorski, Z.E. and Kołakowska, A. 2001. Lipids in fish raw material and smoked products: A review, *Pol. J. Food Nutr. Sci.*, 10/51, 2:3–10.

Sikorski, Z.E., Kołakowska, A., and Sun Pan, B. 1990. The nutritive composition of the major groups of marine food organisms, in: *Seafood: Resources, Nutritional Composition and Preservation*, Z.E. Sikorski (Ed.), pp. 29–54, Chapter 3, CRC Press, Inc., Boca Raton, FL.

Simeonidou, S.,Govaris, A., and Vareltzis, K. 1997. Effect of frozen storage on the quality of whole fish and fillets of horse mackerel (*Trachurus trachurus*) and mediterranean hake (*Merluccius mediterraneus*), *Z. Lebensm Unters Forsch A*, 204:405–410.

Simonetti, M.S., Blasi, F., Bosi, A., Maurizi, A., Cossignani, L., and Damiani, P. 2008. Stereospecific analysis of triacylglycerol and phospholipid fractions of four freshwater species: *Salmo trutta, Ictalurus punctatus, Ictalurus melas and Micropterus salmoides*, *Food Chem.*, 110, 1:199–206.

Sinclair, A.J., Dunstan, G.A., Naughton, J.M., Sanigorski, A.J., and O'Dea, K. 1992. The lipid content and fatty acid composition of commercial marine and freshwater fish and molluscs from temperate Australian waters, *Aust. J. Nutr. Diet.*, 49:77–83.

Siriamornpun, S., Yang, L., Kubola, J., and Li, D. 2008. Changes of omega-3 fatty acid content and lipid composition in canned tuna during 12-month storage. *J. Food Lipids*, 15, 2:164–175.

Sirot, V., Oseredczuk, M., Bemrah-Aouachria, N., Volatier, J.-L., and Leblanc, J.Ch. 2008. Lipid and fatty acid composition of fish and seafood consumed in France: CALIPSO study, *J. Food Compos. Anal.*, 21, 1:8–16.

Sohn, J.-H., Taki, Y., Ushio, H., Kohata, T., Shioya, I., and Ohshima, T. 2005. Lipid oxidation in ordinary and dark muscles of Fish: Influences on rancid off-odor development and color darkening of yellowtail flesh during ice storage, *JFS*, 70, 7:S490–S496.

Stołyhwo, A., Kołodziejska, I., and Sikorski, Z.E. 2006. Long chain polyunsaturated fatty acids in smoked Atlantic mackerel and Balic sprats, *Food Chem.*, 94:589–595.

Tarley, C.R.T., Visentainer, J.V., Matsushita, M., and De Souza, N.E. 2004. Proximate composition, cholesterol and fatty acids profile of canned sardines (*Sardinella brasiliensis*) in soybean oil and tomato sauce, *Food Chem.*, 88, 1:1–6.

Testi, S., Bonaldo, A., Gatta, P.P., and Badiani, A. 2006. Nutritional traits of dorsal and ventral fillets from three farmed fish species, *Food Chem.*, 98:104–111.

Tobin, D., Kause, A., Mantysaari, A., Martin, E.A., Houlihan, S.A.M, Dobly, D.F., Kiessling, A., Rungruangsak-Torrissen, K., Ritola, O., and Ruohonen, K. 2006. Fat or lean? The quantitative genetic basis for selection strategies of muscle and body composition trains in breeding schemes of rainbow trout (*Oncorhynchus mykiss*), *Aquaculture*, 261:510–521.

Tomczyk, K. 2008. The effect of marinading process on fatty acid profile in herring (*Clupea harengus*). PhD thesis, Agricultural University, Kraków, Poland (in Polish).

Tomita, Y. and Ando, Y. 2009. Reinvestigation of positional distribution of tetracosahexaenoic acid in triacyl-*sn*-glycerols of flathead flounder flesh, *Fisheries Sci.*, 75, 2:445–451.

Undeland, I., Ekstrand, B., and Lingnert, H. 1998. Lipid oxidation in herring (*Clupea harengus*) light muscle, and skin, stored separately or as intact fillets, *JAOCS*, 75:581–590.

Undeland, I., Hall, G., and Lingnert, H. 1999. Lipid oxidation in fillets of herring (*Clupea harengus*) during ice storage, *J. Agric. Food Chem.*, 47:524–532.

Usydus, Z., Szlinder-Richert, J., Polak-Juszczak, L., Kanderska, J., Adamczyk, M., Malesa-Ciecwierz, M., and Ruczynska, W. 2008. Food of marine origin: Between benefits and potential risks. Part I. Canned fish on the Polish market, *Food Chem.*, 111, 3:556–563.

Velasco, J., Marmesat, S., Dobarganes, C., and Marguez-Ruiz, G. 2006. Heterogenous aspects of lipid oxidation in dried microencapsulated oils. *Agric Food Chem.*, 54, 5:1722–1729.

Wang, Y.J., Miller, L.A., Perren, M., and Addis, P.B. 1990. Omega-3 fatty acids in lake superior fish, *J. Food Sci.*, 55, 1:71–76.

Weaver, K.L., Ivester, P., Chilton, J.A., Wilson, M.D., Pandey, P., and Chilton, F.H. 2008. The content of favorable and unfavorable polyunsaturated fatty acids found in commonly eaten fish. *J. Am. Diet. Assoc.*, 108:1178–1185.

Weber, J., Bochi, V.C., Ribeiro, C.P., Victório, A.D.M., and Emanuelli, T. 2008. Effect of different cooking methods on the oxidation, proximate and fatty acid composition of silver catfish (*Rhamdia quelen*) fillets, *Food Chem.*, 106, 1:140–146.

Yazdan, M., Jamilah, B., Yaakob, C.M., and Sharifah, K. 2009. Moisture, fat content and fatty acid composition in breaded and non-breaded deep-fried black pomfret (*Parastromateus niger*) fillets, *ASEAN Food J.*, 16, 2:225–231.

Zienkowicz, L. 2009. Effects of fish freshness on lipids changes during frying and storage of fried fish, PhD thesis, West Pomeranian University of Technological, Szczecin, Poland (in Polish).

Zotos, A., Petridis, D., Siskos, I., and Gougoulias, C. 2001. Production and quality assessment of smoked tuna (*Euthynnus affinis*) product, *J. Food Sci.*, 66, 8:1184–1190.

Zunin, P., Boggia, R., and Evangelisti, F. 2001. Identification and quantification of cholesterol oxidation products in canned tuna, *JAOCS*, 78, 10:1037–1041.

16 Milk Lipids

Michael H. Tunick

CONTENTS

16.1 INTRODUCTION

Milk fat has the most complex structure and chemical composition of all food lipids. This chapter deals with lipids found in bovine milk, although the milk of other mammals follows the same general principles. Lipids in milk are typically found at a concentration of 3.5–5.0 g/100 g milk, in the form of emulsified globules dispersed in the aqueous phase. The globules range in diameter from 0.1 to 10 μm and are basically a core of triacylglycerols (TAG) surrounded by a thin (8–10 nm) bilayer milk fat globule membrane (MFGM) that maintains the integrity of the globule. The TAGs are nonpolar molecules consisting of three fatty acid (FA) molecules esterified onto a backbone of glycerol. The TAG core of the globule also contains small amounts of cholesteryl and retinol esters. The MFGM consists of proteins and lipids, with the latter mainly being phospholipids, cholesterol, and glycolipids. There are approximately 1.5×10^{10} fat globules in 1 mL of milk, with a surface area of about $2 \, \text{m}^2$.

16.2 COMPOSITION

16.2.1 LIPID CLASSES

The lipids in milk consist of TAG (96–98 g/100 g milk fat), diacylglycerols (DAG) (0.3–0.6 g/100 g), monoacylglycerols (MAG) (0.2–0.4 g/100 g), and phospholipids (0.2–1.0 g/100 g), along with sterols (0.25–0.45 g/100 g), unesterified FA (0.1–0.4 g/100 g), and minor amounts of other compounds.

16.2.2 FATTY ACIDS

Well over 400 FA residues have been found in milk (Collins et al., 2003), but many are present only in trace amounts. Milk fat contains significant amounts of C_4-C_{18} FA, and is one of the few natural sources of C_4-C_{12} FA (Table 16.1). FA in milk arise from the lipids in the feed and from synthesis in the mammary gland, in roughly equal proportions. All of the C_{18} and some of the C_{16} FA originate in the feed, where the primary lipids are TAG, phospholipids, and glycolipids (MacGibbon and Taylor, 2006). Most of the FA in diet of the cow are linoleic and linolenic acid, and are hydrogenated in the rumen by microorganisms to stearic acid. Shorter-chain FAs are synthesized in the mammary gland, mostly from acetate (CH_3COO^-) and partly from β-hydroxybutyrate, both of which originate in the rumen from the fermentation of carbohydrates such as cellulose. The step wise addition of acetate causes the FA chains to be extended by two carbons at a time, producing an unbranched chain with an even number of carbon atoms. Propionate ($CH_3CH_2COO^-$) or isobutyrate [$(CH_3)_2COO^-$] can also be precursors, resulting in odd-numbered carbon chains or branched chains. The FAs are esterified into TAG, circulate in the bloodstream, and are absorbed by the mammary gland. A desaturase system in that gland converts much of the stearic acid to oleic acid (MacGibbon and Taylor, 2006).

TABLE 16.1

Fatty Acid Composition of Bovine Milk Fat

Abbreviation	Fatty Acid	Range (g/100 g Fat)
4:0	Butyric	2–5
6:0	Caproic	1–5
8:0	Caprylic	1–3
10:0	Capric	2–4
12:0	Lauric	2–5
14:0	Myristic	8–14
15:0	Pentadecanoic	1–2
16:0	Palmitic	22–35
16:1	Palmitoleic	1–4
17:0	Margaric	0.5–1.5
18:0	Stearic	9–14
cis-9 18:1	Oleic	20–30
trans-11 18:1	Vaccenic	3.3–3.8
cis-9,12 18:2	Linoleic	1–3
cis-9,12,15 18:3	Linolenic	0.5–2

Sources: Data from German, J.B. and Dillard, C.J., *Crit. Rev. Food Sci.*, 46, 57, 2006; Molkentin, J., *Br. J. Nutr.*, 84(Suppl. 1), S47, 2000; Jensen, R.G., *J. Dairy Sci.*, 85, 295, 2002.

Note: FAs at levels <0.5 g/100 g fat are not included.

16.2.3 TRIACYLGLYCEROLS

There are 24–54 carbon atoms in milk fat TAG, resulting in molecular weights of 470–890 Da (Collins et al., 2003). The positioning of FA on the glycerol backbone is not random, owing to the specificities of the enzymes involved in biosynthesis. Milk fat is unique in that the 4:0, 6:0, 8:0, and 10:0 FA are preferentially esterified at the *sn*-3 position; two-thirds of the FAs at the *sn*-2 position are 14:0 and 16:0 (Table 16.2). Nearly all of the FAs at *sn*-1 have at least 14 carbon atoms.

With several hundred FAs known to be in milk fat, millions of TAG species are theoretically possible but several thousand combinations of FAs are probably present, mostly in trace amounts (Jensen, 2002). Gresti et al. (1993) separated 223 different TAG that contained FA with even carbon numbers, representing 83.3% of the total TAG. Table 16.3 shows the distribution of the most common of these TAG. The 4:0 and 6:0 FAs are typically found with two FAs in the 14–18 carbon number range. The monounsaturated TAG are much more prevalent than diunsaturated and polyunsaturated TAG.

When consumed, TAG, DAG, and MAG are packaged into vesicles known as chylomicrons, which are synthesized by the intestinal lining and transported to the liver and other parts of the body. The concentration of TAG in the blood has been

TABLE 16.2

Positional Distribution of Fatty Acids in Bovine Milk Triacylglycerols

Abbreviation	Approximate Composition (mol/100 mol)		
	sn-1	*sn*-2	*sn*-3
4:0	1	<0.5	18
6:0	<0.5	1	8
8:0	<0.5	<0.5	5
10:0	1	1	9
12:0	3	5	5
14:0	12	23	8
16:0	47	44	12
16:1	2	2	2
18:0	11	5	5
18:1	22	16	25
18:2	2	2	3
18:3	<0.5	<0.5	<0.5

Source: Data adapted from Blasia, F. et al., *J. Food Compos. Anal.*, 21, 1, 2008.

identified as a risk factor in coronary heart disease, although individual genetic susceptibility may be important (Austin, 1991).

16.2.4 DIACYLGLYCEROLS, MONOACYLGLYCEROLS, AND FATTY ACIDS

The DAG, MAG, and FA result from lipolysis of TAG, but DAG and MAG are also found in freshly secreted milk, suggesting they are intermediates in the formation of TAG (Fagan et al., 2004). DAGs are slightly polar. MAG and FA are more polar, with longer chains being surface-active compounds and shorter chains being partly soluble in water.

16.2.5 STEROLS

Cholesterol is a four-ring, 27-carbon alcohol comprising about 95% of the sterols in milk fat. Cholesterol controls lipid and protein transitions and conformation in membranes, and is a precursor for some hormones (MacGibbon and Taylor, 2006). Milk fat has long been thought to increase the risk of coronary heart disease because of the cholesterol and saturated FA it contains, although some studies indicate that milk may not affect adversely blood lipids as would be predicted from its fat composition (Pfeuffer and Schrezenmeir, 2000).

TABLE 16.3
TAG Present in Bovine Milk Fat

Carbon Number	Fatty Acids	Composition (mol/100 mol)
	Trisaturated TAG	
34	4:0, 14:0, 16:0	3.1
36	6:0, 14:0, 16:0	1.4
36	4:0, 14:0, 18:0	1.3
36	4:0, 16:0, 18:0	3.2
38	6:0, 16:0, 16:0	1.5
38	4:0, 16:0, 18:0	2.5
40	6:0, 16:0, 18:0	1.1
Total[a]		32.4
	Monounsaturated TAG	
36	4:0, 14:0, 18:1	1.8
38	4:0, 16:0, 18:1	4.2
40	6:0, 16:0, 18:1	2.0
40	4:0, 18:0, 18:1	1.6
44	10:0, 16:0, 18:1	1.6
46	12:0, 16:0, 18:1	1.2
48	14:0, 16:0, 18:1	2.8
50	14:0, 18:0, 18:1	1.4
50	16:0, 16:0, 18:1	2.3
52	16:0, 18:0, 18:1	2.2
Total[a]		32.6
	Diunsaturated TAG	
40	4:0, 18:1, 18:1	1.5
50	14:0, 18:1, 18:1	1.3
52	16:0, 18:1, 18:1	2.5
54	18:0, 18:1, 18:1	1.2
Total[a]		13.1[b]
	Polyunsaturated TAG	
54	18:1, 18:1, 18:1	1.0
Total[a]		5.2

Source: Data adapted from Gresti, J. et al., *J. Dairy Sci.*, 76, 1850, 1993.

Note: TAG at levels <1.0 mol/100 mol are not included. Stereospecific numbering positions were not determined.

[a] Includes others at levels <1.0 mol/100 mol not listed.

[b] Includes 2.5 mol/100 mol of TAG containing 18:2.

16.2.6 Compound Lipids

16.2.6.1 Phospholipids

Up to 1% of the total lipid in milk consists of phospholipids. Sphingomyelin, phosphatidyl choline, and phosphatidyl ethanolamine comprise 90% of the phospholipids (Deeth, 1997), with about two-thirds incorporated into the MFGM and one-third present in the aqueous phase (MacGibbon and Taylor, 2006). Phosphatidyl choline and phosphatidyl ethanolamine are composed of two FAs and a phosphate group joined to glycerol, with the phosphate being bound to choline or ethanolamine. Sphingomyelin is similar to phosphatidyl choline except the glycerol is replaced by serine, and the carbonyl group on one of the FAs is replaced by a double-bonded carbon. The MFGM is disrupted during butter churning and migrates into the buttermilk fraction, which thus contains a large proportion of the phospholipids. Phospholipids exhibit hydrophilic and lipophilic properties, which allow them to stabilize oil/water emulsions and make them a desirable additive when vegetable oil emulsions are required for food applications. Phospholipids appear to have a role in liver protection and memory improvement (Thompson and Singh, 2006).

16.2.6.2 Glycolipids

The cerebrosides and gangliosides are sphingolipids, which are bioactive compounds that affect cell regulation. Sphingolipids consist of a long hydrocarbon chain attached to a polar domain containing an amino group. Cerebrosides have monosaccharides attached to the polar area and gangliosides have complex oligosaccharides attached. Glycolipids do not appear to be required in the diet but might confer protection against colon cancer, atherosclerosis, and infections in newborns (Vesper et al., 1999). Some 70% of glycolipids in bovine milk are associated with the MFGM, and their concentrations in milk are small: 26 mmol cerebrosides/kg and 14 mmol gangliosides/kg (Newburg and Chaturvedi, 1992).

16.2.7 Ether Lipids

Ether lipids contain an ether linkage at the *sn*-1 position instead of an ester linkage. These are present in small amounts in milk: phospholipids have 0.16 g/100 g milk of 1-*O*-alkylacylphospholipids and neutral lipids have 0.01 g/100 g milk of 1-*O*-alkyldiacylglycerols. Ether lipids are believed to accumulate in cell membranes and have anticancer properties (Molkentin, 2000).

16.2.8 Flavor Compounds

Over 120 flavor compounds are found in milk fat. These are usually present in small amounts, but improper storage of milk leads to higher concentrations and off-flavors. Many flavor compounds are aldehydes, which are oxidation products of FA and generate pungent flavors. Aldehydes are present in low concentrations, but the threshold of detection is also low (1–10 μg/kg). Spontaneous rearrangement of hydroxyacids leads to formation of four-carbon γ-lactones and five-carbon δ-lactones, at concentrations of 10–30 μg/g. Lactones are responsible for sweet and fruity flavors in milk. Methyl ketones and diacetyl (2,3-butanedione) are also present in milk fat, and together with FA and lactones generate the characteristic flavor of butter, with diacetyl being predominant (Tunick, 2007).

16.2.9   OTHER CONSTITUENTS

Milk fat also contains components such as carotenoids and fat-soluble vitamins. Carotenoids are derived from 40-carbon chains with conjugated double bonds and cyclic end groups. About 95% of the carotenoids are in the form of β-carotene, which ranges from 2.5 to 8.5 μg/g fat, depending on diet and breed of cow (MacGibbon and Taylor, 2006). The fat-soluble vitamins are A (retinol), D, E (tocopherols), and K, with A and D being found only in foods of animal origin. One liter of milk supplies around four-tenths of the recommended daily allowance of vitamins A and K, and about one-tenth the allowance of vitamins D and E, though commercially available milk is fortified with vitamin D. Fat-soluble vitamins are not lost when processing milk into cheese and yogurt, and are not affected by heat treatment of milk or exposure to light (Schaafsma, 2002).

16.2.10   FACTORS AFFECTING COMPOSITION

The most important aspect influencing milk fat composition is the feed, with genetics and stage of lactation also being important. The age and health of the cow and seasonal differences are contributing factors.

Pasteurization apparently has little effect on the lipids (Jensen, 2002). Homogenization of milk strips off the MFGM and reduces the size of the globules by a factor of five. The globules are then partially covered by micelles containing α-, β-, and κ-casein. This covering may be mechanical or charge related and imparts stability to the droplets in milk.

16.3   PHYSICOCHEMICAL PROPERTIES

16.3.1   MILK FAT GLOBULE MEMBRANE

The TAGs are synthesized on the smooth endoplasmic reticulum within the mammary secretory cells. As they migrate to the apical membrane, they form droplets that fuse together and force membrane outward. The membrane surrounds the droplet and finally pinches off to form a spherical globule coated by MFGM. The MFGM contains about two-thirds lipid, primarily TAG and phospholipids (Table 16.4), with the remainder being protein (Singh, 2006). Glycolipids and choline-containing phospholipids (phosphatidyl choline and sphingomyelin) are mostly on the outer surface of the MFGM, and phosphatidyl ethanolamine, which is neutral, is on the inner surface. This asymmetric arrangement is found in red blood cells and other membranes (Deeth, 1997). Aside from emulsification, the MFGM protects globular fat against coalescence, lipolysis, and oxidation.

TABLE 16.4
Composition of Lipids in the Milk Fat
Globule Membrane

Component	g/100 g Membrane Lipids
Triacylglycerols	62
Diacylglycerols	9
Free fatty acids	1–6
Sterols	<2
Phospholipids	26–31
	g/100 g Phospholipids
Phosphatidyl choline	36
Phosphatidyl ethanolamine	27
Sphingomyelin	22

Source: Singh, H., *Curr. Opin. Colloid Interface Sci.*, 11, 154, 2006. With permission.

16.3.2 PHYSICAL CHARACTERISTICS OF MILK FAT

At room temperature, the density of milk fat is around 0.92 g/mL and the solubility is 0.14 g/100 g water. Milk fat does not conduct heat well, as evidenced by its low thermal conductivity of 0.17 W/m·K (Taylor and MacGibbon, 2002). Milk fat has a characteristic melting profile from about −35°C to 38°C; the broad range is due to the diverse TAG it contains. The melting points of individual TAG ranges up to 72°C (for tristearin), but the higher-melting TAG dissolve in the liquid fat. Cooling causes the formation of three polymorphic crystal forms, designated β′ (the most common), β, and α, which are different from each other because of the packing of the various hydrocarbon chains of the TAG. Milk fat has a latent heat of fusion (conversion of solid to liquid) of 70–80 J/g. Homogenization of milk alters melting by causing the globules to be associated with casein (Tunick, 1994).

16.3.3 LIPOLYSIS

Lipoprotein lipase enzymes cause lipids in milk to hydrolyze into DAG, MAG, and FA. The rancid off-flavors arising from butyric, caproic, and capric acids are sometimes described as hydrolytic rancidity. Milk lipase is a component of raw milk found in the aqueous phase, and the MFGM prevents it from reacting with the fat globule. Lipolysis will result if the MFGM becomes permeable, which may take place in late-lactation cows, or is mechanically damaged. The latter occurs during homogenization, so processors pasteurize milk (inactivating the lipase) immediately before or after it is homogenized (Deeth, 2006).

Pseudomonas fluorescens and other bacteria have proteinases that can break down the MFGM, thus allowing lipases to hydrolyze the lipids. Pseudomonads are found in water and soil and may be present in stored milk even after pasteurization.

Bacterial lipolysis is controlled by limiting storage time to three days while maintaining the temperature under 4°C (Sørhaug and Stepaniak, 1997).

16.3.4 OXIDATION

Exposure to light and elevated temperature may cause double bonds in unsaturated FA to be oxidized to peroxides, which leads to formation of undesirable flavors. Phospholipids act synergistically with tocopherols and β-carotene to prevent oxidation. Oxidation may also be caused by contact with copper, iron, and other metals during processing (O'Brien and O'Connor, 2002).

16.4 PROPERTIES OF MILK FAT IN FOOD PRODUCTS

16.4.1 FUNCTIONAL ATTRIBUTES

Milk fat may be concentrated to produce cream or isolated as in butter. It may also be homogenized along with the milk to prevent separation. It is difficult to satisfactorily substitute for milk fat because its unique characteristics contribute to the appearance, flavor, mouthfeel, texture, and other properties of foods containing it.

Milk fat is versatile, and the many attributes that it contributes to food include the following:

- Aeration: the lightening of texture by whipping creams and butter oils.
- Color and flavor: milk lipids impart a pale yellow color and a pleasing buttery flavor to food and serve as a reservoir for other flavors.
- Dispersion: the size and arrangement of discrete particles in a liquid. For example, milk fat globules prevent coalescence of air bubbles in ice cream.
- Emulsification by MFGM material, as previously mentioned.
- Firmness and softness, which depend on the proportion of solid fat.
- Inhibition of fat bloom (undesirable cocoa butter crystallization) in cookies and chocolate products.
- Layering: the ability of milk fat to allow for the formation of layers in laminated pastries such as Danish and croissants.
- Lubricity: the presence of milk fat increases lubricity in the mouth since melting is completed at body temperature.
- Plasticity: milk fat is soft enough not to fracture pastry dough that is being rolled.
- Spreadability: a key function of dairy-based spreads, which depends on the solid fraction of fat at a particular temperature.
- Structure formation: the ability of milk fat to disrupt the gluten matrix in cookies and pastries.
- Viscosity, which relates to imparting a creamy mouthfeel to food.

16.4.2 MODIFICATION OF PROPERTIES

Melting and rheological attributes of milk fat are affected by the biological processes occurring in the cow (especially FA and TAG composition) and may be tailored by processing. The type of fat being secreted into milk is influenced by stage of lactation, season, genetic improvement (Soyeurt et al., 2006), supplementation of the cow's diet (Glasser et al., 2007), or infusion of long-chain FA into the abomasums of lactating cows (Ortiz-Gonzalez et al., 2007). Fractionation, hydrogenation, and interesterification are the most common processing techniques for altering milk fat. Fractionation is performed by supercritical extraction or progressive crystallization from melted fat and leads to fractions with specific properties. Hydrogenation increases the saturated fat content, which improves stability and hardness but also generates trans FA. Interesterification, by inorganic catalysts or lipase enzymes in the presence of certain FA, rearranges the FA residues on the TAG molecule.

16.4.3 Uses in Food

For many years, milk fat and its solid and liquid fractions have been used in baking, confectionery, and spreads because of its functional attributes. Compatibility with other fats, coating of air bubbles in foams, emulsification with water, and coating of solid nonfat particles are all factors that must be considered when using milk fat as an ingredient.

Research on modified milk fat has resulted in new applications. For instance, butyric, caproic, and caprylic acids may be converted to ethyl esters, creating fruity flavors (Lubary et al., 2009). Cholesterol may be removed by short path molecular distillation (evaporation into a vacuum), steam distillation, or complexation with cyclodextrin powder (Han et al., 2007) to produce cholesterol-free milk. The MFGM may be separated through solvent or supercritical fluid extraction (Dewettinck et al., 2008) or by high-pressure homogenization (Thompson and Singh, 2006) and used as a health-promoting additive to certain foods.

16.5 NUTRITIONAL SIGNIFICANCE

16.5.1 Saturated Fatty Acids

The 4:0, 6:0, 8:0, and 10:0 FA are quite digestible. They are hydrolyzed from TAG by gastric lipases, more so than longer-chain FA, and are transferred directly to the blood for oxidation by the liver, and thus do not tend to form adipose tissue (Molkentin, 2000). Stearic acid (18:0) does not affect blood cholesterol levels but the 12:0–16:0 FA have been shown to increase the concentration of total and low-density lipoprotein (LDL) cholesterol in plasma (Molkentin, 2000).

16.5.2 Unsaturated Fatty Acids

16.5.2.1 Oleic, Linoleic, and Linolenic Acids

Oleic acid and stearic acid (which is rapidly converted into oleic acid in the body) decrease the blood pressure and concentration of LDL-cholesterol in humans while leaving the high-density lipoprotein cholesterol level unaffected (Bonanome and Grundy, 1988). A theory by Terés et al. (2008) states that oleic acid's *cis* configuration allows it to pack densely into cell membranes of blood vessels, where it makes the cells more receptive to signals that reduce blood pressure. Linoleic acid and linolenic acid, the primary polyenoic FA in milk, are essential for human metabolism, and linolenic acid may also have an LDL-cholesterol lowering effect (de Lorgeril and Renaud, 1994).

16.5.2.2 Conjugated Linoleic Acids

Conjugated linoleic acids (CLA) are conjugated dienes found chiefly in milk fat, with nine-tenths of the total being the *cis*-9, *trans*-11 isomer, sometimes called rumenic acid. All known physiologic effects of CLA are found in the 9-*cis*, 11-*trans* and 10-*trans*, 12-*cis* isomers (Pariza, 2004). CLA are derived from linoleic acid and possibly linolenic acid. Their concentration in milk tends to increase during summer pasture feeding (Bargo et al., 2006). In young rodents, 10-*trans*, 12-*cis* appears to be responsible for the reduction of body fat gain, 9-*cis*, 11-*trans* enhances growth and feed efficiency, and both inhibit mammary carcinogenesis (Pariza, 2004).

16.5.2.3 Trans Fatty Acids

Vaccenic acid, the 11-*trans* isomer of oleic acid, is the primary *trans* FA in milk. Vaccenic acid occurs naturally in milk and appears to have an anticarcinogenic effect after being converted into the *cis*-9, *trans*-11 isomer, unlike trans FA created by hydrogenation of *cis*-unsaturated fats, which increase the risk of coronary heart disease (Lock et al., 2004).

16.5.3 Overall Picture

A review of the current literature on the effects of dietary fat, and milk fat in particular, points to the intriguing possibility that consuming food containing milk lipids is more helpful than harmful. For years, reducing the consumption of dairy fat has been emphasized as a means of improving the diet and ultimately the health of consumers. Half of the lipids in milk (including human milk) are saturated fats, which are noted targets of nutritionists. However, since the production of saturated fats in milk has not been reduced by evolution, it is possible that there are unseen benefits to saturated fat that cause mammals to continue to make it. Mammary glands already produce unsaturated fats, and natural selection would presumably cause those to be predominant if saturated fats served no purpose. Moreover, milk fat globules are structured differently than other dietary fats, contain a range of FA chain lengths, and serve as carriers of fat-soluble vitamins, phospholipids, and other bioactive compounds (German, 2008). Some scientists reject the idea of reducing milk consumption simply because of the FA and cholesterol in it (Pfeuffer and Schrezenmeir, 2000). It appears that the nutritional benefits of milk lipid consumption outweigh the possible drawbacks of some of its components.

REFERENCES

Austin, M.A. 1991. Plasma triglyceride and coronary heart disease, *Arterioscler. Thromb. Vasc. Biol.*, 11, 2–14.

Bargo, F., Delahoy, J., Schroeder, G., Baumgard, L., and Muller, L. 2006. Supplementing total mixed rations with pasture increase the content of conjugated linoleic acid in milk, *Anim. Feed Sci. Technol.*, 131, 226–240.

Blasia, F., Montesano, D., De Angelis, M., Maurizi, A., Ventura, F., Cossignani, L., Simonetti, M.S., and Damiani, P. 2008. Results of stereospecific analysis of triacylglycerol fraction from donkey, cow, ewe, goat and buffalo milk, *J. Food Compos. Anal.*, 21, 1–7.

Bonanome, A. and Grundy, S.M. 1988. Effect of dietary stearic acid on plasma cholesterol and lipoprotein levels, *New Engl. J. Med.*, 318, 1244–1248.

Collins, Y.F., McSweeney, P.L.H., and Wilkinson, M.G. 2003. Lipolysis and free fatty acid catabolism in cheese: A review of current knowledge, *Int. Dairy J.*, 13, 841–866.

de Lorgeril, M. and Renaud, S. 1994. Mediterranean alpha-linolenic acid-rich diet in secondary prevention of coronary heart disease, *Lancet*, 343, 1454–1459.

Deeth, H.C. 1997. The role of phospholipids in the stability of milk fat globules, *Aust. J. Dairy Technol.*, 52, 44–46.

Deeth, H.C. 2006. Lipoprotein lipase and lipolysis in milk, *Int. Dairy J.*, 16, 555–562.

Dewettinck, K., Rombaut, R., Thienpont, N., Le, T.T., Messens, K., and Van Camp, J. 2008. Nutritional and technological aspects of milk fat globule membrane material, *Int. Dairy J.*, 18, 436–457.

Fagan, P., Wijesundera, C., and Watkins, P. 2004. Determination of mono- and di-acylglycerols in milk lipids, *J. Chromatogr. A*, 1054, 251–259.

German, J.B. 2008. Milk fats: A different perspective, *Sci. Aliments*, 28, 176–186.

German, J.B. and Dillard, C.J. 2006. Composition, structure and absorption of milk lipids: A source of energy, fat-soluble nutrients and bioactive molecules, *Crit. Rev. Food Sci.*, 46, 57–92.

Glasser, F., Doreau, M., Ferlay, A., Loor, J.J., and Chilliard, Y. 2007. Milk fatty acids: Mammary synthesis could limit transfer from duodenum in cows, *Eur. J. Lipid Sci. Technol.*, 109, 817–827.

Gresti, J., Bugaut, M., Maniongui, C., and Bezard, J. 1993. Composition of molecular species of triacylglycerols in bovine milk fat, *J. Dairy Sci.*, 76, 1850–1869.

Han, E.-M., Kim, S.-H., Ahn, J., and Kwak, H.-S. 2007. Optimizing cholesterol removal from cream using β-cyclodextrin cross-linked with adipic acid, *Int. J. Dairy Technol.*, 60, 31–36.

Jensen, R.G. 2002. The composition of bovine lipids: January 1995 to December 2000, *J. Dairy Sci.*, 85, 295–350.

Lock, A.L., Corl, B.A., Barbano, D.M., Bauman, D.E., and Ip, C. 2004. The anticarcinogenic effect of *trans*-11 18:1 is dependent on its conversion to *cis*-9, *trans*-11 CLA by Δ9-desaturase in rats, *J. Nutr.*, 134, 2698–2704.

Lubary, M., ter Horst, J.H., Hofland, G.W., and Jansens, P.J. 2009. Lipase-catalyzed ethanolysis of milk fat with a focus on short-chain fatty acid selectivity, *J. Agric. Food Chem.*, 57, 116–121.

MacGibbon, A.K.H. and Taylor, M.W. 2006. Composition and structure of bovine milk lipids. In *Advanced Dairy Chemistry*, Vol. 2: *Lipids*, 3rd edn., P.F. Fox and P.L.H. McSweeney (Eds.). Springer, New York, pp. 1–42.

Molkentin, J. 2000. Occurrence and biochemical characteristics of natural bioactive substances in bovine milk lipids, *Br. J. Nutr.*, 84(Suppl. 1), S47–S53.

Newburg, D.S. and Chaturvedi, P. 1992. Neutral glycolipids of human and bovine milk, *Lipids*, 27, 923–927.

O'Brien, N.M. and O'Connor, T.P. 2002. Lipid oxidation. In *Encyclopedia of Dairy Sciences*, H. Roginski, J.W. Fuquay, and P.F. Fox (Eds.). Academic Press, London, U.K., pp. 1600–1607.

Ortiz-Gonzalez, G., Jimenez-Flores, R., Bremmer, D.R., Clark, J.H., DePeters, E.J., Schmidt, S.J., and Drackley, J.K. 2007. Functional properties of butter oil made from bovine milk with experimentally altered fat composition, *J. Dairy Sci.*, 90, 5018–5031.

Pariza, M.W. 2004. Perspective on the safety and effectiveness of conjugated linoleic acid, *Am. J. Clin. Nutr.*, 79(Suppl.), 1132S–1136S.

Pfeuffer, M. and Schrezenmeir, J. 2000. Bioactive substances in milk with properties decreasing risk of cardiovascular diseases, *Br. J. Nutr.*, 84(Suppl. 1), S155–S159.

Schaafsma, G. 2002. Vitamins: General introduction. In *Encyclopedia of Dairy Sciences,* H. Roginski, J.W. Fuquay, and P.F. Fox (Eds.). Academic Press, London, U.K., pp. 2653–2657.

Singh, H. 2006. The milk fat globule membrane—A biophysical system for food applications, *Curr. Opin. Colloid Interface Sci.*, 11, 154–163.

Sørhaug, T. and Stepaniak, L. 1997. Psychrotrophs and their enzymes in milk and dairy products: Quality aspects, *Trends Food Sci. Technol.*, 8, 35–67.

Soyeurt, H., Dardenne, P., Gillon, A., Croquet, C., Vanderick, S., Mayeres, P., Bertozzi, C., and Gengler, N. 2006. Variation in fatty acid contents of milk and milk fat within and across breeds, *J. Dairy Sci.*, 89, 4858–4865.

Taylor, M.W. and MacGibbon, A.K.H. 2002. Lipids: General characteristics. In *Encyclopedia of Dairy Sciences,* H. Roginski, J.W. Fuquay, and P.F. Fox (Eds.). Academic Press, London, U.K., pp. 1544–1550.

Terés, S., Barceló-Coblijn, G., Benet, M., Álvarez, R., Bressani, R., Halver, J.E., and Escriba, P.V. 2008. Oleic acid content is responsible for the reduction in blood pressure induced by olive oil, *Proc. Natl. Acad. Sci. USA*, 105, 13811–13816.

Thompson, A.K. and Singh, H. 2006. Preparation of liposomes from milk fat globule membrane phospholipids using a microfluidizer, *J. Dairy Sci.*, 89, 410–419.

Tunick, M.H. 1994. Effects of homogenization and proteolysis on free oil in Mozzarella cheese, *J. Dairy Sci.*, 77, 2487–2493.

Tunick, M.H. 2007. Origins of cheese flavor. In *Flavors of Dairy Products*, K.R. Cadwallader, M.A. Drake, and R.J. McGorrin (Eds.). ACS Books, Washington, DC, pp. 155–173.

Vesper, H., Schmelz, E.-M., Nikolova-Karakashian, M.N., Dillehay, D.L., Lynch, D.V., and Merrill, A.H., Jr. 1999. Sphingolipids in food and the emerging importance of sphingolipids to nutrition, *J. Nutr.*, 129, 1239–1250.

17 The Role of Lipids in Meat

Zdzisław E. Sikorski and Izabela Sinkiewicz

CONTENTS

17.1 INTRODUCTION

In this chapter, the term "meat" has been used as denoting the skeletal, heart, and tongue muscles, as well as the liver and the kidney of slaughtered mammals and birds. The tissues of fish and edible invertebrates have been excluded, since they have been treated in Chapter 15.

Meat is often regarded as rich in fat. This notion, however, is not true in respect to all edible parts and to all species and breeds of slaughter animals. The lipids of muscle food raw materials are either structural lipids, i.e., components of cell and organelle membranes, or depot fat. Just like in many other food raw materials, the lipids

- Contribute to the structure of various tissues and organelles
- Supply substrates for synthesis of different metabolites
- Are the richest source of energy in the human diet
- Accumulate the fat soluble vitamins
- Affect the sensory properties and the shelf life of many commodities

Among the lipids playing structural roles are phospholipids (PL) and cholesterol. They are present in meat in considerably low amounts, although very uniformly distributed, since they constitute the material of cellular membranes, the plasma membrane, and the nervous tissue. Their proportion in the total lipids in meat decreases with the increase of fattiness of the carcass. Energy is stored mainly in triacylglycerols (TAG). Due to the efforts of animal husbandry and processing technology, it is actually possible to breed livestock with a desirable lipid profile and to manufacture meat products of programmed fat content and sensory characteristics.

17.2 THE CONTENTS AND DISTRIBUTION OF LIPIDS IN MEAT

Lipids are accumulated in the animal carcass primarily under the skin as a layer in the subcutaneous connective tissue consisting of cells almost entirely filled with fat. This tissue has in the animal body an insulating and contour-building role and serves as a rich energy store. Furthermore, there is the intramuscular fat that comprises the PL that builds predominantly the cell membranes, the TAG present mainly, in about 80%, in the intramuscular adipocytes and as droplets within the myofiberss' cytoplasm (5%–20%), as white flecks or streaks between the bundles of muscle fibers within the lean muscle tissue that is known as marbling, and cholesterol. The marbling fat is dispersed in the loose networks of the perimysial connective tissue septa. The fat located within the septa, between muscles, is known as intermuscular or seam fat. On top of that, lipids also appear as kidney, leaf, snowball, ruffle, caudal, crotch, and jowl fat.

The fattiness depends on the species, the breed, the maturing rate, the age, and the feeding regime of the slaughter animal, as well as on the cut of the meat (Moloney 2006, Hocquette et al. 2009) (Table 17.1). Lean beef and lean pork cuts as well as white poultry muscles may contain as little as 2%–4% fat wet weight, while fatty pork as much as 30%–45% fat. The beef and the lamb meat produced nowadays contain less fat than what was produced 10 years ago. This has been caused mainly by intentional changes introduced in animal husbandry. The commonly offered cuts of lamb and beef on the market, trimmed of fat, may have not more than 5% fat (McAfee et al. 2010).

In various meat products, the content and the distribution of fat depends on the recipe and is controlled in respect of both nutritional and sensory requirements (Table 17.2). In cooked meats, it depends on the cooking methods, since the loss in the drip as well as the penetration of fat from the frying medium into the meat affect the outcome (Table 17.3).

17.3 THE FATTY ACID COMPOSITION OF MEAT LIPIDS

The proportion of various fatty acid (FA) residues in the lipids of red meats and poultry depends on the species, the breed, the age, the sex, and the diet of the animal and on the cut (Table 17.4). The effect of the composition of the animals' diet within the species of the slaughter animals is very significant. The properties of fats from free-living ruminants reflect the lipid composition of the vegetation on their pasture. The contents of linoleic acid ($C_{18:2}$) (LA) and other n-3 FA tend to be higher

TABLE 17.1
Proximate Contents of Fat in Raw Meats

Meat Kind and Cut	Fat, g/100 g	Meat Kind and Cut	Fat, g/100 g
Beef		*Pork*	
Loin, fat	31	Shoulder butt, medium	42
Loin, medium	25	Spare rib	32
Flank	18	Ham, medium	31
Chuck, medium	16	Loin, medium	25
Round, medium	13		
Tongue	15	*Goat*	
Kidney	8	Meat	2.3
Liver	3		
		Goose	
Lamb		Meat	7.1
Loin	5.9	Liver	4.3
Rib	9.2		
Shoulder	6.8	*Turkey*	
Breast	15	Dark meat	4.4
Foreshank	3.3	Light meat	1.6
Leg	4.5	Liver	16
Liver	5.0	Neck	5.4
Kidney	3.0		

Source: USDA National Nutrient Database for Standard Reference, Release 22, Individual food reports (SR 22-Page Reports), 2009. With permission.

TABLE 17.2
The Formulation for Frankfurter Sausage (United States) with Different Contents of Fat

Ingredients	10% Fat	20% Fat	30% Fat
Beef trimmings 90 CL	29.6	21.0	18.0
Beef trimmings 50 CL	—	—	14.7
Turkey trimmings 85 CL	—	15.0	—
Pork trimmings 75 CL	20.6	—	15.5
Pork trimmings 50 CL	—	26.1	23.8
Water/Ice	36.15	30.0	21.2
Spices and additives	13.65	7.9	6.8

Source: Hoogenkamp, H.W., *Soy Protein & Meat Formulations*, Protein Technologies International, St. Louis, MO, 2001, p. 196. With permission.

TABLE 17.3
Proximate Contents of Fat in Fresh and Cooked Chicken, g/100 g

Meat Cut	Raw	Fried	Roasted	Stewed
Dark meat	4.3	12	9.7	—
Light meat	1.7	5.5	4.5	4.0
Drumstick	3.4	8.1	5.7	5.7
Leg	3.8	9.3	8.4	8.1
Neck	8.8	12	—	8.2
Thigh	3.9	10	11	9.8
Wing	3.5	9.2	8.1	7.2

Source: USDA National Nutrient Database for Standard Reference, Release 22, 2009. With permission.

TABLE 17.4
Proximate Composition of the Fatty-Acid Residues in the Lipids of Red Meat and Poultry

Fatty Acid	Contents (g/100 g Tissue)			
	Beef	Pork	Chicken	Turkey
C14:0	0.07	0.15	0.07	0.07
C14:1	0.02	0	0.02	0
C16:0	0.53	1.44	1.95	1.54
C16:1	0.09	0.27	0.58	0.31
C18:0	0.37	0.74	0.53	0.61
C18:1	0.76	3.11	3.17	1.33
C18:2	0.05	0.39	1.86	1.24
C18:3	0.02	0.04	0.08	0.07
C20:3	0	0	0	0
C20:4	0	0.04	0.01	0.31
C20:5	0	0	0	0.09
C22:5	0	0	0	0.12
C22:6	0	0	0	0.31

Source: Kunachowicz, H. et al., Tables of Nutritive Value of Foods, Food and Nutrition Institute, Warsaw, Poland, 1998. With permission.

in the fat of grass-fed ruminants than in concentrate-fed animals. Increasing the proportion of polyenoic FA (PEFA) in the feed brings about a raise in the content of these acids in the meat lipids. The fat amount, composition, and distribution in animals bred for manufacturing various specialty products is often strictly regulated by the food law, e.g., in the heavy pigs used in Italy for "Prosciutto di Parma" and

"Prosciutto di S. Daniele." The farmer who wants to comply with the regulations should know the FA composition of the feed, since it affects the properties of the animals' lipids (Della Casa et al. 2010). Generally, it is more difficult to change the FA composition of the meat lipids of ruminants than that of monogastric animals because of the biohydrogenation of PEFA in the rumen. Broiler chickens kept on a diet with 5% fish meal have a significant proportion of eicosapentaenoic acid (EPA), docosahexaenoic acid (DHA), and other n-3 PEFA residues in their fat. Their content increased when the fish meal proportion in the feed was raised up to 30% or when 2%–4% red fish oil was added (Jeun-Horng et al. 2002). Feeding broilers with basal corn and soybean meal diets enriched with 4 g/kg of canola oil, fish meal, or fish oils significantly increased the contents of EPA, DHA, and the total n-3 FA in the lipids of chicken thigh and breast meat in respect to those in samples from animals fed non-enriched diets (Jeun-Horng et al. 2002, Nobar et al. 2007).

17.4 THE NUTRITIVE VALUE OF MEAT LIPIDS

The dietary significance of meat lipids is related to their contribution to the energy value of the diet, as well as to their positive and negative effects on human metabolism and health depending on their FA composition.

Many consumers believe that the fat of beef is very rich in saturated FA and thus nutritionally unfavorable. This notion is, however, not correct, since in beef lipids the contents of saturated FA residues is about the same as that of unsaturated FA. About one half of the total amount of FA residues in the intra-muscular fat of beef and lamb is made up of monoenoic species, predominantly oleic acid ($C_{18:1}$ c-9), and of PEFA, mainly LA, and alpha-LA ($C_{18:3}$) (ALA), with smaller amounts of EPA, docosapentaenoic acid ($C_{22:5}$) (DPA), and DHA. Pasture feeding of animals increases the proportion of long chain (LC) n-3 PEFA in beef and lamb. Although the concentration of these FA residues in the fat of red meat is significantly lower than that in fish lipids, they may have nutritional significance in populations consuming a lot of red meat and rarely eating sea fish. Experiments aimed at increasing the biological value of lamb by changing the FA composition of the intramuscular fat have shown that replacing the sunflower oil supplementation of the animals' feed with linseed oil increased the content of LC n-3 PEFA but decreased the proportion of $C_{18:2}$ n-6, total LC n-6 PEFA, and $C_{18:2}$ cis-9, trans-11 FA. However, blends of sunflower oil and linseed oil added to the standard diet resulted in an enrichment of the intramuscular lipids in conjugated LA (CLA) and LC n-3 PEFA (Jerónimo et al. 2009).

Meat lipids contain significant amounts of CLA. This generic term comprises several geometric and positional isomers of LA, which have conjugated cis or trans double bonds in their hydrocarbon chains. In animal experiments, they have been shown to be anticarcinogenic, antidiabetic, and antiatherogenic and to have a positive effect on the immune system, bone metabolism, and body composition. CLA occurs mainly in the meat and milk of ruminants, since its main route of synthesis relies on the metabolism of rumen microflora. The predominant CLA, rumenic acid ($C_{18:2}$ cis-9, trans-11), is formed as a result of the biohydrogenation in the rumen of $C_{18:2}$ n-6. However, it is also generated endogenously by the desaturation of the trans-vaccenic acid ($C_{18:1}$ trans 11), which is the predominant trans FA in red meat. Thus, although the trans-FA in foods is generally regarded as nutritionally harmful, the *trans*-vaccenic acid may play a beneficial role as a substrate for CLA synthesis. The CLA in beef and lamb consists of about 80% of the $C_{18:2}$ *cis*-9, *trans*-11 isomer. The lipids of monogastric mammals and of fish are poor in CLA. Lamb has been reported to contain 4.3–19 mg CLA/g lipid, which is followed by beef, 1.3–10 mg/g, and below 1 mg/g in pork, chicken, and horse meat. The concentration of CLA in meats depends also on the feeding regime of the animals and on the kind of muscle. The contents of CLA in ruminants' lipids do not seem to be affected by processing—in meat products it is similar to that in the raw meat (Schmid et al. 2006).

17.5 THE ROLE OF LIPIDS IN THE SENSORY QUALITY OF MEAT AND MEAT PRODUCTS

17.5.1 INTRODUCTION

Lipids affect all important sensory attributes of meat just by being its inherent component, which has characteristic properties like a physical structure, a melting temperature, rheological behavior, hydrophobicity, color, and the capacity to dissolve various non-polar compounds, including pigments and volatiles. The impact of the amount of fat in meat on the appearance and the eating quality as well as on preferences of consumers in different countries has been discussed by Dransfield (2008). Not less important for the sensory properties of meat are the interactions of other tissue components with fats and predominantly with the secondary products of lipid autoxidation (Sikorski and Sikorska-Wiśniewska 2006). Of utmost significance among these interactions are oxidative processes.

Oxidation in food systems and antioxidants have been expertly treated in Chapters 9 and 10, respectively. Sections 17.5.2 and 17.5.3 present the role of autoxidation in changes of meat color and flavor. Here, only some general aspects of the use of antioxidants in meat products will be addressed.

Antioxidants have been always used in meat processing for retarding the oxidative changes in lipids that lead to the deterioration of the color, flavor, and texture of the products. Initially, they were various spices and herbs, as well as wood smoke applied traditionally in smoking. Later, numerous synthetic antioxidants were proposed and accepted by health authorities and the food law. Many synthetic antioxidants are on the list of additives permitted to be used in the food industry, e.g., butylated hydroxytoluene (BHT) and butylated hydroxyanisole (BHA), in the European Union registered as E320 and E321, respectively. However, some persisting concerns about their safety and the consumers' attitude to the avoidance of synthetic additives in foods have encouraged food producers to search for natural substances to serve the same purpose.

Antioxidant compounds are applied in different forms; as components of dips or curing solutions, as wood smoke or smoke preparations, as fresh or dried original spices, or extracts added to the minced meats as liquids or powders, or as additives to the animals' feed. Various spices, e.g., rosemary, oregano, sage, and thyme, as well as their extracts are known to have antioxidant activity. This is mainly due to the content of a variety of phenolic compounds. Rosemary is commercially available and may be used as an antioxidant in Europe and the United States. Its extract contains carnosic and rosmarinic acids, carnosol, rosmanol, epirosmanol, isorosmanol, methylcarnosate, and rosmaridiphenol; oregano is rich in polyhydroxylbenzoic, cinnamic, rosmarinic, protocatechuic, and caffeic acids. These and other spices and herbs are used as natural antioxidants in different formulations in order to protect meat products from oxidative deterioration during prolonged storage. The antioxidant effectiveness of various spices and herb extracts in numerous food systems has been investigated. Trindade et al. (2010) have shown that rosemary extract, added in an amount of 400 mg/kg ground beef or in combination with either BHT/BHA or oregano extract, was more effective in decreasing the rate of lipid oxidation in ^{60}Co irradiated beef burgers, packaged aerobically in polyethylene bags, for a period of 90 days at −20°C than the oregano extract used individually or in combination with BHT/BHA. Essential oil and extracts from rosemary and other herbs of the *Labiatae* family added to the feed of turkey, hog, chicken, and lamb have been found effective as antioxidants in the animals' meat. Distilled rosemary leaves supplementing the basal diets of pregnant ewes had a carryover antioxidant effect in the lambs from these ewes. The lipids in the meat were less oxidized and the color was more stable after refrigerated storage for 21 days under retail display conditions in an atmosphere containing 70% O_2 and 30% CO_2 than those in the control samples (Nieto et al. 2010).

17.5.2 The Effect of Lipids on Color

Lipids affect the color of meat since their oxidation products may cause oxidative changes in the pigments—myoglobin (MbFe(II)), hemoglobin, and the cytochromes. The natural, cherry-red color of meat on the freshly cut surface is caused by the reduced forms of the heme proteins (Figure 17.1). Oxygenation of these pigments at a high partial pressure of oxygen yields bright-red oxymyoglobin (MbFe(II)O_2) and oxyhemoglobin (HbFe(II)O_2). This change in color after exposure to fresh air, known as blooming, takes only about 30 min. These two desirable colors— cherry-red and bright-red—are, however, unstable and turn to undesirable brown in

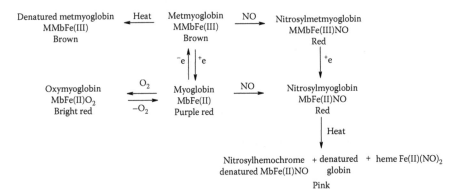

FIGURE 17.1   Reduced and oxidized forms of heme proteins.

conditions favoring oxidation of the pigments. The rate of oxidation depends on the reducing capacity of the meat and the presence of various oxidants and antioxidants. The change in color proceeds gradually as the brown metmyoglobin accumulates in the muscles. Its progress means a loss of the high quality of the fresh meat since the light-red color is highly valued by the consumers, especially in beef. Thus, the market value of fresh beef cuts exhibited in retail display gradually decreases and a discount in price must follow. The time over which the meat retains high color stability is from one to about five days, depending on the reducing capacity of the tissues, the composition of the atmosphere around the cut, the properties of the packaging, and the storage temperature.

The browning reaction of meat heme proteins is catalyzed by oxidized muscle lipids, especially the PEFA residues of the membrane PL, and vice versa. Secondary lipid oxidation products, predominantly unsaturated aldehydes, may covalently bind to $MbFe(II)O_2$ changing the conformation of the pigment molecule and thus facilitating the oxidation of the heme iron (Faustman and Wang 2000). On the other hand, oxidation of $MbFe(II)O_2$ may initiate the chain reactions of lipid oxidation in meat (Monahan 2000). The factor involved in this catalytic activity is the heme iron that may initiate the formation of lipid radicals (Figure 17.2). Thus, any factors and treatments that increase the reducing capacity of the muscles and decrease the rate of oxidative reactions serve the purpose of stabilizing the desirable color of the cut surface of the meat.

The rate of oxidation of $MbFe(II)O_2$ can be decreased by adding various carotenoids to the feed of the livestock. The carotenoids can also reduce the oxidized forms of heme proteins, thus they increase the color stability of meat (Mortensen and Skibsted 2000). A similar stabilizing effect can be achieved also by using α-tocopherol. Its action is probably based on its ability to decrease the rate of generation of primary lipid peroxides and their breakdown to water-soluble free-radical secondary products like $HO^\bullet$ radicals, as well as unsaturated aldehydes (Faustman and Wang 2000). Synthetic vitamin E, DL-α-tocopheryl acetate, is

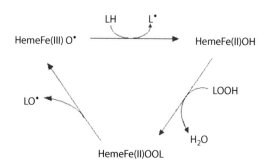

LH – unsaturated fatty acid
LOOH – lipid hydroperoxide
$LO^\bullet$, $L^\bullet$ – lipid radicals

FIGURE 17.2 The catalytic role of heme proteins in lipid oxidation.

effective in reducing lipid oxidation and prolonging the shelf life of fresh pork if added to the feed in the concentration of 200 mg/kg. A similar result can be obtained by adding 40 mg/kg of natural vitamin E, D-α-tocopheryl acetate (Boler et al. 2009).

The economic effectiveness of adding antioxidants to the livestock feed must be evaluated with respect to the cost of the additives and the financial gain due to a possibly achievable prolongation of the high quality life of the product (Smith et al. 2000).

17.5.3 The Effect of Lipids on Flavor

17.5.3.1 Fresh Meat Flavor

Fresh, raw meat has only a slightly perceptible aroma and a blood-like taste. Therefore, these properties are not important in the quality grading of uncooked and unprocessed meats. However, according to Dransfield (2008), the taste of LCFA can be sensed by humans thanks to a specific detection system. During prolonged storage, especially under abuse conditions, numerous volatile compounds are generated due to catabolic reactions in the muscles, bacterial activity, hydrolysis, and oxidation. The primary products of FA oxidation, i.e., the hydroperoxides are odorless. On the other hand, the compounds generated further in the course of oxidation cause undesirable odors known as rancid, which may be tallowy in beef, muttony in mutton, and cheesy, acrylic, or fishy in pork. In turkey kept on diets rich in PEFA, the meat may have a perceptible fishy taint. The rancid flavor can be detected in meat products when the value of the 2-thiobarbituric acid reactive substances (TBARS), the recognized indicator of the concentration of the secondary products of lipid oxidation, is between 0.5 and 2.0 mg/kg.

17.5.3.2 Cooked Meat Flavor

Flavor ranks in the preferences of many consumers as the second most important criterion of meat quality, just after tenderness. Within the range of the fat content in the meat from about 3% to about 7%, the flavor preference tends to increase. Below and above this range, it declines. In cooked meats and meaty dishes the lipids contribute to the aroma in two ways. Firstly, the volatile secondary products of oxidation add their characteristic aroma to the total flavor. Secondly, some reactive compounds participate in the Maillard reaction that is known to yield typical aromas, as well as in other processes generating odorous volatiles.

The lipid degradation products in cooked meat are numerous and very diversified in chemical and sensory properties. Among the total of over 1000 volatile compounds found in meat, the number of products derived from lipids has been estimated to reach several hundred (Shahidi et al. 1986, Mottram 1998). Here belong the short chain carboxylic acids, esters, aliphatic and aromatic hydrocarbons, saturated and unsaturated aldehydes, ketones, lactones, and alkylfurans. Their precursors are both the TAG and the PL of the storage and structural lipids. Therefore, the volatile lipid oxidation products contribute to the aroma of both fatty and lean meats, since even very lean muscles contain at least about 1% lipids, mainly PL. The FA residues in PL are usually more unsaturated than the storage lipids and thus are highly susceptible to oxidation. The contribution of the secondary oxidation products to the perceived cooked meat aroma depends on their concentration, the characteristic odor, and the threshold values. The characteristic flavors/aromas of 62 compounds identified in beef have been presented by Calkins and Hodgen (2007) in their interesting paper on meat flavor. Although the concentration of the secondary lipid oxidation products is usually higher than that of the heterocyclic compounds generated mainly from non-lipid precursors, their contribution to the cooked meat aroma may be relatively low because the odor thresholds of many of them are high (Mottram 1998). Sulfurous and carbonyl compounds are regarded as the most important contributors to meat flavor. The volatile compounds formed due to lipid changes, especially unsaturated aldehydes, may be responsible for the species' characteristics of cooked meat aroma, while the odor developed by heating water-soluble precursors is recognized as having a general meaty note. The qualitative and the quantitative composition of the pool of these volatiles in cooked meats depend also on the species and the feeding regime of the slaughter animals—increasing the proportion of PEFA in the animals' feed brings about a significant rise in the contents of secondary lipid-oxidation products. However, in frankfurters made of meat that was obtained from chicken kept on a diet supplemented with 2% and 4% fish oil and with α-tocopheryl acetate (50 IU/kg), no significant fishy flavor was noted during the 30 days storage in the dark at 0°C (Jeun-Horng et al. 2002).

In reheated dishes, a rancid, fatty, pungent, and other undesirable off-flavor odors known as a warmed-over flavor may appear. These off-flavors are caused by several volatile lipid degradation products. The primary outcome of autoxidation of LA is the 9-hydroxyperoxide, which as a result of β-scission yields 2-octenal, 2-nonenal, 2,4-decadienal, 1-octen-3-ol, 1-octen-3-one, 2,4-nonadienal, and hexanal. The 2,4-decadienal and 2-nonenal have low odor thresholds, comparable to those of the sulfurous compounds contributing to the meaty flavor (Calkins and Hodgen 2007). Hexanal has been recognized as the characteristic aldehyde indicative for the warmed-over flavor. Its content in reheated samples may be about ten times higher than in freshly cooked beef (Belitz et al. 2001). Arachidonic acid oxidizes into 11-hydroxyperoxide. Its secondary oxidation products of intense aroma are trans-4,5-(epoxy)-(E)-2-decenal, 2,4,7-tridecatrienal, and again 2,4-decadienal, 1-octen-3-one, and hexanal. Glycerophospholipids with an LC fatty aldehyde in the sn-1 position are the precursors of 12-methyltridecanal that has a tallowy and beef-like odor with a threshold of 0.1 µg/kg in water (Calkins and Hodgen 2007). PL, rich in PEFA, set free from the tissue membranes by heating, are very prone to oxidation catalyzed by the Fe(II) released due to thermal denaturation from the heme proteins. The liberation of Fe(II) from myoglobin occurs in the same temperature range as the loss in the oxidative stability of cooked muscle lipids (Decker and Xu 1998). In cured, cooked meats the rate of generation of the volatile compounds responsible for the warmed-over flavor is lower, since curing retards the liberation of Fe(II) from the heme proteins. Furthermore, the cured meat pigment MbNO acts as an antioxidant. The formation of the warmed-over flavor may be inhibited by additives capable of binding Fe ions, e.g., polyphosphates, phytin, or ethylene diamine tetra acetic acid (EDTA). According to Belitz et al. (2001) typical antioxidants are less effective.

Carbonyls and alcohols derived from lipids are also responsible for the character-istic odor that develops in meat sterilized by irradiation.

Besides adding their own odor, many secondary products of oxidation of PUFA, especially the unsaturated aldehydes, participate in the formation of a total flavor by reacting with other components of the cooked meat. The 2,4-alkabienals and conjugated alkatrienals react with amino acids yielding a typical fishy off-flavor. Among the compounds formed due to such interactions and identified in cooked meats are numerous alkyl-3-thiazolines, alkylthiazoles, alkylpyridines, alkylthio-phenes, and alkylpyrazines. The concentration of many of these compounds is especially high in roast beef from animals whose feed had been supplemented with fish oils.

17.5.4 LIPIDS AND THE RHEOLOGICAL PROPERTIES OF MEAT AND MEAT PRODUCTS

The results of studies in various countries have shown that the optimum texture of meat from animals of different species and breeds, as perceived by the consumers, depends interalia on the characteristic levels of the fat content. Other factors affecting the texture are related to the age and the growth rate of the animals. Intramuscular fat is regarded as affecting the meat's tenderness, juiciness, and flavor. Marbling is one of several criterions in establishing the quality grades of beef carcass—positively in North America and Asia, but too much of visible fat is less favored in Europe. Marbling is less visible in pork. The beef of different breeds and ages containing from 1.5% to 7% of intramuscular fat is often recognized as having the highest sensory quality (Mojto et al. 2001). Generally, consumers are satisfied with about 3%–4% of intramuscular fat in beef, 5% in sheep meat, and over 2.5% in pork (Hocquette et al. 2009). The extent of marbling increases proportionally to the total amount of intramuscular fat. The advantageous effect of marbling on the texture of cooked meats may be due to the lubricating action of the fat layers during chewing and swallowing. Marbling fat uniformly distributed in the meat cut also improves the flavor and the juiciness of the product.

The adipose fat and the marbling affect the shape of the carcass and the cut. This effect is related to the melting point of the fat that depends on the FA composition of the lipids. A high content of PEFA in the animals' feed brings about a softening of the fatty tissues. This effect occurs especially in pork. The firmness of the meat on the carcass and of the retail cuts during refrigerated storage is related to the physical properties of the TAG at the chilling temperature.

In comminuted-type meat products, small pieces of adipose tissue and fat droplets are embedded in a continuous, gelled matrix composed of water, various salts, and proteins, or else only the fat globules are uniformly distributed, surrounded by the other components of the system. The fat globules are stabilized in the structure of the sausage by interactions with proteins and other surface-active components. These interactions and the gelling of the proteins of the sausage batter are responsible for the texture of such products like frankfurters, wieners, and bologna. The sausage batter is made by cutting and mixing all the components in a proper sequence in a chopper or an emulsifier. The components comprise, generally, lean and fatty meats and adipose tissue in proportions so as to ensure the desirable fat content in the final product, protein preparations if necessary, according to the formula, salts, water or ice, and a mixture of spices and other additives. They are uniformly blended and the sausage batter behaves like a viscoplastic body. The muscle proteins, predominantly myosin, solubilized in the aqueous medium in the presence of salts act in the system as emulsifiers and stabilize the fat globules in the batter by proteinaceous membranes bound to the globules by hydrophobic interactions. This protects the fat droplets from coalescence and the formation of separate lipid phases in the batter in the sausage casings during steaming and smoking. The proteins of the formulation, when heated, form a viscoelastic gel holding in its structure the emulsified fat globules, dispersed adipose tissue fragments, and water. The effectiveness of the cross-linked structure of the proteinaceous gel in arresting the emulsified fat droplets depends on the proportion of lipids/proteins, the emulsifying capacity of the meat proteins and the added protein preparations, the degree of dispersion of the fat, and the conditions of emulsification and heating. The emulsifying capacity of the proteins may decrease due to a very high temperature of the batter during chopping. In over-chopped batters, the total surface of the small fat droplets may be too large to be effectively protected by the proteins, and coalescence leads to the formation of greasy pockets of the separated fat layers. This quality defect may occur in cooked sausages, e.g., liverwurst, in which the emulsifying capacity of the cooked meat used for making the batter is low.

In order to respond to the contemporary market demand for low-fat meat products there is a tendency to decrease the amount of fat in emulsion-type commodities. There is, however, at the same time, the consumer requirement for a proper mouth-feel attributed to fat. These contradictory health and pleasure demands can be to some extent satisfied both by simultaneously decreasing the proportion of the total fat in the formulation, replacing part of the animal fat by vegetable oils, and adding structure-forming and health-promoting dietary fiber. It has been shown (Choi et al. 2009) that a meat batter prepared of ham and pork-back fat, containing 10% pork fat, 10% vegetable oil, 2% rice bran fiber, 1.5% sodium chloride, and 0.15% sodium tripolyphosphate, cooked at 75°C had better sensory properties, a higher emulsion stability, and a lower cooking loss than the control product containing 30% pork fat, no vegetable oil, and no rice bran.

17.6 FINAL REMARKS

Lipids are ubiquitous in foods. They may be a source of sensory pleasure for the consumers of meat and meat products and, if taken in not overlarge quantities, good for the human health. After all, they are not only a source of energy, but also of essential FA and a reservoir of lipid-soluble vitamins. Furthermore, the changes in lipids during storage, processing, and culinary preparations of foods can be desirable or unwelcome. To make good use of the meat lipids, it is necessary to learn more about their chemical and functional properties.

ACKNOWLEDGMENT

The authors highly appreciate the helpful suggestions and information received from Professor Zbigniew Duda, Wrocław University of Life Sciences.

REFERENCES

Belitz, H.-D., Werner, G., and Peter, S. 2001. *Lehrbuch der Lebensmittelchemie*, Fünfte, vollständig überarbeitete Auflage. Berlin, Germany: Springer.

Boler, D.D., Gabriel, S.R., Yang, H., Balsbaugh, R., Mahan, D.C., Brewer, M.S., McKeith, F.K., and Killefer, J. 2009. Effect of different dietary levels of natural-source vitamin E in grow-finish pigs on pork quality and shelf life. *Meat Science* 83:723–730.

Calkins, C.R. and Hodgen, J.M. 2007. A fresh look at meat flavor. *Meat Science* 77:63–80.

Choi, Y.S., Choi, Y.H., Han, D.J., Kim, H.Y., Kim, H.W., Jeong, J.Y., and Kim, C.J. 2009. Characteristics of low-fat meat emulsion systems with pork fat replaced by vegetable oils and rice bran fiber. *Meat Science* 82:266–271.

Decker, E.A. and Xu, Z. 1998. Minimizing rancidity in muscle foods. *Food Technology* 52 (10):54–58.

Della Casa, G., Bochicchio, D., Faeti, V., Marchetto, G., Poletti, E., Rossi, A., Panciroli, A., Mordenti, A.L., and Brogna, N. 2010. Performance and fat quality of heavy pigs fed maize differing in linoleic acid content. *Meat Science* 84(1):152–158.

Dransfield, E. 2008. The taste of fat. *Meat Science* 80:37–42.

Faustman, C. and Wang, K.W. 2000. Potential mechanisms by which vitamin E improves oxidative stability of myoglobin. In *Antioxidants in Muscle Foods. Nutritional Strategies to Improve Quality*, E. Decker, C. Faustman, and C.J. Lopez-Bote (eds.), pp. 135–152. New York: Wiley-Interscience.

Hocquette, J.F., Gondret, F., Baéza, F., Médale, F., Jurie, C., and Pathick, D.W. 2009. Biological markers of intramuscular fat (IMF) content. In *Proceedings of the 62nd American Meat Science Association. Reciprocal Meat Conference*, Rogers, AR, pp. 1–5.

Hoogenkamp, H.W. 2001. *Soy Protein and Meat Formulations*, p. 196. St. Louis, MO: Protein Technologies International.

Jerónimo, E., Susana, P., Alves, S.P., Prates, J.A.M., Santos-Silva, J., and Bessa, R.J.B. 2009. Effect of dietary replacement of sunflower oil with linseed oil on intramuscular fatty acids of lamb meat. *Meat Science* 83(3):499–505.

Jeun-Horng, L., Yuan-Hui, L., and Chun-Chin, K. 2002. Effect of dietary fish oil on fatty acid composition, lipid oxidation and sensory property of chicken frankfurters during storage. *Meat Science* 60:61–167.

Kunachowicz, H., Nadolna, I., Przygoda, B., and Iwanov, K. 1998. *Tables of Nutritive Value of Foods*. Warsaw, Poland: Food and Nutrition Institute (in Polish).

McAfee, A.J., McSorley, E.M., Cuskelly, G.J., Moss, B.W., Wallace, J.M.W., Bonham, M.P., and Fearon, A.M. 2010. Red meat consumption: An overview of the risks and benefits. *Meat Science* 84 (1):1–13.

Mojto, J., Zaujec, K., and Pavlič, M. 2002. Analysis of marbling in meat of slaughter bulls for the purpose of classification and realization. In *Congress Proceedings, 47th International Congress of Meat Science and Technology*, Vol. 1, Kraków, Poland, pp. 130–131.

Moloney, A.P. 2006. Reducing fat in raw meat. In *Improving the Fat Content of Foods*, Ch. Williams and J. Buttriss (eds.), pp. 313–335. Boca Raton, FL: CRC Press.

Monahan, F.J. 2000. Oxidation of lipids in muscle foods: Fundamental and applied concerns. In *Antioxidants in Muscle Foods. Nutritional Strategies to Improve Quality*, E. Decker, C. Faustman, and C.J. Lopez-Bote (eds.), pp. 3–23. New York: Wiley-Interscience.

Mortensen, A. and Skibsted, L.H. 2000. Antioxidant activity of carotenoids in muscle foods, In *Antioxidants in Muscle Foods. Nutritional Strategies to Improve Quality*, E. Decker, C. Faustman, and C.J. Lopez-Bote (eds.), pp. 61–82. New York: Wiley-Interscience.

Mottram, D.S. 1991. Meat, in *Volatile Compounds in Foods and Beverages*, H. Maarse (ed.), p. 129. New York: CRC Press.

Mottram, D.S. 1998. Flavor formation in meat and meat products: A review. *Food Chemistry* 62(4):415–424.

Nieto, G., Díaz, P., Bañón, S., and Garrido, M.D. 2010. Dietary administration of ewe diets with a distillate from rosemary leaves (*Rosmarinus officinalis* L.): Influence on lamb meat quality. *Meat Science* 84(1):23–29.

Nobar, R.S.D., Nazeradl, K., Gorbani, A., Dghadmshahriar, H., and Gheyasi, J. 2007. Incorporation of DHA and EPA fatty acids into broiler meat lipids. *Journal of Animal and Veterinary Advances* 6(10):1199–1203.

Schmid, A., Collomb, M., Sieber, R., and Bee G. 2006. Conjugated linoleic acid in meat and meat products: A review. *Meat Science* 73(1):29–41.

Shahidi, F., Rubin, L., and D'Souza, L.A. 1986. Meat flavor volatiles. A review on the composition, techniques of analysis, and sensory evaluation. *CRC Critical Reviews in Food Science and Nutrition* 24(2):141–243.

Sikorski, Z.E. and Sikorska-Wiśniewska, G. 2006. The role of lipids in food quality. In *Improving the Fat Content of Foods*, Ch. Williams and J. Buttriss (eds.), pp. 213–235. Boca Raton, FL: CRC Press.

Smith, G.C., Belk, K.E., Sofos, J.N., Tatum, J.D., and Williams, S.N. 2000. Economic implications of improved color stability in beef. In *Antioxidants in Muscle Foods. Nutritional Strategies to Improve Quality*, E. Decker, C. Faustman, and C.J. Lopez-Bote (eds.), pp. 397–426. New York: Wiley-Interscience.

Trindade, R.A., Mancini-Filho, J., and Villavicencio, A.L.C.H. 2010. Natural antioxidants protecting irradiated beef burgers from lipid oxidation. *LWT: Food Science and Technology* 43(1):98–104.

18 Egg Lipids

Waldemar Ternes and Astrid M. Drotleff

CONTENTS

18.1 EGG SHELL AND ALBUMEN LIPIDS

The egg contains lipids on a large scale, even in its outer region. The shell contains 0.045% lipids in the cuticula and 1.35% lipids in the shell membrane (Suyama et al., 1977). Neutral and polar lipids occur in proportions of 5:1 and 6:1, respectively. Neutral lipids consist of mono-, di-, and triacylglycerols (TAG), cholesterol and cholesteryl esters, and free fatty acids (FA). The main components in the neutral lipid fraction of the shell and its adjoining layers are cholesterol and cholesteryl esters, in addition to diacylglyerols; the TAG content is low.

The polar lipid fraction contains mainly sphingomyelins, whereas phosphatidylethanolamines (PE) and phosphatidylcholines (PC) occur along with cerebrosides (ceramide mono- and diglucoside) only in very small amounts. Seventeen different FA are involved in the structure of the lipids; in the neutral fraction of the shell, palmitic, stearic, and oleic acids predominate, and within the membrane, the proportion of linoleic acid is higher. The polar lipids in the cuticula and the membrane are very similar in their composition and arachidonic and behenic acids are found in relatively high concentrations in the cuticle layer (Table 18.1). The lipid content of the insoluble egg shell layer after decalcification is 2%–4%. Table 18.1 shows a comparison of different egg shell layers in the FA composition. It is remarkable that in the palisade layer 1 and 2 as well as in the mammillary layer, the concentration of behenic acid is between 10% and 17%.

Albumen contains up to 0.02% lipids, about 15% of which are phospholipids. TAG make up the main part, but a small amount of a similar phospholipid such as that in the egg yolk, can also be found. More than half of the lipids appear to be bound to albumen. The lipid content is thought to increase during storage at 12°C and 30°C, but not at 4°C. At higher temperatures, TAG and cholesteryl esters—but not phospholipids—move from the egg yolk into the albumen (Sato et al., 1973).

TABLE 18.1
Fatty Acid Composition of the Decalcificated Eggshell Layers (in %)

Fatty Acid	Cuticle Layer	Palisade Layer I	Palisade Layer II	Mammillary Layer
$C_{10:0}$	1.26	3.15	1.43	1.29
$C_{12:0}$	1.26	1.57	2.38	0.97
$C_{14:0}$	2.52	3.54	4.28	4.21
$C_{16:0}$	17.13	27.56	26.67	33.66
$C_{16:1}$	1.26	3.15	4.28	4.85
$C_{18:0}$	8.56	13.38	10.95	14.56
$C_{18:1(n-6)}$	1.76	1.57	1.43	1.29
$C_{18:1(n-9)}$	4.78	7.08	7.14	8.74
$C_{18:1(t-9)}$	0.25	0.39	0.48	0.48
$C_{18:2}$	3.02	4.33	5.71	5.50
$C_{18:3}$	2.77	3.15	4.28	3.56
$C_{20:0}$	12.34	5.51	9.47	4.21
$C_{20:4}$	2.52	3.54	3.80	2.91
$C_{20:5}$	0.25	0.39	0.47	0.65
$C_{22:0}$	37.27	16.93	11.90	10.03
$C_{24:6}$	3.02	4.72	5.71	3.24
Total saturated fatty acids	80.35	71.65	66.67	68.93
Total unsaturated fatty acids	19.64	28.35	45.34	31.06
Total fatty acids[a]	39.70	25.40	21.00	30.90

Source: Adapted from Miksik, I. et al., *Electrophoresis*, 24, 843, 2003.
[a] In ng/µg.

During storage, the lipid content in the albumen increases. Pankey and Stadelman (1969) reported an increase in lipid content at room temperature from 0.0037% to 0.0054% damp weight within 16 weeks.

Table 18.2 informs about the fatty acid composition in egg albumen and vitelline membrane with different diet supplementations. Palmitic, oleic, linoleic, stearic, and arachidonic acids are the dominant fatty acids. Negishi et al. (1975) found

TABLE 18.2

Composition of Fatty Acids of Egg Albumen and of Vitelline Membrane after Supplementation of Safflower Oil, Docosahexaenoic Acid (DHA), and Conjugated Linoleic Acid (CLA) (in g/100 g)

	Albumen			Vitelline Membrane		
Fatty Acid	Safflower Oil	DHA	CLA	Safflower Oil	DHA	CLA
$C_{16:0}$	18.30	17.05	27.10	23.23	23.14	26.92
$C_{16:1(n-7)}$	0.81	0.62	0.48	1.15	1.44	0.44
$C_{17:0}$	0.27	0.18	0.30	0.30	0.27	0.36
$C_{18:0}$	7.76	6.09	17.13	11.66	8.74	17.23
$C_{18:1(n-9)}$	16.66	17.05	17.82	27.02	26.95	16.67
$C_{18:2(n-6)}$	20.92	15.41	23.66	22.00	19.79	21.36
$C_{18:3(n-6)}$	0.23	0.19	—	0.05	—	—
$C_{18:3(n-3)}$	0.15	—	0.49	0.26	0.42	0.49
$C_{18:2(c9,t11)}$	—	—	4.82	0.09	—	4.84
$C_{18:2(t10,c12)}$	—	—	1.80	—	—	1.76
$C_{18:2(c,c)}$	—	—	0.43	—	—	0.37
$C_{18:2(t,t)}$	—	—	0.42	—	—	—
$C_{20:1(n-9)}$	0.06	0.41	—	—	—	—
$C_{20:2(n-6)}$	0.23	0.58	0.19	0.13	—	—
$C_{20:3(n-6)}$	0.40	0.46	—	0.15	0.16	—
$C_{20:4(n-6)}$	9.92	7.77	1.30	2.36	1.62	1.47
$C_{20:5(n-3)}$	—	0.50	—	—	—	—
$C_{22:4(n-6)}$	1.14	0.73	0.08	0.17	—	—
$C_{22:5(n-6)}$	0.90	1.10	0.27	0.68	0.34	0.08
$C_{22:5(n-3)}$	0.37	0.56	—	—	—	—
$C_{22:6(n-3)}$	2.07	10.56	0.48	1.00	2.95	0.26
SAT	26.34	23.32	44.52	35.18	32.15	44.51
MONO	17.53	18.08	18.31	28.17	28.39	17.11
PUFAs	36.33	37.87	26.47	26.80	25.27	23.66
n-6	33.74	26.25	25.51	25.54	21.91	22.91
n-3	2.59	11.62	0.96	1.26	3.36	0.75
n-6/n-3	13.84	2.38	30.00	25.36	6.58	31.57
SAT/PUFAs	0.73	0.62	1.70	1.34	1.29	1.91
Total CLA	—	—	7.48	0.09	—	6.97

Source: Data from Watkins, B.A. et al., *J. Agric. Food Chem.*, 51(23), 6870, 2003.

TABLE 18.3
Composition of Fatty Acids of Non-Polar Lipids in Egg Albumen (in %)

Fatty Acid	Lipid Classes		
	Wax (Total)	Alcohols of Wax	Free Fatty Acids
Unknown	—	3.8	—
$C_{14:0}$	1.2	3.8	1.6
Unknown	—	1.5	—
Unknown	—	1.2	—
$C_{16:0}$	28.6	19.1	61.0
$C_{16:1}$	3.4	—	4.5
Unknown	—	1.1	—
$C_{18:0}$	3.2	37.0	4.1
$C_{18:1}$	45.6	—	15.1
$C_{18:2}$	18.1	—	13.7
Unknown	Sp.	1.1	—
Unknown	Sp.	2.3	—
$C_{22:0}$	—	27.9	—
$C_{23:0}$	—	1.6	—

Source: Negishi, T. et al., *J. Zootechnol. Sci.* (*Nihon Chikusan Gakkai-ho*), 46(6), 342, 1975. With permission.

Note: Each lipid class was prepared by thin-layer chromatography.

0.011%–0.014% lipids in fresh egg albumen. The neutral lipids mainly consist of wax, free FA, and free sterols (Table 18.3). Generally, wax is an ester of a long-chain primary alcohol and a fatty acid. In the wax of egg albumen, there are great differences in the chain length of the FA and the alcohol. For example, the stearic acid percentage of wax FA is 3.2%, and the percentage of the wax alcohol with the same chain length is 37%. In the polar lipids, sphingomyelins and cerebrosides are the main fractions (Table 18.4).

18.2 EGG YOLK LIPIDS

18.2.1 Introduction

Egg yolk can be separated with centrifugation into a supernatant phase (plasma) and a precipitate (granules). The plasma consists of a water-soluble protein fraction (the livetins), and lipids that occur in bounded form in the low-density lipoproteins (LDL). The LDL contain on a moisture-free basis 86% lipids, 1% minerals, and 12% proteins; the granules 34% lipids, 69% proteins, and 5% minerals. Only with a minimum ionic strength of 0.3 M NaCl (pH 7.0), can granules become solubilized. The release of the apolipoproteins and the polar lipids from granules and LDL is responsible for the emulsifying action of egg yolk.

TABLE 18.4

Composition of Fatty Acids, Sphingosine Bases, and Sugars of Sphingolipids and Cerebrosides in Egg Albumen (in %)

	Lipid Classes	
Fatty Acids	**Sphingomyelins**	**Cerebrosides**
Unknown	—	1.7
$C_{14:0}$	4.6	4.0
Hydroxy-$C_{14:0}$	—	2.0
Unknown	—	2.2
Hydroxy-unknown	—	1.1
$C_{16:0}$	52.7	18.4
$C_{16:1}$	4.2	2.1
$C_{17:0}$	—	0.7
Hydroxy-$C_{17:0}$	—	2.8
$C_{18:0}$	4.0	5.1
$C_{18:1}$	5.8	0.7
Hydroxy-$C_{19:0}$	—	7.8
$C_{20:0}$	3.4	—
Hydroxy-$C_{21:0}$	—	16.2
$C_{22:0}$	21.5	2.8
Hydroxy-$C_{23:0}$	—	15.3
$C_{24:0}$	3.8	15.3
Hydroxy-$C_{25:0}$	—	1.8
Normal acid/hydroxy acid	100/0	53/47
Sphingosine Base		
3-*O*-Methylsphingosine	27.5	—
3-*O*-Methyldihydroxysphingosine	—	22.1
Sphingosine	63.7	—
Dihydroxysphingosine	8.8	30.3
Dehydrophytosphingosine	—	47.6
Sugar		
Galactose	—	17.8
Glucose	—	56.6
Unknown	—	9.3
Unknown	—	11.4
Unknown	—	4.9

Source: Data from Negishi, T. et al., *J. Zootech. Sci. (Nihon Chikusan Gakkai-ho)*, 46(6), 342, 1975.

The fat content of egg yolk is approximately 31.8%–35.5%; that is, about a third of the whole egg yolk. Lipids occur in different particles (granules, LDL-micelles) and are almost completely associated with proteins. An overview is given by Ternes (2001). About 70% of the dry mass consists of lipids. Differences in the lipid composition are not only genetically determined, but also occur due to the animals' age and diet. The contents of long-chain (C_{20} and C_{22}) polyunsaturated FA (PUFA) of n-6 and n-3 were 20%–25% higher in the lipids of egg yolk from younger hens (Nielsen, 1998) (Table 18.5).

At 56%–64%, the main components are TAG, while the phospholipids represent 21%–31% in the lipids (Table 18.6). Mono- and diacylglycerols can be found in concentrations of 1.5%–2.2% in the lipid fraction. The content of cholesterol is about 1.6% of dry matter (Littmann-Nienstedt, 1996).

TABLE 18.5
Fatty Acid Composition of Whole Lipids and Several Fractions in Yolk Lipids (% weight)

Fatty Acids	Whole Lipids (1)	Triacyl-glycerols (1)	Mono- and Diacylglycerols (2)	Diphosphatidyl-glycerols (6)	Phospho-lipids (3)	Cholesteryl Esters (4)	(5)
$C_{14:0}$	0.4	0.5	1.1–2.5	7.1	—	1.8	4.0
$C_{16:0}$	23.4	23.8	10–20	33.8	26.6	—	32.5
$C_{16:1}$	3.4	4.0	—	7.6	1.3	—	6.5
$C_{17:0}$	0.4	0.4	—	—	—	—	—
$C_{18:0}$	10.2	7.6	2.8	16.8	14.8	2.8	10.5
$C_{18:1}$	44.5	49.4	44–61	19.9	25.7	38.9	35.0
$C_{18:2}$	12.3	11.2	11–24	6.6	18.6	32.7	11.5
$C_{18:3}$	0.6	0.6	Traces	—	0.2	5.0	—
$C_{20:0}$	Traces	—	—	—	—	—	—
$C_{20:1}$	0.6	0.6	—	—	—	—	—
$C_{20:2}$	0.1	0.1	—	—	0.2	—	—
$C_{20:3}$	0.2	—	—	—	—	—	—
$C_{20:4}$	2.5	0.4	—	2.2	6.8	—	—
$C_{20:?}$	Traces	3.5	—	—	—	—	—
$C_{22:0}$	Traces	—	—	—	—	—	—
$C_{22:1}$	2.5	Traces	—	—	—	—	—
$C_{22:4}$	Traces	0.4	—	—	—	—	—
$C_{22:5}$	0.3	0.1	—	—	1.3	—	—
$C_{22:6}$	0.6	Traces	—	—	2.2	—	—
$C_{24:0}$	Traces	—	—	—	—	—	—
$C_{24:4}$	0.1	0.1	—	—	—	—	—
$C_{26:1}$	0.5	0.1	—	—	—	—	—

Source: Kukis, A., *Biochim. Biophys. Acta*, 1124, 205,1992; Nakane, S. et al., *Lipids*, 36, 413, 2001.

Note: (1) Kaufmann and Mankel (1967); (2) Holopainen (1972); (3) Burghelle-Mayeur et al. (1989); (4) Swaczyna and Montag (1984); (5) Sezille and Biserte (1964); (6) Noble and Moore (1965).

TABLE 18.6
Composition of Lipids in Egg Yolk in %

Lipids	Holopainen (1972)	Evans et al. (1967)
Hydrocarbons	0.3	—
Triacylglycerols	56.1	65
Mono-, diacylglycerols	6.7	1.5–2.2
Free fatty acids	0.7	—
Cholesteryl esters	0.1	0.3–1
Cholesterol	6.0	5.2
Phospholipids	30.8	28.3

Source: Navarro, J.G., et al., *J. Sci. Food Agric.*, 23, 1287, 1972. With
permission.

Ushakova et al. (1979) examined the hydrocarbon content of egg yolk, finding 15–30 mg alkanes and cycloalkanes/kg egg yolk, 1–3 mg/kg n-alkanes, 15–17 mg/kg squalenes, and about 1 mg monocyclic arenes/kg egg yolk. Table 18.6 shows the FA composition of whole lipids and several fractions.

After freezing/thawing, egg yolk assumes a gel-like behavior with rheological characteristics typical of physical gels, in which the network is held together by linkages weaker than covalent ones. The LDL-micelles are mainly involved in the gelation. Jaekel et al. (2008) developed a freeze-drying process for egg yolk, which is able to reduce the freeze-induced gelation of LDL-micelles.

Pasteurized egg yolk showed no appreciable difference in comparison to native egg yolk. With phospholipase (PLA1), fermented pasteurized egg yolk showed a rise in apparent viscosity after 70%. Apparently, fermentation of pasteurized egg yolk inhibits a decisive stage in heat-induced yolk gelation. The formed lysophospatidylcholin can stabilize LDL-micelles (Jaekel and Ternes, 2009). Therefore, the production of temperature-stable egg yolk for use in convenience products such as "sauce hollandaise," is possible.

18.2.2 Triacylglycerols of Egg Yolk

The TAG composition of egg yolk and the FA composition are influenced by diet. For example, with a basal diet (Gornall and Kuksis, 1971), the C_{52} TAG are the main components at 58.4%, whereas C_{50} and C_{54} TAG are present in nearly equal amounts (19.5% and 17.3%, respectively). C_{48} and C_{46} TAG occur as minor components at 4.8% and 2.0%, respectively (see also Tables 18.11 through 18.13). In their FA composition, the TAG correspond to the hen's depot fat. A comparison of the FA composition of egg yolk lipids and TAG lipids is shown in Table 18.5.

The TAG show higher proportions of palmitoleic and oleic acids. The FA composition of TAG in the high-density lipoprotein (HDL) fraction in the granules and LDL fraction of the yolk plasma show similar composition (Tables 18.12 and 18.13).

In the TAG fraction, the major species were recognized in the monoenes as *sn*-1: $C_{16:0}$, *sn*-2: $C_{18:1}$, *sn*-3: $C_{16:0}$ and *sn*-1 $C_{16:0}$, *sn*-2: $C_{18:1}$, *sn*-3: $C_{18:0}$ and their racemates.

The dienes consist almost entirely of sn-1: $C_{16:0}$, sn-2: $C_{18:1}$, sn-3: $C_{18:1}$ and their racemate. The trienoic TAG were mainly sn-1: $C_{16:0}$, sn-2: $C_{18:2}$, sn-3: $C_{18:1}$ and their racemate as well as trioleate. In the polyenoic fraction, sn-1: $C_{18:1}$, sn-2: $C_{18:2}$, sn-3: $C_{18:1}$ are mostly present (Gornall and Kuksis, 1971).

18.2.3 MONO- AND DIACYLGLYCEROLS OF EGG YOLK LIPIDS

Mono- and diacylglycerols are present in egg yolk lipids at 1.5%–2.2% (Evans et al., 1967). Holopainen (1972) cited contents of 6.7% but this seems to be too high. Mono- and diacylglycerols primarily occur in the HDL fraction of the granules (5.8%–8.5% in the lipids of HDL fraction). In the lipids of the LDL fraction from plasma, only low concentrations of 0.4%–0.7% were found (Tsutsui and Obara, 1979); but more than three-fourths of the lipids can be found in the LDL fraction, giving a total concentration of 2% in the lipids.

The FA composition takes an intermediate position between the TAG and the polar lipids, but diacylglycerols correspond somewhat more to the TAG, whereas the monoacylglycerols correspond to the polar lipids (Acker and Ternes, 1994).

Noble and Moore (1965) have reported on the FA composition of diphosphatidyl-glycerols (DPG) in yolk (Table 18.5). However, there is no follow-up report on the presence of DPG in egg yolk. In mammalian tissues, the DPG composition differs greatly as linoleic acid is dominant there.

18.2.4 PHOSPHOLIPIDS

The main component of the phospholipids is the PC fraction at 60%–73%, followed by the PE fraction at approximately 15%–26%. The proportion of sphingomyelins is 2.5%–4.8%. Phosphatidylinositols and other lipids such as glycolipids occur only in trace quantities (Table 18.7).

In careful examinations, Rhodes and Lea (1957) obtained the composition of phospholipids. Here, for the first time, lysophospholipids were shown to be the natural components of egg lipids and not artifacts produced during sample preparation, as the authors explicitly point out. The PC fraction contains 0.9% plasmalogens; the cephalin fraction contains 0.2% phosphatidylserines. Table 18.7 gives an overview of the composition of phospholipids in egg yolk. There is high agreement between the results of all individual authors. Only the results of Holopainen (1972) for sphingomyelins appear to be too high.

A comparison of the FA spectrum of TAG from the whole egg yolk with the phospholipid fraction shows that the concentration of stearic acid is reduced, whereas the proportion of more highly unsaturated FA in the phospholipid fraction increases markedly. Oleic acid (20%–27% in the phospholipid fraction) is present only half as frequently as in the TAG. The concentration of palmitic acid (approximately 26.6%) is almost as high as that of oleic acid, with an average of 25.7%. The content of stearic and linoleic acid in the phospholipid fraction (18.6%) is notably higher than that in the TAG fraction. While the palmitoleic acid in the phospholipid fraction occurs at lower levels (average = 1.3%), docosahexaenoic acid ($C_{22:6}$) (DHA) occurs in higher concentrations at an average of 2.2% (Table 18.5).

TABLE 18.7

Composition of Phospholipids in Egg Yolk in %

Phospholipids (PL)

Phosphatidylcholines (PC)	66.0	70
Phosphatidylethanolamines (PE)	18.4	24
Lysophophatidylcholines (LPC)	4.1	—
Lysophosphatidylethanolamines	—	—
Sphingomyelins	4.8	4
Plasmalogen	—	0.9
Phosphatidylinositol (PI)	0.6	1
Phosphatidylserines (PS)[a]	—	<1
Phosphatidylglycerol	—	5 µg in 120 mg PL
Lysophosphatidylserines (LPS)	—	5 µg in 120 mg PL
Lysophosphatidic acid (LPA)	—	0.02%
Lysoplasmanic acid	—	0.004%
Cerebroside A	—	1.4%
Cerebroside B	—	1.6%

Note: Several authors: Kuksis (1992) and Nakane et al. (2001).

[a] Probably yolk does not contain PI, because of coelutions with PS.

In contrast to the TAG, the proportion of unsaturated FA is higher in the phosphatides. Numerous feeding tests have shown that only the quantity and distribution of unsaturated FA within the phospholipid fraction depend on the quantity of monoenoic FA in chicken feed.

There are significant differences in the FA compositions of PC and PE (Table 18.8). In the PE fraction, the concentrations of stearic acid ($C_{18:0}$) and unsaturated C_{20} and C_{22} FA are considerably higher. In contrast, palmitic acid, oleic acid, and linoleic acid are present with increased frequency in the PC fraction.

This unequal distribution of FA in PE and PC is also present in the single lipoprotein fractions (HDL and LDL) (see Tables 18.12, 18.13, and 18.22).

PE are responsible for the antioxidative properties of commercial preparations known as "egg lecithin," which consist of whole egg phosphatides. Due to the NH_2 group, they can participate in Maillard reactions, which occur not only during heating (boiling), baking, and frying, but also during storage of dried egg products, and that lead to brown-stained, mostly fluorescent compounds (Lea, 1957).

PE are very sensitive to oxygen and autoxidize faster than PC, a process accompanied by an intense brown discoloration. Studies to reduce the oxidation of the PUFA with dietary supplementation of rosemary extract and tocopherols showed a higher oxidative stability of n-3 FA (Cherian et al., 1996; Galobart et al., 2001).

The egg lipid sample that initially seems to have a simple structure, appears more complex when modern chromatographic separation methods are used to determine sphingolipids and glycolipids, that occur in small amounts (Table 18.5). In this way, Fujino et al. (1971a) shed some light on the jumble of complex lipid

TABLE 18.8

Fatty Acid Composition of Phosphatidylethanolamine (PE) Fraction, Phosphatidylcholine (PC) Fraction, Phosphatidylserine Fraction, and Diphosphatidylglycerols in Yolk (% Weight)

Fatty Acids	PE Fraction[a]	PC Fraction[a]	PS Fraction[b]
$C_{14:0}$	—	—	5.2
$C_{16:0}$	21.9	37.1	29.6
$C_{16:1}$	2.2	3.0	4.8
$C_{18:0}$	30.7	7.2	24.1
$C_{18:1}$	20.5	28.0	14.0
$C_{18:2}$	7.5	14.3	6.4
$C_{18:3}$	0.4	0.5	—
$C_{20:4\,(n-6)}$	9.2	3.1	8.4
$C_{22:5\,(n-6)}$	1.3	0.5	—
$C_{22:6}$	5.3	3.0	—

[a] Navarro et al. (1972).
[b] Noble and Moore (1965).

groups. They detected sphingomyelins with 11 different FA (mainly lignoceric, stearic, and behenic acids), cerebrosides with 14 different FA—mainly hydroxy FA—as well as glucose and galactose, and ceramides with 16 different FA (mainly lignoceric acid). An essential N-containing segment in this case is sphingosine, in addition to small amounts of dihydrosphingosine. Apart from this, very low concentrations of gangliosides (glycosphingolipids, containing sialic acid) have been found (Keenan and Berridge, 1973). Tests reveal that mainly monosialosyl-N-tetraglycosylceramid with 14 different FA $C_{16:0}$, $C_{18:0}$, $C_{18:1}$, and $C_{22:0}$ are available. Glucose, galactose, hexosamin, and sialic acid are present in proportions of 1.0: 1.25: 1.95, and 1.19, respectively. In contrast to ceramids as tested by Fujino and Momma (1971), only small amounts of $C_{23:0}$, $C_{24:0}$, $C_{24:1}$ occur in the gangliosides, but no hydroxy FA. The carbohydrates glucose, galactose, and hexosamin are present, as is sialic acid. Even these small amounts contribute to the emulsifying characteristics of egg yolk. As early as 1957, Carter et al. (1957) noted the occurrence of plasmalogens in the egg (as alkali-stable ether lipids), and Lea (1957) found 0.9% plasmalogens in the PC fraction. Do and Ramachandran (1980) characterized those connections in more detail, identifying the 1-O-alkyl-sn-glycero-3-phosphoethanolamine as the most important ether lipid, apart from a very small amount of 1-O-alkyl-sn-glycero-3-phosphocholine.

Moschidis et al. (1984) found small amounts of phosphonolipids in the egg yolk. The comparison of the FA composition of minor components shows that in sphingomyelins, the amount of palmitic acid is twice that in the egg yolk lipids, whereas there is almost 40% less oleic acid. Palmitoylsphingosin is with 85% the important sphingomyelin. Linoleic acid at 30% is found at concentrations about three times higher in lysophosphatidylcholines than in egg yolk lipids, and twice as much as in the PC fraction. At 28%, docosanoic acid ($C_{20:0}$=behenic acid) is present in considerable amounts in gangliosides, as opposed to lower concentrations of oleic and linoleic acids in comparison with the FA spectrum of egg yolk lipids. Momma et al. (1972) also found a high proportion of approximately 67% hydroxy FA in the cerebrosides. Many of the hydroxy FA have a chiral center. In any case, Momma et al. (1972) determined a relatively high concentration (15.7%) of tetracosanoic acid ($C_{24:0}$=lignoceric acid), while smaller concentrations of the main FA of the egg yolk lipids were found in the cerebrosides. Table 18.9 shows an overview of the FA spectrum of the minor components.

18.2.5 Fatty Acid Position in the Glycerol Molecule

In the TAG molecule, a chiral center develops when both primary hydroxy groups of the glycerol are esterified with two different FA, so that—owing to the position of the FA—a distinction can be made between *sn*-1, *sn*-2, and *sn*-3. Interestingly, the *sn*-1- and *sn*-2-preferential location of palmitic-, stearic-, oleic-, linoleic, and linolenic acids in the polar lipids (as well as in PC and PE) is the same. The 1-position in the TAG of egg yolk is primarily occupied by palmitic acid, the 2-position by oleic and linoleic acids, while there are oleic acid and saturated FA at the 3-position (Christie and Moore, 1970). In the PC fraction, the 2-position is occupied up to 87% by unsaturated FA (Privett et al., 1962; Gornall and Kuksis, 1971) (Table 18.10).

$C_{16:0}$–$C_{18:1}$ (38%–47%) belongs to the main molecular species composition of PC in egg yolk. $C_{18:0}$–$C_{18:2}$ (8%–11%), $C_{18:1}$–$C_{19:0}$ (9%–19%) are also significant species in PC (Kuksis and Marai, 1967).

The major species of PE in monoenes are $C_{16:0}$–$C_{18:1}$ and $C_{18:0}$–$C_{18:1}$; in dienes $C_{16:0}$–$C_{18:2}$ and $C_{18:0}$–$C_{18:2}$ are the main molecular species. In PE tetraenes $C_{18:0}$–$C_{20:4}$, and in hexaenes $C_{16:0}$–$C_{22:6}$ and $C_{18:0}$–$C_{22:6}$ are important. Species with DHA occur at 6.3% in $C_{16:0}$–$C_{22:6}$ and at 4.29% in $C_{18:0}$–$C_{22:6}$. Arachidonic acid $C_{20:4}$ is also present mainly in the PE fraction, for example, $C_{18:0}$–$C_{20:4}$ with 28.46% located in PE (Holub and Kuksis, 1969).

18.2.6 The Lipid Component of Lipoproteins

Lipid components in the lipoproteins of the egg yolk differ markedly (Table 18.11). Whereas in the HDL fraction of the granules, the lipids constitute 22%–27% of the dry matter, the LDL fraction constitutes 85% of the plasma. Phospholipids can be found in higher concentrations (approximately 50%) in the HDL fraction than in the LDL fraction of the plasma (approximately 20%). Cholesterol and cholesteryl esters are present, at approximately 5% in the HDL fraction and 4% in the LDL fraction of plasma (Gornall and Kuksis, 1971).

By ultracentrifugation ($d<1.01$ and $d<1.006$), the very low-density lipoproteins (VLDL) fraction (Evans et al., 1977; Burghelle-Mayeur et al., 1989) can be separated from the LDL fraction. The VLDL fraction contains approximately 92.1% ± 2.7% of

TABLE 18.9

Fatty Acid Composition of Sphingomyelins, Lysophosphatidyl Cholines, Gangliosides, Cerebrosides and Ceramides in Yolk (% Weight)

Fatty Acids	Sphingomyelins (1) Fed with			Lysophosphatidyl Cholines (1) Fed with		Gangliosides (2)	Cerebrosides (3)	Hydroxy-Acids	(4)	Ceramides	
	(5)	Basal Diet	Fish Oil 1.64%	Basal Diet	Fish Oil 1.64%					(6)a Non-Hydroxy	(6)a α-OH
$C_{12:0}$	—	—	—	—	—	0.3	—	—	—	—	—
$C_{14:0}$	5.8	—	—	—	—	1.4	—	—	0.08	1.0	0.1
$C_{15:0}$	—	—	—	—	—	0.5	—	—	0.05	—	—
$C_{16:0}$	37.9	52.0	23.7	20.6	24.4	26.6	9.6	—	3.54	84.3	0.5
$C_{16:1}$	5.9	—	3.9	—	6.2	2.3	0.4	—	0.43	—	—
$C_{17:0}$	—	—	—	—	—	5.7	—	—	0.05	—	—
$C_{18:0}$	16.0	—	11.1	6.5	23.6	9.2	2.5	3.5	2.01	5.2	0.2
$C_{18:1}$	21.5	32.1	21.8	29.6	30.6	22.9	1.8	—	2.55	—	—
$C_{18:2}$	8.2	15.2	3.1	30.3	4.7	1.8	—	—	0.59	—	—
$C_{18:3}$	—	—	1.6	3.2	1.2	—	—	—	—	—	—

	1	2	3	4	5	6	7	8	9	10	11
$C_{19:0}$	—	—	—	—	—	—	0.3	—	—	0.2	—
$C_{20:0}$	—	—	34.8	—	—	0.2	0.8	0.6	1.18	1.0	98.5
$C_{20:2}$	—	—	—	—	1.2	—	—	—	—	—	—
$C_{20:3}$	—	—	—	9.8	3.9	—	—	—	—	—	—
$C_{20:4}$	—	—	—	—	—	—	0.4	—	0.77	—	—
$C_{21:0}$	1.5	—	—	—	—	—	—	—	—	—	—
$C_{22:0}$	—	—	—	—	—	28.3	7.2	19.3	22.73	1.5	0.6
$C_{22:1}$	—	—	—	—	1.2	—	—	—	—	0.5	—
$C_{22:4}$	—	—	—	—	—	—	—	—	—	—	—
$C_{23:0}$	—	—	—	—	—	0.4	3.7	12.2	15.05	0.6	—
$C_{24:0}$	3.5	—	—	—	—	0.4	15.7	22.0	35.66	0.7	0.1
$C_{24:1}$	—	—	—	—	—	0.3	—	—	13.46	4.5	—
$C_{26:0}$	—	—	—	—	3.1	—	—	—	1.37	—	—
Others	—	—	—	—	—	—	—	—	—	0.3	—

Note: (1) Navarro et al. (1972); (2) Keenan and Berridge (1973); (3) Momma et al. (1972); (4) Fujino and Momma (1971); (5) Noble and Moore (1965); (6) Do and Ramachandran (1980).

[a] mol%.

TABLE 18.10
Stereospecific Distribution of Fatty Acids in Glycerols of Yolk

Fatty Acids	Phospholipids (1) sn-1	Phospholipids (1) sn-2	Triacylglycerols (1) sn-1	Triacylglycerols (1) sn-2	Triacylglycerols (1) sn-3	Triacylglycerols (2) sn-1	Triacylglycerols (2) sn-2	Triacylglycerols (2) sn-3
$C_{14:0}$	—	—	—	—	—	1	1	1
$C_{16:0}$	71.8	2.0	71.5	5.1	6.5	62	5	13
$C_{16:1}$	—	—	6.0	2.6	5.7	7	3	7
$C_{18:0}$	18.1	0.5	4.1	1.4	8.7	6	3	8
$C_{18:1}$	7.3	44.2	14.7	56.0	74.6	21	57	64
$C_{18:2}$	0.6	27.8	2.3	33.5	3.5	3	32	6
$C_{18:3}$	—	6.9	0.8	1.2	0.6	—	—	—
$C_{20:4}$	—	1.5	—	—	—	—	—	—
$C_{22:6}$	—	13.8	—	—	—	—	—	—

Note: (1) Privett et al. (1962); (2) Gornall and Kuksis (1971).

TABLE 18.11
Composition of HDL Fraction of Granules and LDL Fraction of Egg Yolk Plasma

	Lipoproteins in % HDL Fraction	Lipoproteins in % LDL Fraction
Proteins	72.2	14.4
Lipids	27.3	85.6
Neutral lipids	*ca.* 14.0	64.1
Triacylglycerols	91.0	80.0
Mono- and diacylglycerols	8.5	0.7
Sterols	9.2[a]	5.4[a]
Sterylesters	1.7	1.6
Hydrocarbons	0.6	0.6
Phospolipids	*ca.* 13.3	21.5
Phosphatidylcholines	85.7	78
Phosphatidylethanolamines	9	15.8
Lysophosphatidylcholines	5.3	6.0

Source: Data from Tsutsui, T. and Obara, T., *Nippon Shokuhin Kogyo Gakkaishi*, 26, 81, 1979.

[a] The concentration is lower in the paper of Gornall and Kuksis (1971).

TABLE 18.12

Fatty Acid Composition of Triacylglycerols, Phosphatidylcholines, and Phosphatidylethanolamines in the HDL Fraction of Egg Yolk Granules (% Weight)

Fatty Acids	Fraction d > 1.2		Triacylglycerols		Phosphatidylcholines		Phosphatidyl-ethanolamines	
	(1)	HDL	(2)	(3)	(2)	(3)	(2)	(3)
$C_{14:0}$	—	—	0.3–0.5	0.4	0.3	—	0.2	—
$C_{14:1}$	—	—	—	—	—	—	—	—
$C_{16:0}$	27.0	30.0	24.0–25.5	24.2	31.0–35.0	31.6	12.6–18.5	21.8
$C_{16:1}$	3.2	3.7	4.4–5.9	4.5	1.3–2.7	2.2	0.4–1.4	0.8
$C_{18:0}$	14.5	12.8	5.9–7.3	7.8	13.5–16.4	16.0	28.5–32.3	41.2
$C_{18:1}$	40.1	47.6	45.0–50.0	51.5	26.0–44.0	29.6	20.0–30.9	26.0
$C_{18:2}$	15.2	6.0	5.0–10.0	11.6	6.4–20.1	17.9	5.8–17.3	8.6
$C_{20:4}$	—	—	—	—	3.4–3.9	2.8	12.2	0.4
$C_{18:3}$	—	—	—	—	—	—	—	—
$C_{20:0}$	—	—	—	—	—	—	—	0.7

Note: (1) Evans et al. (1977); (2) Pankey and Stadelman (1969); (3) Tsutsui and Obara (1979).

the lipoproteins of the plasma and basically corresponds to the LDL fraction reported by other authors (cp. Table 18.13). The VLDL fraction contains 87% lipids, of which 68% are TAG, 26% phospholipids, and 6% cholesterol and cholesteryl esters (Evans et al., 1977). Burghelle-Mayeur et al. (1989) give similar values for cholesterol, but approximately 10% of the total cholesterol occurs as cholesteryl esters.

There is no significant difference between the FA distribution of TAG in the HDL and LDL fractions (Tsutsui and Obara 1979; cp. Tables 18.12 and 18.13). When comparing the FA spectrum of egg yolk lipids with the FA spectra of TAG in the HDL and LDL fractions, it can be established that oleic acid is present in 10% higher concentration. In the TAG fraction, the PC and PE fractions differ significantly. Palmitic acid in the PC fraction is present in higher concentrations, whereas TAG concentrations in the PE fraction are correspondingly lower. On the other hand, the stearic acid in the PE fraction is found in higher concentrations. Palmitoleic acid mainly occurs in the TAG fraction and, in lesser quantities, in the PC fraction of HDL and LDL lipids. In the PE fraction of both proteins, the palmitoleic acid content is low. Arachidonic acid is to be found almost exclusively in the phospholipid fraction. In 1969, Pankey and Stadelman found an extremely high arachidonic acid content in the PE fraction, but Tsutsui and Obara (1979) could not confirm this.

18.2.7 Influence of Diet on the Lipid Composition

Dietary fat supplements are frequently used to increase the energy content in the hens' ration. Increased dietary fat does not lead to an increase in the fat content in

TABLE 18.13

Fatty Acid Composition of VLDL- and LDL Fraction, Triacylglycerols, Phosphatidylcholines, and Phosphatidylethanolamines in the LDL Fraction of Egg Yolk Plasma (% Weight)

Fatty Acids	Fraction VLDL/LDL (1)	(1)	Triacylglycerols (2)	(3)	(4)	Phosphatidyl-cholines (2)	(3)	Phosphatidyl-ethanolamines (2)	(3)
$C_{14:0}$	—	—	0.5	0.4	—	0.3	—	—	—
$C_{14:1}$	—	—	0.4	—	—	—	—	—	—
$C_{16:0}$	24.6	25.2	23.0–24.7	24.0	22.3	3.2	32.4	5.0–20.1	21.6
$C_{16:1}$	3.4	3.5	3.8–6.2	4.6	3.8	1.3–2.1	2.2	0.3–1.1	0.9
$C_{18:0}$	10.3	11.4	5.2	7.7	6.6	13.1–18.8	16.4	25.7–32.0	40.5
$C_{18:1}$	48.1	46.6	44.4–57.9	51.6	50.4	24.0–42.3	30.3	24.5–30.0	23.9
$C_{18:2}$	13.1	13.4	5.4–10.0	11.7	14.6	7.4–12.3	15.2	7.9–11.0	9.5
$C_{18:3}$	—	—	—	—	0.6	—	—	—	—
$C_{20:0}$	—	—	—	—	—	—	—	—	1.1
$C_{20:4}$	—	—	—	—	0.2	2.2–2.8	2.9	12.7–15.9	2.4

Note: (1) Evans et al. (1977); (2) Pankey and Stadelman (1969); (3) Tsutsui and Obara (1979); (4) Burghelle-Mayeur et al. (1989).

the egg yolk, but the FA and TAG spectra in egg yolk lipids do change in relation to the fat composition of the feed. There is nearly always a connection between diet and the egg yolk FA spectrum, but the amounts of FA do not completely correspond.

The digested fat is absorbed as FA and monoacylglycerols through the intestinal epithelium and esterified back to TAG. The TAG produced in the hens' bloodstreams are associated with lipoproteins. The lipoprotein and TAG synthesis occurs *de novo* in the liver and is, among other things, necessary for the formation of egg yolk lipid in the ovaries. Due to this synthesis, the FA distribution in the 1-, 2-, and 3-positions in the TAG are similar in the liver and egg yolk (Hirata et al., 1987). Hens, like many other vertebrates, are not able to synthesize PUFA, especially linoleic acid, from the "non-fats" in the feed (such as carbohydrates and proteins). This explains the conspicuously high proportion of saturated and monoenoic FA (such as $C_{16:0}$, $C_{18:0}$, $C_{18:1}$) in the egg yolk fat after a low-fat diet (Tables 18.14 and 18.15).

When feed contains higher quantities of PUFA, as in fish or algae oil supplements, the content of linoleic acid and other PUFA increases significantly. Even the linoleic acid content of the feed is involved in the amount of unsaturated FA in the egg yolk. Supplements of oils containing linoleic acid, such as safflower, flax, soybean, sunflower, and corn (maize) oil, lead to high enrichment of this acid in the egg lipids (Biedermann et al., 1971).

According to Sim et al. (1972), there is a marked increase of oleic acid content in the egg yolk after supplementing the feed with sunflower oil. The increase of oleic acid is higher than in other feed mixtures with rape oil, soybean oil, or beef tallow.

TABLE 18.14
Distribution of Fatty Acids in Dietary Fat and Yolk Lipids Depending on Dietary Fat (% Weight)

Fatty Acids	Fish Oil (1) a	b (3%)	Soybean Oil a	b (12%)	Coconut Oil (2) a	b (10%)	Sunflower Oil (2) a	b (10%)
$C_{10:0}$	—	—	—	—	7.9	Traces	—	—
$C_{12:0}$	—	—	—	—	40.0	1.0	1.8	Traces
$C_{14:0}$	—	—	0.1	0.2	18.5	7.5	1.8	0.3
$C_{16:0}$	23.0	26.8	11.4	24.0	12.5	25.5	7.5	20.8
$C_{16:1}$	7.9	4.9	Traces	1.6	—	4.6	—	1.8
$C_{17:0}$	—	—	0.1	0.4	—	—	—	—
$C_{18:0}$	7.0	18.2	5.0	8.6	3.6	8.1	4.3	9.2
$C_{18:1}$	15.8	31.7	24.5	38.1	10.9	39.3	26.7	35.9
$C_{18:2}$	4.9	11.3	50.3	33.1	4.9	9.0	58.6	27.4
$C_{18:3}$	4.4	0.4	7.5	1.4	0.3	0.2	0.9	0.2
$C_{20:4}$	4.9	1.3	—	1.2	—	—	—	—
$C_{20:5}$	9.5	0.4	—	—	—	—	—	—
$C_{20:1}$	—	—	0.6	0.2	—	—	—	—
$C_{21:1}$	—	—	—	—	—	—	—	—
$C_{22:4}$	6.2	0.2	—	—	—	—	—	—
$C_{22:5}$	7.5	0.5	—	—	—	—	—	—
$C_{22:6}$	8.0	4.1	—	0.5	—	—	—	—
Others	—	0.2	0.5	0.8	—	2.0	—	4.3

Note: (1) Navarro et al. (1972); (2) Biedermann et al. (1971).
[a] Distribution of fatty acids in dietary fat.
[b] Distribution of fatty acids in yolk lipids (percentage of dietary fat in the diet).

It is remarkable that with other diets low in linoleic acid, a reduction in linoleic acid content in the egg yolk does not conform to the decrease in linoleic acid in the feed. The living organism can produce linoleic acid from certain precursors (e.g., *cis*-2-octenoic acid), where the synthesis required for chain augmentation must be repeated five times for the production of a double bond, although linoleic acid and linolenic acid are precursors for the production of arachidonic acid ($C_{20:4}$), docosapentaenoic acid ($C_{22:5}$), eicosapentaenoic acid ($C_{20:5}$) (EPA), and DHA. According to Reiser (1950), hens can produce eicosatrienoic acid ($C_{20:3}$).

In contrast to other investigators, Chen et al. (1965) were unable to detect an increase of linoleic acid with a dietary supplement of 10% flax oil or cottonseed oil. The discrepancy can be explained by the fact that at 44%, the content of linoleic acid in the animals' diet was already high, so that even in the eggs of control animals, the concentration in the yolk was high at 15%. Obviously, there is an upper limit for the interchange of FA with linoleic acid. Even Kaufmann and Mankel (1967) thought that a selective influence might play a role during the biosynthesis of egg lipids in contrast to the depot fat of laying hens.

TABLE 18.15
Distribution of Fatty Acids in Triacylglycerols of Dietary Fat and Egg Yolk

Fatty Acids	Soybean Oil (1)		Coconut Oil (1)		Lard (1)		Tallow (1)		Safflower Oil (2)	
	a	b (10%)	a	b (10%)	a	b (10%)	a	b (10%)	a	b (30%)
$C_{8:0}$	—	—	10.3	—	—	—	—	—	—	—
$C_{10:0}$	—	—	7.4	—	—	—	—	—	—	—
$C_{12:0}$	—	0.3	49.1	2.2	—	Traces	—	0.4	—	—
$C_{14:0}$	Traces	1.8	17.4	10.0	1.8	0.7	3.7	1.2	—	0.3
$C_{14:1}$	—	0.2	—	3.5	0.2	Traces	0.1	0.2	—	0.1
$C_{16:0}$	11.5	22.1	7.7	27.9	25.9	25.2	25.7	27.1	7.0	19.9
$C_{16:1}$	—	2.6	—	7.7	3.6	4.7	3.1	4.9	—	1.4
$C_{17:0}$	0.1	0.2	—	0.1	0.4	0.2	1.2	0.5	—	—
$C_{17:1}$	—	0.2	—	—	0.3	0.3	0.6	0.6	—	—
$C_{18:0}$	3.7	3.7	22.1	3.7	12.1	3.8	18.0	4.2	3.0	5.1
$C_{18:1}$	22.2	37.4	4.9	37.8	43.4	55.3	40.7	54.1	13.0	23.6
$C_{18:2}$	53.6	28.8	1.1	6.4	10.5	9.0	3.6	5.6	76.0	48.7
$C_{18:3}$	8.6	2.1	—	0.4	0.7	0.1	0.5	0.2	1.0	0.7
$C_{20:1}$	0.3	0.2	—	0.3	1.0	0.3	0.6	0.3	—	—
$C_{20:2}$	—	0.2	—	Traces	—	0.1	—	Traces	—	—
$C_{20:4}$	—	0.3	—	0.1	—	0.3	—	Traces	—	—

Note: (1) Hirata et al. (1987); (2) Naber and Biggert (1989). Traces, less than 0.1%.
[a] Distribution of fatty acids in feed fat.
[b] Distribution of fatty acids in yolk (percentage of dietary fat in the diet).

While dietary saturated FA have only a slight effect on the FA in the samples, trans-FA can have a considerable influence. With a diet containing 5.5 g trans-FA daily in the form of hardened fats, after 16 days, the trans-FA increased up to 10% at the expense of some proportion of oleic acid in the neutral lipids. Fewer trans-FA are transferred to the phospholipid fraction than into the TAG fraction (Kaufmann and Mankel, 1967). From rapeseed oil (which contains erucic acid), erucic acid is transferred to the egg (Biedermann et al., 1971). In addition to the increase in $C_{18:2}$ acid in egg lipids, the physical characteristics change, thus increasing the tendency toward oxidation.

Naber and Biggert (1989) showed in their feeding tests that as the fat content in the feed increases, there is a reduction in lipogenesis in the liver; and the fat is stored in the liver and egg yolk with the same FA as in the dietary fat.

Metabolic changes in PUFA such as linoleic acid to less highly saturated acids such as oleic and stearic acids affect the amount of linoleic acid. Dietary fats high in oleic acid should reduce the transformation of stearic and linoleic acids to other FA. The reduction of linoleic acid to palmitic and palmitoleic acids can also be assumed.

Huang et al. (1990) show further connections between dietary and egg yolk fat on closer examination of n-3 FA from menhaden oil. It was demonstrated that the amounts of n-3 FA, EPA, and DHA in egg yolk fat are not proportional to those in the feed, although EPA is metabolized in a high proportion. The quantities of docosapentaenoic acid and DHA in the egg yolk react in inverse proportions to the content in menhaden oil. Differences in the absorption rate of both acids in the intestine are assumed to be responsible.

Linoleic acid is thought to be not only a precursor of arachidonic acid, but also of EPA, DHA, docosapentaenoic acid, and linolenic acid.

Navarro et al. (1972) found that an increase of fish oil in the feed leads to an increase in n-3 FA as compared with the content in the feed, whereas the quantity of n-6 FA was reduced. This corresponds to earlier reports that n-3 FA are preferentially transported into the egg in lipid metabolism.

Caston and Leeson (1990) reported that the feeding of flaxseed increases n-3 FA and slightly increases n-6 FA as well. The more flaxseed oil is in the feed, the higher the amounts of linoleic and linolenic acids, as well as EPA and DHA, present in the egg yolk lipids.

Hirata et al. (1987) conducted extensive tests comparing the FA composition of egg yolk after the feed supplements of vegetable oil and animal fat. The FA sample of the dietary fat supplement is reflected considerably more clearly in TAG than in polar lipids. Table 18.15 reveals that compared to the supplement of animal fats, the proportion of linoleic acid in the TAG of the egg yolk is at its highest after a 10% soy oil supplementation, and the proportion of oleic acid at its lowest; this also applies to a coconut oil supplementation. Furthermore, after coconut oil supplementation, the TAG contain substantially more myristic, myristoleic, and palmitoleic acids. Since myristoleic acid ($C_{14:1}$) and palmitoleic acid ($C_{16:1}$) are not present in coconut oil, but do occur in egg yolk lipids, the production of a double bond in myristic and palmitic acids by fat metabolism can be assumed. Although coconut oil consists of 50% of lauric acid ($C_{12:0}$), only small amounts of this acid can be found in the egg yolk. It has been assumed that hens can use only very small amounts of FA with shorter chains for the egg yolk lipid production. Shorter FA chains in the coconut oil do not contribute to the egg yolk lipid production. The lipid spectrum in the egg yolk lipids differs significantly after supplementation with animal and vegetable fats. The addition of beef tallow or lard supplements results in only minor differences. Numerous reports have confirmed the data of Hirata et al. (1987).

The isomers of conjugated linoleic acids (CLA) have anticancer activity, immune-enhancing-weight reductions, and antiatherogenic properties. CLA are found in food produced from ruminant animals. The development of CLA-enriched eggs showed a reduction in monoenoic FA and non-CLA PUFA after CLA feeding, and an increase in EPA and DHA. The incorporation rates of different CLA isomers into the whole lipids of egg yolk and TAG, PC- and PE lipid classes were different. *cis*-9, *trans*-11 and *cis*-10, *trans*-12 CLA were deposited more in TAG, but *cis*-11, *trans*-13 CLA was less. There are large differences in the concentrations of *cis*-8, *trans*-10 CLA in PC and PE (Table 18.16).

TABLE 18.16

Influence of Dietary Conjugated Linoleic Acid (CLA; 2.5%) on the Fatty Acid Composition of Total Lipids, Triacylglycerols, Phosphatidylcholine (PC), and Phosphatidylethanolamine (PE)

Fatty Acids	Total Lipids	Triacylglycerols	PC	PE
$C_{16:0}$	24.85	26.2	27.17	16.42
$C_{16:1}$ (n-9)	0.63	1.16	0.22	—
$C_{16:2}$ (n-9,12)	0.23	—	—	0.69
$C_{16:2}$ (1)[a]	0.12	0.28	—	—
$C_{16:2}$ (2)[a]	0.13	0.26	—	—
$C_{16:2}$ (3)[a]	0.06	0.08	—	—
$C_{17:0}$	0.29	0.28	0.25	0.45
$C_{18:2}$ (n-9,12)	19.25	22.45	18.14	15.41
$C_{18:1}$ (n-9)	25.21	33.88	26.45	20.86
$C_{18:0}$	16.48	10.17	17.60	28.59
CLA (cis-9, trans-11)	2.74	3.17	2.19	2.16
CLA (trans-10, cis-12)	2.54	3.50	1.76	1.50
CLA (cis-8, trans-10)	0.60	0.57	0.36	0.74
CLA (cis-11, trans-13)	0.94	0.80	0.98	1.32
$C_{20:4}$ (n-5,8,11,13)	3.91	0.18	2.78	11.34

Source: Data from Du, M. et al., *Poult. Sci.*, 78, 1639, 1999.

[a] Corresponding to three hexadecadienoic isomer peaks identified by gas chromatograph-mass analysis.

18.2.8 POSITIONS OF FATTY ACIDS IN THE GLYCEROL MOLECULE

Hirata et al. (1987) showed modifications of the TAG spectrum, depending on the feeding fat (Table 18.17). Coconut oil with a high proportion of C_{32}–C_{42} TAG, which develops due to the high concentration of $C_{8:0}$, $C_{10:0}$, and $C_{12:0}$ FA, displaces the TAG composition in favor of low-molecular-weight TAG. A large part of these low-molecular-weight FA does not reach the egg yolk because they are metabolized in the liver. The feeding of soybean oil, with its high proportion of C_{52} and C_{54} TAG, leads to increased C_{52} and C_{54} TAG in the egg yolk lipids.

At lower concentrations, linoleic acid in the feed is preferentially transported into the PC fraction. At higher concentrations, linoleic acid is also transported in higher amounts into the TAG and PC fractions (Tables 18.18 through 18.21).

Hirata et al. (1987) analyzed the position of FA in the TAG depending on dietary fats (Tables 18.19 and 18.20). It is clear that preference is also given to the sn-2 position for linoleic acid. Cossignani et al. (1994; Table 18.21) show the stereospecific distribution of FA in the PC and PE fractions, in reference to the FA on sn-1 and sn-2 positions of lysoPC and lysoPE (Table 18.22). Amate et al. (1999) observed a great difference in the FA composition of lysoPC and lysoPE in comparison to PC and PE fractions.

TABLE 18.17

Composition of Triacylglycerols in Yolk after Feeding Different Fats (% Weight)

Number of C-Atoms in Triacylglycerols	Soybean Oil		Coconut Oil		Tallow		Lard	
	a	b	a	b	a	b	a	b
$C_{28} + C_{30}$	—	—	4.3	—	—	—	—	—
C_{32}	—	—	13.5	—	—	—	—	—
C_{34}	—	—	17.3	—	—	—	—	—
C_{36}	—	—	19.8	—	—	—	—	—
C_{38}	—	—	17.3	—	—	—	—	—
C_{40}	—	—	10.7	—	—	—	—	—
C_{42}	—	0.5	7.5	1.5	—	—	—	—
C_{44}	—	1.0	4.1	4.2	0.4	—	1.0	—
C_{46}	—	1.9	2.1	9.5	1.6	0.5	1.9	—
C_{48}	0.6	4.4	1.6	21.7	7.0	3.9	4.8	1.7
C_{50}	3.9	9.0	0.9	24.2	19.8	17.5	16.6	12.2
C_{52}	28.8	56.4	0.7	33.9	48.0	63.4	55.5	67.8
C_{54}	66.7	26.8	0.2	5.0	23.2	14.9	20.2	18.3

Source: Hirata, A. et al., *Nippon Shokuhin Kogyo,* 34(5), 320, 1987. With permission.

[a] Distribution of fatty acids in feed fat.

[b] Distribution of fatty acids in yolk.

18.3 CHOLESTEROL

18.3.1 INTRODUCTION

Among the sterols of the egg, which make up approximately 4% of egg lipids, cholesterol predominates. It occurs only in the egg yolk, mainly in a free form. The cholesterol content of yolk plasma is approximately three times greater than that of the granules. Only 4%–15% are esterified with FA (Chung et al., 1965). The FA composition of the cholesteryl esters is shown in Table 18.5.

The cholesterol values in individual eggs differ considerably, but on average, the proportions of cholesterol in relation to the dry matter of the egg yolk are relatively constant. Currently, reliable tests indicate a ratio of 1.6% egg yolk to dry matter (Littmann-Nienstedt, 1996). In this ratio, the egg surpasses all other foods, so that cholesterol content can be used qualitatively and quantitatively as an indicator for the amount of egg supplement in food.

Sterols accompanying cholesterol make up only approximately 4% of the total sterol content. Cholesterol, 7-cholesterol, and campesterol are found most frequently, but β-sitosterol, 24-methylencholesterol, desmosterol, and lanosterol could also be detected in very small amounts (Dresselhaus, 1974) (Table 18.23).

Tu et al. (1970) identified the same accompanying sterols as 1.8% of the total sterols. Estrogen steroids or other steroid sterols could not be detected. β-Sitosterol, which is found most frequently, obviously derives from vegetable feed.

TABLE 18.18

Fatty Acid Composition in Triacylglycerols, Phosphatidylcholine (PC) Fraction, and Phosphatidylethanolamine (PE) Fraction in the HDL and LDL Fraction Depending on Dietary Fat (% Weight)

Fatty Acids	Fatty Acid Composition of Dietary Fat		Triacylglycerols (2)/(3)			
	Saff. Oil	Olive Oil	LDL Saff. Oil	LDL Olive Oil	HDL Saff. Oil	HDL Olive Oil
$C_{14:0}$	—	—	0.5	0.3	0.5	0.3
$C_{16:0}$	0.7	11.5	22.8	21.6	23.4	20.8
$C_{16:1}$	—	1.5	3.4	4.2	3.2	3.8
$C_{18:0}$	0.3	2.5	6.5	3.6	6.7	4.1
$C_{18:1}$	18.0	75.5	37.9	63.1	39.0	62.9
$C_{18:2}$	76.0	7.5	28.9	7.2	27.2	8.2
$C_{18:3}$	1.0	1.0	—	—	—	—
$C_{20:0}$	—	0.5	—	—	—	—
$C_{20:4}$	—	—	—	—	—	—

Fatty Acids	PC Fraction				PE Fraction			
	LDL Saff. Oil	(2)/(3) LDL Olive Oil	HDL Saff. Oil	HDL Olive Oil	(2)/(3) LDL Saff. Oil	LDL Olive Oil	HDL Saff. Oil	HDL Olive Oil
$C_{14:0}$	—	—	—	—	—	—	—	—
$C_{16:0}$	29.6	30.2	31.4	31.0	18.6	19.0	16.5	16.8
$C_{16:1}$	—	—	—	—	—	—	—	—
$C_{18:0}$	15.5	15.0	17.8	13.4	27.3	25.0	32.5	30.3
$C_{18:1}$	21.2	38.7	21.5	40.6	17.5	28.5	15.3	27.8
$C_{18:2}$	28.0	12.3	26.8	10.9	21.5	10.6	20.7	8.4
$C_{18:3}$	—	—	—	—	—	—	—	—
$C_{20:0}$	—	—	—	—	—	—	—	—
$C_{20:4}$	5.6	3.8	3.7	4.1	15.2	16.0	15.0	16.8

Note: (2) Naber and Biggert (1989); (3) Pankey and Stadelman (1969). Saff. Oil: Safflower Oil.

TABLE 18.19

Position of Fatty Acids in Triacylglycerols of Yolk Depending on Dietary Fat

Fatty Acids	Triacylglycerols of Coconut Oil (Dietary Fat)			Triacylglycerols of Yolk (Diet with Coconut Oil)		
	Position 1 (sn-1)	Position 2 (sn-2)	Position 3 (sn-3)	Position 1 (sn-1)	Position 2 (sn-2)	Position 3 (sn-3)
$C_{8:0}$	8.3	0.7	21.9	—	—	—
$C_{10:0}$	6.5	3.4	10.3	—	—	—
$C_{12:0}$	35.1	80.8	31.4	0.7	2.0	3.9
$C_{14:0}$	20.4	8.0	23.8	5.8	5.9	18.3
$C_{14:1}$	—	—	—	0.2	2.9	7.4
$C_{16:0}$	12.3	1.5	9.3	63.2	9.4	10.1
$C_{16:1}$	—	—	—	5.1	6.6	11.4
$C_{18:0}$	3.6	0.6	2.1	3.9	3.7	3.5
$C_{18:1}$	12.6	3.7	0.6	17.7	52.1	43.6
$C_{18:2}$	1.3	1.3	0.7	2.3	16.2	0.7
$C_{18:3}$	—	—	—	—	0.6	0.6
$C_{20:1}$	—	—	—	0.1	0.3	0.5

Fatty Acids	Triacylglycerols of Soybean Oil (Dietary Fat)			Triacylglycerols of Yolk (Diet with Soybean Oil)		
	Position 1 (sn-1)	Position 2 (sn-2)	Position 3 (sn-3)	Position 1 (sn-1)	Position 2 (sn-2)	Position 3 (sn-3)
$C_{12:0}$	—	—	—	0.7	0.1	0.1
$C_{14:0}$	—	—	—	1.1	1.8	2.5
$C_{16:0}$	24.8	6.9	2.8	52.5	9.2	4.6
$C_{16:1}$	—	—	—	0.7	2.0	5.1
$C_{18:0}$	9.7	1.4	—	7.4	3.9	—
$C_{18:1}$	31.2	19.0	16.4	24.0	26.3	61.9
$C_{18:2}$	32.2	64.8	63.8	12.0	53.9	20.3
$C_{18:3}$	1.4	7.5	16.6	0.8	2.3	3.2
$C_{20:1}$	0.3	0.2	—	0.3	0.2	0.1

Source: Data from Hirata, A. et al., *Nippon Shokuhin Kogyo Gakkaishi*, 34(5), 320, 1987.

18.3.2 AUTOXIDATION OF CHOLESTEROL

During processing, not only the unsaturated FA and the phospholipids of the egg are influenced by autoxidation, but also the cholesterol. Apart from oxygen, the effects of moisture, low pH, temperature, heating time, addition of H_2O_2, light and a photosensitizer, a long storage period, and packaging are important. Riboflavin can serve as a photosensitizer; riboflavin occurs in substantial amounts in the egg, both in dried egg yolk and in pasta (Acker and Greve, 1963). However, under extreme conditions such as storage in direct sunlight, cholesterol hydroperoxide and a hydroxy

TABLE 18.20

Position of Fatty Acids in Triacylglycerols of Yolk Depending on Dietary Fat

Fatty Acids	Triacylglycerols of Yolk (Basal Diet)			Triacylglycerols of Yolk (Diet with 7% Fish Oil)		
	Position 1 (sn-1)	Position 2 (sn-2)	Position 3 (sn-3)	Position 1 (sn-1)	Position 2 (sn-2)	Position 3 (sn-3)
$C_{14:0}$	0.9	0.2	0.3	1.5	0.5	0.8
$C_{16:0}$	65.9	2.5	6.7	66.8	3.0	2.8
$C_{16:1(n-7)}$	3.1	—	3.0	0.3	—	—
$C_{16:1(n-5)}$	3.5	1.4	5.4	4.3	2.0	5.9
$C_{18:0}$	5.8	2.0	8.3	6.2	2.5	11.0
$C_{18:1(n-9)}$	12.9	42.4	67.7	10.5	40.9	63.8
$C_{18:1(n-7)}$	1.4	1.2	2.9	1.7	1.7	4.3
$C_{18:2(n-6)}$	4.2	45.8	7.3	2.8	42.8	5.4
$C_{18:3(n-3)}$	1.4	1.2	0.6	1.4	1.6	0.3
$C_{20:4(n-6)}$	—	0.9	0.1	—	0.5	0.1
$C_{20:5(n-3)}$	—	—	—	—	0.4	—
$C_{22:5(n-3)}$	0.3	—	—	—	1.2	0.2
$C_{22:6(n-3)}$	0.5	—	—	—	1.9	0.2

Source: Data from Cossignani, I. et al., *Ital. J. Food Sci.*, 293, 1994.

compound derived from it (7-hydroxycholesterol) are formed. Chicoye et al. (1968) confirmed these results and indicated the presence of additional oxidation products. Smith (1987) also suggested that the hydroperoxides of PUFA formed during lipid oxidation may play a role as initiator of cholesterol oxidation. As reviewed by Hur et al. (2007), the concentrations of cholesterol oxidation products in egg products can frequently reach 200 mg/kg.

Oxidation products from cholesterol are of physiological significance. For example, 5α-cholestan-5,6-α-epoxy-3-β-ol can be found in the skin after UV radiation and has been suggested to be involved in the development of skin cancer (Black and Chan, 1976). The compound is also thought to be involved in arteriosclerotic deposits (Benditt, 1977). *In vitro* studies show that 5α-cholestan-5,6-α-epoxy-3-β-ol is associated with DNA (Blackburn et al., 1979). 7β-Hydroxycholesterol and 7-keto-cholesterol also have toxic properties. Smith (1981, 1987) gives an overview of the literature and reports on about 80 identified oxidation products. The oxidized cholesterols are subdivided into primary and secondary product groups. Primary products are those that are oxidized at C_7 and in side-chains from C_{20} to C_{25}. Secondary oxycholesterols are mainly epoxides on C_5 and C_6, as well as cholestantriols. In some publications, oxidation products of the A-ring can be found on C_3 and C_4. Figure 18.1 shows the corresponding structures. Even after 18 months of storage at 4°C, no cholesterol oxides could be detected by Pike and Peng (1985) in the eggs shells.

TABLE 18.21

Structural Analysis of Phosphatidylcholine (PC) Fraction and Phosphatidylethanolamine (PE) Fraction in Yolk Depending on Dietary Fat (7% Fish Oil) and Long Chain ω-3-PUFA Diet

Fatty Acids	PC Fraction (Basal Diet)		PC Fraction (7% Fish Oil)		PC Fraction (ω-3-Enriched)		PE Fraction (Basal Diet)		PE Fraction (7% Fish Oil)		PE Fraction (ω-3-Enriched)	
	Position		Position		Position		Position		Position		Position	
	1 (sn-1)	2 (sn-2)	1 (sn-1)	2 (sn-2)	1 (sn-1)	2 (sn-2)	1 (sn-1)	2 (sn-2)	1 (sn-1)	2 (sn-2)	1 (sn-1)	2 (sn-2)
$C_{14:0}$	0.4	0.1	0.6	0.1	0.49	0.11	0.1	0.2	0.2	0.3	1.14	Traces
$C_{15:0}$	—	—	—	—	0.32	—	—	—	—	—	0.14	—
$C_{16:0}$	64.1	1.1	67.5	2.0	67.03	6.36	33.6	1.1	40.8	6.9	38.30	4.77
$C_{16:1(\omega-7)}$	0.5	—	0.4	—	1.84	1.38	—	—	0.2	0.4	0.34	29.34
$C_{16:1(\omega-5)}$	1.5	0.6	1.6	0.7	—	—	0.1	0.2	0.2	0.8	—	—
$C_{18:0}$	26.7	0.2	23.9	0.6	22.96	2.24	59.5	0.7	53.4	2.5	52.87	5.39
$C_{18:1(\omega-9)}$	4.1	43.2	3.6	40.0	nd	nd	5.2	22.7	2.9	24.2	nd	nd
$C_{18:1(\omega-7)}$	1.1	0.9	1.0	1.5	nd	nd	—	0.7	1.1	1.0	nd	nd
$C_{18:1(\omega-9+7)}$	—	—	—	—	5.79	53.85	—	—	—	—	5.62	0.41
$C_{18:2(\omega-6)}$	0.8	36.1	0.6	32.3	0.70	21.90	1.1	22.8	0.5	18.5	0.35	12.41
$C_{18:3(\omega-6)}$	—	0.3	—	0.2	Traces	Traces	—	0.2	—	—	Traces	Traces
$C_{18:3(\omega-3)}$	—	—	—	—	—	0.17	—	—	—	0.4	—	0.14
$C_{18:4(\omega-3)}$	—	—	—	—	—	0.42	—	—	—	—	—	0.48
$C_{20:1(\omega-9)}$	—	—	—	—	0.21	0.27	—	—	—	—	0.26	Traces
$C_{20:4(\omega-6)}$	—	8.7	—	5.5	—	2.50	0.2	27.2	—	8.3	Traces	11.21

(continued)

TABLE 18.21 (continued)

Structural Analysis of Phosphatidylcholine (PC) Fraction and Phosphatidylethanolamine (PE) Fraction in Yolk Depending on Dietary Fat (7% Fish Oil) and Long Chain ω-3-PUFA Diet

Fatty Acids	PC Fraction (Basal Diet) Position		PC Fraction (7% Fish Oil) Position		PC Fraction (ω-3-Enriched) Position		PE Fraction (Basal Diet) Position		PE Fraction (7% Fish Oil) Position		PE Fraction (ω-3-Enriched) Position	
	1 (sn-1)	2 (sn-2)	1 (sn-1)	2 (sn-2)	1 (sn-1)	2 (sn-2)	1 (sn-1)	2 (sn-2)	1 (sn-1)	2 (sn-2)	1 (sn-1)	2 (sn-2)
$C_{20:5(n-3)}$	—	—	—	1.6	—	1.17	—	—	—	1.7	—	2.63
$C_{22:4(n-6)}$	—	0.5	—	0.3	—	—	—	—	—	—	—	—
$C_{22:5(n-6)}$	—	1.3	—	0.4	—	—	—	3.6	—	0.5	—	—
$C_{22:5(n-3)}$	—	0.4	—	1.1	—	0.43	—	—	—	1.1	—	1.62
$C_{22:6(n-3)}$	—	5.7	—	12.6	Traces	8.84	—	17.1	—	33.4	Traces	30.64

Source: Data from Cossignani, I. et al., *Ital. J. Food Sci.*, 3, 293, 1994; Schreiner, M. et al., *J. Food Lipids*, 13, 36, 2006.

Note: nd, not detected; only the sum of n-9 and n-7 from $C_{18:1}$.

TABLE 18.22

Fatty Acid Composition of the *sn*-1 and *sn*-2 Positions Obtained from the Corresponding Lysophospholipid Fraction (Lysophosphatidylcholines [lysoPC] and Lysophosphatidylethanolamines [lysoPE] after Phospholipase A$_2$ Hydrolysis, the Fatty Acid Composition of PC and PE Fractions Is Given for a Comparison (mol%))

	Fatty Acid Composition		Position *sn*-1[a]		
	PC Fraction	PE Fraction	LysoPC	LysoPE	*sn*-2[b]
$C_{16:0}$	40.27	23.04	70.49	36.18	9.8
$C_{16:1(n-9)}$	1.03	0.7	0.34	—	1.64
$C_{18:0}$	13.11	26.6	26.26	58.62	0.63
$C_{18:1(n-9)}$	27.1	19.17	2.52	2.88	48.65
$C_{18:2(n-6)}$	14.75	13.43	0.19	0.77	28.29
$C_{18:3(n-3)}$	0.26	0.38	0.1	0.36	0.41
$C_{20:2(n-6)}$	0.25	0.34	—	—	0.43
$C_{20:3(n-6)}$	0.18	0.42	—	—	0.31
$C_{20:4(n-6)}$	1.96	9.26	—	0.28	5.5
$C_{22:4(n-6)}$	—	1.06	—	—	0.25
$C_{22:6(n-3)}$	1.08	5.6	0.1	0.91	2.89

Source: Data from Amate, L. et al., *Lipids*, 34, 8, 1999.

[a] Fatty acid composition of the *sn*-1 position obtained from the corresponding lysophopholipid fraction (lysoPC and lysoPE) after phospholipase A$_2$ hydrolysis.

[b] Fatty acid composition of the *sn*-2 position obtained as follows: |FFA| = |2 PL| − |LysoPL| where PL = total phospholipids.

TABLE 18.23

Attendant Sterols of Cholesterol in Egg Yolk

	% Related to Whole Sterols	mg/100 g Dry Matter	
		(1)	(2)
Cholestanol	1.10	26.5	9.7
7-Cholestenol	0.66	15.9	6.3
Campesterol	0.51	12.3	—
β-Sitosterol	0.23	5.5	6.7
24-Methylencholesterol	0.26	6.3	—
7-Dehydrocholesterol	0.32	7.7	6.3
Desmosterol	0.16	3.9	15.1
Lanosterol	0.11	2.7	3.1
Ergosterol	—	—	7.3
Total	3.89	93.8	64.5

Note: (1) Dresselhaus 1974; (2) Tu et al. 1970.

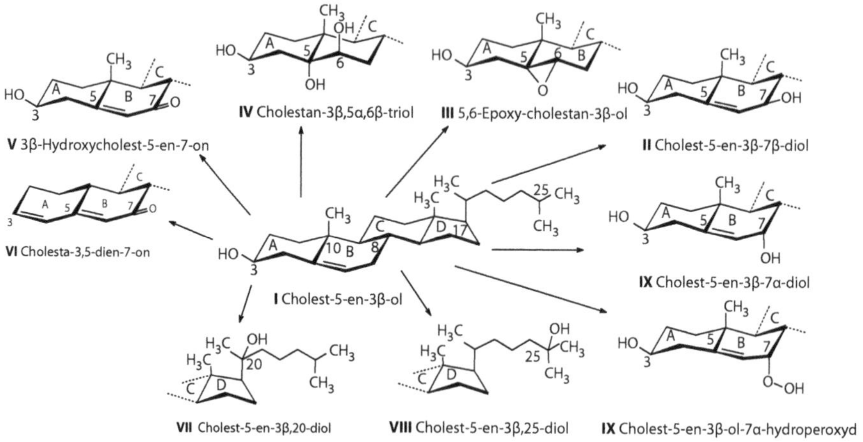

FIGURE 18.1   Cholesterol and oxidized cholesterol.

In fresh eggs, only 7α and 7-ketocholesterol occur in small amounts (Nourooz-Zadeh, 1990). Both compounds can be found in the metabolism of cholesterol; they occur naturally in the egg yolk in concentrations of approximately 0.6 mg/kg.

During thermic processing (e.g., spray drying for the production of egg yolk powder), but scarcely during the preparation of egg dishes, oxidation products do develop, with primary oxidation products developing first. Up to 10% of the cholesterol can be oxidized during drying and storage under unfavorable conditions. Freezer storage (−20°C) leads to an increase in secondary oxidation products. The analysis of these compounds leads, in part, to faulty results, especially when cholesterol oxides are determined after saponification. In spray-dried egg products, Fischer et al. (1985) found the amounts of oxycholesterol cited in Table 18.24. Apart from those three main products, traces of cholestan-3β,5α,6β-triol and cholest-5-en-3β,25-diol have been identified in some samples.

TABLE 18.24
Amounts of Oxycholesterols (µg/g) in Spray-Dried Egg Products

	Whole Egg Powder	Yolk Powder
Cholest-5-en-3β,7α-diol	8.6; 18.8	7.9; 8.3
	20.2	13.0
Cholest-5-en-3β,7β-diol	7.6; 21.3	5.6; 7.1
	24.6	14.0
5,6-Epoxy-cholestan-3β-ol	5.8; 20.2	4.3; 5.9
α- and β-Epoxyd	24.9	8.7

Source: Data from Fischer, K.-H. et al., *Z. Lebensm. Unters. Forsch.*, 181, 14, 1985.

18.3.3 TECHNOLOGICAL DEPENDENCIES FOR THE FORMATION OF OXIDATION PRODUCTS OF CHOLESTEROL

These compounds develop spontaneously in the presence of oxygen, light, and high temperature. H_2O_2, which is partly added to the enzymatic decarbohydration process before drying, also represents a potential type of oxidation, so that mainly epoxides develop.

During storage, the contents of oxidation products increase, for example, from 0 to 20 µg/g after 7 months of storage. When H_2O_2 was added, a rapid increase in cholesterol-5,6-epoxides was noted (from 16 to 123 µg/g). Generally, the amounts of oxidized cholesterol can be reduced significantly by adding antioxidants (Morgan and Armstrong, 1987; Huber et al., 1995; Li et al., 1996). During extended ambient storage of liquid and spray-dried yolk applying equimolar amounts of ascorbyl palmitate, tocopherols, and butylated hydroxyanisole (BHA), the latter two substances could be identified as inhibiting the formation of 7-ketocholesterol, but other cholesterol oxides were not affected (Brinkerhoff et al., 2002). 2,6-Di-tertiarbutyl-4-methylphenol (BHT) and tertiary butylhydroquinone (TBHQ) behave under certain test conditions as the most effective synthetic antioxidants, whereas alpha- and gamma-tocopherol, rosemary extracts, and the flavonoid quercetin as natural antioxidants are most efficient in inhibiting thermal-induced oxidation of cholesterol (Valenzuela et al., 2004).

The amounts of cholesterol oxides in egg powder from eggs dried with direct procedures (i.e., by the use of burning gas) are obviously higher than from drying with indirect heating. Nitric oxides, which develop during the burning of organic material, are thought to stimulate the oxidation of cholesterol (Missler et al., 1985), and N_2O_4 is proposed as the chief initiator for the oxidation process (Table 18.25).

Table 18.26 shows the occurrence of oxidized cholesterols depending on the storage period. There is a significant increase in oxidation products in dried egg

TABLE 18.25
Oxidized Cholesterol Derivates after Direct and Indirect Heating

	Mixed Eggs (Dried)	
Sterols	Direct Heating (µg/g)	Indirect Heating (µg/g)
7α-Hydroxycholesterol	7.0	1.8
7β-Hydroxycholesterol	18.5	1.5
Cholesterol-5α,6α-epoxyd	50.0	21.5
Cholesterol-5β,6β-epoxyd	37.4	1.9
25-Hydroxycholesterol	5.1	1.4
7-Ketocholesterol	37.0	2.0
5α-Cholestane-3β,5,6β-triol	13.0	11.6

Source: Data from Missler, S.R. et al., *J. Food Sci.*, 50, 595, 646, 1985.

TABLE 18.26

Oxidized Cholesterols in Fresh Yolk, and in Spray-Dried Yolk, Depending on Storage Time and Water Content of Yolk Powders

Product	Water Content	Storage Time	Oxidized Cholesterols (in µg/g Egg Lipid)								
			5α,6α-Epoxy	5β,6β-Epoxy	7-Keto	7α-Hydroxy	7β-Hydroxy	20α-Hydroxy	25-Hydroxy	5,6-Dihydroxy	
Fresh yolk		0	n.n.	n.n	n.n.	n.n.	n.n.	n.n.	n.n.	n.n	(1)
Yolk powder	3%, (n.m.)	0	0.18	0.12	0.13	0.08	0.06	tr.	tr.	—	(2)
Yolk powder	3%, (n.m.)	3 months	0.66	0.21	0.12	0.16	0.15	tr.	tr.	—	(2)
Yolk powder	7%–8%	0	0.09	0.15	0.09	0.16	0.05	tr.	tr.	—	(2)
Yolk powder	7%–8%	3 months	0.38	0.12	0.72	0.12	0.11	tr.	tr.	—	(2)
Yolk powder	11%–12%	0	0.19	0.08	0.07	0.10	0.03	tr.	tr.	—	(2)
Yolk powder	11%–12%	3 months	0.34	0.10	0.07	0.12	0.10	tr.	tr.	—	(2)
Yolk powder	n.m.	6 months	1.3	6.5	3.5	2.2	2.5	0.8	n.n.	n.n.	(1)
Yolk powder	n.m.	12 months	2.5	12.0	2.9	8.9	9.4	0.7	n.n.	tr.	(1)
Yolk powder[a]	n.m.	8 years	9.4	dp.	5.7	27.5	46.8	6.6	10.4	27.6	(1)

Note: (1) Data from Nourooz-Zadeh and Appelqvist (1987); (2) Data from Obara et al. (2006). n.m., not moisturized; n.n., less than 0.2 ppm (not detectable); tr., traces; —, not researched; dp., disturbing peak, not evaluated.

[a] Partial unknown storage conditions.

yolk during storage (Nourooz-Zadeh and Appelqvist, 1987). Obara et al. (2006) investigated the influence of water activity on cholesterol oxidation in spray- and freeze-dried egg powders. Generally, the oxysterol accumulation of storage in moisturized powders (8%–12% water content) is lower than in nonmoisturized ones, with 5,6-epoxycholesterol isomers being produced in the highest amounts after 3 months of storage at room temperature. However, the largest increase during storing could be observed in 7β-hydroxycholesterol, probably because of the transformation of 7α-hydroxycholesterol into the more thermally stable 7β-isomer (Table 18.26).

Furthermore, in spray-dried powders, the level of cholesterol oxidation products is higher than in freeze-dried ones.

Most of the cholesterol oxidation products are highly bioavailable and are carried rapidly to tissues where they can exert their adverse physiological effect (Caboni et al., 2005). Hur et al. (2007) concluded in their review that the prevention of cholesterol oxidation in processed food should be similar to procedures to prevent lipid oxidation. Low processing temperatures, antioxidant addition to the food, oxygen-proof packaging, optimally under protective atmosphere, as well as low-temperature storage in the dark or dietary antioxidants to laying hens (Galobart et al., 2002) minimize the formation of cholesterol oxidation products.

18.3.4 Strategies for Lowering Cholesterol Content of Egg Yolk

18.3.4.1 Influencing Cholesterol Content by Genetic Selection, Hens' Diet, or Drugs

At present, chicken eggs provide an average of 190 mg cholesterol/52 g egg substance and are thus considered to be a high-cholesterol food. A problem, however, associated with the yolk is that its high cholesterol and possibly cholesterol oxide content has been connected in the consumers' mind as a causative agent for heart disease and various other diseases. As cholesterol in eggs accounts for more than 50% of daily cholesterol intake (Wang and Pan, 2003), low-cholesterol eggs meet the consumers' demands and are therefore interesting from an economic point of view.

Since more than 40 years, many efforts have been made to reduce the cholesterol content in shell eggs. Recently, Elkin gave excellent overviews about genetic and nutritional strategies (Elkin, 2006), as well as further approaches utilizing non-nutritive dietary factors or drugs (Elkin, 2007). Species, breed or strain, age, and rate of egg production influence egg cholesterol content. Genetic selection programs (Hargis, 1988) achieved a reduction of 7%, and other authors also obtained only small changes in egg cholesterol content of 5%–7%. The cholesterol content per gram decreases proportionally with the increasing age of the hens, while the proportion of yolk increases. Since 1935, the laying capacity of a hen has increased from 120 to 257 eggs/year (Westermann, 1991). Due to this increased laying capacity, the cholesterol content has decreased by about 25% since the 1950s.

Table 18.27 summarizes studies that described successful attempts in lowering egg yolk cholesterol by dietary factors. Generally, most of the conducted experimental approaches could achieve only minimal changes of less than 10%. Best results, namely a reduction of yolk cholesterol levels of >30%, could be achieved when garlic paste or supra-optimal amounts of copper were administered to laying hens.

TABLE 18.27

Studies Describing Successful Approaches in Lowering Egg Yolk Cholesterol by Layers' Dietary Factors

Dietary Factor	Supplemental Dietary Level of Layers Feed /Experiment Duration	Reduction in Yolk Cholesterol Concentration (mg/g yolk) vs. Unsupplemented Control	Effect on Laying Performance	References
Omega-3 fatty acids	≤10% menhaden fish oil	≤13%	Not reported	Oh et al. (1991)
	7%, 14%, 21%, and 28% whole chia seed (*Salvia hispanica* L.)/90 d	~20% (dietary level 28% chia seeds)	Egg production decreases with increasing omega-3 fatty acid incorporation into the egg	Ayerza and Coates (2000)
Conjugated linoleic acid (CLA)	≤5%/5 weeks	≤6%	Not reported	Hur et al. (2003)
	≤2%/12 weeks	No reduction, but 13% lower egg cholesterol content due to CLA-mediated reduction in yolk size	Egg mass was uniformly lower in hens fed the CLA-enriched diets	Szymczyk and Pisulewski (2003)
Copper (Cu)	0.005%, 0.010%, 0.015%, and 0.020% Cu (as CuSO₄.5H₂O)/90 d	≤20% (dietary level 0.015% Cu)	None	Balevi and Coskun (2004)
	0.005%, 0.015% and 0.025% (as sulfate or acetate)/84 d	14% (dietary level 0.025%)	Slightly lower egg production	Al Ankari et al. (1998)
	0.0125% and 0.0250% (as CuSO₄·5H₂O/8 weeks	26% and 32%, respectively	Egg weights were not consistently affected; egg production increased; small amounts of Cu accumulated in yolks (0.80 µg/g), whites (0.56 µg/g) and shells (5.46 µg/g) of eggs (dietary Cu supplement 0.0250%)	Pesti and Bakalli (1998)
Garlic (*Allium sativum*)	≤1.5% garlic powder	≤23%[a]	Not reported	Mottaghitalab and Taraz (2002)

	Treatment/dose	Value	Effect	Reference
	2%, 4%, 6%, 8%, and 10% Sun-dried garlic paste/6 weeks	5%, 9%, 14%, 20%, and 24%, respectively; ≤32%[a]	No effect on egg weight, egg mass, feed consumption, feed efficiency, and BW gain; higher egg production (only Babcock strain); yolk weight responded quadratically with increasing levels of dietary garlic	Chowdhury et al. (2002)
	0.5% and 1% garlic powder/22 weeks	5% and 10%, respectively; ≤6%[a]	Egg weight increased ($P<0.01$) with garlic powder supplementation; no effect on, e.g., egg production, and egg yolk weight	Yalcin et al. (2006)
Garlic and copper (Cu)	1%, 3%, and 5% garlic powder (GP); 0.020% Cu; 3% GP+0.020% Cu/5 weeks	5.1%[b], 10.0%, and 6.8%, respectively for GP; 7.7% (Cu); 11.3% (GP+Cu)	No effect observed	Lim et al. (2006)
Flavonoids, or pectin	0.05% Hesperitin, 0.05% naringenin, and 0.5% pectin/2 months	16.9% and 11.7%[a] (hesperitin), 20.3% and 15.1%[a] (naringenin), and 22.4% and 21.8%[a] (pectin)	Flavonoids (hesperetin and naringenin) increased the yolk weight and the ratio of yolk weight/egg weight; pectin increases cholesterol excretion; hesperitin and naringenin reduce activity of liver-HMGR and acyl-CoA:cholesterol acyltransferase	Lien et al. (2008)
Probiotics	0.010% and 0.015% Probiolac®/10 weeks	23.2% and 22.6%, respectively; ≤22%[a]	Egg production was improved by 5% (dietary level 0.010%)	Mohan et al. (1995)
	0.010% and 0.020% Probiolac/47 weeks	14% and 13%, respectively	Egg production increased, egg weight was not influenced	Panda et al. (2003)
	0.025%, 0.050%, and 0.075% BioPlus 2B®/90d	16.0%, 35.4%, and 37.2%, respectively; ≤37.8%[a]	Egg production increased by 4% (dietary level 0.075%); egg and egg yolk weight were not affected	Kurtoglu et al. (2004)
	Lactobacillus acidophilus/16 weeks	23%; 22%[a]	Egg production was improved	Abdulrahim et al. (1996)
	Lactobacillus acidophilus, up to four million viable cells per gram of feed/48 weeks	≤18.8%	Levels of egg production and feed conversion were higher (8% and 14.8%, respectively)	Haddadin et al. (1996)

(continued)

TABLE 18.27 (continued)
Studies Describing Successful Approaches in Lowering Egg Yolk Cholesterol by Layers' Dietary Factors

Dietary Factor	Supplemental Dietary Level of Layers Feed /Experiment Duration	Reduction in Yolk Cholesterol Concentration (mg/g yolk) vs. Unsupplemented Control	Effect on Laying Performance	References
Green tea powder (GTP)	1% and 2% (Chinese, Korean and Japanese GTP, respectively)	17.7% and 17.1% (Chinese GTP) (Korean and Japanese GTP caused no significant decrease)	Egg production rates of the layers were increased with all supplements up to 10%; the egg weight was reduced by 1.8% in layers fed 1% Chinese green tea	Uuganbayar et al. (2006)
β-Cyclodextrin	2%, 4%, 6%, and 8%	5.1%[b], 21.1%, 28.2%, and 29.9%, respectively	Egg production and egg weight were 1.9% and 2.9%, respectively, lower in the hens fed on 8% beta cyclodextrin	Park et al. (2005)
HMGR inhibitors ("statins")	Lovastatin, simvastatin, or pravastatin at 0.03% and 0.06%/4 weeks	5.3%[a] and 13.0%[a] (lovastatin), 18.7%[a] and 14.2%[a] (simvastatin), 11.1%[a] and 19.6%[a] (pravastatin)	Egg weight was lowered with all statin treatments; egg production was relatively unaffected	Kim et al. (2004)
	Atorvastatin 0.03 and 0.06%/5 weeks	31.9%[a] and 46.5%[a], respectively	Egg production decreased by 19% (dietary level 0.06%)	Elkin et al. (1999)
	2%, 5%, and 8% red mold rice, containing 0.0145%, 0.035% and 0.056% monacolin K (lovastatin), respectively, as red yeast metabolite/8 weeks	14%[a], 13%[a], and 5.7%[a], respectively	No differences in egg production and yolk weight	Wang and Pan (2003)

[a] Reduction (%) in egg cholesterol content (mg/yolk) vs. unsupplemented control.
[b] Not significant.

Garlic (*Allium sativum*) and its products have been shown to exhibit hypocholesterolemic properties in broiler chickens, probably because it decreases the activity of hepatic 3-hydroxy-3-methylglutaryl coenzyme A reductase (HMGR), a rate limiting enzyme in cholesterol biosynthesis. Garlic contains a variety of organosulfur compounds, some of which may be responsible for the hypocholesterolemic action and egg cholesterol-lowering effect, but the active substances are still to be identified (Yalcin et al., 2006).

Copper has been investigated in terms of its egg cholesterol-lowering effect in numerous studies after it could be demonstrated that feeding copper (Cu^{2+}) in excess of nutritional requirement markedly reduced plasma and muscle cholesterol concentrations in broiler chickens. However, Patterson et al. (2004) failed to confirm this effect. The variation between the studies has not been able to be explained so far.

Considerable attempts have been made to influence the cholesterol synthesis of laying hens by drugs (non-nutritive dietary factors or pharmacologic agents) in order to reduce egg cholesterol levels. Best results in lowering yolk cholesterol levels could be achieved when HMGR inhibitors (so-called *statins*) were orally administered to chicken. Kim et al. (2004) suggested that "pravastatin, unlike other classes of statins, may be a good candidate for commercial production of low cholesterol eggs" because of its unique limited impact on hen physiology and egg production. Beyond that, atorvastatin is the most potent compound investigated so far and has been shown to lower egg cholesterol (mg/yolk) by 46.5% (Elkin et al., 1999). A subsequent study from the same laboratory demonstrated that egg yolks from atorvastatin-treated hens showed favorable nutrient compositional changes (e.g., greater amounts of amino acids and reduced levels of total FA and cholesterol), whereas 0.07% of the total administered dose of $[^{14}C]$-atorvastatin transferred to the yolks of eggs laid during the first 10 days postdosing (Elkin et al., 2003).

Lovastatin, also called monacolin K, is a secondary metabolite of red yeast (*Monascus* species) and has been widely used in the human diet ("red mold rice") and as folk remedies for over 1200 years in Asia. Wang and Pan (2003) fed laying hens a diet supplemented with up to 8% red mold rice containing 0.056% lovastatin and observed that egg cholesterol content was lowered by ~14%.

There are also several hypocholesterolemic "nonstatin" compounds, but attempts to block cholesterogenesis with azasterol and triparanol resulted in an extensive (up to 87%), but unacceptable replacement of yolk cholesterol by desmosterol, causing a number of serious side effects including cataracts, both in humans and animals (Elkin, 2007).

Of course, in the case of commercial application, supplemented layers feed has to meet legal requirements and the obligate garnering of government regulatory approval for not "generally recognized as safe" substances has to be taken into account.

18.3.4.2 Reduction of Egg Yolk Cholesterol through Enzymatic Methods

Recent approaches to reduce cholesterol through microbes, as recently summarized (Li-Chan and Kim, 2007), succeeded for example in a cholesterol bioconversion of 85.6% into cholest-4-en-3-one, caused by extracellular cholesterol oxidase from mutant *Brevibacterium* sp. ODG-007 (Lv et al., 2002). Christodoulou et al. (1994)

found that up to 93.4% cholesterol was degraded by cholesterol oxidases from *Pseudomonas fluorescens* and *Streptomyces* species after 72 h of incubation at 37°C. At a low temperature (4°C, 48 h), *P. fluorescens* cholesterol oxidase could still degrade up to 64.9% cholesterol. Enzymatic methods are thought to have the advantage to degrade cholesterol selectively, but, they are difficult to apply and bioconversation products have to be identified.

18.3.4.3 Removing Cholesterol by Physical Treatments and Effects on Functional Properties of Egg Yolk

Physical treatments seemed to be most promising in the cholesterol removal of egg yolk, but changes in its functional properties have to be taken into account. Egg yolk contributes important functional constituents such as emulsifier or foaming and gelling agent when used for food. Phospholipids, lipoproteins (LDL and HDL), and nonassociated proteins (livetins and phosvitins) are responsible for the emulsifying capacity, with LDL being the most important contributor. Mine and Bergougnoux (1998) found that cholesterol is an important component in the stabilization of LDL emulsions. In general, the functional activity of egg yolk is highly influenced by the rheological properties of the egg yolk.

For approximately 30 years, a lot of studies have been conducted, applying subcritical or supercritical CO_2-extraction in order to remove cholesterol selectively from the egg yolk matrix (Levi and Sim, 1988; Froning et al., 1990, 1998; Warren et al., 1991; Sun et al., 1995; Bringe et al., 1996; Miranda et al., 2002). Promising are procedures by which the cholesterol and lipid (primarily TAG) content can be reduced by approximately 60%–80% with CO_2-extraction. With this method, it is also possible to remove neutral lipids. As an advantage, this technical procedure exhibits product safety, low toxicity, lipid extraction selectivity, and high maintenance of yolk functionality. Due to this process, phospholipid and protein contents are concentrated. It has been shown that rheological parameters increase with the level of cholesterol reduction by up to 40–80 wt%, as the concentration of surface-active agents increases with cholesterol removal. Higher cholesterol reduction, however, results in a lipoprotein structural rearrangement and the parameters decrease (Moros et al., 2002). Unfortunately, CO_2-extraction is only successful with a prior spray-drying step, which thermally denatures the (lipo)proteins, thus destroying the noncovalently stabilized lipid–protein complexes that impart high functionality. A dramatic change from fluid to gel-like behavior can be observed (Miranda et al., 2002).

The cholesterol contents can also be reduced by means of extraction in the form of complexes with cyclodextrins (Smith et al., 1995; Awad et al., 1997; Chiu et al., 2004; Jung et al., 2005), or polysorbat 80 (Paraskevopoulou and Kiosseoglou, 1995a,b). Nanofiltration in ethanolic solution allows cholesterol removal of more than 60% (Allegre et al., 2006). Recently, high methoxyl pectins have been tested regarding their cholesterol-removing ability (Rojas et al., 2007). However, but only a maximum decrease of 14.4% was reached, whereas Streamline Phenyl® resin as a hydrophobic adsorbent was reported to cause a 70% reduction of yolk plasma cholesterol, obtaining a low-cholesterol yolk fraction (Rojas et al., 2006). All adsorption techniques named above unfortunately need dilution of the egg yolk. β-Cyclodextrin absorbs no TAG, but free and esterified cholesterol to an extent of 91.6% and 94.4%, respectively. Awad et al. (1997) concluded that such cholesterol-reduced egg yolk has similar compositional and functional properties as the original egg yolk and can be used, for example, in sponge cake production.

Solvent extraction has been performed using acetone (Borges et al., 1996; Martucci and Borges, 1997), or other organic solvents such as petroleum ether or petroleum ether-ethanol (Warren et al., 1991; Paraskevopoulou and Kiosseoglou, 1994). The chemical extraction techniques have the disadvantage that, in addition to cholesterol, all the lipids and phospholipids, mainly responsible for the functional properties, are removed as well. Beyond that, also pigments and flavor components are extracted, causing a deterioration in quality. The extraction of cholesterol using petroleum ether affects the protein quality of the remaining egg yolk, resulting in very weak gelling characteristics (Paraskevopoulou et al., 2000), whereas the emulsifying properties are reported to be comparable to those of native egg yolk (Paraskevopoulou and Kiosseoglou, 1994).

REFERENCES

Abdulrahim, S.M. et al. 1996. The influence of *Lactobacillus acidophilus* and bacitracin on layer performance of chickens and cholesterol content of plasma and egg yolk, *Br. Poult. Sci.*, 37, 341.

Acker, L. and Greve, H. 1963. Über die Photooxidation des Cholesterins in eihaltigen Lebensmitteln, *Fette, Seifen, Anstrichm.*, 65, 1000.

Acker, L. and Ternes, W. 1994. Chemische Zusammensetzung des Eies, in: *Ei und Eiprodukte*, Ternes, W., Acker, L., and Scholtyssek, S. (Eds.), Verlag Paul Parey, Berlin, Germany, Chapter 6.

Al Angari, A., Najib, H., and Al Hozab, A. 1998. Yolk and serum cholesterol and production traits, as affected by incorporating a supraoptimal amount of copper in the diet of the leghorn hen, *Br. Poult. Sci.*, 39, 393.

Allegre, C. et al. 2006. Cholesterol removal by nanofiltration: Applications in nutraceutics and nutritional supplements, *J. Membr. Sci.*, 269, 109.

Amate, L., Ramirez, M., and Gil, A. 1999. Positional analysis of triglycerides and phospholipids rich in long-chain polyunsaturated fatty acids, *Lipids*, 34, 8.

Awad, A.C., Bennink, M.R., and Smith, D.M. 1997. Composition and functional properties of cholesterol reduced egg yolk, *Poult. Sci.*, 76, 649.

Ayerza, R. and Coates, W. 2000. Dietary levels of chia: Influence on yolk cholesterol, lipid content and fatty acid composition for two strains of hens, *Poult. Sci.*, 79, 724.

Balevi, T. and Coskun, B. 2004. Effects of dietary copper on production and egg cholesterol content in laying hens, *Br. Poult. Sci.*, 45, 530.

Benditt, E.P. 1977. Implications of the monoclonal character of human arteriosclerotic plaques, *Am. J. Pathol.*, 86, 693.

Biedermann, R., Rabucki, A.L., and Schürch, A. 1971. Über den Einfluß des Futterfettes auf das Fettsäuremuster der Hühnereilipide, Diss. 4400, Zürich, Switzerland, 72.

Black, H.S. and Chan, J.T. 1976. Etiologic related studies of ultraviolet light-mediated carcinogenesis, *Oncology*, 33, 119.

Blackburn, G.M., Rashid, A., and Thompson, M.H. 1979. Interaction of 5,6-cholesterol oxide with DNA and other nucleophiles, *J. Chem. Soc. Chem. Commun.*, 420.

Borges, S.V., Martucci, E.T., and Müller, C.O. 1996. Optimization of the extraction of cholesterol from dehydrated egg yolk using acetone, *Lebensm. Wiss. Technol.*, 29, 687.

Bringe, N.A., Howard, D.B., and Clark, D.R. 1996. Emulsifying properties of low-fat, low-cholesterol egg yolk prepared by supercritical CO_2 extraction, *J. Food Sci.*, 61, 19.

Brinkerhoff, B.E. et al. 2002. Effect of antioxidants on cholesterol oxidation in spray-dried egg yolk during extended ambient storage, *J. Food Sci.*, 67, 2857.

Burghelle-Mayeur, C., Demarne, Y., and Merat, P. 1989. Influence of the sex-linked dwarfind gene (dw) on the lipid composition of plasma, egg yolk and abdominal fat pad in white leghorn laying hens: Effect of dietary fat, *Am. Inst. Nutr.*, 1361.

Caboni, M.F. et al. 2005. Effect of processing and storage on the chemical quality markers of spray-dried whole egg, *Food Chem.*, 92, 293.

Carter, H.E., Smith, D.B., and Jones, D.N. 1957. A new ethanolamine-containing lipid from egg yolk, *Biol. Chem.*, 232, 681.

Caston, L. and Leeson, S. 1990. Dietary flax and egg composition, *Poult. Sci.*, 69, 1617.

Chen, P.H. et al. 1965. Some effects of added dietary fats on the lipid composition of hen's egg yolk, *J. Food Sci.*, 30, 838.

Cherian, G., Wolfe, F.H., and Sim, J.S. 1996. Feeding dietary oils with tocopherols: Effects on internal qualities of eggs during storage, *J. Food Sci.*, 61, 15.

Chicoye, E., Powrie, W.D., and Fennema, O. 1968. Photooxidation of cholesterol in spray-dried egg yolk upon irridation, *J. Food Sci.*, 33, 581.

Chiu, S.H. et al. 2004. Immobilization of beta-cyclodextrin in chitosan beads for separation of cholesterol from egg yolk, *Food Res. Int.*, 37, 217.

Chowdhury, S.R., Chowdhury, S.D., and Smith, T.K. 2002. Effects of dietary garlic on cholesterol metabolism in laying hens, *Poult. Sci.*, 81, 1856.

Christie, W.W. and Moore, J.H. 1970. The structure of egg yolk triglycerides, *Biochim. Biophys. Acta*, 218, 83.

Christodoulou, S. et al. 1994. Enzymatic degradation of egg-yolk cholesterol, *J. Food Prot.*, 57, 908.

Chung, R.A., Rogler, J.C., and Stadelmann, W.J. 1965. The effect of dietary cholesterol and different dietary fats on cholesterol content and lipid composition of egg yolk and various body tissues, *Poult. Sci.*, 44, 221.

Cossignani, L. et al. 1994. Incorporation of n-3 PUFA into hen egg yolk lipids. II: Structural analysis of triacylglycerols, phosphatidylcholines and phosphatidylethalomines, *Ital. J. Food Sci.*, 3, 293.

Do, U.H. and Ramachandran, S. 1980. Mild alkali-stable phospholipids in chicken egg yolks: Characterization of 1-alkenyl and 1-alkyl-sn-glycero-3-phosphoethanolamine, sphingomyelin, and 1-alkyl-sn-glycero-3-phosphocholine, *J. Lipid Res.*, 21, 888.

Dresselhaus, M. 1974. Die gaschromatographische Bestimmung des Cholesterins in Lebensmitteln: Ein selektives Verfahren zur Ermittlung des Eigehaltes, Dissertation, Univ. Münster, Munster, Germany.

Du, M., Ahn, D.U., and Sell, J.L. 1999. Effect of dietary conjugated linoleic acid on the composition of egg yolk lipids, *Poult. Sci.*, 78, 1639.

Elkin, R.G. 2006. Reducing shell egg cholesterol content. I. Overview, genetic approaches, and nutritional strategies, *Worlds Poult. Sci. J.*, 62, 665.

Elkin, R.G. 2007. Reducing shell egg cholesterol content. II. Review of approaches utilizing non-nutritive dietary factors or pharmacological agents and an examination of emerging strategies, *Worlds Poult. Sci. J.*, 63, 5.

Elkin, R.G. et al. 1999. Select 3-hydroxy-3-methylglutaryl-coenzyme A reductase inhibitors vary in their ability to reduce egg yolk cholesterol levels in laying hens through alteration of hepatic cholesterol biosynthesis and plasma VLDL composition, *J. Nutr.*, 129, 1010.

Elkin, R.G., Furumoto, E.J., and Thomas, C.R. 2003. Assessment of egg nutrient compositional changes and residue in eggs, tissues, and excreta following oral administration of atorvastatin to laying hens, *J. Agric. Food Chem.*, 51, 3473.

Evans, R.J., Bandemer, S.L., and Davidson, J.A. 1967. Lipids and fatty acids in fresh and stored shell eggs, *Poult. Sci.*, 46, 151.

Evans, R.J. et al. 1977. The influence of crude cottonseed oil in the feed on the blood and egg yolk lipoproteins of laying hens, *Poult. Sci.*, 56, 468.

Fischer, K.-H., Laskawy, G., and Grosch, W. 1985. Quantitative Analyse von Autoxidationsprodukten des Cholesterols in tierischen Lebensmitteln, *Z. Lebensm. Unters. Forsch.*, 181, 14.

Froning, G.W. et al. 1990. Extraction of cholesterol and other lipids from dried egg-yolk using supercritical carbon-dioxide, *J. Food Sci.*, 55, 95.

Froning, G.W. et al. 1998. Moisture content and particle size of dehydrated egg yolk affect lipid and cholesterol extraction using supercritical carbon dioxide, *Poult. Sci.*, 77, 1718.

Fujino, Y. and Momma, H. 1971. The lipids of egg yolk. 5. Presence of ceramide, *J. Food Sci.*, 36, 1125.

Fujino, Y. et al. 1971a. Studies on the lipids of egg yolk, *Agric. Biol. Chem.*, 35, 134.

Fujino, Y., Negishi T., and Momma, H. 1971b. Lipids of egg yolk. II. Nature of sphingolipids, *Agric. Biol. Chem.*, 35, 140.

Galobart, J. et al. 2001. Effect of dietary supplementation with rosemary extract and α-tocopheryl acetate on lipid oxidation in eggs enriched with ω3-fatty acids, *Poult. Sci.*, 80, 460.

Galobart, J. et al. 2002. Influence of dietary supplementation with α-tocopheryl acetate and canthaxanthin on cholestrol oxidation in ω3 and ω6 fatty acid-enriched spray-olried eggs, *J. Food Sci.*, 67, 2460.

Gornall, D.A. and Kuksis, A. 1971. Molecular species of glycerophosphatides and triglycerides of egg yolk lipoproteins, *Can. J. Biochem.*, 49, 51.

Haddadin, M.S.Y. et al. 1996. Effect of *Lactobacillus acidophilus* on the production and chemical composition of hen's eggs, *Poult. Sci.*, 75, 491.

Hargis, P.S. 1988. Modifying egg yolk cholesterol in the domestic fowl: A review, *World's Poult. Sci.*, 44, 17.

Hirata, A. et al. 1987. Effects of dietary fats on triacylglycerol composition and structure of egg yolk lipids, *Nippon Shokuhin Kogyo Gakkaishi*, 34, 320.

Holopainen, M. 1972. Kananmunan lipidikoostumuksesta, *Arvi a. Karisto oy:n kirjapaino*, 125, 5.

Holub, B.J. and Kuksis, A. 1969. Molecular species of phosphatidyl ethanol amin from egg yolk, *Lipids*, 4, 466.

Huang, Z. et al. 1990. Effect of dietary fish oil on ω-3 fatty acid levels in chicken eggs and thigh flesh, *Agric. Food Chem.*, 38, 743.

Huber, K.C., Pike, O.A., and Huber, C.S. 1995. Antioxidant inhibition of cholesterol oxidation in a spray-dried food system during accelerated storage, *J. Food Sci.*, 60, 909.

Hur, S.J. et al. 2003. Effect of dietary conjugated linoleic acid on lipid characteristics of egg yolk, *Asian-Australas. J. Anim. Sci.*, 16, 1165.

Hur, S.J., Park, G.B., and Joo, S.T. 2007. Formation of cholesterol oxidation products (COPS) in animal products, *Food Control*, 18, 939.

Jaekel, T., Dautel, K., and Ternes, W. 2008. Preserving functional properties of hen's egg yolk during freeze-drying, *J. Food Eng.*, 87, 522.

Jaekel, T. and Ternes, W. 2009. Changes in rheological behaviour and functional properties of hen's egg yolk induced by processing and fermentation with phospholipases, *J. Food Sci. Technol.*, 44, 567.

Jung, T.H., Park, H.S., and Kwak, H.S. 2005. Optimization of cholesterol removal by cross-linked beta-cyclodextrin in egg yolk, *Food Sci. Biotechnol.*, 14, 793.

Kaufmann, H.P. and Mankel, A. 1967. Über Trans-Lipoide: Die Lipide des Hühnerei-Dotters, *Fette, Seifen, Anstrichm.*, 69, 107.

Keenan, T. W. and Berridge, L. 1973. Identification of gangliosides as constituents of egg yolk, *J. Food Sci.*, 38, 43.

Kim, J.H. et al. 2004. Oral administration of pravastatin reduces egg cholesterol but not plasma cholesterol in laying hens, *Poult. Sci.*, 83, 1539.

Kuksis, A. 1992. Yolk lipids (Review), *Biochim. Biophys. Acta*, 1124, 205.

Kuksis, A. and Marai, L. 1967. Determination of complete structure of natural lecithins, *Lipids*, 2, 217.

Kurtoglu, V. et al. 2004. Effect of probiotic supplementation on laying hen diets on yield performance and serum and egg yolk cholesterol, *Food Addit. Contam.*, 21, 817.

Lea, C.H. 1957. Deteriovative reactions involving phospholipids and lipoproteins, *J. Sci. Food Agric.*, 8, 1.

Levi, S. and Sim, J.S. 1988. Selective removal of cholesterol from egg-yolk products by supercritical CO_2-fluid extraction, *Can. Inst. Food Sci. Technol. J.-Journal de l Institut Canadien de Science et Technologie Alimentaires*, 21, 369.

Li, S.X., Cherian, G., and Sim, J.S. 1996. Cholesterol oxidation in egg yolk powder during storage and heating as affected by dietary oils and tocopherols, *J. Food Sci.*, 61, 4.

Li-Chan, E.C.Y. and Kim, H.-O. 2007. Structure and chemical composition of eggs, in: *Egg Bioscience and Biotechnology*, Mine, Y. (Ed.), Hoboken, NJ: John Wiley & Sons, pp. 1–97.

Lien, T.F., Yeh, H.S., and Su, W.T. 2008. Effect of adding extracted hesperetin, naringenin and pectin on egg cholesterol, serum traits and antioxidant activity in laying hens, *Arch Anim. Nutr.*, 62, 33.

Lim, K.S. et al. 2006. Effects of dietary garlic powder and copper on cholesterol content and quality characteristics of chicken eggs, *Asian-Australas. J. Anim. Sci.*, 19, 582.

Littmann-Nienstedt, S. 1996. Neuere Beobachtungen zum Cholesterin- und Trocken-massengehalt von Schaleneiern, *Lebensmittelchemie*, 50, 61.

Lv, C.F. et al. 2002. Bioconversion of yolk cholesterol by extracellular cholesterol oxidase from *Brevibacterium* sp., *Food Chem.*, 77, 457.

Martucci, E.T. and Borges, S.V. 1997. Extraction of cholesterol from dehydrated egg yolk with acetone: Determination of the practical phase equilibrium and simulation of the extraction process, *J. Food Eng.*, 32, 365.

Miksik, I. et al. 2003. Insoluble eggshell matrix proteins: Their peptide mapping and partial characterization by capillary electrophoresis and high-performance liquid chromatography, *Electrophoresis*, 24, 843.

Mine, Y. and Bergougnoux, M. 1998. Adsorption properties of cholesterol-reduced egg yolk low-density lipoprotein at oil-in-water interfaces, *J. Agric. Food Chem.*, 46, 2153.

Miranda J. et al. 2002. Rheological characterization of egg yolk processed by spray-drying and lipid-cholesterol extraction with carbon dioxide, *J. Am. Oil Chem. Soc.*, 79, 183.

Missler, S.R., Wasilchuk, B.A., and Merritt, C., Jr. 1985. Separation and identification of cholesterol oxidation products in dried egg preparations, *J. Food Sci.*, 50, 595, 646.

Mohan, B. et al. 1995. Effect of probiotic supplementation on serum yolk cholesterol and on egg-shell thickness in layers, *Br. Poult. Sci.*, 36, 799.

Momma, H., Nakano, M., and Fujino, Y. 1972. Studies on the lipids of egg yolk. IV. Cerebroside in egg yolk, *Jpn. J. Zootechnol. Sci.*, 43, 198.

Morgan, J.N. and Armstrong, D.J. 1987. Formation of cholesterol-5,6-epoxides during spray-drying of egg yolk, *J. Food Sci.*, 52, 1224.

Moros, J.E., Franco, J.M., and Gallegos, C. 2002. Rheological properties of cholesterol-reduced, yolk-stabilized mayonnaise, *J. Am. Oil Chem. Soc.*, 79, 837.

Moschidis, M.C., Demopoulos, C.A., and Kritikou, L.B. 1984. Isolation of hens' egg phospholipids by thinlayer chromatography, their identification and silicic acid column chromatographic separation, *J. Chromatogr.*, 292, 473.

Mottaghitalab, M. and Taraz, Z. 2002. Effects of garlic powder (*Allium sativum*) on egg yolk and blood serum cholesterol in Aryan breed laying hens, *Br. Poult. Sci.*, 43, S42.

Naber, E.C. and Biggert, M.D. 1989. Patterns of lipogenesis in laying hens fed a high fat diet containing safflower oil, *Am. Inst. Nutr.*, 119, 690–695.

Nakane, S. et al. 2001. Hen egg yolk and white contain high amounts of lysophosphatidic acids, growth factor-like lipids: Distinct molecular species composition, *Lipids*, 36, 413.

Navarro, J.G. et al. 1972. Influence of dietary fish meal on egg fatty acid composition, *J. Sci. Food Agric.*, 23, 1287.

Negishi, T. et al. 1975. The lipid class and its composition in egg white, *Jpn. J. Zootechnol. Sci. (Nihon Chikusan Gakkai-ho)*, 46, 342.

Nielsen, H. 1998. Hen age and fatty acid composition of egg yolk lipid, *Br. Poult. Sci.*, 39, 53.

Noble, R.C. and Moore, J.H. 1965. Metabolism of yolk phospholipids by developing chick embryo, *Can. J. Biochem.*, 43, 1677.

Nourooz-Zadeh, J. 1990. Determination of the autoxidation products from free or total cholesterol: A new multistep enrichment methodology including the enzymatic release of esterified cholesterol, *J. Agric. Food Chem.*, 38, 1667.

Nourooz-Zadeh, J. and Appelqvist, L.A. 1987. Cholesterol oxides in Swedish foods and food ingredients: Fresh eggs and dehydrated egg products, *J. Food Sci.*, 52, 57.

Obara, A., Obiedzinski, M., and Kolczak, T. 2006. The effect of water activity on cholesterol oxidation in spray- and freeze-dried egg powders, *Food Chem.*, 95, 173.

Oh, S.Y. et al. 1991. Eggs enriched in omega-3-fatty-acids and alterations in lipid concentrations in plasma and lipoproteins and in blood-pressure, *Am. J. Clin. Nutr.*, 54, 689.

Panda, A.K. et al. 2003. Production performance, serum/yolk cholesterol and immune competence of white leghorn layers as influenced by dietary supplementation with probiotic, *Trop. Anim. Health Prod.*, 35, 85.

Pankey, R.D. and Stadelman, W.J. 1969. Effect of dietary fats on some chemical and functional properties of eggs, *J. Food Sci.*, 34, 312.

Paraskevopoulou, A. and Kiosseoglou, V. 1994. Cholesterol and other lipid extraction from egg-yolk using organic-solvents: Effects on functional-properties of yolk, *J. Food Sci.*, 59, 766.

Paraskevopoulou, A. and Kiosseoglou, V. 1995a. Effect of cholesterol extraction from dried yolk with the aid of polysorbate-80 on yolks functional-properties, *Food Hydrocoll.*, 9, 205.

Paraskevopoulou, A. and Kiosseoglou, V. 1995b. Use of polysorbate-80 to reduce cholesterol in dehydrated egg-yolk, *Int. J. Food Sci. Technol.*, 30, 57.

Paraskevopoulou, A. et al. 2000. Small deformation measurements of single and mixed gels of low cholesterol yolk and egg white, *J. Texture Stud.*, 31, 225.

Park, B.S., Kang H.K., and Jang A. 2005. Influence of feeding beta-cyclodextrin to laying hens on the egg production and cholesterol content of egg yolk, *Asian-Australas. J. Anim. Sci.*, 18, 835.

Patterson, P.H., Cravener, T.L., and Hooge, D.M. 2004. The impact of dietary copper source and level on hen performance, egg quality and egg yolk cholesterol, *J. Dairy Sci.*, 87, 435.

Pesti G.M. and Bakalli R.I. 1998. Studies on the effect of feeding cupric sulfate pentahydrate to laying hens on egg cholesterol content, *Poult. Sci.*, 77, 1540.

Pike, O.A. and Peng, I.C. 1985. Stability of shell egg and liquid egg yolk to lipid oxidation, *Poult. Sci.*, 64, 1470.

Privett, O.S., Blank, M.L., and Schmit, J.A. 1962. Studies on the composition of egg lipid, *J. Food Sci.*, 27, 463.

Reiser, R. 1950. Fatty acid changes in egg yolk of hens on a fat-free and a cottonseed oil ration, *J. Nutr.*, 40, 429.

Rhodes, D.N. and Lea, C.H. 1957. Phospholipids. 4. On the composition of hen's egg phospholipids, *Biochem. J.*, 65, 526.

Rojas, E.E.G., Coimbra, J.S.D., and Minim, L.A. 2006. Adsorption of egg yolk plasma cholesterol using a hydrophobic adsorbent, *Eur. Food Res. Technol.*, 223, 705.

Rojas, E.E.G. et al. 2007. Cholesterol removal in liquid egg yolk using high methoxyl pectins, *Carbohydr. Polym.*, 69, 72.

Sato, Y., Watanabe, K., and Takahashi, T. 1973. Lipids in egg white, *Poult. Sci.*, 52, 1564.

Schreiner, M., Moreira, R.G., and Hulan, H.W. 2006. Positional distribution of fatty acids in egg yolk lipids, *J. Food Lipids*, 13, 36.

Sezille, G. and Biserte, G. 1964. Composition en acides gras des lipides du jaune d'œuf, *C. r. Seanc. Soc. Biol.*, 158, 1092.

Sim, J.S., Bragg, D.B., and Hodgson, G.C. 1972. Effect of dietary animal tallow and vegetable oil on fatty acid composition of egg yolk, adipose tissue and liver of laying hens, *Poult. Sci.*, 52, 51.

Smith, D.M. et al. 1995. Cholesterol reduction in liquid egg-yolk using beta-cyclodextrin, *J. Food Sci.*, 60, 691.

Smith, L.L. 1981/1987. *Cholesterol Autoxidation*, Plenum Press, New York.

Sun, R., Sivik, B., and Larsson, K. 1995. The fractional extraction of lipids and cholesterol from dried egg-yolk using supercritical carbon-dioxide, *Fett Wiss. Technol.*, 97, 214.

Suyama, K. et al. 1977. Lipids in the exterior structures of the hen egg, *J. Agric. Food Chem.*, 25, 799.

Swaczyna, H. and Montag, A. 1984. Beitrag zur analytischen Erfassung der Cholesterylfettsäureester in biologischem Material, *Fette, Seifen, Anstrichmittel*, 11, 436.

Szymczyk, B. and Pisulewski, P.M. 2003. Effects of dietary conjugated linoleic acid on fatty acid composition and cholesterol content of hen egg yolks, *Br. J. Nutr.*, 90, 93.

Ternes, W. 2001. Egg proteins, in: *Chemical and Functional Properties of Food Proteins*, Sikorski, Z.E. (Ed.), Technomic Publishing, Lancaster, PA, Chapter 12.

Tsutsui, T. and Obara, T. 1979. Lipid and fatty acid composition of low density lipoprotein and high density lipoprotein of egg yolk, *Nippon Shokuhin Kogyo Gakkaishi*, 26, 81.

Tu, C., Powrie, W.D., and Fennema, O. 1970. Steroids in egg yolk, *J. Food Sci.*, 35, 601.

Ushakova, T.M. et al. 1979. Hydrocarbons of hens'egg yolk, *Voprosy-Pitaniya*, 3, 69.

Uuganbayar, D., Shin, I.S., and Yang, C.J. 2006. Comparative performance of hens fed diets containing Korean, Japanese and Chinese green tea, *Asian-Australas. J. Anim. Sci.*, 19, 1190.

Valenzuela, A., Sanhueza, J., and Nieto, S. 2004. Cholesterol oxidized products in foods: potential health hazards and the role of antioxidants in prevention. *Grasas y Aceites*, 55, 312.

Wang, J.J. and Pan, T.M. 2003. Effect of red mold rice supplements on serum and egg yolk cholesterol levels of laying hens, *J. Agric. Food Chem.*, 51, 4824.

Warren, M.W. et al. 1991. Lipid-composition of hexane and supercritical carbon-dioxide reduced cholesterol dried egg-yolk, *Poult. Sci.*, 70, 1991.

Watkins, B.A. et al. 2003. Conjugated linoleic acids alter the fatty acid composition and physical properties of egg yolk and albumen, *J. Agric. Food Chem.*, 51, 6870.

Westermann, H. 1991. Beitrag zum Cholesteringehalt im Ei (Literaturübersicht), *Arch. Geflügelkunde*, 55, 49.

Yalcin, S. et al. 2006. Effect of garlic powder on the performance, egg traits and blood parameters of laying hens, *J. Agric. Food Chem.*, 86, 1336.

19 Modified Triacylglycerols and Fat Replacers

Marek Adamczak and Włodzimierz Bednarski

CONTENTS

19.1 INTRODUCTION

Fats and oil are major nutritional compounds and the main source of energy (38 kJ/g of fat) for the human body and cell compounds. They also provide regulatory substances that play important physiological roles. For a long time, polyenoic fatty acids (PEFA) *n*-3 and *n*-6 were identified as important nutritional compounds and later as being nutraceutical compounds with biomedical applications (Moghadasian 2008). They are also precursors of hormone-like compounds known as eicosanoides (Shahidi and Wanasundara 1998).

Unbalanced fat consumption can induce many diseases, i.e., cardiovascular disease (CVD) or cancer. Factors that increase the risk of coronary heart disease, e.g., distribution of plasma cholesterol over the low-density and high-density lipoproteins (LDL and HDL), oxidizability of LDL, and hemostasis, can be modified by changing the sources and amount of fats in a diet (Tarrago-Trani et al. 2006).

Naturally occurring fats and oils are not ideal products but modification of their composition, regio- and stereochemical structures can improve their properties and nutritional value.

Diets high in trans fatty acids (TFA) and saturated FA can cause many health problems, e.g., CVD and metabolic syndrome. The negative influence on health associated with "bad" fat consumption could be changed by modifying the fat composition. It is estimated that replacing 5% of the daily energy intake from saturated fats with carbohydrates and mono- or PEFA would reduce the risk of CVD by 22%–37%.

The first important problem is the definition and the regulations concerning TFA. In most countries, the definition of TFA is not only based on the chemical structure but it also distinguishes between different dietary sources of TFA and functional and metabolic aspects.

The evidence presented in epidemiological and clinical studies has shown that TFA can cause health problems, and these findings have resulted in regulations to limit the consumption of TFA. In Europe, the TFA content in margarine and spreads should be lower than 1% while in the United States information on the label about TFA content is required and products containing less than 0.5 g TFA per serving can be declared as "zero trans" (Bezelgues and Dijkstra 2009). There are also efforts (mainly in the United States) to make even more restricted regulations to reduce daily energy intake of TFA and saturated FA to less than 10%.

Nutrition experts have been encouraging consumers to consume unsaturated FA for a long time. The benefits of long chain n-3 and n-6 FA in promoting human health, especially children's growth and development, have been confirmed in many studies, e.g., a correlation was found between the amount of docosahexaenoic acid (DHA) in mother's milk and the baby's cognitive ability. DHA is now added to over 75% of U.S. infant formulas.

An important issue is also low-calorie food, which is mostly recognized as a low-fat food. The most important low-fat foods are dairy products, i.e., cheese, fermented products, and cream (Solan 2000). Low-calorie structured triacylglycerols (sTAG) are currently available on the market, e.g., Salatrim (Benefat™), which is an acronym for short- and-long-chain acyl TAG molecules, which contributes 25 kJ/g.

The fats in food products are responsible for many properties influencing the physicochemical characteristics of the product as well as consumer acceptance (Figure 19.1). The main trend in changing lipids has been their modification by different methods, lowering the TFA content, and removing or substituting fat in food products. However, following the trends to lower the energy levels in food products by removing fat can cause problems that have to be solved by using substitutes or fat mimetics possessing designed properties (Table 19.1). The multifunctional characteristics of fat make substitution a challenging task and it is required to determine the role fat plays in food products.

Fat substitutes are macromolecules whose physical and chemical properties are similar to TAG. The substitutes can be used in cooking or frying because they are stable under these conditions. They can usually replace the fat in food products based on a one-to-one ratio. Fat-based fat replacers are also either chemically or enzymatically modified acylglycerols, e.g., Olestra. Fat mimetics, i.e., protein- or

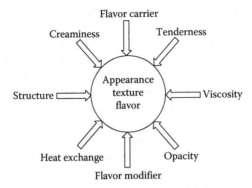

FIGURE 19.1 The possible functions of fats in food products.

TABLE 19.1
Examples of Fat Roles in Food Products and Changes after Its Removal

Fat Function in the Food Products	Food Product without Fat
Cheese	
Structure formation	Firmer
Specific flavor	Reduced flavor
Softness	Soft and pasty (high-moisture cheese)
Creaminess	Rubbery (low-moisture cheese)
Opacity	Translucent
Salad dressings and mayonnaise	
Specific viscosity	Reduced viscosity
Cling	Reduced cling
Thickness	Less body
Emulsion stability	Watery
Specific flavor	Reduced flavor and more sour
Opacity	Translucent
Baked goods	
Tenderness	Tougher and drier
Flakiness	Less flaky
Cell structure and crumb	Decreased volume
Aeration and volume	Decreased volume
Springiness	Denseness
Surface texture	Stickiness and gumminess
Moistness	Increased rate of staling

Source: Adapted from Miraglio, A.M., *Am. J. Clin. Nutr.*, 62, 1175, 1995.

carbohydrate-based fat replacers, imitate the properties of TAG but cannot substitute fat in equal amounts (Akoh 1998).

The third group of fat replacers consists of fat barriers that are used to block the absorption of fat during the cooking process when fat is used as a heat-exchange medium. Fried foods are often cooked with fat barriers to reduce the absorption of fat during frying. In some cases, fat extenders are also used which are responsible for optimization of fat functionality and decrease the amount of fat in a product.

Fat extenders are also classified as fat replacers. They are described as substances that optimize the functionality of fat, thus allowing a decrease in the usual amount of fat in the products. They provide a barrier for products that use fat as a heat-exchange medium, e.g., carbohydrate-based fat replacers such as cellulose, microcrystalline cellulose, modified starches, and gums.

Cost-effective and improved methods of fat modification and synthesis of fat replacers must meet nutritional and consumer requirements and enable production of a wider assortment of food products. The nutritional and functional properties of fats can be modified by using physical, chemical, and biotechnological methods.

19.2 CHARACTERISTIC OF FAT REPLACERS

Removing fat from food products is not always possible. Sometimes it is acceptable to replace water or air for fat. However, in many cases, a so-called systems approach should be applied which uses a combination of ingredients and processing methods to compensate for specific functions of the fat being replaced.

A common fat substitute is Olestra (a mixture of hexa-, hepta-, and octaesters of saturated and unsaturated FA and sucrose), which is a zero-energy product, stable under ambient and high temperatures (Table 19.2). The main concern regarding its safe use is about the reduction of vitamins A, D, E, and K and absorption and depletion of carotenoids. This substitute has also restricted application to savory snacks because of a slight aftertaste and a potential adverse effect on the wastewater treatment process.

The sTAG are a group of products that provide the physical properties of fat but also defined FA (including PEFA) and lower energy. The properties of short- (C_2–C_6, SCFA) and medium- (C_6–C_{12}, MCFA) FA influence its application as low-energy products. The small size of these molecules and solubility in water determine their properties, i.e., SCFA are faster absorbed in the stomach than other FA and provide low-energy (15–32 kJ/g), MCFA are preferentially transported via the portal vein to the liver (Straarup and Hoy 2000) and for entry into the mitochondria of all tissues, and MCFA are not carnitine-dependent (Babayan 1987). In addition to the properties of the FA, the design of the new sTAG should take into consideration the specificity of human pancreatic lipase.

The most popular fat-based replacers from this group are Salatrim and Carpenin. Salatrim is obtained by replacing long-chain FA in highly hydrogenated vegetable oils using a mixture of SCFA specifically distributed on the glycerol backbone. The properties of Salatrim can be modified by changing the FA composition and position at the glycerol backbone. Carpenin contains caprylic, capric, and behenic acid, and provides 17 kJ/g because of just partial adsorption of behenic acid by the body and

TABLE 19.2

The Characteristics of Selected Fat Substitutes and Their Possible Application

Product	Characteristic	Application
Sucrose FA polyesters	Sucrose polyesters of 6–8 FA/nondigestible	Food for ill people, savory snacks, benefits people at high risk of illness, e.g., coronary diseases, obese, colon cancer
Sucrose FA esters	Mono-, di-, and tri-esters of sucrose with FA/digestible	Emulsifiers and stabilizers in a variety of food products, lubricant, anticaking, antimicrobial, and thinning agent
sTAG and medium chain TAG (MCT), Jojoba oil	Designer TAG, TAG containing predominantly saturated C_8 and C_{10} and traces C_6, C_{12} FA/partially digestible	Health benefits, low-energy food, stable at high-temperature fat
Diacylglycerols	Diacylglycerols/digestible	Cooking and salad oils, health benefits, low-energy food, emulsifier
Other carbohydrate FA esters and polyol FA esters	Sorbitol, trehalose, raffinose polyesters, alkyl glucoside polyesters/nondigestible	Low-energy food, heat stable products, dressings, baked goods
Emulsifiers	Carbohydrate FA esters, acylglycerols, polyglycerol esters/partially digestible	Replace fat, form and stabilize aeration, lubricants, used in combination with other ingredients
Esterified propoxylated glycerols	Polyether polyol esterified with FA	Low-energy frozen desserts, salads dressings, spreads
Trialkoxytricarballylate, trialkoxycitrate, trialkoxyglyceryl ether	Polycarboxylic acids esterified with saturated or unsaturated alcohols (straight or branched 8–30 carbon atoms)	Fat-free or fat-reduced margarine, mayonnaise

metabolism of MCFA which is similar to that of carbohydrates. MCFA are also not accumulated as stored fat because they are not easily re-esterified into TAG.

Fat mimetics, which are protein- or carbohydrate-based fat replacers, provide from 0 to 17 kJ/g (Table 19.3). These products have the ability to absorb a substantial amount of water but they are not useful for frying at high temperatures because of protein denaturation and carbohydrate carmelization. Protein-based replacers are recommended for use at temperatures up to 15°C and usually in amounts up to 3% (w/w). These products usually act in food products as fat extenders and texture and viscosity enhancers (Leman 2007). The possibility of the use of specific enzymes for protein modification has led to the development of attractive fat mimetics. It is

TABLE 19.3
Examples of Protein- and Carbohydrate-Based Fat Mimetics

Product Category	Product	Properties
Protein-based	Microparticulated protein (Simpless)	Made from whey proteins or milk and egg proteins, low-energy product, many applications
	Modified whey protein concentrate (Dairy-Lo)	Controlled thermal denatured whey proteins, many applications
	Egg white and milk proteins, e.g., K-Blazer	Properties and application similar to microparticulated proteins
	Corn protein, e.g., Ultra-Bake	Low-energy product from corn protein
Carbohydrate-based	Cellulose, e.g., Avicell	Different forms of cellulose, replace partially of all fat in the product
	Dextrins, e.g., N-Oil	Obtained from, e.g., tapioca, applied in many food products
	Inulin, e.g., Raftiline	Low-calorie fat product, used in many products
	Maltodextrins, e.g., CrystaLean, Paselli, Lycadex	Used as fat mimetic (4 kcal/g), texture modifier or bulking agent, many applications
	Z-Trim	Zero-calorie product obtained from insoluble fiber from oat, soybean, pea, and rice hulls, corn or wheat bran, used in baked goods, cheese, ice cream
	Oatrim	Hydrolyzed, water-soluble oat flour containing beta-glucan, reduced energy product, many applications

important to analyze in detail the structural properties of the new products obtained through enzymatic modification.

19.3 NATURALLY MODIFIED FATS

19.3.1 MODIFIED VEGETABLE OILS

From a nutritional point of view, vegetable oils generally are the healthiest but they do not meet all technological requirements such as oxidative stability, melting point, and titer. In order to improve the oxidative stability, the oils should contain less PEFA and larger amounts of oleic acid. Oils containing a larger amount of oleic acid are especially suited for frying and baking (Hammond 2000a,b).

The modification methods for FA composition and vegetable oil properties are selected in regards to crop variety, strain, and location of crop cultivation. Oils of desirable technological and nutritional properties can be produced by selecting the varieties and strains of oil plants. Developments in breeding methods have produced new varieties and strains of oil plants that produce oils of modified and custom-designed FA composition and of desired properties. The breeding methods have been applied, first of all, to improve soybean, rape, sunflower, and flax (Hammond 2000a). Through breeding, a new double-improved rapeseed variety was introduced in the industry, which initiated widespread production of nutritionally valuable oil. Canola seeds are a source of erucic-acid-free oil and meal low in the undesirable sulfur-containing glucosinolates. Canola was developed through conventional plant breeding from rapeseed, but genetically modified canola, which is resistant to herbicide, was also introduced to Canada in 1995.

Desirable modifications of soybean oil obtained with conventional methods include: a reduction in lipoxygenase activity and linolenic and palmitic acid concentrations to the level of 3.0%–3.5% and 4.0%–8.4%, respectively (MacKenzie 1999). The above modifications serve the reduction of undesirable changes in oil flavor and odor, which are partially influenced by the activity of lipoxygenase.

The *in vitro* methods also include chemical mutation techniques with the use of *N*-ethyl-*N*-nitrosourea (ENU) and ethyl-methane sulfonate (EMS). These mutagenic agents have led to increases in the oleic acid concentration in *Brassica napus* seed oil to over 80% (MacKenzie 1999) (Table 19.4).

One way of modifying vegetable oil FA composition is to increase the concentration of both stearic and palmitic acids. Oils enriched in these acids crystallize in

TABLE 19.4
Examples of Transgenic and Mutant Plants with Modified FA Composition

Species/Oil	Description
Natreon canola oil	Mid-oleic canola oil. High stability oils for industrial applications
Soybean	Reduced linolenic acid in seed, which is determined by temperature
Mid-oleic sunflower (NuSun)	High stability oil for industrial frying, baking, and blending
	Saturated fat less than 10%, high oleic acid level (55%–75%) and linoleic acid (15%–35%)
Soybean	Increased oleic acid to 80% and reduced linolenic acid to less than 3%
Nutrium Low Lin soybean oil	Less than 3% linolenic acid soybean oil
Soybean	Increased stearic acid to 30%
Soybean/*n*-3 desaturase antisense	Reduced linoleic acid in seed to <1.5%
Soybean and canola/*n*-3 desaturase antisense	Reduced linoleic acid and increased oleic acid in seed (>80%)
Rapeseed/MCFA-specific thioesterase and β-ketoacyl-ACP	Increased medium-chain FA (>50%)

Source: Adapted from MacKenzie, S.L., Chemistry and engineering of edible oils and fats, in *Molecular Biotechnology for Plant Food Production*, Paredes-Lopez, O. (ed.), Technomic Publishing Co. Inc., Basel, Switzerland, 1999, 525; Kinney, A.J., *Curr. Opin. Biotechnol.*, 5, 144, 1994.

Note: ACP, acyl carrier protein.

cool storage without the need for prior hydrogenation. In this way, it is possible to obtain plastic fats at room temperature. Breeding methods allow soybean strains to be obtained, whose oil contains 28% stearic acid, 17% palmitic acid, and a total concentration of saturated acids of about 45% (Horejsi et al. 1994, Hartmann et al. 1996). By decreasing the level of linolenic acid and increasing the amount of stearic acid in soybean oil, oxidative stability improvement was observed (Kinney 1996). The decrease in unsaturated FA in soybean oil and an increase in saturated acids improve oxidative stability and enable production of plastic fats. This oil enables the elimination or limitation of hydrogenated oils and decreases the level of TFA in food.

New sunflower and safflower (Sunola and Saffola) varieties were also obtained as the result of breeding. Modified sunflower oil contains less PEFA, over 80% oleic acid, 7% saturated FA, and a significantly reduced amount of linolenic acid. The ratio of these FA is thus similar to that in olive oil. Such modified sunflower oils are especially suited for frying and baking and do not form any foreign flavors, which occur when oils containing large amounts of PEFA are used for this purpose (Hammond 2000a).

19.3.2 Fat Modified by Feeding of Animals

Although nutritionists object to the excessive consumption of animal fat, this component makes up a considerable percentage of global food consumption. In order to improve its nutritional value, it is modified with regard to animal lipid composition and properties. The fat present in milk, eggs, and poultry meat is of particular interest. Nutritional modification of the amount and quality of cow milk fat plays an important role and can be done efficiently. However, fat modification by feeding is limited due to biohydrogenation in rumen, resulting in a lower supply of unsaturated FA from feed. In addition, animal fats are modified in respect to sterol composition and physical properties.

In spite of the biohydrogenation of unsaturated acids in rumen, administering direct doses of vegetable and fish oils (supplemented to feed) results in larger concentrations of unsaturated FA such as $C_{18:1}$, $C_{18:2}$, $C_{18:3}$ in milk fat and lower-saturated acids.

In order to improve the effectiveness of milk fat modification and limit biohydrogenation, animals are fed encapsulated fats in the form of, e.g., full oil seeds of flax, rape, vegetable oil preparations encapsulated in casein denatured with formaldehyde, FA calcium salts, or FA amides.

Conjugated linoleic acid (CLA), also referred to as rumen acid (*cis*-9, *trans*-11-octadecadienoic acid), is a nutritionally important fat component. If present in the human diet, it prevents obesity, acts as an anti-atheromatosic and anti-neoplastic compound, and also stimulates the immune system. Its concentration in milk fat is determined by the feeding method. For example, its level can be increased in milk twofold in comparison to that produced in winter by letting cows graze on a pasture field. Milk products can also be enriched in CLA by introducing lactic bacteria into the cow's diet. Feeding strategies that promote CLA production in the rumen and milk depend on the composition of lipid substrates and plant oils containing linoleic and linolenic acid, e.g., sunflower, soybean, and rapeseed are very efficient at increasing milk CLA content (Lawson et al. 2001).

Perfecting the biosynthesis of fats by specially selected microorganisms has led to the effective production of fungi and sea-weed biomass containing γ-linolenic acid (GLA), eicosapentaenoic acid (EPA), and DHA. Animal feed supplemented by this biomass produces increased amounts of these acids in the lipids of milk, eggs, and meat (Hridinka et al. 1996). The effect of feed composition on fat modification was tested in feeding pigs and chickens, and was shown to be highly effective in the modification of animal fat composition with respect to the incorporation of essential PEFA. Enriching animal fat with oleic acid has a beneficial effect on the levels of cholesterol and nutritionally undesirable TAG in the blood plasma of the consumers. It was shown that a supplement of soybean oil in the feed of chickens (broilers) resulted in higher levels of linoleic acid in lipids of the leg muscles. To a smaller extent, this effect in poultry muscle lipids can be obtained by using 9% rapeseed oil and 2% soybean oil in the chicken feed. This resulted in an increase in the concentration of linolenic acid in the breast muscle (3.5%) and thigh muscle (6.2%) lipids. In addition, by adding vitamin E to the feed, this modification was more effective and the level of n-3 PEFA was higher than in the muscle lipids of chickens fed diets not supplemented with vitamin E. In the nutritional modification of the poultry muscle lipid composition and properties, it is important to adequately select the fats to be incorporated in the feed, their composition, and properties, as well as to adjust the diet size to maximize the effect on the oxidative stability and sensory properties of the meat that incorporates the modified lipids.

The nutritional modification of egg yolk lipids mainly involves an increase in the concentration of mono- and PEFA acids and a decrease in the cholesterol concentration. Feeding layer diets with an increased concentration of olive oil rich in $C_{18:1}$ acid resulted in an increase of this acid in the egg yolk lipids. However, the greatest increases in the concentration of PEFA in egg yolk lipids were obtained by supplementing feed with fish oils or sea-weed biomass rich in these acids.

Nutritional methods can also be applied to enriching egg yolk lipids with CLA by administering it in adequate doses with the feed. The effectiveness of nutritional methods in the reduction of cholesterol in egg yolk lipids is low. The use of 3-hydroxy-3-methylglutarylo-CoA reductase for this purpose reduced the cholesterol level in egg yolk lipids by 30%; however, this method reduced productivity (Pszczoła et al. 2000).

19.4   LIPIDS MODIFIED BY PHYSICAL AND CHEMICAL METHODS

19.4.1   BLENDS

The physical methods used for fat modification include the following techniques: blending, fat fractionation at different temperatures, extraction with supercritical liquids (mainly supercritical CO_2), microwave treatment, or conventional extraction with organic solvents.

Blending is the simplest method of fat modification and involves mixing two or more different kinds of fats to produce fat with greater oxidative stability, higher nutritive value, and required melting behavior with a low or zero-level of TFA.

Blending vegetable oils of different origins or their combination with animal fats, e.g., with milk fat, is justified for technological, nutritional, and economic reasons. If the FA composition of different vegetable oils is known, they can be combined in certain proportions to obtain products of improved composition and better functional properties, e.g., improved oxidative stability or shifted melting point. Blending sunflower oil with soybean, rapeseed, or corn oil can serve this purpose (Hammond 2000a,b). Similar effects can be obtained by combining soybean oil with palm olein or flax oil with triolein.

In order to modify the physical and nutritional properties of milk fat, butter is enriched with vegetable oils such as sunflower, soybean, rapeseed, and palm oils. These oils can be added separately or in combinations to the cream or directly to the final product. In this way, milk fat can be enriched with essential PEFA to improve the nutritional value of dairy products. In addition, changes in the unsaturated–saturated FA proportions can result in favorable rheological modifications of butter-like dairy products. Supplementing cream before churning or butter with oils is also beneficial for the standardization of the rheological properties of butter-like products all year round. The effectiveness of the modification of physical properties, mainly the rheological properties of butter-like products, is greatly determined by the type of oil used (Shen et al. 2001). The selection of oils and their proportions also determines the nutritional value and functional properties of margarine, which also consists of blended fats.

19.4.2 FRACTIONATION

Fat fractionation involves fat separation into different fractions, by structure, molecular size, melting point, titer, and solubility in different solvents or volatility. The selection of the method and process conditions is based on the knowledge of the fat's structure, its behavior at different temperatures, and the mechanism of β and β' fat crystal formation, which determines the melting point value. Fats can be fractionated directly after melting or after dissolving in acetone, ethanol, or hexane. The composition of the fractions obtained with the use of solvents is different from the fraction composition obtained as the result of dry fractionation without solvents. Fractionation with organic solvents is less time consuming and the separation of crystals from liquid fractions is easy. The drawback of this method is that compounds responsible for flavor and odor are removed with solvents during the fat-refining process.

The technology of milk fat fractionation does not use solvents but crystallization from melted fat. Crystalline fractions of different chemical composition and different physical properties are formed at controlled temperature (Bhaskar et al. 1998). The results of fractionating are solid and liquid fractions with melting points of 48°C and 25°C, respectively. The melting point of unfractionated fat was 41.6°C. The liquid fraction in comparison to the solid fraction had 17% more $C_{8:0}$–$C_{10:0}$ FA and 11% less $C_{12:0}$–$C_{16:0}$ FA and 32% less $C_{18:0}$ FA but 41% more $C_{18:1}$, $C_{18:2}$, and $C_{18:3}$ FA.

By fractionating, products of different physical properties and different nutritional values can be obtained.

Progress in fat fractionation has resulted in the use of extraction by CO_2 in supercritical conditions. This is applied to obtain milk fat fractions enriched with the high-melting-point TAG with reduced amounts of cholesterol, vitamins, and some flavor- and odor-responsible components. After extraction with CO_2, the resulting fraction has a higher melting point and 2.5-fold more TGA, including $C_{10:0}$–$C_{12:0}$ acids than the initial anhydrous fat, and the unsaturated-to-saturated-acid ratio is 0.75 and 0.57 for the resulting fat and initial fat, respectively. This method allows the separation of fractions differing in respect to their FA composition from milk fat. Solid fractions are enriched with TAG and with a large number of carbon atoms that contain larger amounts of long-chain FA. The separated liquid fractions are enriched with low-molecular TAG with large amounts of SCFA that play an important role in nutrition. Extraction with CO_2 also allows the differentiation of the cholesterol amounts in the resulting fractions.

Another possibility for the modification of fat composition and physical properties, e.g., milk fat, is the application of molecular distillation. This method involves the evaporation of more volatile molecules and their condensation. After the process of milk fat distillation, fractions rich in short- and long-chain TAG and with a reduced amount of cholesterol are obtained. These fractions are used for the production of dietetic butter for people suffering from disturbances in fat assimilation.

19.4.3 HYDROGENATION

The mechanism and a key parameters influencing TFA formation during hydrogenation have been discovered during recent decades. It is now accepted that the mechanism of hydrogenation of oils follows a Horiuti–Polanyi mechanism.

Fat hydrogenation, also referred to as fat hardening, is the main, classical process of vegetable oil modification. In this process, double bonds of the FA are saturated with hydrogen. This is a typical chemical process in which, by using different catalysts and changing process parameters, mainly temperature, it is possible to obtain products with a wide variety of properties and FA composition. Through partial hydrogenation, it is possible to obtain liquid frying oils. The process of hydrogenation is usually performed in a three-phase system consisting of: gaseous hydrogen, liquid oil, and a solid catalyst (generally nickel), in semi-continuous reactors at temperatures up to 250°C with a hydrogen pressure of up to 5 bar. Most of the hydrogenated fats are characterized by a melting point at about 35°C and are used in the production of margarine, confectionery, and baking fats. Partially hydrogenated fats contain up to a few dozen percent of TFA and final products based on such fats used for frying and baking contain from 6% to 37% TFA.

TFA always occur in oil-fried products as they are formed at high temperatures. High temperatures applied during fat refining, e.g., deodorization at 180°C–260°C, can also cause double-bond isomerization. Advances in fat hydrogenation, which optimize the process and selection of catalysts, have resulted in a gradual reduction of TFA in partially hydrogenated fats. A new hydrogenation procedure has been developed in Bunge Oils, Inc. that involves the use of a conditioned nickel catalyst that selectively prevents the formation of *trans* isomers. Another modification of hydrogenation is presented that uses supercritical carbon dioxide or a mixed catalyst (nickel and palladium) at lower temperatures. A new electrochemical hydrogenation method with nickel and palladium catalysts and formate as an electro-catalyst was also introduced. However, all of presented procedures have resulted in hydrogenated fat that still contains less than 10% TFA.

The results of complete hydrogenation, in which all bonds of unsaturated FA are saturated by hydrogen, are hardened fats that undergo further modification by interesterification. Highly hydrogenated fats have a low concentration of unsaturated FA, TFA, and higher oxidative and thermoxidative stability than the initial oils. However, the production of zero-trans fats by hydrogenation is a nonachievable goal, mainly because of the cost involved. The other alternatives for the production of zero-TFA are: the application of plant breeding and genetic engineering techniques to produce vegetable oils with modified composition, the use of tropical oils, and the application of interesterification to modify the properties and FA composition of food products. The application of these methods has led to the marketing of the first group of products containing zero-TFA, e.g., snacks (Frito-Lay, UTZ), spreadable butter (Land O'Lakes), margarines (Lipton, Fleischmann's), and a wide range of products from Kraft.

19.4.4 CHEMICAL INTERESTERIFICATION

Chemical interesterification has been known since the mid-1800s and is used to change FA moieties within and among TAG. It has also been used in the food industry since the 1940s, and in the 1970s there was renewed interest in this process for synthesizing zero-trans-margarines.

Interesterification is an interchange of acyl groups between two esters, ester and alcohol (alcoholysis) or ester and acid (acidolysis). The latter two reactions are called transesterification. However, the nomenclature used in the literature to describe each reaction is different, and "transesterification" is a synonym of "interesterification." Interesterification is most widely applied in fat modification because it relocates FA in TAG of one or several modified fat components. In this process, acyl groups are repositioned both within TAG molecules (intramolecular) and between different molecules (intermolecular). The exchange of FA reaches a thermodynamic equilibrium, which can be predicted by the laws of probability. By selecting reaction components, catalysts, and process temperature, modified fat products with custom-made functional properties can be obtained.

The most often used catalyst for chemical interesterification is sodium methoxide (0.2%–0.3%), which is more active than other base, metal, and acid catalysts (0.1%–9.2%). Sodium alkylate catalysts are easy to use, inexpensive, active at relative low temperatures, and are used in small concentrations. Two mechanisms for chemical interesterification are proposed: a carbonyl addition mechanism or an enolate intermediate (Claisen condensation) mechanism. These two mechanisms are basically very similar, and the main role in the reaction is played by the ester carbonyl group ($C=O$) of TAG.

The oil blends before the reaction need to be dried under vacuum at a higher temperature than 90°C for 0.5–1 h (Xu et al. 2006). Interesterification is frequently used for modifying blends of fats of different melting points. In this case, an appropriate reaction temperature must be selected, usually 60°C–70°C, because a liquid state of all the substrates facilitates an even distribution of a particular FA within TAG and consequently produces a very advantageous structure resulting from a fine, strong crystal structure. The reaction is terminated by the addition of water, and subsequent processes encompass the purification steps of randomized fat (Figure 19.2). During chemical interesterification, a substantial loss of oil is observed (about 30%) because of the formation of soap and FA methyl esters.

The fact that fine crystalline β′ form prevails in the products containing interesterified fats also reduces the range of undesirable polymorphic changes in fats during their storage (Rousseau et al. 1998). Such modified fats are used in the production of margarine with a high content of PEFA. Merely 18% of interesterified fat in margarine allows the incorporation of 82% of oil in a product matrix, while margarine itself remains solid. The application of interesterified fats facilitates the production of nutritionally valuable margarines and special-purpose confectionery fats and shortenings.

Despite the lack of selectivity, some structured lipid-like products have been synthesized by chemical methods (Table 19.5). For example, Benefat is prepared by interesterification of triacyl-acetin, propionin, or butyrin, or their mixtures with either hydrogenated canola, soybean, cottonseed, or sunflower oil. TAG with three SCFA are removed in the process and typical molecules contain 30–67 mol% SCFA and 33–70 mol% long-chain FA (LCFA); stearic acid is the predominant LCFA (Akoh and Kim 2008).

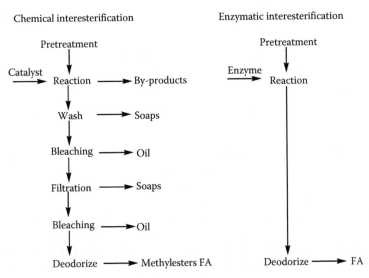

FIGURE 19.2 Simplified comparison of chemical and enzymatic interesterification. (Modified from Graham, I.A., et al., *Curr. Opin. Biotechnol.*, 18, 142, 2007.)

TABLE 19.5

Selected Examples of the Products Obtained by Chemical or Enzymatic Interesterification

Process	Product	Characteristic
Chemical interesterification or esterification	Benefat (Salatrim)	Low-energy product of interesterification of SCFA and LCFA (fully hydrogenated oil)
	Enova	Low-energy product of esterification of soybean/canola unsaturated FA with glycerol (contain 80% DAG)
	Zero-TFA fat	Obtained by interesterification of soybean oil and fully hydrogenated soybean oil
	Neobee MCT and MLT-B	Shortenings made by esterification of MCFA with glycerol or by interesterification of MCT with fully hydrogenated soybean oil
	Caperin	Interesterified coconut, palm kernel, and rapeseed oil
Enzymatic reactions	NovaLipid (the first fat created using the enzymatic interesterification)	Shortenings made from interesterified soybean oil with fully hydrogenated soybean and cottonseed oil
	Human milk substitute	Product of interesterification catalyzed by Lipozyme TL IM between lard and soybean oil
	Cocoa butter analog	Obtained by interesterification of camel hump fat in supercritical carbon dioxide Reaction catalyzed by Lipozyme TL IM
	Fat for table margarines and spreads or dressings	Interesterification between TAG rich in PEFA and palm/palm kernel oils. Reaction catalyzed by *Candida parapsilosis* lipase
	sTAG with caprylic acid located at *sn*-1 and *sn*-3 and DHA at *sn*-2	Product of acidolysis catalyzed by Novozym 435
	sTAG, plastic fat for various culinary purposes	Obtained by acidolysis of stearic acid and rice bran oil

Note: MCT, medium-chain triacylglycerol.

Mono- and diacylglycerols (MAG, DAG) are important products that can be synthesized by chemical glycerolysis processing for 30–60 min at 220°C–260°C with inorganic catalysts (NaOH or $Ca(OH)_2$). To remove the impurities from the reaction mixture, distillation and molecular distillation should be used. The progress of the reaction can be controlled to a certain extent by the glycerol-to-oil ratio and the physical parameters of the reaction. DAG are used in the composition of cooking oils because of their role in reducing the levels of serum TAG. The first DAG oil has been marketed in Japan under the brand name Healthy Econa Cooking Oil since 1999. The same product (named Econa oil) has been available in the United States since January 2005. MAG and DAG have been approved by the EU as food grade additives and they have the status of Generally Recognized as Safe (GRAS). Information about the properties and the application of acylglycerols is presented in Chapter 20.

19.5 BIOTECHNOLOGICAL MODIFICATION OF FAT

19.5.1 ENZYMATIC MODIFICATION OF FATS AND OILS

The structure and composition of sTAG has resulted, besides the low-calorie intake and functional properties (plastic fats), in many benefits for human health: superior nitrogen retention, enhanced absorption of the FA at the *sn*-2 position, reduction in serum TAG, LDL-cholesterol and total cholesterol, improved immune function, prevention of thrombosis, improved absorption of other fats, etc. sTAG are often compared and described as "nutraceuticals" or "pharmafoods" and from a nutritional point of view as "functional lipids" (Bornscheuer 1999, Bornscheuer et al. 2003). The "true" sTAG can be best obtained through enzymatic modification of existing products or by sTAG synthesis from glycerol and FA as the result of esterification, acidolysis, alcoholysis, or interesterification (Lee and Akoh 1998, Xu 2000, Osborn and Akoh 2002) (Figure 19.3).

Lipases are the basic tool for the biotechnological modification of fats and oils and are used in interesterification or to obtain specific sTAG, i.e., TAG of a defined

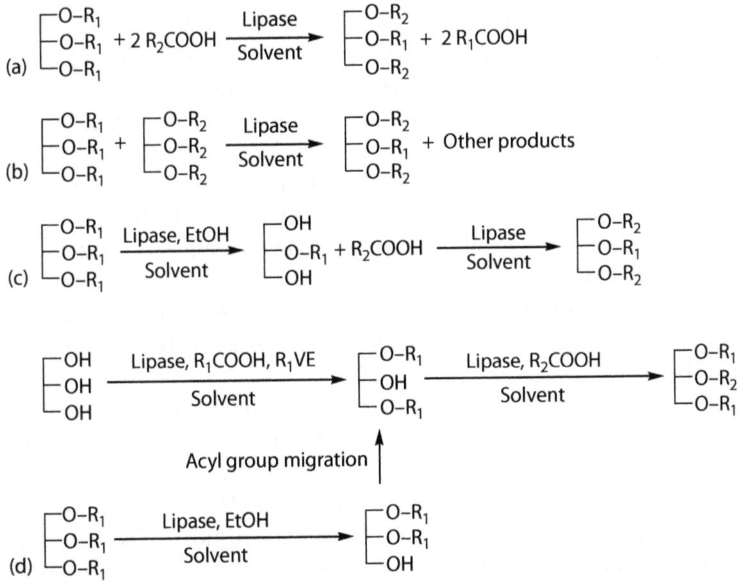

FIGURE 19.3 Possible methods of enzymatic synthesis of sTAG by (a) acidolysis, (b) interesterification, and (c, d) two-step processes with mono- or diacyl-glycerols, respectively, as semi-products (the reactions can be also performed in a solvent-free system). (Modified from Bornscheuer, U.T. et al., Lipase-catalyzed synthesis of modified lipids, in *Lipids for Functional Foods and Nutraceuticals*, Gunstone, F.D. (ed.), The Oil Press, Bridgwater, U.K., 2003, p. 149.)

stereochemical structure, modified qualitative and quantitative FA composition compared to the natural TAG. Progress in the enzymatic modification of fats and oils has resulted from research into lipases gathered within the last 15 years (EC 3.1.1.3) and made commercial enzymes available (Kazlauskas and Bornscheuer 1998, Bornscheuer 2000). The application of lipases in lipid modification is attractive because they show three main types of selectivity:

1. Regio-selectivity
   a. *sn*-1,3-Regio-selectivity
   b. Non-selectivity
2. FA selectivity
   a. LCFA
   b. Saturated FA
   c. *cis*-ω-9 unsaturated FA
   d. SCFA
3. Acylglycerol selectivity
   a. MAG
   b. MAG and DAG
   c. TAG

The selectivity of lipases is influenced by many parameters described by medium engineering: solvent polarity, water activity coefficient, substrates, lipase, carrier used for enzyme immobilization, and procedure of immobilization, etc. (Adamczak 2004, Adamczak and Krishna 2004). The new possibilities of obtaining tailor-made lipases for lipid modification are: directed evolution, rational protein design, and metagenome exploration (Bornscheuer 2008, Urban and Adamczak 2008).

The mild reaction conditions and the selectivity of lipases make the process easier mainly because of the lack of a need to purify the final product (Figure 19.3). If the chemical reaction is performed with ABA and BBB type TAG (A, B–acyl group), 18 isomers can be obtained, and if the reaction is catalyzed by *sn*-1,3-specific lipase, 8 TAG isomers can be detected in the reaction mixture (Xu et al. 2006). The advantage of lipase selectivity is also visible when the structure of the obtained products is analyzed compared to that of the TAG obtained by interesterification (Table 19.4).

For the synthesis of sTAG, enzymes that are immobilized to increase their stability in extreme temperature, pH, organic solvents and are more easily recovered are usually used. The most popular commercial immobilized lipases used for lipid modification are as follows: *Candida antarctica* lipase B (Novozym 435), *Rhizomucor miehei* lipase (Lipozyme RM IM, bonded to the ionic resin), or *Thermomyces lanuginosa* lipase (Lipozyme TL IM, silica granulated enzyme). The synthesis of sTAG can be performed by either one- or two-step methods (Figure 19.3a and b). The one-step method is relatively easy to perform but it can be difficult to control the selectivity of the reaction and obtain high-yield sTAG. A well-known example of a one-step process is the synthesis of cocoa-butter equivalent. Cocoa butter is predominantly 1,3-disaturated-2-oleyl-glycerol, containing more than 95% palmitic, stearic, and oleic acids. The lipase-catalyzed synthesis of cocoa-butter equivalent from palm oil and stearic acid uses sn-1,3-selective lipases to replace palmitic acid with stearic acid at the sn-1- and sn-3-positions.

The two-step enzymatic synthesis of sTAG seems to be an alternative to one-step reactions. Although these methods are time consuming and require purification of intermediates, they can result in tailor-made sTAG (Figure 19.3c and d). In practice, however, the acyl group migration occurring in an aqueous medium leads to the formation of undesired by-products. The first step in the reactions involves alcoholysis of appropriate TAG to obtain 2-MAG. In the second step after separation, this product is esterified with a FA. The key to the successful development of the two-step synthesis of sTAG was the discovery that high concentrations of 2-MAG can be obtained by alcoholysis of TAG in aprotic solvents. The process should be carried out in conditions preventing migration of acyl groups and must be catalyzed by sn-1,3-selective lipases (Bornschuer et al. 2003).

Lipases can be used for the synthesis of concentrates of PEFA or an acylglycerol fraction rich in these FA, as not so many enzymes selective toward PEFA have been discovered. However, FA selective enzymes can theoretically also be used for removing TFA from food fat. There are still basic problems to be solved in order to successfully use the enzymatic method of sTAG synthesis, e.g., process scaling-up. Further studies are also necessary to describe sTAGs' nutritional and other properties such as oxidative stability (Bornschuer et al. 2003).

The mechanism of enzymatic glycerolysis is similar to a chemical reaction. A mild-condition of the reaction and use of solvents enable the application of procedures to quickly reach the equilibrium of the reaction. The efforts of "fishing" the product include crystallization, adsorption, membrane separation, extraction, etc. and also improve the progress of the reaction (Bornscheuer 1995, Peng et al. 2000). Other approaches to improve the yield of MAG and/or DAG synthesis involve the application of protected glycerol, FA vinyl esters, or adsorption of glycerol on silica gel. Recent progress in the yield of synthesis of acylglycerols was obtained in a reaction catalyzed by *C. antarctica* lipase B, using solvent partitioning in an ionic liquid:*tert*-butanol or *tert*-butanol:*tert*-pentanol system (Guo and Xu 2005, Damstrup et al. 2007). In an industrial installation operating for about 90 days, a high capacity of lipase with 2000 L of pure MAG per 1 kg of enzyme was obtained.

The idea of biohydrogenation, the development of a natural hydrogenation process, has been known for many years and concerns rumen and lactic acid bacteria. However, there are at least three properties of the biohydrogenation reaction that limit its application: (1) migration of double bounds can occur, (2) the enzymes of the system act on free FA and not TAG, and (3) the enzymes are rather unstable and require cofactors. Enzymes can be also useful for the synthesis of PEFA by using desaturases as discussed below. In addition, isomerases and hydratases, e.g., from lactic acid bacteria, can be used to modify the FA composition of the fats.

Lipases are used in lipophilization reactions involving the attaching of a lipophilic moiety to a hydrophilic one, such as water-soluble vitamins and phenolic compounds (Villeneuve 2007, Adamczak and Bornscheuer 2009). The obtained amphiphiles are soluble in lipophilic media and their biological activity is unchanged.

Phospholipases are also used for the modification of phospholipids. These enzymes are not only used in the degumming of oils but are also able to modify and synthesize specific biologically active phospholipids (De Maria et al. 2007). The details of phospholipase application are discussed in Chapter 20.

19.5.2 LIPIDS FROM GENETICALLY MODIFIED ORGANISMS

In addition to previously developed methods involving culture procedures that promote desirable modification of vegetable oils, genetic engineering techniques are also becoming common (Kinney 1996, Zou et al. 1997, Smith et al. 2007). The application of genetic engineering to perfecting the qualitative and quantitative properties of FA in oils has proved to be a difficult task. Knowledge of the metabolic paths of FA synthesis has been considerably enriched recently. Additionally, several new enzymes participating in seed oil biosynthesis, as well as new key genes determining the synthesis of certain FA and TAG, have been discovered. Particular progress, however, was achieved in defining desaturases and related diiron-oxo proteins in plants. It seems that plants possess two main families of diiron-oxo enzymes: (1) located in plastids, dissolved acyl-acyl carrier protein (ACP) desaturases reacting with C_{14}, C_{16}, and C_{18} saturated acyl-ACP substrates (class II of the diiron-oxo enzymes); and (2) membrane-bound desaturase using complex lipids or acyl-CoA as substrates (class III of the diiron-oxo enzymes) (Murphy 1999). The characteristics of some desaturases already enabled the application of the point mutation to obtain, for example, palmitoyl-ACP-specific enzyme from Δ^9 stearoyl-ACP desaturase. It now seems only a matter of time until a method of rational design of desaturases will be possible for controlling enzyme selectivity. Enzymatic processes will then replace chemical synthesis completely and the insertion of conjugated double bonds will be possible without the use of chemicals.

New opportunities are also offered by transgenic oil crops containing large amounts of DHA, EPA, and GLA. Plant desaturases with an amino-terminal cytochrome b_5 domain have been selected. These are enzymes that can participate in the introduction of a double bond into an acyl chain between an existing double bond and the carbonyl terminal (Murphy 1999).

Modifications by genetic engineering techniques can lead to quantitative changes of natural FA in a vegetable oil or can be used to introduce new, unnatural FA. Frequently, in the modification of FA composition in vegetable oils, there is a problem of reaching a compromise between, on the one hand, obtaining the required product and, on the other, solving crop cultivation problems (reduced crop yield, reduced resistance to diseases). Backcrossing seems to be a solution. It involves crossing a transgenic plant with material with favorable agrotechnical properties.

In spite of the considerable progress in FA synthesis by transgenic oilseed plants, few genetically recombined commercial products are available. One of them is rapeseed oil modified through cloning lauroyl-specific acyl-ACP thioesterase from *Umbellularia californica* (California Bay) to *B. napus* (Del Vecchio 1996). The resulting oil, Laurical®, contains about 40% lauric acid. It seems that the only visible possibility of a further increase in the concentration of lauric acid to 50%–60% is an additional cloning of *sn*-2 acyltransferase gene.

An increase in the lauric acid concentration in the *sn*-1 and *sn*-3 positions induces, among others, β-oxidation and the glyoxylate cycle, leading to a reduction in the lauric acid concentration for the benefit of acetyl-CoA and sucrose. All these additional treatments generate additional costs, and, moreover, multiple cloning may cause gene expression instable.

The isolation of a *B. rapa* stearoyl-ACP desaturase cDNA (accession no. X60978) and expression of antisense stearoyl-ACP desaturase constructs in both *B. napus* and *B. rapa* has led to the inhibition of oleic acid and increased the stearic acid concentration in oil by up to 40% (Knutzon et al. 1992).

The modification of vegetable oil composition involves a reduction in the saturated FA concentration in soybean oil from 15% to less than 4% by suppressing the activity of a type II acetyl-ACP thioesterase (Kinney 1996). The other direction of genetic modification is toward increasing the concentration of unsaturated FA in oils. For example, higher unsaturated FA concentrations in canola seed oil (from 68% to 83%) were obtained by suppressing the activity of the Δ^{12} fatty acid desaturase gene (Kinney 1996). High oxidative and temperature stability of genetically modified high oleic sunflower oil with a linoleic acid reduced from 71.6% to 5.5% has been confirmed (Smith et al. 2007).

Vegetable oil modification not only involves qualitative and quantitative FA manipulations, but also the manipulation of such components as tocopherols and phytosterols. In order to fully characterize oils from genetically modified plants, their stereochemical structure must be determined because it determines the properties and oil suitability for different applications.

The first examples of the production in transgenic plants of C_{20} PEFA were published just a few years ago (Graham et al. 2007). It was demonstrated that transgenic *Arabidopsis* had arachidonic acid (AA) and EPA levels of ~7% and 3% of total FA, respectively. However, the expression was only reported for leaf tissue and not in seed oils (expression controlled by cytomegalovirus, CMV). The application of seed-specific promoters to express C_{20} PEFA in transgenic tobacco and linseed resulted in low levels of AA and EPA and a high content of intermediates (Figure 19.4). Detailed biochemical analysis indicated that the activity of heterologous Δ^6-elongase was inhibited *in planta* because of the absence of substrate FA in the acyl-CoA pool.

It is inevitable that new oils will be produced from genetically engineered plants. This, however, will require in-depth studies into the mechanisms of FA and oil synthesis and especially into the interactions between particular stages of their synthesis and the controlling mechanisms. Genetic engineering techniques are also being applied in the modification of FA concentration and composition in, e.g., milk, eggs, and meat.

19.5.3 ALTERNATIVE FATS AND OILS

For centuries, fat shortages have led to a wider interest in alternative lipid sources. Microbiological lipids received a lot of attention when they were found to be attractive due to their very desirable FA composition and the advantageous stereochemical structure of the TAG (Leman 1997, Ratledge 2001) (Table 19.6). Microbiological lipids are also attractive thanks to their wide range and advantageous yield of microbial biomass in media supplemented with a wide range of substrates that are frequently waste products (Leman 1997). Many oleaginous microorganisms were characterized

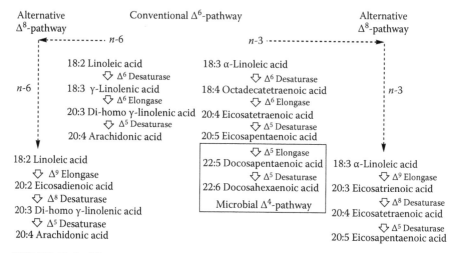

FIGURE 19.4 The generalized scheme of aerobic long chain-PEFA synthesis by predominant Δ^6-pathway.

TABLE 19.6
Lipids in Biomass of Lipid-Accumulating Organisms

Organism	Lipid Content (% w/w of Biomass)	Main FA (% of Total)
Chlorella vulgaris	39	$C_{18:1}$—58
Apiotrichum curvatum	58	$C_{18:1}$—44
Candida lipolytica	32–36	$C_{18:2}$—51
Mucor ramannianus	56	$C_{18:3}$ $\omega6$—31
Aspergillus terreus	57	$C_{18:2}$—40
Tolyposporium ehrenbergii	41	$C_{18:1}$—81

Source: Adapted from Weber, N. and Mukherjee, K.D., Lipid biotechnology, in *Food Lipids: Chemistry, Nutrition, and Biotechnology*, Akoh, C.C. and Min, D.B. (eds.), CRC Press, Boca Rota, FL, 2008, 707.

in respect to their lipid accumulation capacity. Oleaginous microorganisms are defined as those whose biomass contains over 20% lipids in dry matter.

The synthesis of microbial fat by bacteria is often ignored because the average fat concentration in dry biomass does not exceed 10%. However, there are strains of *Arthrobacter* sp., *Mycobacterium*, and *Corynebacterium*, which are able to accumulate from 30% to 80% lipids in dry matter (Ratledge 2001). Unfortunately, there are other problems related to low growth rate and yield of bacteria, lipid extraction, and possible allergenity and toxicity of resulting lipids. Microalgae, e.g., *Botryococcus braunii* and *Chlorella pyrenoidosa*, serve as attractive sources of PEFA, and the content of fat in dry biomass can reach 85% (Kay 1991, Ward and Singh 2005).

Moreover, microalgae are a very attractive source of EPA and DHA. For example, in oil from *Crypthecodinium cohnii* the DHA concentration is about 40% (Gunstone 1998). Lipid synthesis by microalgae is difficult to justify for economic reasons. In spite of solutions decreasing the costs of their synthesis, microalgae require special growth conditions and costly and difficult methods for lipid separation from microalgae.

One of the key achievements in expanding the pool of DHA and EPA is the isolation of the gene encoding synthesis of Δ^5 desaturase from the yeast *Mortierella alpina* (Michaelson et al. 1998). Obtaining PEFA from microbial biomass is justified for economic reasons. Moreover, microbiological oil is cholesterol-free and not contaminated with heavy metals or pesticides.

GLA synthesized by the *Mucor javanicus* (synonym: *Mucor circinelloides*) strain was commercially available in Great Britain as Oil of Javanicus. Unfortunately, in spite of its relatively high GLA concentration of 15%–18%, this product could not resist the competition of cheaper sources of GLA: evening primrose, borage, and blackcurrant oils. In Japan, oil from *Mortierella isabellina* is a commercially available GLA source.

M. alpina is one of the most valuable sources of PEFA. Many firms use microorganisms to synthesize oil with over 40% of AA. Controlling the synthesis of certain PEFA by *M. alpina* became possible after obtaining mutants with varied desaturase activity (Certik et al. 1998). By modifying the culture conditions for this fungus, the PEFA composition was considerably altered. For example, the reduction of the temperature and supplementing the culture liquid with α-linolenic acid stimulated the synthesis of EPA. A mixed culture of *M. alpina and Pythium irregulare* cultured on medium with processed canola flake and cake produced modified lipid with increased amount of γ-linoleninic acid, AA, and EPA. The ratio of PEFA-to-saturated FA increased 50% compared to the original canola oil (Dong and Walker 2008).

CLA refers to a mixture of positional and geometrical linoleic acid isomers containing conjugated double bounds with 9,11- and 10,12-octadecadienoic acid ($C_{18:2}$) as the main isomers. At each position, the double bond can be either in the *cis*- or the *trans*-configuration. Modern analytical methods have identified 17 CLA isomers in food products, but most research concentrates on the two most abundant isomers, *cis*-9, *trans*-11 and *trans*-10, *cis*-12. The most biologically active form, which is also the major CLA component in milk (90% of total CLA) and beef (75% of total CLA), is rumenic acid, *cis*-9, *trans*-11-octadecadienoic acid. More recently, biological activity was also proposed for other CLA isomers, e.g., *trans*-10, *cis*-12 and *cis*-11, *trans*-13. Ruminants, i.e., cows, goats, sheep, can produce CLA and thus ruminant products show the highest concentration of CLA in total FA (Adamczak et al. 2008). The CLA isomers are produced by the strains of *Lactobacillus, Propionibacterium,* or *Bifidobacterium* (Ogawa et al. 2005). The new, green technology of CLA synthesis consists of ultraviolet photo-isomerization, but predominantly all-trans isomers are produced.

Under appropriate conditions for the cultivation of microorganisms, it is possible to obtain microbiological cocoa-butter-like fat (cocoa-butter substitutes). This can be achieved through supplementation of the cultivation medium with FA, whose increased concentration in the resulting microbial oil is required. The stearic acid concentration in bio-oil can also be increased by the inhibition of the stearic to oleic acid conversion by repressing the Δ^9-desaturase gene.

Microbiological lipid synthesis undoubtedly offers vast opportunities in regard to the synthesis of PEFA and special TAG such as cocoa-butter substitutes. However, problems related to production economics and the perfection of the bio-oils' composition have to be solved first.

19.6 MODIFIED TRIACYLGLYCEROLS AND FAT REPLACERS AND THEIR EFFECTS ON FOOD QUALITY

The available data indicate that consumption of sTAG is safe and beneficial for consumers, including postoperative patients. Pure medium-chain triacylglycerols (MCT) consumed in high amounts can cause the formation of ketone bodies. On the other hand, the lack of toxicity of MCT was demonstrated during studies with beagles (Matulka et al. 2009). The problem of the application of fat substitutes was presented by Akoh and Kim (2008). It is necessary to establish uniform digestibility and absorbability coefficients, especially for low-energy fats. The current guidelines cover total fat, i.e., saturated fat and not the digestibility coefficient of the fat.

The wide interest in the recently discovered nutraceutical, CLA, and sTAG containing high amounts of PEFA, has drawn attention to the oxidative stability of such new products (Sieber et al. 2004). Some studies indicated that after downstream processing of sTAG, natural antioxidants are removed and the stability of the products is low. It seems that the addition of antioxidants, preferentially natural products, is required and can effectively protect the new products, but more studies in this subject are needed. There are also reports about similar oxidative stability of sTAG, obtained from vegetable and fish oils, to substrates used for their production.

DAG oil is an edible oil with taste and usability characteristics comparable to those of naturally occurring oils. The safety of unheated or heated DAG oil at levels up to 5.5% in diets for 90 days as well as in chronic studies has been confirmed in studies with rats. Genotoxicity studies into unheated and heated DAG oil did not reveal any genotoxic effects (Morita and Soni 2009).

One of the most important problems with low-fat products is the changes concerning their structure and rheological properties. Extensive research should be performed not only to demonstrate the nutritional value of new designer lipids, but also the functionality and physical properties of new products. A chemically randomized blend of a 30% stearic-acid-rich and oleic-acid-rich can be used as a zero-trans shortening with improved health properties because of the increased content of oleic acid. Comparable plasticity to the commercial shortening was achieved by chemical interesterification of the mixture followed by crystallization and shearing of the fat. This high-stability shortening can be used in deep fat frying and many all-purpose bakery applications (Ahmadi and Marangoni 2009).

The reduction or replacement of fat in food products requires modification of the structure and rheological properties of their standard product counterparts.

Extruded low-fat ice cream is an example of a low-energy food product in which the fat reduction was compensated for by modification of the microstructure (Palzer 2009). Low-fat mayonnaise emulsion is characterized by lower viscosity compared to that of the full-fat product. The desired viscosity of the low-fat product can be increased by using a viscous starch solution, but the properties of the obtained emulsion are not identical with those of the original product (Ma et al. 2006). The modification of the structure of the low-fat spreads can also be obtained by adding modified starch or hydrocolloids (Liu et al. 2007).

The elastic properties of food, i.e., snap, mouth-feel, and hardness, which have been changed by removing or modifying the fat-phase, can be improved by using organogels and organogelators such as phytoserols and oryzanols, ceramides, monoacylglycerols, and waxes (Rogers 2009). This approach could lead to zero-trans, zero-saturated food products characterized by traditional physical properties. However, to date just one system, phytoserols plus oryzanol, has been accepted as a food grade oil-structuring ingredient not obtained from saturated FA.

The main products recognized by consumers as possible reduced-fats are dairy products. However, these products are good examples of how many problems are involved in removing and replacing the fat in food products. The main problems concern the structure, texture, aroma, and flavor of low-fat cheese. There are some suggestions as to how to improve the quality of the product, e.g., maintain the same moisture in the non-fat solids ratio as in full-fat cheese, increase the salt-in-moisture level, curd washing and ripening acceleration (addition of enzymes or cultures), but the final product differs from the original.

The modification of fat properties in food needs additional basic experiments and the example of the application of wax technology for obtaining plastic fat confirms it. There is still a chance to apply exotic oils like Allanblackia seed oil, which is solid at room temperature, in the production of novel food products.

REFERENCES

Adamczak, M. 2004. The application of lipases in modifying the composition, structure and properties of lipids—A review, *Pol. J. Food Nutr. Sci.*, 13(54):3–10.

Adamczak, M., Bornscheuer, U.T. 2009. Improving ascorbyl oleate synthesis catalyzed by *Candida antarctica* lipase B in ionic liquids and water activity control by salt hydrates, *Process Biochem.*, 44:257–261.

Adamczak, M., Krishna, S.H. 2004. Strategies for improving enzymes for efficient biocatalysis, *Food Technol. Biotechnol.*, 42:251–264.

Adamczak, M., Bornscheuer, U.T., Bednarski, W. 2008. Properties and biotechnological methods to produce lipids containing conjugated linoleic acid, *Eur. J. Lipid Sci. Technol.*, 110:491–504.

Akoh, C.C. 1998. Fat replacers, *Food Technol.*, 52:47–53.

Akoh, C.C., Kim, B.H. 2008. Structured lipids, in *Food Lipids: Chemistry, Nutrition, and Biotechnology*, 3rd edn., eds. C.C. Akoh and D.B. Min, Boca Raton, FL: CRC Press, pp. 841–872.

Ahmadi, L., Marangoni, A.G. 2009. Functionality and physical properties of interesterified high oleic shortening structured with stearic acid, *Food Chem.*, 117:668–673.

Babayan, V.K. 1987. Medium chain triglycerides and structure lipids, *Lipids*, 22:417–420.

Bezelgues, J.-B., Dijkstra, A.J. 2009. Formation of *trans* fatty acids during catalytic hydrogenation of edible oils, in *Trans Fatty Acids in Human Nutrition*, eds. F. Destaillats, J.-L. Sebedio, F. Dionisi, and J.-M., Chardigny, Bridgwater, U.K.: The Oily Press, pp. 43–63.

Bhaskar, A.R., Rizvi, S.S.H., Bertoli, C., Fayl, B., Hug, B. 1998. A comparison of physical and chemical properties of milk fat fractions obtained by two processing technologies, *J. Am. Oil Chem. Soc.*, 75(10):1249–1264.

Bornscheuer, U.T. 1995. Lipase-catalyzed synthesis of monoacylglycerols, *Enzyme Microb. Technol.*, 17:578–586.

Bornscheuer, U.T. 1999. Recent advances in the lipase-catalyzed biotransformation of fats and oils, *Recent Res. Dev. Oil Chem.*, 3:93–106.

Bornscheuer, U.T. 2000. *Enzymes in Lipid Modification*, Weinheim, Germany: Wiley-VCH.

Bornscheuer, U.T. 2008. Alteration of lipases properties by protein engineering methods, *Oléagineux Corps Gras, Lipides*, 15:1–5.

Bornscheuer, U.T., Adamczak, M., Soumanou, M.M. 2003. Lipase-catalysed synthesis of modified lipids, in *Lipids for Functional Foods and Nutraceuticals*, ed. F.D. Gunstone, Bridgwater, U.K.: The Oil Press, pp. 149–182.

Certik, M., Sakuradani, E., Shimizu, S. 1998. Desaturase-defective fungal mutants: Useful tools for the regulation and overproduction of polyunsaturated fatty acids, *Trends Biotechnol.*, 16:500–505.

Damstrup, M.L., Kill, S.Z., Jensen, A.D., Sparso, F.V., Flemming, V., Xu, X. 2007. Process development of continuous glycerolysis in an immobilized enzyme-packed reactor for industrial monoacylglycerol production, *J. Agric. Food Chem.*, 55:7786–7792.

De Maria, L., Vind, J., Oxenbøll, K., Svendsen, A., Patkar, S. 2007. Phospholipases and their industrial applications, *Appl. Microbiol. Biotechnol.*, 74:290–300.

Del Vecchio, A.J. 1996. High-laurate Canola. How Calgene's program began, where it's headed, *Inform*, 7:230–232.

Dong, M., Walker, T.H. 2008. Addition of polyunsaturated fatty acids to Canola oil by fungal conversion, *Enzyme Microb. Technol.*, 42:514–520.

Graham, I.A., Larson, T., Napier, J.A. 2007. Rational metabolic engineering of transgenic plants for biosynthesis of omega-3 polyunsaturates, *Curr. Opin. Biotechnol.*, 18:142–147.

Gunstone, F.D. 1998. Movements towards tailor-made fats, *Prog. Lipid Res.*, 37(5):277–305.

Guo, Z., Xu, X. 2005. New opportunity for enzymatic modification of fats and oils with industrial potentials, *Org. Biomol. Chem.*, 3:2615–2619.

Hammond, E.G. 2000a. Genetic alteration of food fats and oils, in *Fatty Acids and Their Health Implications*, ed. Ch.K. Chow, New York: Marcel Dekker Inc., pp. 357–374.

Hammond, E.G. 2000b. Sources of fats and oils, in *Introduction to Fats and Oils Technology*, eds. R.D. O'Brien, W.E. Farr, P.T. Wan, Champaign, IL: AOCS Press, pp. 49–62.

Hartmann, R.B., Fehr, W.R., Welke, G.A., Hammond, E.G., Duvick, D.N., Cianzio, S.R. 1996. Association of elevated palmitate content with agronomic and seed traits of soybean, *Crop Sci.*, 36:1466–1470.

Holm, H.C., Cowan, D. 2008. The evolution of enzymatic interesterification in the oils and fats industry. *Eur. J. Lipid Sci. Technol.*, 110:679–691.

Horejsi, T.F., Fehr, W.R., Welke, G.A., Duvick, D.N., Hammond, E.G., Cianzio, S.R. 1994. Genetic control of reduced palmitate content in soybean, *Crop Sci.*, 34:331–334.

Hridinka, C., Zollitisch, W., Knaus, W., Lettner, F. 1996. Effect of dietary fatty acid pattern on melting point and composition of adipose tissues and intramuscular fat of broiler carcasses, *Poultry Sci.*, 75:208–211.

Kay, R.A. 1991. Microalgae as food and supplement, *Crit. Rev. Food Sci. Nutr.*, 30(6):555–573.

Kazlauskas, R.J., Bornscheuer, U.T. 1998. Biotransformation with lipases, in *Biotechnology*, vol. 8a, *Biotransformation I*, eds. H.-J. Rehm, G. Reed, A. Pühler, P. Stadler, Weinheim, Germany: Wiley-VCH, pp. 37–191.

Kinney, A.J. 1994. Genetic modification of the storage lipids of plants, *Curr. Opin. Biotechnol.*, 5:144–151.

Kinney, A.J. 1996. Development of genetically engineered soybean oils for food applications, *J. Food Lipids*, 3:273–292.

Knutzon, D.S., Thompson G.A., Radke, S.E., Johnson, W.B., Knauf, V.C., Kridl, J.C. 1992. Modification of *Brassica* seed oil by antisense expression of a stearoyl-acyl carrier protein desaturase gene, *Proc. Natl. Acad. Sci. U. S. A.*, 89:2624–2626.

Lawson, R.E., Moss, A.R., Givens, D.I. 2001. The role of dairy products in supplying conjugated linoleic acid to man's diet: A review, *Nutr. Res. Rev.*, 14:153–172.

Lee, K.-T., Akoh, C.C. 1998. Structured lipids: Synthesis and applications, *Food Rev. Int.*, 14(1):17–34.

Leman, J. 1997. Oleaginous microorganisms: An assessment of the potential, *Adv. Appl. Microbiol.*, 43:195–243.

Leman, J. 2007. Enzymatically modified whey protein and other protein-based fat replacers, in *Novel Enzyme Technology for Food Application*, ed. R. Rastall, Cambridge, U.K.: Woodhead Publishing Limited, pp. 140–159.

Liu, H., Xu, X.M., Guo, Sh.D. 2007. Rheological, texture and sensory properties of low fat–mayonnaise with different fat mimetics, *LWT Food Sci. Technol.*, 40:946–954.

Ma, Y., Cai, C., Wang, J., Sun, D.-W. 2006. Enzymatic hydrolysis of corn starch for production fat mimetics, *J. Food Eng.*, 73:297–303.

MacKenzie, S.L. 1999. Chemistry and engineering of edible oils and fats, in *Molecular Biotechnology for Plant Food Production*, ed. O. Paredes-Lopez, Basel, Switzerland: Technomic Publishing Co. Inc., pp. 525–620.

Matulka, R.A., Larry Thompson, D.V.M., Burdock, G.A. 2009. Lack of toxicity by medium chain triglycerides (MCT) in canines during a 90-day feeding study, *Food Chem. Toxicol.*, 47:35–39.

Michaelson, L.V., Lazarus, C.M., Griffiths, G., Napier, J.A., Stobart, A.K. 1998. Isolation of a Δ^5-fatty acid desaturase gene from *Mortierella alpina*, *J. Plant. Biochem.*, 30:19055–19059.

Miraglio, A.M. 1995. Nutrient substitutes and their energy values in fat substitutes and replacers, *Am. J. Clin. Nutr.*, 62:1175–1179.

Moghadasian, M.H. 2008. Advances in dietary enrichment with n-3 fatty acids, *Crit. Rev. Food Sci. Nutr.*, 48:402–410.

Morita, O, Soni, M.G. 2009. Safety assessment of diacylglycerol oil as an edible oil: A review of the published literature, *Food Chem. Toxicol.*, 47:9–21.

Murphy, D.J. 1999. Production of novel oils in plants, *Curr. Opin. Biotechnol.*, 10:175–180.

Ogawa, I., Kishino, S., Ando, A., Sugimoto, S., Mihara, K., Shimizu, S. 2005. Production of conjugated fatty acids by lactic acid bacteria, *J. Biosci. Bioeng.*, 100:355–364.

Osborn, H.T., Akoh, C.C. 2002. Structured lipids-novel fats with medical, nutraceutical, and food applications, *Comp. Rev. Food Sci. Food Saf.*, 3:93–103.

Palzer, S. 2009. Food structures for nutrition, health and wellness, *Trends Food Sci. Technol.*, 20:194–200.

Peng, L., Xu, X., Tan, T. 2000. Enzymatio production of high quality monoacylglycerols, in *Research Advances in Oil Chemistry*, ed. R.M. Mohan, Calcutta, India: GRN, pp. 53–78.

Pszczoła, D.E., Katz, F., Giese, J. 2000. Research trends in healthful foods, *Food Technol.*, 54(10):45–52.

Ratledge, C. 2001. Microorganisms as sources of polyunsaturated fatty acids, in *Structured and Modified Lipids*, ed. F.D. Gunstone, New York: Marcel Dekker, Inc., pp. 351–399.

Rogers, M.A. 2009. Novel structuring strategies for unsaturated fats—Meeting the zero-trans, zero-saturated fat challenge: A review, *Food Res. Int.*, 42:747–753.

Rousseau, D., Marangoni, A.G., Jeffrey, K.R. 1998. The influence of chemical interesterification on the physicochemical properties of complex fat systems. 2. Morphology and polymorphism, *J. Am. Oil Chem. Soc.*, 75(12):1833–1839.

Schörken, U., Kempers, P. 2009. Lipid biotechnology: Industrially relevant production processes, *Eur. J. Lipid Sci. Technol.*, 111:627–645.

Shahidi, F., Wanasundara, U.N. 1998. Omega-3 fatty acid concentrates: Nutritional aspects and production technologies, *Trends Food Sci. Technol.*, 9:230–240.

Shen, Z., Birkett, A., Augustin, M.A., Dungey, S., Versteeg, C. 2001. Melting behaviour of blends of milk fat with hydrogenated coconut and cottonseed oils, *J. Am. Oil Chem. Soc.*, 78(4):387–394.

Sieber, R., Collomb, M., Aeschlimann, A., Jelen, P., Eyer, H. 2004. Impact of microbial cultures on conjugated linoleic acid in dairy products—A review, *Int. Dairy J.*, 14:1–15.

Smith, S.A., King, R.E., Min, D.B. 2007. Oxidative and thermal stabilities of genetically modified high oleic sunflower oil, *Food Chem.*, 102:1208–1213.

Solan, A.S. 2000. The top ten functional food trends, *Food Technol.*, 54(4):33–62.

Straarup, E.M., Hoy, C.-E. 2000. Structured lipids improve fat absorption in normal and malabsorbing rats, *J. Nutr.*, 130:2802–2808.

Tarrago-Trani, M.T., Phillips, K.M., Lemar, L.E., Holden, J.M. 2006. New and existing oils and fats used in products with reduced *trans*-fatty acid content, *J. Am. Diet. Assoc.*, 106:867–880.

Urban, M., Adamczak, M. 2008. Exploration of metagenomes for new enzymes useful in food biotechnology—A review, *Pol. J. Food Nutr. Sci.*, 58:11–22.

Villeneuve, P. 2007. Lipases in lipophilization reactions, *Biotechnol. Adv.*, 25:515–536.

Ward, O.P., Singh, A. 2005. Omega-3/6 fatty acids: Alternative sources of production, *Process Biochem.*, 40:3627–3652.

Weber, N., Mukherjee, K.D. 2008. Lipid biotechnology, in *Food Chemistry: Chemistry, Nutrition, and Biotechnology*, 3rd edn., eds. C.C. Akoh and D.B. Min, Boca Raton, FL: CRC Press, pp. 707–765.

Xu, X. 2000. Production of specific-structured triacylglycerols by lipase-catalyzed reactions: A review, *Eur. J. Lipid Sci. Technol.*, 102:287–303.

Xu, X., Guo, Z., Zhang, H., Vikbjerg, A.F., Damstrup, M.L. 2006. Chemical and enzymatic interesterification of lipids for use in food, in *Modifying Lipids for Use in Food*, ed. F. Gunstone, Abington, U.K.: Woodhead Publishing Limited, pp. 234–272.

Zou, J., Katavic, V., Gibblin, E.M., Barton, D.L., Mackenzie, S.L., Keller, W.A., Hu, A., Taylor, D.L. 1997. Modification of seed oil content and acyl composition in the *Brassicacea* by expression of a yeast *sn*-2 acyl transferase gene, *Plant Cell*, 9:909–923.

20 Lipids with Special Biological and Physicochemical Activities

Marek Adamczak and Włodzimierz Bednarski

CONTENTS

20.1 INTRODUCTION

The increasing cost of health care, increasing life expectancy, and the need to improve life quality have stimulated a growing interest in functional food (also referred to as nutraceuticals or pharmaceutical food) developments. The new food categories require new legislation to describe products through their properties (Figure 20.1). Functional food is defined as food that is "satisfactorily demonstrated to affect beneficially one or more target functions in the body, beyond adequate nutritional effects, in a way that is relevant to either improved state of health and well-being and/or reduction of risk of disease" (Diplock et al. 1999). Using the knowledge concerning lipids' biological activity, health, and physical properties, functional products can be also obtained. It has been suggested that the quality and quantity of dietary lipids may be an important factor influencing the risks associated with metabolic syndrome. Metabolic syndrome has the following symptoms: abdominal obesity, hypertriglyceridemia, a low level of high-density-lipoprotein (HDL)-cholesterol,

FIGURE 20.1 The position of functional food with a modified lipid phase.

hypertension and high fasting-glucose level, increased cardiovascular morbidity, and mortality (Grundy et al. 2005).

Several important issues that should be analyzed when functional food is introduced to the market are legislation, consumer knowledge, and understanding of functional food nutritional and health properties. Information on legislation in the European Union and the United States may be found in the recent publications of Gulati and Ottaway (2008), Ruckman (2008), Gulati and Ottaway (2006), and Cheftel (2005).

Examples of functional food with a modified lipid phase can be found in all types of functional food, i.e., fortified food with polyenoic fatty acids (PEFA); enriched food with added new fatty acids (FA) and altered products—low-fat, low-cholesterol products; or enhanced commodities—eggs with increased n-3 FA content achieved by changes in feed composition (Table 20.1). Consumer acceptance of the concept of functional food depends on consumer education and understanding of the basic principles of human nutrition. There are also some other factors, e.g., sociodemographic parameters, that influence the acceptance of functional products by the consumer. Generally, Europeans are more critical of new products and technologies than American or Japanese consumers. The results of trials for empirical prediction of functional product acceptance by consumers differ, depending on the methodology employed and the analyzed parameters. Obviously, new food acceptance is a multistage process and requires input from scientists and industry (Siró et al. 2008).

TABLE 20.1
Examples of Functional Food with Modified Lipids

Product	Modified Lipids/Changes in Lipids' Composition
Functional drinks	Cholesterol-lowering drinks containing n-3 FA and soy
"Eye-health" drinks	Drinks with increased lutein concentration
Becel pro-active	Margarine with phytostanol esters (cholesterol-lowering food)
Benecol	A spread containing camelina oil
Balade cheese, cream	Low-cholesterol butter, cheese, and cream (90% cholesterol removed by the addition of crystalline β-cyclodextrin to the molten milk fat)
Columbus	Eggs enriched in n-3 FA and vitamin E
Functional meats	Products enriched with n-3 FA

TABLE 20.2
Functional Lipids That Can Be Obtained from Microalgae

Functional Ingredients	Function, Activity	Extraction Method	Microalgae
Lutein	Antioxidant	SFE	*Chlorella pyrenoidos* *Haematococcus pluvialis*
β-Carotene	Antioxidant	SFE	*Dunaliella salina* *H. pluvialis*
Astaxanthin	Antioxidant, immunomodulation, and cancer prevention	SFE	*H. pluvialis* *Chlorella vulgaris*
Cantaxanthin	Antioxidant, immunomodulation, and cancer prevention	SFE	*C. vulgaris* *H. pluvialis*
EPA	Reduce risk of CVD	LL	*Phaeodactylum tricornutum* *Monodus subterraneus* *Porphyridium cruentum*
Oleic acid	Antioxidant	LL, SFE, PLE	*C. vulgaris* *H. pluvialis* *D. salina* *Spriulina platensis*
DHA	Reduce risk of CVD	LL	*S. platensis*
Palmitic acid	Antimicrobial	PLE	*D. salina*
Palmitoleic acid	Reduce risk of CVD	SFE	*S. platensis*
Tocopherol	Antioxidant	Ultrasounds, SFE	*Porphyridium* spp. *S. platensis*

Source: Adapted from Plaza et al., *J. Agric. Food Chem.*, 57, 7159, 2009.

Note: SFE, supercritical fluid extraction; LL, liquid-liquid extraction; PLE, pressurized liquid extraction.

One of the important factors that determine the development of functional food is research concerning new functional ingredients. Traditional sources of lipophilic ingredients are fish byproducts (PEFA), plants (phytosterols) and new, nonconventional resources are needed, e.g., microorganisms, algae (microalgae) (Plaza et al. 2008) (Table 20.2).

20.2 BIOLOGICAL ACTIVITY OF LIPIDS CONTAINING POLYENOIC AND CONJUGATED FATTY ACIDS

It is very difficult to analyze the effect of PEFA, especially essential fatty acids (EFA), and arrive at clear conclusions concerning their effect on human health. An analysis of the literature in this field results in rather confusing conclusions, and there is no clear and unquestionable evidence of positive or negative interactions between lipids and human health. Experiments on model animals and clinical trials indicate different effects of biological active lipids (i.e., *n*-3 PEFA, conjugated FA,

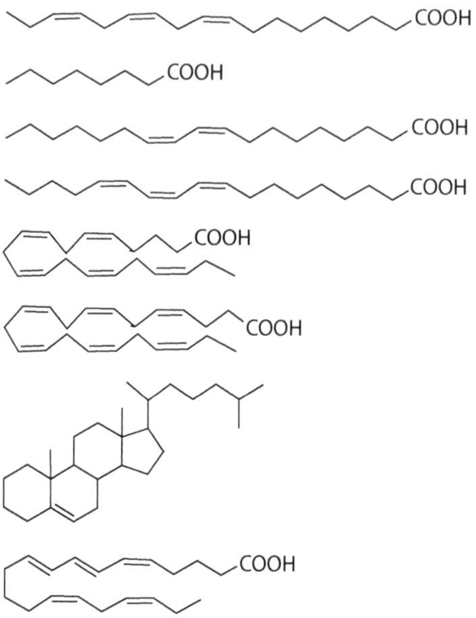

$C_{18:3}$, *n*-3, LNA

$C_{8:0}$, CA

$C_{18:2}$, CLA

$C_{18:3}$, CLNA

$C_{20:5}$, *n*-3, EPA

$C_{22:6}$, *n*-3, DHA

Campestenone

$C_{20:5}$, *n*-3, CEPA

FIGURE 20.2 The main biological-active food lipids. Acids: LNA-linoleic, CA-caprylic, CLA-conjugated linoleic, CLNA-conjugated linolenic, EPA-eicosapentaenoic, DHA, docosahexaenoic, CEPA-conjugated eicosapentaenoic.

sterols, medium chain fatty acids (MCFA), diacylglycerols (DAG), and phospholipids) on human health (Figure 20.2).

It is known that the consumption of *n*-3 PEFA (α-linoleic acid, $C_{18:3}$, eicosapentaenoic acid $C_{20:5}$ (EPA), docosahexaenoic acid $C_{22:6}$ (DHA)) is correlated with a reduced risk of cardiovascular disease (CVD) and cancer in clinical and animal studies (Fernandez et al. 1999, Holub and Holub 2004).

Pariza and Hargraves (1985), during their studies on the synthesis of heterocyclic aromatic amines known as substances with mutagenic and carcinogenic activities, discovered organic compounds that inhibited mutation induction in bacteria in meat subjected to thermal treatment. These compounds were identified as isomers of linoleic acid (18:2), characterized by double conjugated bonds. Conjugated linoleic acids (CLA) were known much earlier, yet they had not been assumed to exhibit different activity than that of typical linoleic acid containing solely *cis,cis*-unsaturation (*c*9,*c*12). Subsequent experiments carried out by the Pariza research group confirmed the anticarcinogenic properties of CLA in the *in vivo* tests conducted on mice with neoplasmic lesions induced with 7,12-dimethylbenzene (Pariza and Hargraves 1985). Eventually, their investigations led to the determination and isolation of the *c*9,*t*11 isomer, considered to be the most biologically active form of CLA.

Later, multioriented investigations of numerous research groups confirmed that conjugated dienes of linoleic acid also possess other, valuable health-promoting properties. Beyond all doubt, they have been demonstrated to exhibit antiatherogenic, immunostimulatory, antioxidative, and bacteriostatic activities against

TABLE 20.3
Pros and Cons of CLA Consumption

Pros	Cons
Alter LDL/HDL cholesterol ratio in rabbits	—
Fat-to-lean agent	—
Inhibition of fatty acid synthase *in vitro*	—
Incorporation of CLA isomers into tissue lipids of chicken	Impaired growth performance
A CLA-rich source for humans	
Anti-carcinogenic properties: epidermal, gastrointestinal carcinomas (AM)	—
Inhibition of initiation of mouse skin carcinogenesis	—
Inhibition of initiation of mouse forestomach carcinogenesis	—
Inhibition of rat mammary tumorigenesis	—
Stimulation of immune function, e.g., in rats	—
Anti-obesity effects in diet-induced obese mice	—
Inhibition human colon cancer cells (cancer cell line)	—
—	Increase in reactive oxygen species
Inhibition of proliferation of several human cancer cell lines *in vitro*	—
Attenuation of inflammation and preventing colony cancer (*in vitro* and *in vivo* in mice)	—
No effect on blood pressure or isobaric arterial elasticity in men	—
Improved insulin sensitivity in rats	—

Listeria monocytogene (Collomb et al. 2006). In addition, they have been shown to prevent aggregation of blood cells (thus preventing coronary heart disease) and to support the treatment of diabetes (Table 20.3) (Adamczak et al., 2008). It is possible, therefore, to apply CLA both for prophylactic and therapeutic purposes. CLA regulate the metabolism of lipids and inhibit the accumulation of adipose tissue, which has been used by pharmaceutical companies in the production of a number of slimming preparations and supplements for bodybuilders.

The anticarcinogenic action of CLA has been confirmed in many studies (Igarashi and Miyazawa 2001, Belury 2002, Wahle and Heys 2002, Chujo et al. 2003). Although its mechanism is still not fully understood, there are some clues. For instance, it may be linked with antioxidative properties that protect against the effect of free radicals on cellular membranes. One cannot exclude CLA oxidation in cancer cells to free radicals that exhibit strong cytotoxic activity or that CLA might display a capacity for modulating the pathway of signal conduction that is incorporated into the cycle of cell replication or inhibit the synthesis of nucleotides and proteins in cancer cells (Wahle and Heys 2002). Another theory claims they can modulate cell defense systems by affecting the activity of lymphocytes and macrophages. Specific modulation of lipid metabolism is also probably included, e.g., competence with linoleic acid in biosynthesis of AA is a precursor of eicosanes for stimulating cell development (Igarashi and Miyazawa 2001).

The mechanism of the action of CLA on tumors may vary depending on the location and stadium of the neoplasm as well as on age and exposure time. CLA may also affect the process of carcinogenesis at its various stages (initiation, promotion, progression) (Watkins and Li 2001). Although other FA have also been claimed to possess anticarcinogenic potential, CLA dienes have attracted special interest due to the fact that they also exhibit such activity at lower concentrations of fat in a diet. On the contrary, the application of other substances requires several times higher doses.

It has been demonstrated that the intake of a diet rich in CLA (according to some studies–mainly of $t10,c12$ isomer) may diminish the level of total cholesterol and its LDL fraction without changing the level of HDL-cholesterol, while improving the ratio of LDL/HDL-cholesterol and decreasing the amount of cholesterol deposits in aortas. The effect is opposite with the intake of CLA rich in $c9,t11$ CLA, which decreases them (Tricon et al. 2004).

In vitro experiments have demonstrated that CLA have higher antioxidative potential than α-tocopherol, but weaker than that of β-carotene. Owing to this property, it may protect the cell membrane against the action of free radicals. Presumably, it affects the antiatherogenic and anticarcinogenic activity of conjugated dienes of linoleic acid (Watkins and Li 2002).

CLA is also capable of regulating the metabolism of adipose tissue (Salas-Salvado et al. 2006) as it inhibits its formation by blocking enzymes responsible for fat deposition. In addition, it intensifies the process of lipolysis (fat burning) (Park and Pariza 2007). A possible explanation of this phenomenon is the stimulation of the activity of carnitine-palmitic transferase, which regulates the process of β-oxidation of FA by CLA (Whigham et al. 2000).

It has been demonstrated that lipid metabolism is determined, to the greatest extent, by CLA that contain $t10,c12$. The effect of CLA on the regulation of lipid metabolism in the body has also been confirmed in some experiments carried out on animals. A diet with 0.5% addition of CLA administered for a period of 4–8 weeks caused a distinct decrease in the body weight of the examined mice, by 75% in females and by 57% in males. In rats, the corresponding decrease in body weight accounted for 23% and in chickens for 22% (Sebedio and Christie 1998).

The interesting results of experiments on animals have prompted scientists to undertake analogous investigations in humans. Their results are, however, divergent. A test carried out on Norwegian volunteers demonstrated that in a group of subjects of both sexes with similar parameters (age, body mass index (BMI), body fat content), persons consuming CLA were characterized by a slight decrease of body mass, with simultaneous considerable reduction in the content of adipose tissue as compared to the control group. Further studies by Norwegian scientists on a group of obese people (BMI ranging from 25 to 35) showed that the intake of CLA at a level of 3.4 and 6.8 g/day for 2 weeks evoked a tangible decrease in the level of adipose tissue. However, the authors quote examples of experiments that had failed to demonstrate such an effect (Rainer and Heiss 2004). This indicates that the effects of CLA on the body as well as the dose effectively diminishing the level of adipose tissue are probably individual characteristics (Bessa et al. 2000).

CLA has also been demonstrated to positively affect the immune system of rabbits and chickens. It increases the immunological potential in macrocytes of mice and rats, which is reflected in enhanced phagocytosis and bactericidal activity and cytotoxicity of T lymphocytes as well as enhanced phagocytosis of leukocytes. This results in the proliferation of lymphocytes in the body. CLA has been postulated to be helpful in the case of patients with AIDS or tumors (Whigham et al. 2000). In addition, its presence neutralizes the effect of *Escherichia coli* endotoxin.

CLA may also alleviate symptoms of diabetes, e.g., they may neutralize the blood level of insulin and modify the functional properties of cellular membranes through their incorporation into phospholipid (PL) membranous structures, e.g., their permeability (mainly *c*9, *t*11 isomer) (Riserus 2006, Park and Pariza 2007).

Recently, Considine et al. (2007) showed in a model system consisting of β-lactoglobulin (β-LG) and CLA that it has the ability to inhibit the transition of β-LG from one form to another, during heat- and high–pressure-induced unfolding and aggregation of β-LG.

20.3 ANTIMICROBIAL AND SURFACE ACTIVITY OF LIPIDS

The antimicrobial, microbicidal activity of lipids have been known for many years (Kabara et al. 1972, Kabara 1980, Thormar and Hilmarsson 2007). Initially, the bactericidal effects of soaps and FA were analyzed. Later, it was also demonstrated that monoacylglycerols (MAG) can kill many pathogens, i.e., viruses, bacteria, and fungi. Lauricidin®, monolaurin (monolaurate glycerol) is currently available as a nutritional supplement for better human health (Table 20.4).

TABLE 20.4
Examples of Antimicrobial Activity of Lipids

Lipids	Antimicrobial Activity	References
Bioconverted EPA and DHA	Antifungal activity against plant pathogens and foodborne pathogenic bacteria	Bajpai et al. (2008), Shin et al. (2007)
10-Hydroxystearic acid 7,10-Dihydroxy-8-octadecenoic acid 12,13,17-Trihydroxy-9-octadecenoic acid	Inhibition of plant phatogenic fungi (*Erysiphe graminis*, *Puccinia recondita*, *Phytophthora infestans*, *Botrytis cinerea*)	Hou and Forman (2000)
9,10-Epimino octadecane	Antimicrobial activity toward gram-positive and yeasts (not gram-negative)	Kabara et al. (1977)
Selene substituted FA Branched fatty acids	Fungistatic and bacteriostatic activity	Larsson et al. (1975a,b)
CA, MAC, SC	Inactivation of fish pathogens (*Edwardsiella ictaluri*, *E. tarda*, *Streptococcus iniae*, *Yersinia ruckeri*)	Kollanoor et al. (2007)
Monolaurin	Anti-listerial activity	Sprong et al. (2001)

Note: EPA, eicosapentaenoic acid; $C_{20:5}$, CA, caprylic acid; MAC, monoacylcaprylate; SC, sodium caprylate.

Antimicrobial lipids are commonly found on skin, mucosa, and in natural products such as milk.

General rules that can explain the relations between lipid structure and antibacterial activity are as follows: (1) lauric acid is the most active saturated FA, palmitoleic acid is the most active monounsaturated acid, and linoleic acid is the most active PEFA against gram-positive bacteria; (2) MAG of MCFA are more active than FFA (the most active is monolaurylate); (3) FA have very low activity against gram-negative bacteria but the exception are very short chain FA (<6 carbons); (4) short chain FA (C_{10} to C_{12}) affect the growth of yeasts. The general rules also hold that MAGs are more active than DAGs and triacylglycerols (TAG).

It seems that gram-positive bacteria are more susceptible to the antibacterial effect of lipids than gram-negative bacteria. However, there are many exceptions to this general information, e.g., *Chlamydia trachomatis*, *Neisseria gonorrhoeae*, and *Helicobacter pylori* are susceptible to FFA and MAG, and *Campylobacter jejuni* is easily killed by capric acid and 1-monocaprin glycerol. Interestingly, *E. coli* and *Salmonella* spp. are resistant to antibacterial lipids at neutral pH, but are killed at acid pH (Bergsson et al. 2002, Thormar et al. 2006, Thormar and Hilmarsson 2007).

Hydroxy fatty acids (HFA) are present in mammals and are metabolites of PEFA (Figure 20.3). HFA are produced by plants and the specific properties of these FA, i.e., high viscosity and reactivity compared to the other FA, are because of the presence of hydroxyl groups. Some HFA are known to have antifungal and cytotoxic activity. Value-added products can also be obtained by microbial biotransformation of unsaturated FA. Microorganisms are able to produce three main groups of HFA: mono-, di-, or trihydroxy FA.

Biosurfactants are a structurally-diverse group of surface-active molecules synthesized by microorganisms or by enzymatic synthesis (Nitschke and Costa 2007). The following biosurfactants are classified as lipids: glycolipids, i.e., sophorolipids (synthesized by *Candida bombicola*, *Candida apicola*), rhamnolipids (*Pseudomonas aeruginosa*), trehalose lipids (*Rhodococcus erythropolis*), mannosylerythritol lipids (*Pseudozyma antarctica*), as well as PL (*Acinetobacter* sp., *Corynebacterium lepus*), FA (*Corynebacterium indibasseosum* synthesizing corynomicolic acids), and neutral lipids. Most of the biosurfactants are anionic or neutral and are synthesized by bacteria; but yeasts are also valuable producers. The high surface activity of biosurfactants, its tolerance to pH, temperature, ionic strength, biodegradability, low

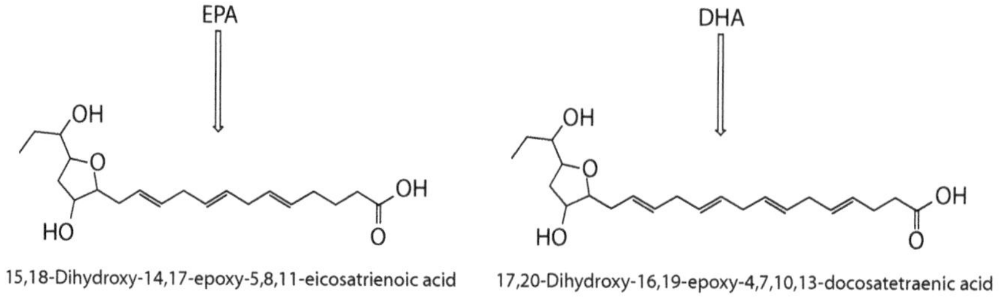

15,18-Dihydroxy-14,17-epoxy-5,8,11-eicosatrienoic acid   17,20-Dihydroxy-16,19-epoxy-4,7,10,13-docosatetraenic acid

FIGURE 20.3 Products of EPA and DHA bioconversion by *Bacillus megaterium* ALA2 (former *Clavibacter* sp. ALA2).

toxicity, and antimicrobial activity all make these compounds very attractive for application in the food industry. Rhamnolipids inhibited the growth of harmful algae at concentrations from 0.4 to 10.0 mg/dm^3 (Wang et al. 2005). Gram-positive bacteria were inhibited by mannosylerythritol lipids (Kitamoto et al. 1993, Adamczak and Bednarski 2000). The most important properties of biosurfactants could be the control of biofilm formation and adhesion of pathogens on different surfaces.

Carbohydrate esters of FA (CEFA) are biodegradable, nontoxic, nonionic surfactants used in the food, pharmaceutical, cosmetics, and detergent industries. In many studies, the antimicrobial properties of CEFA have been analyzed (Blaszyk and Holley 1998, Ferrer et al. 2005). Novel mono-substituted CEFA and ether were more active (minimum inhibitory concentration, MIC = 0.04 mM) against gram-positive bacteria than monolaurin (Nobmann et al. 2009). CEFA can be synthesized chemically and enzymatically by interesterification, transesterification, and direct esterification. The antimicrobial action of CEFA is not fully understood but the cytoplasmic membrane is thought to be the primary site of action, affecting the respiratory activity through inhibition of enzymes involved in oxygen uptake.

The most important problem of widespread biosurfactant application is an economical issue. The price of surfactants synthesized by biosurfactants could be attractive if waste products are used as components of the cultivation medium.

20.4 MODIFICATION OF PHOSPHOLIPIDS AND THEIR ROLE

PL are ubiquitous in all organisms and, because of their amphiphilic properties, they can form micelles, reverse micelles, and bilayer vesicles. The physiological functions of PL are the formation of biomembranes, support of membrane proteins, and they also play an important role in transmembrane signaling (Nagao and Yanagita 2008).

PL like carotenoids, tocopherols, ascorbic acid, flavonoids, and sterols are natural antioxidants, but the most important properties of PL, appropriate for industrial application, are its physical properties and its biocompatibility. There has recently been evidence presented that the nutritional value of PL is higher than that of TAG. It was demonstrated that phosphatidylethanolamine (PE) lowered cholesterol level (Imaizumi et al. 1983), phosphatidylcholine (PC) decreased lymphatic cholesterol absorption and reduced hepatic FA synthesis in rabbits and rats (Jiang et al. 2001), and phosphatidylinositol (PI) increased the level of HDL-cholesterol in rabbits and humans (Stamler et al. 2000, Burgess et al. 2005). The most common industrial applications of PL are in emulsifiers, cosmetics components, medical formulations, and liposome preparations.

The modification of PL properties is possible through the application of the following phospholipases: A_1, A_2, B, C, and D (PLA$_1$, PLA$_2$, PLB, PLC, and PLD) (Figure 20.4). The most important are the applications of PLA$_2$ and PLD that aimed at the following:

1. Obtaining homogenous PL
2. Improving the emulsification properties of PL
3. Obtaining PL that are not present in natural sources, e.g., phosphatidylserine, which is only present in animal brains

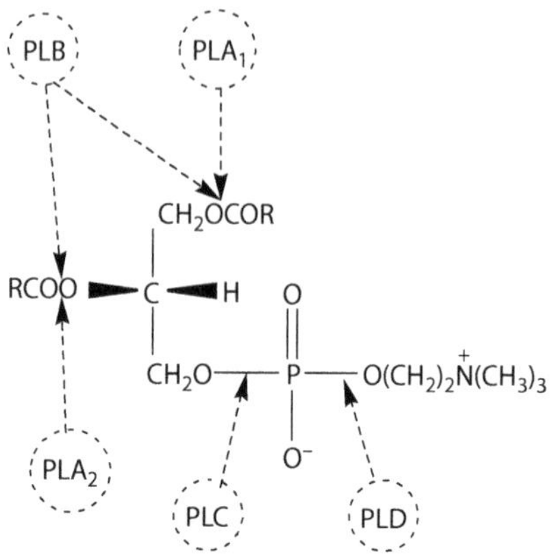

FIGURE 20.4 Specificity of phospholipases on PC.

4. Synthesizing structured sPL, which are characterized by specific composition of acyl and polar head groups (Adamczak 2004, Ulbrich-Hofmann et al. 2005, De Maria et al. 2007)

Modification of acyl groups of PL can be obtained by using PLA_1, PLA_2, and lipases. However, because PLA_1 is not commercially available for biocatalysis, PLA_2 and PLD are mainly in use. The main application of these enzymes in biocatalysis with PL is in the synthesis of lysophospholipids, which have superior emulsifying properties and in the introduction of specific FA into PL. The lysophospholipids that are soluble in water and form stable oil-in-water emulsions can be obtained by using three procedures (Figure 20.5). PLA_2, one of the catalysts of these reactions, is

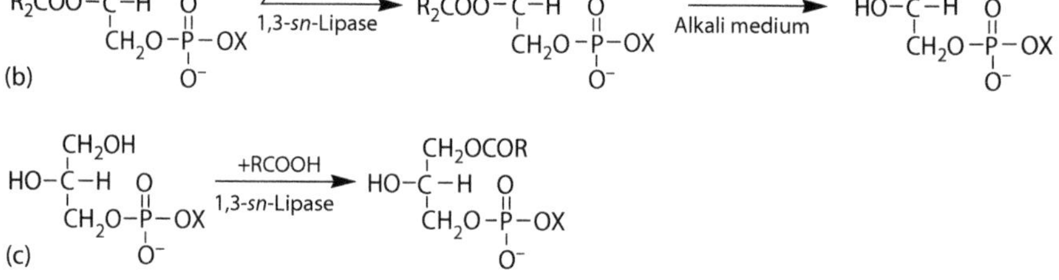

FIGURE 20.5 Possible enzymatic methods of lysophospholipids synthesis: (a) PLA_2 catalyzed hydrolysis; (b) two-step synthesis by hydrolysis or alcoholysis catalyzed by 1,3-*sn*-lipase, followed induction of acyl migration by alkali; (c) 1,3-*sn*-lipase catalyzed esterification.

mostly obtained from animal sources and can be a high molecular weight intracellular enzyme or low molecular weight extracellular enzyme. Many of these animal PLA$_2$s were cloned, and this stopped the research on plant and microbial enzymes that *Streptomyces* or *Aspergillus* are known to contain.

The industrial synthesis of lysophospholipids is performed by hydrolysis of soybean lecithin in an aqueous solution containing Ca^{2+} ions as a cofactor, catalyzed by porcine pancreatic PLA$_2$ (Figure 20.5a). After the FA removal by solvent washing and dehydration under reduced pressure, a powder of lysophospholipids is obtained (Iwasaki and Yamane 2002).

In the second strategy, *sn*-1,3-selective lipases can by used for deacylation of acyl groups from *sn*-1-position (Figure 20.5b). The reaction can also be performed as alcoholysis and, in this condition, a homogenous reaction system is formed, which makes it easy to recover the product. The reaction is performed in a membrane reactor, and lysophospholipids are subsequently obtained by acyl migration catalyzed by ammonia vapor.

The third method of lysophospholipids synthesis is also based on the use of 1,3-*sn*-lipase, but the esterification occurs with glycerophosphorylcholine or glycerol phosphate (Figure 20.5c). Based on this procedure, lysophosphatidic acid (a biologically active compound and constituent of cell membranes) can be produced in a reaction catalyzed by *sn*-1,3-lipase. However, the yield of the reaction may be low because of water formation and changes in reaction are directed toward hydrolysis.

PLD used in biocatalysis is usually obtained from plants: cabbage, carrots, peanuts, and castor bean, but also from microorganisms, mainly *Streptomyces* strains. The most important reaction catalyzed by PLD is not hydrolysis, but transphosphatidylation, i.e., a reaction that involves the replacement of polar head groups with other hydroxyl compounds (Figure 20.6). Substrates for synthesis of specific PL are usually natural compounds like PC (lecithin), which are also subjected to different methods of modification (Amit et al. 2006). Examples of head groups introduced by PLD into glycerophospholipids are presented in Table 20.5. Surprisingly, the reaction of transphosphatidylation can also be performed in water-rich media (high water activity value). Usually, this reaction is processed in two-phase systems containing an organic phase, lipids in organic solvents, and a water phase of enzyme and hydroxyl compounds. PLs above their critical micellar concentration (CMC) are

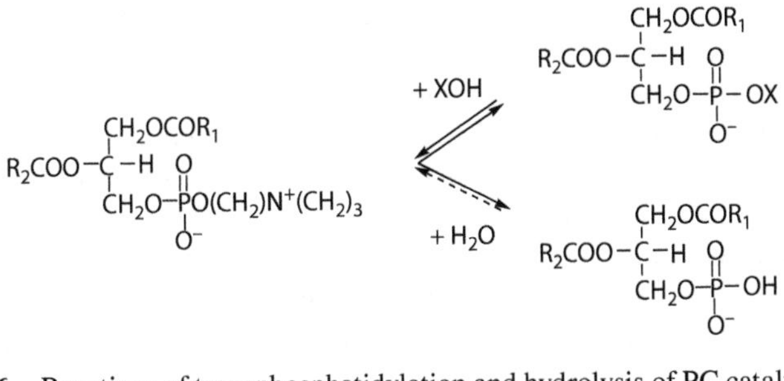

FIGURE 20.6 Reactions of transphosphatidylation and hydrolysis of PC catalyzed by PLD.

TABLE 20.5

Examples of the Head Group Introduced by PLD into Glycerophospholipids

Head Group Introduced	Head Group Structure
Aliphatic secondary alcohols	R_1 $-CH_3, -CH_2CH_3$ R_2 $-CH_3, -CH_2CH_3, -(CH_2)_3-CH_3$
(R)- and (S)-Isopropylideneglycerol	
Cyclic nonaromatic alcohols	
Arbutin	
Koji acid	
Arsenocholin	

present in the form of different aggregates, used *in vivo* as a substrates for phospholipase. During the *in vitro* reactions, PLs are prepared in the form of liposomes or micelles with the help of different surfactants.

The industrial application of phospholipases includes removal of PL during the oil refining process. In the EnzyMax® procedure, Lecitase® 10L (porcine PLA$_2$) was the only enzyme used. Recently, Lecitase Novo, PLA$_1$ from *Fusarium oxysporum* was proposed as an appropriate alternative. Lecitase Novo is an enzyme that is active toward PL in the oil phase (Lecitase 10L toward hydrophilic PL), the optimal parameters of the enzyme fit the conditions of oil degumming ($T=30°C-70°C$, pH 4–5) and finally the process is cheaper because of the smaller amount of water used for phosphorous removal (Claussen 2001).

The new enzyme used for degumming is Lecitase Ultra. It is a protein-engineered carboxylic ester hydrolase extracted from *Thermomyces lanuginosus/F. oxysporum* and is produced by the submerged fermentation of a genetically modified *Aspergillus oryzae*. This is an acidic enzyme complex, which exhibits maximal activity at pH = 5.0, and it has an inherent activity toward both PL and TAG. When the temperature is over 40°C, the phospholipase activity predominates and the lipase activity is partly suppressed. It is intriguing to observe that the enzyme is also able to identify the PL exclusively as a substrate in the oil degumming process, and the yield loss due to TAG hydrolysis is negligible (Yang et al. 2008).

PLCs from bacteria are usually metalloenzymes, e.g., PLC from *Bacillus cereus* is characterized by the presence of three Zn^{2+} ions that form a metal cluster as part of an active site and attracts the phosphate group of the substrate. Three classes of PLCs can be selected, i.e., PC-specific (PC-PLC), PI-specific (PI-PLC), and sphingomyelin-specific (SM-PLC). The most prominent application of this enzyme is the synthesis of enantiomerically-pure 1,2-DGA and phosphate monoesters by hydrolysis of the corresponding glycerophospholipids (Anthonsen et al. 1999). The other application of PLC could be the synthesis of organic phosphates (glycerol-3-phosphate) and dihydroxyacetone. The oil degumming can also be obtained by using PLC. Some of the enzymes (due to their high structural and catalytic similarities to mammalian PLC) could be of interest in medical research. Moreover, this enzyme was used in basic research in the areas of lipid metabolism, chemistry of lipids, blood coagulation, and of eukaryotic cell membranes (Kamberov and Ivanov 1990).

20.5 PRODUCTION AND PROPERTIES OF FATTY ACIDS AS FOOD SUPPLEMENTS

20.5.1 INTRODUCTION

The physiological effects of FA depend on their structural properties, e.g., chain length, degree of unsaturation, the positional and geometric configuration of the unsaturation, and their position in TAG as well as the total composition and structure of TAG. The nutritional properties of fats and oils are important and designed in the formulation of sTAG. Physical properties are important in spreads, coating lipids, and creams.

PEFA have various physiological functions, e.g., EPA and DHA play roles in the prevention of a number of diseases (Gill and Valivety 1997a,b). Gamma-linolenic acid (GLA) shows the physiological functions of modulating atopic eczema and rheumatoid arthritis, and AA is important for pre-term infants.

MCT at *sn*-1,3-position and long-chain at *sn*-2-position of sTAG (MLM-type of sTAG) are absorbed into the intestinal mucosa more rapidly than natural oils composed of only LCFA.

20.5.2 GAMMA-LINOLENIC ACID

A very important group of vegetable oils consist of seed oils rich in GLA ($C_{18:3}$, *n*-6). GLA prevents a wide variety of human diseases, it is an important cosmetic component and is the intermediate in the synthesis of AA ($C_{20:4}$, *n*-6). The main sources of GLA are seed oils from: *Oenothera biennis* (evening primrose), *Borago officinalis* (borage), and *Ribes nigrum* (black currant). A list of plant species rich in GLA acid is presented in Table 20.6.

TABLE 20.6
Examples of Plant Species Rich in γ-Linolenic Acid

Family	Genus and Species	Oil Content of Seed (%)	GLA Content in Oil (%)
Boraginaceae	Borago officinalis	28–31	17–25
	Anchusa azurea	21	11–13
	A. capensis	29	10
	Buglossoides arvensis	11	6
	Echium asperrimum	19	10
	E. sabulicola	20	11
	E. vulgare	15–22	12–14
	Myosotis nemorosa	20	20
	M. secunda	23	12
	Nonea vesicaria	19	9
	Symphytum officinale	21	27
Cannabaceae	Cannabis sativa	38	3–6
Onagraceae	Oenother biennis	17–25	7–10
	Oenothera grandiflora	4	9
Saxifragaceae	Ribes alpinum	19	9
	R. nigrum	30	15–19
	R. rubrum	25	4–6
Scrophulariaceae	Scrophularia marilandica	38	10
	S. sciophila	22	10

Source: Adapted from Guil-Guerrero, J.L. et al., *J. Am. Oil Chem. Soc.*, 78, 677, 2001; Clough, P.M., Specialty vegetable oils containing γ-linolenic acid and stearidonic acid, in *Structured and Modified Lipids*, Gunstone, F.D. (eds.), Marcel Dekker, Inc., New York, 2001, 75.

The consumption of oils rich in GLA can cause vasoconstriction, platelet aggregation, or immunosuppression. Incorporation into the diet of n-3 PEFA is necessary to shift the physiological balance in the direction of vascodilation and antiaggregation. In addition, SCFA and/or MCFA are effective means of delivering of sTAG. MCFA are more rapidly cleared from the blood because of their small size and higher solubility, they are also a source of quick energy because they are rapidly hydrolyzed in lumen and oxidized in the liver. Restriction of evening primrose oil and borage oil is aimed at obtaining a special, universal rich source of both n-3 PEFA and GLA. Using lipases catalyzed acidolysis, MCFA and n-3 PEFA were incorporated into evening primrose oil (Akoh et al. 1996) and borage oil (Namal Senanayake and Shahidi 1999).

High-purified GLA is required for medical and cosmetic applications, and the enzymatic method is used for purification and concentration of GLA from vegetable oils. Shimada et al. (1998) used *Rhizopus delemar* lipase, which discriminates against GLA to enrich the GLA concentration in FFA fraction to about 97%.

Syed Rahmatullah et al. (1994) found Lipozyme discriminated against GLA in an esterification reaction and lipase from *C. cylindracea*, discriminated against GLA in the hydrolysis of borage oil and evening primrose oil. GLA was enriched in FFA fraction (about 70% GLA) or acylglycerols (about 50% GLA).

20.5.3 FISH OILS AS A MATERIAL FOR SYNTHESIS OF sTAG

Fish oil is the main source of EPA and DHA and other long-chain *n*-3 PEFA (Haraldsson and Hjaltason 2001). The FA composition of fish oils has been thoroughly presented in Chapter 15.

The modification of fish oils is done in a two-direction synthesis of sTAG and obtaining concentrates of DHA and EPA. Synthesized sTAGs based on the fish oil contain MCFA at *sn*-1 and *sn*-3-position and LC-PEFA at the *sn*-2 position. Table 20.7 shows the FA composition of tuna oil and sTAG obtained after acidolysis.

The variety of FA in fish oils makes it difficult to obtain pure concentrates of DHA and EPA. PEFA can be concentrated in the form of TAG or monoesters or FFA. Lipases useful for concentration and purification of EPA and DHA can be divided into two categories: discriminating against other *n*-3 FA used for the concentration of DHA and EPA, and lipases discriminating against EPA or DHA used

TABLE 20.7
**Fatty Acid Composition of 1(3)- and
2-Positions in Tuna Oil and sTAG
Obtained by Acidolysis[a]**

| | Fatty Acid Content (mol%) | | | |
| | Tuna Oil | | sTAG | |
Fatty Acid	**1(3)-*sn***	**2-*sn***	**1(3)-*sn***	**2-*sn***
$C_{8:0}$	ND	ND	41.9	0.5
$C_{14:0}$	1.9	1.8	0.9	1.8
$C_{16:0}$	14.7	7.2	3.7	7.1
$C_{16:1}$	4.9	0.8	1.1	1.1
$C_{17:0}$	0.8	0.4	0.3	0.5
$C_{17:1}$	1.1	0.4	0.3	0.3
$C_{18:0}$	5.0	0.4	0.7	0.4
$C_{18:1}$	15.4	3.3	2.5	3.2
$C_{18:2}$	1.0	0.4	0.3	0.3
$C_{20:4}$	2.0	0.7	0.7	0.6
$C_{20:5}$	3.5	2.2	1.6	2.1
$C_{22:6}$	8.8	11.9	7.8	12.4

Note: ND, not detectable.

[a] Reaction of acidolysis between tuna oil and caprylic acid, catalyzed by *R. delemar* lipase.

for concentration of EPA or DHA. The first category, e.g., includes lipases from *Pseudomonas* sp., *P. fluorescens*, *C. antarctica*, and lipase from *Geotrichum candidum*. Lipases from *C. rugosa*, *R. delemar*, and *Rhizomucor miehei* discriminate between EPA and DHA, and all in favor of EPA. However, there is also some evidence that *C. rugosa* lipase does not discriminate against DHA or EPA.

REFERENCES

Adamczak, M. 2004. The application of lipases in modifying the composition, structure and properties of lipids: A review. *Pol. J. Food Nutr. Sci.*, 13/54, 3–10.

Adamczak, M. and Bednarski, W. 2000. Properties and yield of synthesis of mannosyleryth-ritol lipids by *Candida antarctica*, in: *Food Biotechnology*, S. Bielecki, J. Tramper, and J. Polak (eds.), Amsterdam, the Netherlands: Elsevier, pp. 229–234.

Adamczak, M., Bornscheuer, U.T., and Bednarski, W. 2008. Properties and biotechnological methods to produce lipids containing conjugated linoleic acid. *Eur. J. Lipid Sci. Technol.*, 110, 491–504.

Akoh, C.C., Jennings, B.H., and Lillard, D.A. 1996. Enzymatic modification of evening prim-rose oil: Incorporation of n-3 polyunsaturated fatty acids. *J. Am. Oil Chem. Soc.*, 73, 1059–1062.

Amit, J., Swaroopa, G.P., and Bhaskar, N.T. 2006. Modification of lecithin by physical, chemical and enzymatic methods. *Eur. J. Lipid Sci. Technol.*, 108, 363–373.

Anthonsen, T., D'Arrigo, P., Pedrocchi-Fantoni, G., Secundo, F., Servi, S., and Sundby, E. 1999. Phospholipids hydrolysis in organic solvents catalyzed by immobilized phospholipase C. *J. Mol. Catal. B: Enzymes*, 6, 125–132.

Bajpai, V.K., Shin, S.Y., Kim, H.R., and Kang, S.C. 2008. Anti-fungal action of bioconverted eicosapentaenoic acid (bEPA) against plant pathogens. *Ind. Crops Prod.*, 27, 136–141.

Belury, M.A. 2002. Inhibition of cancerogenesis by conjugated linoleic acid: Potential mechanisms of action. *J. Nutr.*, 132, 2995–2998.

Bergsson, G., Steingrímsson, Ó., and Thormar, H. 2002. Bactericidal effects of fatty acids and monoglycerides on *Helicobacter pylori*. *Int. J. Antimicrob. Agents*, 20, 258–262.

Bessa, R.J.B., Santos-Silva, J., Ribeiro, J.M.R., and Portugal, A.V. 2000. Reticulo-rumen bio-hydrogenation and the enrichment of ruminate edible products with linoleic conjugated isomers. *Livestock Prod. Sci.*, 63, 201–211.

Blaszyk, M. and Holley, R.A. 1998. Interaction of monolaurin, eugenol and sodium citrate on growth of common meat spoilage and pathogenic organisms. *Int. J. Food Microbiol.*, 39, 175–183.

Burgess, Y., Wang, Y.M., Cha, J.Y., Nagao, K., and Yanagita, T. 2005. Dietary phosphatidyl-choline alleviates fatty liver induce by orotic acid. *Nutrition*, 21, 867–873.

Cheftel, J.C. 2005. Food and nutrition labeling in the European Union. *Food Chem.*, 93, 531–550.

Chujo, H., Yamasaki, M., Nou, S., Koyanagi, N., Tachibana, H., and Yamada, K. 2003. Effect of conjugated linoleic acid isomers on growth factor-induced proliferation of human breast cancer cells. *Cancer Lett.*, 202, 81–87.

Claussen, K. 2001. Enzymatic oil-degumming by a novel microbial phospholipase. *Eur. J. Lipid Sci. Technol.*, 103, 333–340.

Clough, P.M. 2001. Specialty vegetable oils containing γ-linolenic acid and stearidonic acid, in: *Structured and Modified Lipids*, F.D. Gunstone (ed.), New York: Marcel Dekker, Inc., pp. 75–117.

Collomb, M., Schmid, A., Sieber, R., Wechsler, D., and Ryhanen, E.-L. 2006. Conjugated linoleic acids in milk fat: Variation and physiological effects. *Int. Dairy J.*, 16, 1347–1361.

Considine, T., Patel, H.A., Singh, H., and Creamer, L.K. 2007. Influence of binding conjugated linoleic acid and myristic acid on the heat- and high-pressure-induced unfolding and aggregation of β-lactoglobulin B. *Food Chem.*, 102, 1270–1280.

De Maria, L., Vind, J., Oxenbøll, K., Svendsen, A., and Patkar, S. 2007. Phospholipases and their industrial applications. *Appl. Microbiol. Biotechnol.*, 74, 290–300.

Diplock, A.T., Aggett, P.J., Ashwell, M., Bornet, F., and Fern, E.B. 1999. Scientific concepts of functional foods in Europe: Consensus document. *Br. J. Nutr.*, 81, S1–S27.

Fernandez, E., Chatenoud, L., La Vecchia, C., Negri, E., and Franceschi, S. 1999. Fish consumption and cancer risk. *Am. J. Clin. Nutr.*, 70, 85–90.

Ferrer, M., Soliveri, J., Plou, F.J., López-Cortés, N., Reyes-Duarte, D., Christensen, M.W., Copa-Patiño, J.L., and Ballesteros, A. 2005. Synthesis of sugar esters in solvent mixtures by lipases from *Thermomyces lanuginosus* and *Candida antarctica* B, and their antimicrobial properties. *Enzyme Microb. Technol.*, 36, 391–398.

Gill, I. and Valivety, R. 1997a. Polyunsaturated fatty acids, Part 1: Occurrence, biological activities and applications. *Trends Biotechnol.*, 15, 401–409.

Gill, I. and Valivety, R. 1997b. Polyunsaturated fatty acids, Part 2: Biotransformation and biotechnological applications. *Trends Biotechnol.*, 15, 470–478.

Grundy, S.M., Cleeman, J.L., Daniels, S.R., Donato, K.A., Eckel, R.H., Franklin, B.A., Gordon, D.J. et al. 2005. Diagnosis and management of metabolic syndrome. *Circulation*, 112, 2735–2752.

Guil-Guerrero, J.L., Gracia Maroto, F.F., and Gimenez Gimenez, A. 2001. Fatty acid profiles from forty-nine plant species that are potential new source of γ-linolenic acid. *J. Am. Oil Chem. Soc.*, 78, 677–684.

Gulati, O.P. and Ottaway, P.B. 2006. Legislation relating to nutraceuticals in the European Union with a particular focus on botanical-sourced products. *Toxicology*, 221, 75–87.

Gulati, O.P. and Ottaway, P.B. 2008. Botanical nutraceuticals (Food supplements, fortified and functional foods) in the European Union with main focus on nutrition and health claims regulation, in: *Nutraceutical and Functional Food Regulations in the United States and Around the World*, B. Debasis (ed.), San Diego, CA: Academic Press, pp. 199–219.

Haraldsson, G.G. and Hjaltason, B. 2001. Fish oils as sources of important polyunstaturated fatty acids, in: *Structured and Modified Lipids*, F.D. Gunstone (ed.), New York: Marcel Dekker, Inc., pp. 313–350.

Holub, D.J. and Holub, B.J. 2004. Omega-3 fatty acids from fish oils and cardiovascular disease. *Mol. Cell Biochem.*, 263, 217–225.

Igarashi, M. and Miyazawa, T. 2001. The growth inhibitory effect of conjugated linoleic acid on a human hepatoma cell line, HepG2, is induced by a change in fatty acid metabolism, but not the facilitation of lipid peroxidation. *Biochim. Biophys. Acta*, 1530, 162–171.

Imaizumi, K., Mawatari, K., Murata, M., Ikeda, I., and Sugano, M. 1983. The contrasting effect of dietary phosphatidylethanolamine and phosphatidylcholine on serum lipoproteins and liver lipids in rats. *J. Nutr.*, 113, 2403–2011.

Iwasaki, Y. and Yamane, T. 2002. Phospholipases in enzyme engineering of phospholipids for food, cosmetics, and medical application, in *Lipid Biotechnology*, T.M. Kuo and H.W. Gardner (eds.), New York: Marcel Dekker, Inc., pp. 417–431.

Jiang, Y., Noh, S.K., and Koo, S.I. 2001. Egg phosphatidylcholine decreases the lymphatic absorption of cholesterol in rats. *J. Nutr.*, 131, 2358–2363.

Kabara, J. 1980. Lipids as host-resistance of human milk. *Nutr. Rev.*, 38, 65–73.

Kabara, J.J., Swieczkowski, D.M., Conley, A.J., and Truant, J.P. 1972. Fatty acids and derivatives as antimicrobial agents. *Antimicrob. Agents Chemother.*, 2, 23–28.

Kabara, J.J., Vrable, R., and Lie Ken Jie, M. 1977. Antimicrobial lipids: Natural and synthetic fatty acids and monoglycerides. *Lipids*, 12, 753–759.

Kamberov, E. and Ivanov, A. 1990. Purification of phospholipase-C from *Bacillus cereus* by affinity chromatography on 2-(4-aminophenylsulphonyl) ethyl-cellulose. *J. Chromatogr.*, 525, 307–318.

Kitamoto, D., Yanagishita, H., Shinbo, T., Nakane, T., Kamisawa, C., and Nakahara, T. 1993. Surface active properties and antimicrobial activities of mannosylerythritol lipids as biosurfactants produced by *Candida antarctica*. *J. Biotechnol.*, 29, 91–96.

Kollanoor, A., Vasudevan, P., Nair, M.K.M., Hoagland, T., and Venkitanarayanan, K. 2007. Inactivation of bacterial fish pathogens by medium-chain lipid molecules (caprylic acid, monocaprylin and sodium caprylate). *Aquacult. Res.*, 38, 1293–1300.

Larsson, K., Noren, B., and Odham, G. 1975a. Antimicrobial effect of simple lipids and the effect of pH and positive ions. *Antimicrob. Agents Chemother.*, 8, 733–736.

Larsson, K., Noren, B., and Odham, G. 1975b. Antimicrobial effect of simple lipids with different branches at the methyl end group. *Antimicrob. Agents Chemother.*, 8, 742–750.

Nagao, K. and Yanagita, T. 2008. Bioactive lipids in metabolic syndrome. *Prog. Lipid Res.*, 47, 127–146.

Namal Senanayake, S.P.J. and Shahidi, F. 1999. Enzymatic incorporation of docosahexaenoic acid into borage oil. *J. Am. Oil Chem. Soc.*, 76, 1009–1015.

Nitschke, M. and Costa, S.G.V.A.O. 2007. Biosurfactants in food industry. *Trends Food Sci. Technol.*, 18, 252–259.

Nobmann, P., Smith, A., Dunne, J., Henehan, G., and Bourke, P. 2009. The antimicrobial efficacy and structure activity relationship of novel carbohydrate fatty acid derivatives against *Listeria* spp. and food spoilage microorganisms. *Int. J. Food Microbiol.*, 128, 440–445.

Pariza, M.W. and Hargraves, W.A. 1985. A beef-derived mutagenesis modulator inhibits initiation of mouse epidermal tumors by 7,12-dimethylbenz[α]anthracene. *Carcinogenesis*, 6, 591–593.

Park, Y. and Pariza, M.W. 2007. Mechanisms of body fat modulation by conjugated linoleic acid (CLA). *Food Res. Int.*, 40, 311–323.

Plaza, M., Cifuentes, A., and Ibanez, E. 2008. In the search of new functional food ingredients from algae. *Trends Food Sci. Technol.*, 19, 31–39.

Rainer, L. and Heiss, C. 2004. Conjugated linoleic acid: Health implications and effects on body composition. *J. Am. Diet. Assoc.*, 104, 963–968.

Riserus, U. 2006. *Trans* fatty acids and insulin resistance. *Atheroscler. Suppl.*, 7, 37–39.

Ruckman, S.A. 2008. Regulations for nutraceuticals and functional foods in Europe and the United Kingdom, in: *Nutraceutical and Functional Food Regulations in the United States and Around the World*, B. Debasis (ed.), San Diego, CA: Academic Press, pp. 221–238.

Salas-Salvado, J., Marquez-Sandoval, F., and Bullo, M. 2006. Conjugated linoleic acid intake in humans: A systematic review focusing on its effect on body composition, glucose, and lipid metabolism. *Crit. Rev. Food Sci. Nutr.*, 46, 479–488.

Shimada, Y., Sakai, N., Sugihara, A., Fujita, H., Honda, Y., and Tominaga, Y. 1998. Large-scale purification of γ-linolenic acid by selective esterification using *Rhizopus delemar* lipase. *J. Am. Oil Chem. Soc.*, 75, 1539–1543.

Shin, S.Y., Bajpai, V.K., Kim, H.R., and Kang, S.C. 2007. Antibacterial activity of bioconverted eicosapentaenoic (EPA) and docosahexaenoic acid (DHA) against foodborne pathogenic bacteria. *Int. J. Food Microbiol.*, 113, 233–236.

Siró, I., Kápolna, E., Kápolna, B., and Lugasi, A. 2008. Functional food. Product development, marketing and consumer acceptance: A review. *Appetite*, 51, 456–467.

Sprong, R.C., Hulstein, M.F.E., and Van der Meer, R. 2001. Bactericidal activities of milk lipids. *Antimicrob. Agents Chemother.*, 45, 1298–1301.

Stamler, C.J., Breznan, D., Neville, T.A., Viaua, F.J., Camlioglua, E., and Sparks, D.L. 2000. Phosphatidylinositol promotes cholesterol transport *in vivo*. *J. Lipid Res.*, 41, 1214–1221.

Syed Rahmatullah, M.S.K.S., Shukla, V.K.S., and Mukherjee, K.D. 1994. γ-Linolenic acid concentrates from borage and evening primrose oil fatty acids via lipase-catalyzed esterification. *J. Am. Oil Chem. Soc.*, 71, 563–567.

Thormar, H. and Hilmarsson, H. 2007. The role of microbicidal lipids in host defense against, pathogens and their potential as therapeutic agents. *Chem. Phys. Lipids*, 150, 1–11.

Thormar, H., Hilmarsson, H., and Bergsson, G. 2006. Stable concentrated emulsions of the 1-monoglyceride of capric acid (monocaprin) with microbicidal activities against the food-borne bacteria *Campylobacter jejuni*, *Salmonella* spp., and *Escherichia coli*. *Appl. Environ. Microbiol.*, 72, 522–526.

Tricon, S., Burdge, G.C., Kew, S., Banerjee, T., Russell, J.J., Jones, E.L., Grimble, R.F., Williams, C.M., Yaqoob, P., and Calder, P.C. 2004. Opposing effects of *cis*-9, *trans*-11 and *trans*-10, *cis*-12 conjugated linoleic acid on blood lipids in healthy humans. *Am. J. Clin. Nutr.*, 80, 614–620.

Ulbrich-Hofmann, R., Lerchner, A., Oblozinsky, M., and Bezakova, L. 2005. Phospholipase D and its application in biocatalysis. *Biotechnol. Lett.*, 27, 535–543.

Wahle, K.W.J. and Heys, S.D. 2002. Cell mechanism, conjugated linoleic acids (CLAs) and anti–tumorigenesis. *Prostaglandins, Leukot. Essent. Fatty Acids* 67, 183–186.

Wang, X., Gong, L., Liang, S., Han, X., Zhu, C., and Li, Y. 2005. Algicidal activity of rhamnolipid biosurfactants produced by *Pseudomonas aeruginosa*. *Harmful Algae* 4, 433–443.

Watkins, B.A. and Li, Y. 2001. Conjugated linoleic acid: The present state of knowledge, in: *Handbook of Nutraceuticals and Functional Foods*, R.E.C. Wildman (ed.), Boca Raton, FL: CRC Press, pp. 445–476.

Watkins, B.A. and Li, Y. 2002. Conjugated linoleic acid: Nutrition and biology, in: *Food Lipids: Chemistry, Nutrition, and Biotechnology*, C.C. Akoh and D.B. Min (eds.), New York: Marcel Dekker Inc., pp. 637–661.

Whigham, M.E., Cook, M., and Atkinson, R.L. 2000. Conjugated linoleic acid: Implications for human health. *Pharmacological Res.*, 42, 503–510.

Yang, B., Zhou, R., Yang, J.-G., Wang, Y.-H., and Wang, W.-F. 2008. Insight into the enzymatic degumming process of soybean oil. *J. Am. Oil Chem. Soc.*, 85, 421–425.

21 Frying Fats

Dimitrios Boskou

CONTENTS

21.1   INTRODUCTION

Frying is one of the most widely used practices in the preparation and manufacture of food. There is an increased consumption of fried food today and this is related to the preference of modern consumers for convenience food, the growing of the snack food industry, and the increasing number of fast-food restaurants and takeaway outlets.

During frying, the oil is exposed continuously to a high temperature in the presence of air and moisture. Under such conditions many complex reactions occur, which can be classified as oxidation, polymerization, and hydrolysis. Some of these reactions result in the desirable flavor, color, and texture of the fried food. Others are undesirable from the perspectives of quality, nutrition, and toxicology.

One of the factors that contribute significantly to the quality of the final product of frying is the frying medium. Its degradation depends on the fatty acid (FA) composition and initial state, as well as the presence of prooxidants or antioxidants. Although most of the modifications are generally known, it is difficult to foresee the rate of fat degradation, because of the high number of factors involved. These factors are not only linked to the medium but also to the frying process itself (temperature, duration of heating, mode of heating, rate of oil turnover) or to the food subjected to heating.

Frying in shallow fat is called pan frying or sautèing. Shallow frying means that the oil is used only once and its role as a heat transfer medium is minor. Its action is mainly to prevent sticking to the pan or to give glaze.

Deep frying uses enough fat to cover the food. The oil is hot enough to seal the surface of the food to form a crust. Deep-fat frying is the most widely used method domestically, industrially, and in catering outlets.

Roasting in an oven where temperatures may be as high as 220°C can be also considered as frying (Bognar, 1998).

When acceptable frying practices are applied and the oil is discarded periodically, the oil remaining in the fried food retains its nutritive value and a significant part of liposoluble vitamins. The fried product retains water-soluble vitamins, minerals, and other nutrients to a greater extent than in other cooking methods.

Frying is one of the oldest methods of food preparation. It is believed that it was used as early as 1600 BC by the ancient Egyptians. Most probably it was discovered by roasting a piece of meat in a pot over a hot fire. The fat would be rendered from the meat, the moisture would boil away, and the result would be pan fried meat. In the Old Testament, rules for sacrifices make a distinction between bread "baked in the oven" and bread cooked in the "griddle" or "in the pan." Pliny (Gaius Plinius Secundus,

27–79 AD) recorded a prescription that required eggs steeped in vinegar and then fried in oil. Authors in the middle ages (Chaucer, Cervantes) described in detail the cooking in oil (Morton, 1998).

In the modern era, the practice of frying has both declined and flowered. Two important steps in the technological advance of frying were the development of continuous fryers and the introduction into the food market of partially fried French fries (par-fries). The industrial production of the latter led to further improvement of frying equipment. The consumption of fried food is increasing today due to the preference of modern consumers for convenience food. Modern frying practices have contributed to the production of better products that combine, flavor, color, crust, and texture with reduced loss of nutritive value.

21.2 FRYING MEDIA

21.2.1 "IDEAL" FRYING MEDIUM

It is difficult to give an accurate definition for the perfect medium because there are so many applications and so many factors that affect the frying process. Some general guidelines and characteristics of good frying oil are the following:

- A deep-frying medium should be selected not only on the basis of a good Rancimat value, but also on a specified FAs composition.
- To avoid polymerization, the frying medium should not contain too much linolenic acid (several countries set limits as low as 2% for linolenic acid).
- High levels of saturated FAs (palmitic, stearic) should be avoided because of their effect on the sensory properties of the fried food.
- Blends should be avoided, unless they are properly examined and found not to cause foaming (blending of long-chain FAs oils with short-chain FA oils are known for their foaming effect).

21.2.2 TYPES OF OILS USED FOR FRYING

The fatty materials used for frying may be unhydrogenated fats and oils or hydrogenated products, specifically prepared. Liquid frying fats are made from: (1) refined, bleached, and deodorized oils such as palm olein, corn oil, sunflower oil, soybean oil, safflower oil, cotton seed oil, peanut oil, rape seed, and canola oil; and (2) fractionated hydrogenated oils. In the Mediterranean countries, olive oil in its various forms (virgin, refined, olive residue oil) is also used. Solid fats are palm stearine, coconut fat, palm kernel oil, lard, tallow, and various shortenings with different melting points and solid fat indices. Beef dripping fat is also used because it provides a distinctive flavor to potato chips.

A high intake of saturated and *trans*-FAs from frying fats is considered nutritionally undesirable. To reduce the intake of such FAs, products of genetic modification such as high oleic sunflower oil, low linolenic sunflower, modified canola, soybean oil, and other oils have been proposed as good alternatives to partially hydrogenated oil (Dobarganes et al., 1993, Warner and Mounts, 1993, Ulmansoy et al., 2008).

In manipulating the FA composition to increase monounsaturated FA content, changes in the level of natural antioxidants, which may accompany FA alteration, should be taken into consideration. Normand et al. (2001) demonstrated that stability of modified canola oil is affected far more by the rate of tocopherol degradation than changes in FA composition. This finding may explain why only a small improvement was found by a number of researchers who compared the frying stability of regular and modified oils (Petukhov et al., 1999).

21.3 DEEP FRYING

21.3.1 INTRODUCTION

Deep frying follows defined principles of heat and mass transfer. Convective heat transfer from the hot oil and conductive heat transfer through the food are the driving forces that lead to the chemical and other changes in both the oil and the food. Mass transfer includes dehydration (evolution of steam), distillation (volatile materials), extraction, and leaching from the food to the oil. The physical changes and chemical reactions during deep-fat frying are briefly described below.

21.3.2 PHYSICAL CHANGES

Steam escapes from the surface of the oil and also removes volatile compounds. An increase in volatiles causes lowering of the smoke point. The moisture creates a blanket above the oil that minimizes contact of atmospheric oxygen and food. The formation of steam prevents the oil from getting into the interior of the fried product. Thus, as long as steam is generated, the temperature of the food is always 100°C, regardless of the temperature of the oil.

The formation of polymers leads to an increase in viscosity. The various lipids that can leach into the frying oil change the properties and the performance of the frying oil. Colored lipids solubilized in the oil contribute to the darkening. Phospholipids are emulsifiers. Traces of liposoluble metal compounds may act as prooxidants. Liposoluble vitamins and phenolic compounds are antioxidants. Volatile compounds (e.g., from fish or onions) contribute to off flavors.

Oil penetration depends on the shape of food, its textural properties, porosity, viscosity of the frying medium, and the temperature and duration of frying. Increased viscosity results in larger volumes of absorbed oil. Food that is high in initial fat content does not absorb oil. On the contrary, fat is leached from the food into the oil.

21.3.3 CHEMICAL REACTIONS

At the temperatures of frying (160°C–200°C) the moisture escaping from the food hydrolyzes triacylglycerols (TAG) to FAs, glycerol, and mono- and diacylglycerols.

Oxidation causes the formation of hydroperoxides and conjugated compounds, which by cleavage give aldehydes, alcohols, ketones, lactones, acids, esters, and hydrocarbons. Radical mechanisms lead to the formation of dimer, other oligomers, and oxidized TAG. The latter have one or more acyl group with an extra oxygen (hydroxy, keto, epoxy derivatives). Other oxidation products are TAG with short-chain fatty acyl and n-oxo fatty acyl groups.

Most important volatiles for the quality of frying oil are saturated aldehydes C_6–C_9, enals (e.g., 2-decenal), dienals (e.g., 2,4-heptadienal), and hydrocarbons (hexene, hexane, heptane, octane, nonane, decane). The presence of volatile oxidation products formed during the frying process has been discussed by Perkins (1996), Nawar (1998), and Choe and Min (2007).

The formation of nonvolatile decomposition products is due to the oxidation and polymerization of unsaturated FA. These products act as catalysts in the radical reactions and accelerate the deterioration. Oils heated at intervals have higher degradation rates than oils heated continuously. Aldehydes affect the flavor of deep-fried food. 2-*trans*-4-*trans*-decadienal contributes to the fat flavor, while other aldehydes produce unpleasant aromas. *Trans*-4,5-epoxy-*trans*-2-decenal indicates a prolonged period of heating (Belitz and Grosch, 1999).

Dimers and other oligomers are good markers of oil degradation and their level is regulated in many countries. The polymers are usually divided into polar dimers, nonpolar dimers, and oligomers having a higher molecular weight than dimers. The structures of oxygenated polymers are not fully elucidated because of the heterogeneity of this class of compounds (Dobarganes and Marquez-Ruiz, 1996, Gasparoli, 1998, Mahungu et al., 1999). The structure of an oxidation polymer with ether and peroxide bond has been presented by Belitz and Grosch (1999).

$$
\begin{array}{c}
\text{OH} \\
| \\
\text{R--CH--CH=CH--CH--R} \\
| \\
\text{O} \\
| \\
\text{R--CH--CH=CH--CH--R} \\
| \\
\text{O} \\
| \\
\text{O} \\
| \\
\text{R--CH=CH--CH--R}
\end{array}
$$

Hydrolysis, oxidation, and polymerization reactions in deep-fat frying have been recently discussed by Choe and Min (2007). In the absence of air, hydrolysis, isomerization, and polymerization are the main reactions. Carbon–carbon bonds link the thermal dimers and trimers. A Diels–Alder reaction is proposed for their formation, but dimerization of unsaturated FAs is also explained by a cationic mechanism (Gertz et al., 2000). This cationic mechanism is probably related to the formation of cyclic monomeric FAs. Polymerization is one of the causes of viscosity increase. It is also the cause of gumming.

Cyclic products are important from the health point of view because they are of some toxicological concern and there are a lot of studies on metabolic processes and their ingestion (Sebedio and Chardighy, 1996, Mahungu, 1999). Monomeric cyclic FAs are formed from linolenic, linoleic, and oleic acids during heating of vegetable oils at temperatures above 200°C. Since these temperatures are reached in the deodorization step of refining, refined bleached and deodorized oils may also contain cyclic monomers (Christi and Dobson, 2000, Henon et al., 2001). During cyclization, one double bond is lost. Monocyclic monoenoic and dienoic FAs derive from linolenic and linoleic acid. Cyclization of oleic acid gives saturated FAs. Structures that have been identified are C_{18} disubstituted cyclohexyl dienes with one double bond in the ring and the other in the side chain containing the carboxyl group or the aryl group (10,15 cyclization) and similar cyclopentenyl dienes (11,15 or 10,14 cyclization). From linoleic acid, various isomers are formed containing a cyclopentenyl ring at position C_8–C_{12}, C_{10}–C_{14}, C_{13}–C_{17} and C_5–C_{10}. From oleate, 5- and 6-membered saturated ring compounds are formed (Le Quere and Sebedio, 1996, Mahungu, 1999, Christi and Dobson, 2000). Cyclic FA monomers are practically absent in virgin olive oil and their level is not significantly increased after repeated fryings, i.e., no more than 1 µg/g (Romero et al., 2000). A cyclohexyl FA formed during frying has the following structure:

Trans isomers are formed during frying, whether in the kitchen or in the industry (Chardigny et al., 2001). Such geometrical isomers are also present in refined oils, especially physically refined oils, because of the high temperatures applied. The major isomers found in both cases are 9,12-*cis-trans* and 9,12-*trans-cis*-octadecadienoic acids, followed by conjugated *cis-trans, trans-cis,* and *trans-trans* dienes. Several isomeric trienes are formed from linolenic acid during thermal treatment of oil. Grandgirard et al. (1984) identified di-*trans* and mono-*trans* isomers such as 18:3 9c, 12c, 15t, 18:3 9t, 12c, 15c, 18:3 9c, 12t, 15c and 18:3 9c, 12t, 15t. Determination of geometrical and positional isomers is a tedious work because of the complexity of the mixture and the lack of reference materials. The results of a collaborative study supported by the European Union Commission was published in 2001 (Buchgraber and Ulberth, 2001). The objective of this study was the intercomparison of methods for the determination of *trans* FAs in edible oils and the certification of three materials. One of the reference materials is for *trans* FAs formed by heating either during refining or frying.

21.3.4 INTERACTIONS BETWEEN FRYING OIL AND FOOD COMPONENTS

Maillard reactions between components of the fried food (sugars-amino acids) are responsible for the golden color of fried products. Lipid oxidation products contribute also to the main reactions. Reactive oxidation products are aldehydes, epoxides, hydroxyketones, and dicarbonylic compounds, which react with lysine, proline, and other amino acids.

Some of the typical fried flavors are due to degradation products of the frying oil. Other compounds contributing to the flavor originate from the food itself. Flavor compounds are also formed by interactions of lipid oxidation products and food components. A typical example is 2,4,6-trimethyl-1,3,5-dithiazine (see structure), a major product of reaction of 2,4-decadienal with cysteine (Pokorny, 1999):

Saccharides participate in Maillard reactions. Polysaccharides form a compact film on the surface in the beginning of frying and produce a crispy crust. The level of dietary fiber and resistant starch may change, but not significantly.

Numerous flavor compounds are formed by caramelization, Maillard or Strecker reactions, or the oxidation of phenolic compounds and terpenes. Some classes of flavor compounds found in frying foods are furan derivatives, pyrrole derivatives, pyrazines, thiols, sulfides, aldehydes, and aromatic compounds from phenol oxidation (Pokorny, 1999).

21.3.5 DESIRABLE AND UNDESIRABLE CHANGES DUE TO DEEP-FAT FRYING

Deep frying

- Adds color, flavor, and texture to the fried food.
- Provides economy and speed of cooking.
- Provides consistent quality of the final product.
- Causes lower losses of nutrients and water-soluble vitamins in relation to those caused by other cooking methods. The proteins, carbohydrates, and minerals are almost fully retained.
- Improves digestibility and bioavailability of nutrients in the digestive system.
- Improves the hygienic quality of the food by inactivation of microorganisms.
- Causes reduction of fat in animal high fat foods and improves nutritionally FAs profile.

Undesirable changes due to frying are

- A higher content of fat in the final product
- Unfavorable changes in the sensory properties when prolonged heating is applied
- Increased oxidation and polymerization products in abused oils
- Losses of heat and oxidation-susceptible vitamins

21.3.6 Chips

Deep fat fried food such as hot chips, chicken, and fish are convenience food in many countries. These products are high in fat and they have been implicated in studies relating fat consumption to certain diseases (e.g., cardiovascular disease, obesity, and some forms of diabetes and cancer).

To obtain lower fat, higher quality chips, good deep-frying techniques are needed. An important factor is the size and shape of chips. Thick-cut chips, bigger than 12 mm in diameter, absorb less fat than thinner chips. Cracks and rough surfaces increase fat absorption. Chips may have a 6%–20% fat content, while crisps may have a fat content as high as 35%. Other important factors in addition to cut thickness are the heat balance and the volume of frying oil. Heat balance is the ability of the frying system to maintain a constant temperature (Mehta and Swinburn, 2001).

21.4 QUALITY OF INDUSTRIAL FRYING OILS AND FATS

21.4.1 Introduction

For a better performance of the oil to be used in deep-frying, several factors should be taken into consideration. These factors include the following: (1) the nature of the frying oil and the presence of minor constituents (antioxidants or prooxidants), (2) the specifications of the oil before use, (3) the rate of oil turnover, and (4) the formation of detrimental materials during the frying operation.

21.4.2 The Importance of FA Composition

Saturated FAs, such as palmitic and stearic, are stable against oxidation and polymerization. Because they have a high melting point they add structure to certain products. However, due to this characteristic, the appearance of a fried food may be affected adversely (e.g., waxy mouth-feel, dry surface of stored fried food). Monoenoic FAs, mainly oleic acid, are considered to be beneficial from a health standpoint. Frying oils rich in such FAs do not add to the structure; they are stable against oxidation and provide a light taste. Polyenoic FAs (PEFAs) deteriorate more rapidly than monoenic FAs and the shelf life of products fried in oils rich in these acids is shorter. Oxidation products formed from PEFA vary widely depending on the structure of the FAs and the relative concentration of linoleic and linolenic acids. The percentage of linolenic acid in heated oils should be very low. Biotechnology allows for the creation of oils with specific FA profiles for particular food applications and vitamin E fortified oil seeds, which prevent oxidation. Marmesat et al. (2008) studied the thermal stability of genetically modified sunflower oil with high content of oleic and palmitic acid and containing alpha- and gamma-tocopherols. The importance of the degree of unsaturation and the presence of tocopherols for the protection against thermal degradation of cotton seed oil was studied by Steel et al. (2006).

21.4.3 Specifications

Frying oil with a high FAs content will smoke at a lower temperature. FFAs content increase rapidly if heating is poorly controlled; if the frying rate is insufficient to remove water from the food; and, of course, if the turnover rate is kept above a certain hours.

The main requirements for a high quality frying oil are FA 0.05%–0.1%; peroxide value 0.5–1 mEq/kg; moisture <0.05%; color: red max. 3, yellow max. 30 (Lovibond cell 5.25 in.); smoke point >220°C; taste and flavor: bland.

Foaming is a safety hazard. If excessive foaming occurs, the heating source must be switched off to avoid risk of fire. Foaming may be due to oil breakdown products, contact with metals and traces of detergents from the cleaning of fryers. Lecithins and mono- or diacylglycerols cause foaming. An effective antifoaming agent is dimethyl polysiloxane. Proteinaceous residues have to be removed by filtration or skimming because they give rise to unpleasant flavors. Debris of the fried food is normally responsible for such residues although traces may be also found in unrefined oils such as cold pressed oils or beef drippings used in frying.

Copper, iron, and manganese catalyze the decomposition of hydroperoxides to free radicals. Copper is a strong prooxidant, almost ten times more active than iron. During the frying operation, contamination of the oil with such metals should be avoided. Otherwise, a premature decomposition will take place causing unpleasant rancid notes in the product.

Alkali contaminant materials (ACM) may be due to the fried food and the degradation, or due to residues left in the oil after refining. ACM are traces of soaps, which give rise to the pH. For the determination of ACM bromothymol blue is used. In restaurant frying operations when discard point is reached, levels of ACM may be as high as 40 mg/kg (expressed as sodium oleate).

21.4.4 Stabilization of Frying Oils with Synthetic and Natural Antioxidants

Various researchers have suggested that modification and the addition of various compounds or mixtures can protect frying oils against thermal and oxidative deterioration. However, there is a lack of agreement among researchers on the effectiveness of antioxidants in retarding deterioration of oils during frying. This is partly due to the fact that using actual frying conditions to study antioxidant effectiveness is not always easy and extrapolation of data on the action of antioxidants from storage or heating experiments in model systems may be misleading. The rate of the initiation reactions of autoxidation is very high at frying temperatures and antioxidants, added usually at levels of 0.01%–0.02%, are rapidly decomposed. In addition, degradation reactions take place; these are not encountered when the oil is autoxidized at room temperature. Certain additives such as butylated hydroxytoluene (BHT) and butylated hydroxyanisole (BHA) are lost through volatilization, and the desired level of the additive is not maintained during the frying operation.

Among the various synthetic antioxidants used for the protection of frying oils against thermal oxidation, the most extensively studied are BHA, BHT, tertiary butyl hydroquinone (TBHQ), and propyl gallate (PG). The addition of TBHQ has been claimed to have a good "carry through" into potato chips, deep fried in oil containing 0.02% antioxidant. Other substances that have been tested for their effect on the stability of frying oils are silicone oil, secondary antioxidants such as citric acid, EDTA, tartaric acid, and ascorbyl palmitate and polymeric antioxidants (Berger, 2005).

In the past three decades, there has been an increasing trend toward natural ingredients in food and many attempts have been made to replace chemicals in the stabilization of lipids. The emphasis given to natural antioxidants is due to toxicity concerns about some synthetic antioxidants but mainly to findings, which relate active ingredients in the diet and radical mechanisms in the human body.

Researchers interested in the properties and occurrence of natural antioxidants concentrate on vitamins E and C, carotenoids, phenolic acids, flavonoids, sesame lignans, phytosterols, extracts from the leaves of the plants belonging to the *Lamiaceae* family, olive leaves extracts, tea leaves extracts, phosphatides, oryzanols, olive oil phenols, pomegranate peel extracts, squalene, and propolis. The effect of natural antioxidants on the stability of frying oils has been discussed by Boskou (1999), Blekas and Boskou (1999), Kochhar (2000a), Koketsu and Satoh (1997), Irwandi et al. (2000), Berger (2005), Farag et al. (2006), Chiou et al. (2008), and El Anany (2007). The antioxidative effects and degradation of alpha- and delta-tocopherols at levels 0.01% and 0.1% on the oxidation of a commercial frying fat at 160°C was studied by Nogala-Kalucka et al. (2005). Under mild frying conditions, the less effective antioxidant, alpha-tocopherol, was found to undergo a more quick disintegration than delta-tocopherol.

21.4.5 Olive Oil as Frying Medium

21.4.5.1 Introduction

Virgin olive oil is not only a light and delicate addition to salads and various dishes but it is now considered one of the most health-promoting types of oils available. Olive oil, rich in monounsaturated FAs, has a remarkable stability during domestic deep-frying or in other uses that require frying temperatures (Boskou, 1999, Sanchez-Gimeno et al., 2008). In comparison to sunflower, cottonseed, corn, and soybean oil, olive oil has a significantly lower rate of alteration. This increased stability to thermal oxidation explains why the oil can be used for repeated frying operations. The resistance of olive oil to rapid deterioration at elevated temperatures is attributed to its FA composition and the presence of natural antioxidants such as squalene, alpha-tocopherol, and delta-5-avenasterol (Blekas and Boskou, 1999, Boskou, 1999). The properties mentioned earlier are well known to people who traditionally use olive oil in cooking and prefer olive oil as a means of shallow frying.

Frying in olive oil offers a means to improve the profile of lipid intake. During the frying process, changes occur in the fat composition since the oil penetrates into the fried food. Western diets using vegetable oils and animal fats are very often rich in saturated FAs and also n-6 FAs. When meat is cooked in olive oil, there is a favorable change in saturated to polyunsaturated FA ratio. A better combination is to fry fish. In sardine frying, for example, the nutritional benefits of the oil are combined with those of the n-3 FAs from the fish.

Another important aspect of frying in olive oil was examined by Persson (2003), who fried beef burgers in various oils. The burgers were analyzed for the levels of 12 different heterocyclic amines (HA), such as 2-amino-3,8-dimethylimidazo (4,5-7) quinoxaline and 2-amino-1-methyl-6-phenyl-imidazol (4,5-6) pyridine. The intake of HA has been associated to the development of cancer in some epidemiological studies. During cooking of animal tissue, these amines are formed at low levels via the Maillard reactions and a free radical mechanism. Frying in virgin olive oil reduced the formation of HA and this was related to the presence of seco-iridoid phenols. Loss of these phenols by storage or heating caused a decrease in the HA-reducing capacity of the oil.

21.4.5.2 Heating of Olive Oil and Phenolic Compounds

In the last decade, researchers have focused on the level of phenolic compounds such as hydroxytyrosol in heated olive because these compounds contribute to the stability of the oil against autoxidation. In addition, these compounds are considered components with an important biological role. Most of the published reports indicate that phenolics in virgin olive oil deteriorate relatively rapidly. Andrikopoulos et al. (2002) determined total phenols during successive pan-frying and deep-frying operations of virgin olive oil under conditions applied in domestic cooking. The loss of polar phenols and tocopherols were significant. Brenes et al. (2002) investigated the changes occurring in virgin olive oil subjected to simulated domestic frying, microwave heating, and boiling with water in a pressure cooker. Heating at 180°C caused a significant loss of tocopherols and hydroxytyrosol derivatives; lignans (pinoresinol and 1-acetoxypinoresinol) were relatively stable. Microwave heating caused lower losses of phenolic compounds. Boiling in a pressure cooker caused rapid hydrolysis of the secoiridoid aglycons. The hydrolysis products were diffused in the water phase. Gomez-Alonso et al. (2003) studied the losses of hydroxytyrosol and its seco-iridoid derivatives in olive oil used in repeated frying operations for the preparation of French fries. The reduction of hydroxytyrosol and the other diphenols was found to be 50%–60% in relation to the original value.

Pellegrini and his coworkers (2001) used the ABTS radical decolorization assay of antioxidant activity to study the effect of heating on the total antioxidant activity (TAA) of extra virgin olive oil and alpha-tocopherol content in the presence of 14 polar phenolic compounds occurring in the oil. Their results indicate that heating causes a significant loss of olive polyphenols, which act as stabilizers of alpha-tocopherol during olive oil heating. Gomez-Alonso et al. (2003) found that the antioxidant activity of the phenolic extract, measured with the DPPH• test, diminishes during the first six frying processes (each frying process: 10 min at 180°C). A rapid loss was observed mainly in the concentration of hydroxytyrosol and its secoiridoid derivatives (aldehydic forms).

A loss of the antioxidant capacity of olive oil and other vegetable oils due to heating at frying temperatures were also reported by Quiles et al. (2002), who used electron spin resonance, and also by Carlos-Espin et al. (2000) and his coworkers, who studied the kinetics for the disappearance of total free radical scavenging capacity (RSC) using the DPPH• test. Kalantzakis and others (2006) examined the loss of antioxidant capacity (evaluated by the DPPH test) and the polar transformation products formed from various vegetable oils heated at 180°C for 10h. It was observed that olive oil lost its RSC at a shorter time of heating in relation to soybean, sunflower, cottonseed oil, and any commercial frying oil. However, olive oil reached the level of 25% total polar compounds (TPC) (rejection point) after prolonged heating, while all the other oils reached this upper limit earlier (10h of heating). It can be concluded that olive oil as a frying medium has a remarkable stability and a resistance to oxidative polymerization due to frying. When, however, health effects are expected from the presence of natural antioxidants, the number of heating operations should be restricted to a minimum.

In a recent report Carrasco-Pancorbo et al. (2007) provided information for the levels of four classes of phenols present in olive oil, simple phenols, lignans, complex phenols, and phenolic acids after heating at 180°C. Hydroxytyrosol, decarboxymethyl oleuropein aglycon, and oleuropein aglycon were found to be more sensitive to heating in relation to hydroxytyrosol acetate and lignans. The stability of lignans was also reported by Allouche et al. (2007).

21.4.6 OTHER STABLE OILS

S. P. Kochhar explained the concept of a stable frying oil in his article "Stable and healthful frying oil for the twenty-first century" (Kochhar, 2000b). The idea is not to use hydrogenation to obtain high solid fat frying oil but high oleic acid oil whose stability is further enriched by sesame seed, rice bran, and olive oil. These oils are rich in sterols such as delta-5-avenasterol and citrostadienol that are antioxidant at elevated temperatures. A range of other antioxidants present in these oils are tocopherols, oryzanol (sterol ferulate), sesamolin (a precursor of an antioxidant), sesaminol, sesamol dimers, and others. (Blekas and Boskou, 1999, Boskou, 1999, Nasirullah and Rangaswamy, 2005).

Fractions obtained by hexane or acetone from nontraditional tree seed oils such as sal, mango, and mahua were also shown to increase the stability when incorporated into base frying oils (Nasirullah, 2001). A study published in 2006 (Warner and Dunlap, 2006) evaluated the stability of expeller-pressed soybean oil used in potato frying that had slightly more tocopherols and phytosterols compared to the refined oil.

21.5 EVALUATION OF DETERIORATION AND ASSESSMENT OF THE QUALITY OF USED OIL

21.5.1 INTRODUCTION

The methods used to determine the effect of heat on frying oils can be divided into two categories. The first category includes relatively simple standardized methods such as color, peroxide value, anisidine value, free FAs, foam, smoke point, viscosity, and/or quick tests based on dielectric constant measurements and color indicators.

The methods of this category are convenient for routine analysis. The second category includes more sophisticated methods, which are based on gas–liquid or high performance liquid chromatographic techniques, or other more or less complex procedures necessary to obtain more information about specific hydrolytic or thermal oxidation products. Furthermore, sensory analysis of the fried products can be used to evaluate the suitability of the frying oils. This is carried out by panelists who examine the color, crispness, flavor, and overall acceptance of chips fried in the oil.

21.5.2 Traditional and Standardized Methods

The development of free FA parallels, to a certain extent, other degradation changes, but the measurement alone does not correlate well to the quality of the frying oil.

The iodine value is a measure of unsaturation. A fall in iodine value is consistent with a decrease in the number of double bonds in heated oil as it becomes oxidized.

The color of the oil becomes darker during frying. The change in color can be measured with a Lovibond tintometer. For a quick assessment of color, comparison kits are available. As in the case of free FAs, color alone does not express the quality of the used oil.

The peroxide value is a normal oxidation index, which has to be carefully used because at high temperatures hydroperoxide decomposition may be more rapid than their formation; thus degradation is not reflected by this index.

The smoke point is an important characteristic for restaurants in deciding when to change the frying oil. This value depends on the development of volatiles.

Excessive foam in the oil during frying is a nuisance and in extreme cases a safety hazard that necessitates changing of the oil. Foaming is due to viscosity changes, emulsifiers, and stabilizers formed from the oil and the food being fried. The foam height test is not standardized. During frying, the viscosity is increased because of the formation of high molecular weight compounds. Oil with an increased viscosity is absorbed by the food more readily and the food becomes soggy and greasy. Viscosity is regulated in Belgium. The frying oil should be discarded if viscosity reaches 27 mPa s at 50°C.

The anisidine value measures aldehydes, which are less easily decomposed. Thus, the test is more meaningful than the peroxide value.

The totox value is defined as $2 \times$ peroxide value + anisidine value. The induction period is usually measured by Rancimat apparatus and expressed in hours. Induction period of nonused oils is approximately 10–15 Rancimat h. This period may be reduced to 1–2 h in one or two days of commercial frying.

Conjugated dienes and trienes measure the shift of double bonds. The shift is measurable by the absorbance at 232 and 268 nm. Similar to the peroxide value, the absorbance at 232 nm measures the level of reactive intermediates. $E_{1\,cm}^{1\%}$ values, which are very low in nonheated oils, may become as high as 15 within a few days of commercial frying. $E_{1\,cm}^{1\%}$ at 268 nm also follows changes in iodine value, viscosity, PEFA, and polymer content.

The carbonyl value measures secondary oxidation products and is determined by the 2,4-dinitrophenyl-hydrazine method. Initial carbonyl values are very low but they may increase to 100–150 mEq/kg when the oil is heated.

Refractive index is a simple means to monitor deterioration of frying oils. This parameter is usually measured at 40°C.

Quantification of FAs composition by gas–liquid chromatography of methyl esters is useful for the monitoring of unsaturated FAs. Losses of PEFA follow the increase in TPC. Near the rejection point (27% TPC), 10%–15% losses of PEFA may be observed.

These tests are not equally meaningful in evaluating the level of deterioration of the oil used in frying and each one alone may place the oil in a different order of merit. The most meaningful tests are polar compounds, polymer content, and anisidine value. Typical scores for various oils used for frying potatoes in electric fryers at 180°C, based on determination of color, free FAs, smoke point, reduction of unsaturation, peroxide value, induction period, totox value, iodine value, anisidine value, foam height, polymer content, and polar compounds were reported by Pantzaris (1999).

21.5.3 QUICK TESTS

Quick tests may be based on chemical reactions and physical properties. Chemical quick tests in the industry may measure color, foam, or free FAs. These tests are helpful when the initial composition of oil and fried food are the same and the conditions of frying do not change.

The two most important quick tests based on physical parameters measure the change in viscosity and dielectric constant. Several studies have been published for the marketed sensors and the correlation to total polar material. The advantages and drawbacks of measurements have been discussed by Dobarganes and Marquez-Ruiz (1998) and Gertz (2000).

21.5.4 TOTAL POLAR COMPOUNDS AND POLYMERS

Quantification of TPC is the basis of legislation in many countries for the control of used frying fats. The method is gravimetric and it is based on the adsorption of polar compounds onto a silica gel column. Limitations up to 27% have been set for TPC in countries where frying oils are regulated. For the determination there are approved methods (IUPAC, AOAC). TPC can also be determined with Sep-Pak cartridges.

Methods for the determination of TPC are an improvement in the methodology, but they have some serious drawbacks. Cyclic FA monomers are not eluted in the polar fraction. These alteration products may be present both in the polar and nonpolar fraction. FAs and partial glycerides are eluted in the polar fraction but they may be either naturally present or due to hydrolysis. Samples with similar total levels of alteration cannot be differentiated in relation to the type of degradation (hydrolytic or thermal oxidation). To overcome some of these difficulties, analysis of methyl esters instead of the oils has been suggested. Marmesat et al. (2007) proposed a rapid method for the determination of polar compounds in used frying fats and oils that uses only milligrams of a sample. Methyl oleate is used as an internal standard, and the nonpolar fraction is separated by solid phase extraction and analyzed by high performance size exclusion chromatography (HPSEC). Response factors for intact TAG and FAs methyl esters are calculated. The polar fraction is estimated by the difference in weight.

21.5.5 POLYMERIZED TAG AND OTHER APPROACHES

Polymer TAG is an important fraction of polar compounds for a reliable monitoring of the changes in the oil during frying. It is determined by size-exclusion high performance liquid chromatography and there are approved IUPAC and AOAC methods.

A completely different approach based on hydrolysis of thermally oxidized oils by pancreatic lipase was proposed by Sanchez-Muniz and his coworkers (2006). According to the authors, more work is needed in this direction to better understand the major stages of the digestion of the oxidized oil, the role of minor components (sterols, phenols), and to better assess the level of alteration at which lipase is completely inhibited.

Gloria and Aguilera (1998) proposed a DSC technique to follow changes in the thermal characteristics of commercial frying oils during heating. DSC tracings indicate that heating of the oil results in a progressive shift of the crystallization peak at 43°C–48°C to lower temperatures and reduced enthalpies of crystallization. These characteristics are well correlated with the increase in viscosity, polar compounds, and color.

Xu (2000) developed a quick spectrophotometric method to evaluate deep-frying oil quality. The oil samples are scanned from 350 to 650 nm. Each type of oil is found to show systematic changes in spectrophotometric absorbance; these changes are more evident between 470 and 500 nm. According to the author, there is a strong correlation between the absorbance and total polar constituents, and the method offers itself as a reliable routine test for assessing frying oil quality.

Deyawatee and coinvestigators (2001) studied the deterioration of soybean oil during the frying of potato chips by Fourier Transform Infra Red (FTIR) spectroscopy. The investigators claim that a decrease in the ester linkages of TAG and in the iodine value, and the increase in *trans* unsaturation and formation of unsaturated aldehydes, as measured by FTIR, correlated well to chemical methods for monitoring deterioration. FTIR spectroscopy combined with a partial least square approach was also used by Che Man and Setiowaty (1999) to predict anisidine value. The latter is dependent on the concentration of 2-alkenals, 2,4-dienals, and to a lesser extent to saturated aldehydes. The method is proposed to replace the laborious wet-chemical method. The technique is based on the influence of the carbonyl group on the C-H frequency (range 2747–2619 cm^{-1}) and the carbonyl region (1715–1763 cm^{-1}). Nuclear magnetic resonance (NMR) techniques have also been developed by Guillen and Uriarte (2008), who tried to correlate the information provided by ^{1}H NMR with the limit of 25% TPC in olive oil used for frying. Various antiradical tests using the 2,2-diphenyl-1-picrylhydrazyl radical (DPPH•) or other radicals were used for the evaluation of any oil heated at frying conditions (Pellegrini et al., 2001, Kalantzakis et al., 2006, Van Loon et al., 2006, see also Section 21.4.5).

21.5.6 PREDICTION OF HEAT STABILITY

Gertz and coworkers (2000) proposed a new method to estimate the oxidative stability at elevated temperatures (OSET). The fatty material is heated at 170°C for 2 h after the addition of water-conditioning silica gel. The polymeric TAG (mainly nonpolar dimers and oligomers) are then measured using size-exclusion HPLC and then the OSET value is calculated.

21.6 NUTRITIONAL AND HEALTH ASPECTS

21.6.1 INTRODUCTION

There are many studies in the scientific literature that examine the nutritional value and safety of heated oils. Some of the most important aspects of the problem are thoroughly discussed in the reviews of Clark and Serbia (1991), Sebedio and Chardigny (1996), Marquez-Ruiz and Dobarganes (1996), Paul and Mittal (1997), Billek (2000), and Ringesis and Eder (2008).

It is now established that some compounds formed during heating under nonrealistic conditions have antinutritional properties. These compounds may inhibit enzymes and in the absorption of nutrients, destroy vitamins, oxidize lipids, and/or potentially cause gastric distress or potentially undergo mutations. When overheated oils or fractions containing degradation products were fed to experimental animals, loss of weight, delay in growth, diarrhea, skin problems, growth of liver, ulcer, and other detrimental effects were observed. These harmful degradation products are absent or present in minute quantities when good frying practices are applied. From the plethora of reviews and articles found in the literature, it is reasonable to assume that when a proper monitoring of discarding is practiced the formation of substances that are toxic is of no practical significance. Nutritional and toxicological effects were detected only in discarded frying oils or extensively fried oils (Juarez and Samman, 2006). A moderate consumption of fried food prepared under normal culinary practices is completely safe. Fats and oils reach harmful levels of degradation long after they become unacceptable from the point of view of flavor. This excludes possible risks for the consumer. Besides, the regulated or suggested limits for TPC, oxidized FAs, and TAG oligomers add additional security. However, many scientists and regulators still have worries about some possible detrimental effects of oils used in frying. Undoubtedly, a better knowledge of changes occurring during frying and metabolism of heated fats can help in the selection of oils and food materials with a minimum fat decomposition, a minimum delivery of fat calories, and a healthful balance of FAs.

21.6.2 INTAKE OF FAT

The intake of too much saturated fat, too many calories, and cholesterol from animal fats used in frying is a problem already addressed by many medical and nutritional bodies. The public is now fully aware of the risks in relation to atherosclerosis and other diseases. The growing demand of the consumer for healthful fats has caused a switch from animal fat to vegetable frying oils and this trend is now general not only among fast-food industries and catering outlets but also in domestic applications. According to Varela and Ruiz-Roso (1998) the fat intake can be improved through frying. For example, when lean meat is fried in oil rich in monoenoic FAs, saturated FAs pass into the frying oil and there is an increase in monoenic acids and a decrease in saturated FAs in the fried meat.

21.6.3   MACRONUTRIENTS, METALS, AND WATER-SOLUBLE VITAMINS

From a nutritional point of view, frying is beneficial in relation to other cooking methods (e.g., boiling or stewing) because proteins, saccharides, and minerals are nearly fully retained (Bognar, 1998). Vitamins B_1, B_2, B_6, and C are better retained than in boiling, steaming, or stewing. The absorption and metabolism of elements (Ca, P, Fe, Cu) and their nutritive utilization does not alter significantly (Vaquero, 1998). According to William Artz and his coworkers (2005), there is iron accumulation in the oil during the deep frying in meat, but this is mainly a technological problem as the metal is a prooxidant and may contribute to the acceleration of hydroperoxides decomposition during the storage period. The nutritive value of proteins is not significantly affected, as indicated by net protein utilization (NPU) values obtained for fried fish and other products (Henry, 1998, Pokorny, 1999). Changes of digestibility of proteins, losses of amino acids, and alterations in dietary fiber and resistant starch are of little nutritional importance.

An advantage of deep-frying is that temperatures within the food do not exceed the temperature of the steam under the crust. Frying time, on the other hand, is short in relation to other cooking procedures. Therefore, fried foods provide important nutrients. Only a small loss of protein availability may be observed in the outer layer of the fried food where oxidation and hydrolytic products from the oil are absorbed on protein and saccharides (Pokorny et al., 2001).

21.6.4   FRYING OILS AS SOURCES OF NATURAL ANTIOXIDANTS

Important natural antioxidants in frying oils are tocopherols and tocotrienols. Other antioxidants, such as derivatives of p-hydroxy-benzoic and hydroxycinnamic acid, hydroxytyrosol, and lignans, are also present in certain oils (Boskou, 1999, Carlos-Espin et al., 2000).

Tocopherols are primarily antioxidants, which are consumed by reacting with lipid radicals. At frying temperatures, losses of tocopherols depend on the type of the oil, the nature of the tocochromanol mixture, and the presence of more active primary antioxidants. It has been demonstrated that a significant alpha-tocopherol equivalent remains after repeated frying in many frying oils. However, it is not clear if alpha-tocopherol is the least stable in relation to the beta-, gamma-, and delta-homologues, and the relative literature is rather conflicting. Simmone and Eitenmiller (1998) reported that alpha-tocopherol is the most stable during simulated frying in soybean oil, corn oil, and palm olein compared to the beta-, gamma-, and delta-homologues and tocotrienols. Tocopherol stability can be increased if frying takes place at low oxygen atmospheres (Fusijaki et al., 2000) or in the presence of more active primary antioxidants. The retention of tocopherols in heated oils is evidenced by the resistance of the oils to the loss of their radical scavenging activity as determined by the DPPH• radical test at 515 nm. A practical application of this stability is the delivery of liposoluble vitamins. The antioxidant activity of tocopherols and tocotrienols provides protection against thermal oxidation of added retinyl palmitate. An increased level of the ester in the oils and the fried foods indicates that fortification of retinyl palmitate to frying oils can be useful for delivery of vitamin A activity (Simmone and Eitenmiller, 1998). Losses of tocopherols in refined rapeseed oil due to frying at 180°C were reported by Aleksander et al. (2008). Alpha-tocopherol was found to be less stable in relation to the delta- homologue. Andrikopoulos and his coworkers (2002) studied the deterioration of tocopherols of vegetable oils during the deep- and pan-frying of potatoes. The retention of tocopherols was found to range from 85%–90% (first frying) to 15%–40% (eighth frying); beta- and gamma-tocopherols of sunflower oil were the most sensitive to high temperatures.

Other antioxidants retained in heated oils enhance the stability of alpha-tocopherol and reduce losses of the free radical scavenging activity. Such antioxidants are sinapic acid in rapeseed oil, sesamol, sesamol dimers, and sesaminol in sesame seed oil, and hydroxycinnamic acids and hydroxytyrosol in olive oil (Boskou, 1999, Pellegrini, 2001). Losses of hydroxytyrosol and its aglycones during deep- and pan-frying of virgin olive oil were reported by Andrikopoulos et al. 2002 (See also Section 21.4.5).

21.6.5 CLASSES OF COMPOUNDS OF TOXICOLOGICAL CONCERN

21.6.5.1 Thermally Oxidized Fats and Oxidative Stress

In a used medium, the distinguishable fraction are intact TAG, hydroperoxy and hydroxyl FAs, volatile breakdown products, nonvolatile breakdown products, polar TAG, polar polymers, and nonpolar polymers. The research work carried out thus far indicates that among the various products formed during frying, those capable of creating adverse physiological conditions in experimental animals and considered harmful to humans include cyclic FAs monomers, polymeric compounds, and oxidation products. The effect of the intake of fried oils on the induction of oxidative stress in rat liver microsomes was studied by Quiles et al. (2002). Highly unsaturated oils were less resistant to the oxidative stress produced by frying and led to a higher degree of lipid peroxidation in liver microsomes.

Feeding experiments using experimental animals have revealed that ingestion of thermally oxidized fats in comparison to fresh fat may provoke a series of biological effects, including oxidative stress lipid metabolism alterations mediated by changes in gene expression. Cyclic FAs, oxidation products such as 9-hydroxyoctadecadienoic, and 13-hydroperoxyoctadecadienoic acid have been shown to be potent ligands and activators of peroxisome proliferators-activated receptors. This property is linked to a positive regulation of gene expression (a reduced concentration of TAG and cholesterol in liver due to the oxidized fat) but also a negative regulation, which is related to the effect of oxidized fat on expression of genes involved in FA synthesis and cholesterol homeostasis (Ringseis and Eder, 2008).

21.6.5.2 Sterols

There is an increasing interest in the oxidation products of sterols because of the possible adverse effect on human health. The bulk of literature, however, focuses on cholesterol oxidative products. The latter have been investigated for possible mutagenic, cancerogenic, cytotoxic, and atherogenic properties. Phytosterols form oxidation products that are similar in structure to those of cholesterol (7-hydroxy epimers, epoxides, triols and 7-keto-sterols, and others, Blekas and Boskou, 1989, Dutta et al., 1996, Grandgirard et al., 2002, Rudzinska et al., 2005, Yan and White, 2007, Kmiecik et al., 2008).

Rudzinska et al. (2005) reported that multiple frying of French fries with rapeseed oil caused a 64% loss of phytosterols. Final levels in the oil and the fries were found to be 200 and 150 µg/kg, respectively.

Vegetable oil spreads containing phytosterol-esters are found in the market as cholesterol-lowering functional foods in many countries. Phytosterols are very stable molecules but oxidation may occur at low levels under extreme heating conditions, resulting in phytosterol oxides. As there is some suggestion of adverse biological effects in the literature for the related cholesterol oxidation products, safety data have been generated for phytosterol oxides. A phytosterol oxide concentrate (POC) was generated by prolonged heating of phytosterol-esters in the presence of oxygen. The genotoxicity and subchronic toxicity of this mixture was assessed in a series of *in vitro* genotoxicity assays and a subchronic feeding study in the rat. A POC containing approximately 30% phytosterol oxides did not possess genotoxic potential and no obvious evidence of toxicity when administered in the diet of the rat for 90 consecutive days. A no observed adverse effect level (NOEL) was established at an estimated dietary level of phytosterol oxides of 128 mg/kg/day for males and 144 mg/kg/day for females (Lea et al., 2004).

Carcinogenic substances were found in drippings, charred pieces of batter, or overheated pan-fried foods. Their content in deep fried food is very low. Chemically, they are heterocyclic compounds belonging to the classes of imidazopyridines, imidazoquinolines, and imidazoquinoxalines (Pokorny, 1999).

Another class of compounds exhibiting mutagenic properties are aldehydes such as 4-hydroxynonenal and 4-hydroxy hexenal (Marquez-Ruiz and Dobarganes, 1996). Recently, it was shown by Guillen and Uriate (2008) that there must be some concern about the formation of 4-hydroxy-trans-2-nonenal and chemically related toxic α,β-unsaturated aldehydes that were detected in heated sunflower oil.

21.6.5.3 Polar Compounds

Polar compounds and TAG oligomers are the basis of legislation for frying oil discarding in many European countries. Some of these countries (e.g., Italy, Belgium, and France) have set a limit of 25% for polar compounds while Austria, Germany, and the Netherlands a limit of 27%. For the content of TAG oligomers, limits ranging from 10% to 16% have been established.

Sanchez-Muniz and Bastida (2001) demonstrated that the two cut-points for polar compounds and TAG polymers are not correlated in the same manner in various oils. Certain compounds defined and determined as polar such as acylated glycerols are not necessarily products of alterations due to heating. On the other hand, the polymers formed are of some toxicological concern and express more precisely the level of changes in the heated oil. The authors stress the need to unify the criteria in the different countries and propose a single upper limit of 10% for TAG oligomers, instead of two, one for the TPC and the other for the oligomers.

21.6.5.4 Acrylamide

Acrylamide is a contaminant that can be formed during the preparation of foods rich in carbohydrates at high temperatures and can be found in potato crisps, French fries, crispy bread, bakery products, breakfast cereals, and coffee. It is formed from the carbonyl compounds formed in lipid oxidation and the amino acids, especially asparagines, through Maillard reactions. The acrolein formed from oil reacted with asparagines was found to produce acrylamide in deep-fat frying (Choe and Min, 2007).

Acrylamide first hit the headlines in 2002, due to a report from the Swedish Food Administration. The compound was found to have a potential to cause genetic damage and cancer in laboratory animals, but at concentrations much higher than those seen in foods. The FAO/WHO Expert Committee on Food Additives (JECFA) and the Scientific Panel on Contaminants of the European Food Safety Authority (EFSA) have agreed that there may be human health concerns (EFSA, 11th Scientific Colloquium, Acrylamide carcinogenicity, Parma, Italy, 22–23 May 2008). EFSA recommended that efforts should be made to reduce acrylamide levels in food, and stated that this effort remains a priority. New research data are expected in order to decide whether new advice on acrylamide is needed. An important step in tackling the acrylamide problem is the HEATOX study. HEATOX is the acronym for a European Union-funded project, Heat-Generated Food Toxicants: the project includes research on formation, chemistry, food technology, analytical methods, hazard characterization, and exposure assessment. The results are useful in the risk assessment and risk management advice.

Acrylamide is typically found in plant-based foods that were cooked with high heat (frying, baking, roasting). Larger sources of acrylamide are potato products such as French fries and chips, coffee, breakfast cereals, and toasts.

More acrylamide accumulates in the fried food when cooking is done for longer periods at higher temperatures. Some rules to modify the profile of frying temperature may result in a decrease of acrylamide formation (Fiselier et al., 2006).

The origin of the frying vegetable oil does not seem to affect acrylamide formation in potatoes during frying (Mestdagh et al., 2005). The presence of antioxidants such as ascorbic acid and ferulic acid seems to reduce the amounts of acrylamide in potato chips. Oxidative processes, free radical intermediates, and the effect of natural antioxidants have been studied by Hadegaard et al. (2008).

Napolitano et al. (2008) investigated the relationship between virgin olive oil phenol compounds and the formation of acrylamide in potato chips. Acrylamide concentration increases during frying, but the formation of acrylamide is less rapid when frying takes place in olive oil with a high content of orthodiphenols, and when the frying operation is carried out under mild or moderate conditions. The authors propose the use of olive oil as a reliable strategy to reduce acrylamide formation in domestic frying.

Many procedures were proposed to reduce acrylamide but most of them affect the food taste. The industry's efforts now focus on the use of some of acrylamide reducing enzymes (asparaginases), which convert the precursor, asparagine, into other amino acids that do not form acrylamide (Konings et al., 2007).

21.6.5.5 3-Chloropropane-1,2-diol(3-MCPD) Esters

3-Chloropropane-1,2-diol (3-MCPD) and some other chloropropanols are known as food processing contaminants for almost 30 years. These compounds are formed at high temperatures from glycerol and acylglycerols and chloride ions and their carcinogenic maximum levels of 0.02 mg/kg have been laid down for hydrolyzed vegetable protein and sauces from soybean. Free 3-MCPD is well known as a food processing contaminant since 1978 and its formation is attributed to a reaction between glycerol or acylglycerols and chloride anions. More recent studies based on better analytical methods showed that in most food products the major part is ester-linked with FAs.

Since there are no specific toxicological data on esters so far, EFSA proposed that risk assessment is based on free 3-MCPD, assuming a 100% release in the digestive system (EFSA Statement, October 2008).

In contrast to acrylamide, which is formed during the frying process, formation of 3-MCPD does not take place due to frying. Its presence is a result of the "carry-over" from the oil used for frying to the fried food. In nonprocessed oils only traces or no 3-MCPD are detected. In refined oil, however, concentrations may range from 0.2 to 20 mg/kg (Weisshaar, 2008). Dolezal et al. (2008) reported levels of 3-MCPD esters that range approximately from 0.1 to 0.6 mg/kg for potato chips fried at 170°C.

21.6.5.6 Exposure to Cooking Fumes

When food is prepared under high temperatures in restaurant kitchens, degradation products may be formed. Aldehydes such as formaldehyde, acetaldehyde, *trans-trans*-decadienal-, and acrolein are generated that should not be inhaled because they are irritating to the lung tissue. Therefore, special cabinets and specific devices should be used for frying in hotel and restaurant kitchens to minimize exposure to aerosols and aldehydes formed during frying (Svendsen et al., 2002, Boskou et al., 2006).

21.6.6 Conclusions

Frying oils and fried food are becoming more important in the diet, owing to the increased consumption of fried food and number of convenience food restaurants and take-away outlets that serve food prepared by frying. Oils and fats are rich sources of essential FAs and other nutrients. When they are used for frying for prolonged periods, they deteriorate by darkening, smoking, and foaming. Their biological value is decreased as unsaturated FAs and important phytonutrients or other minor constituents are partially lost. However, when good frying practices are applied, frying oils can retain a significant part of their initial nutritive value and still carry vitamins and functional ingredients. It has to be stressed also that through the years a lot of information has been accumulated concerning the chemical changes in the frying oil. This information can now be used to improve the quality of frying food and the lipid intake profile by defining more strictly the point at which the oil is no longer fit and by extending the regulations to the oil absorbed by the food. Further research is required to identify areas of limited information and to better understand the role of oxidation retardants in the quality of the oil and the presence of contaminants, the interactions between frying oil and food components, and the effect of preprocessing techniques and post frying practices.

REFERENCES

Aleksander, S., Kmiecik, D., Nogala-Kalucka, M., Korczak, J., and Kobus, J., 2008. The influence of heating on the content of native tocopherols in rapeseed oil after various stages of processing, *6th Eurofed Lipid Congress*, Athens, Greece, *Book of Abstracts*, FRY-003.

Allouche, Y., Jimenez, A., Gaforio, J.J., Uceda, M., and Beltran, G., 2007. How heating affects extra virgin olive oil quality indexes and chemical composition, *J. Agric. Food Chem.*, 55, 9646–9654.

Andrikopoulos, N.K., Dedoussis, G.V.Z., Falirea, A., Kalogeropulos, N., and Hatzinikola, H.S., 2002. Deterioration of natural antioxidant species of vegetable edible oils during the domestic deep-frying and pan-frying of potatoes, *Int. J. Food Sci. Nutr.*, 52, 351–363.

Artz, W., Osidacz, P.C., and Coscione, A.R., 2005. Iron accumulation in oil during the deep-fat frying of meat, *J. Am. Oil Chem. Soc.*, 82, 249–254.

Belitz, H.D. and Grosch, W., 1999. *Food Chemistry*, Berlin, Germany: Springer, pp. 210–212.

Berger, K.G., 2005. *The Use of Palm Oil in Frying, Frying Oil Series*, Selangor, Malaysia: Malaysian Palm Oil Promotion Council.

Billek, G., 2000. Health aspects of thermoxidized oils and fats, *Eur. J. Lipid Sci. Technol.*, 102, 587–593.

Blekas, G. and Boskou, D., 1999. Phytosterols and stability of frying oils, in *Frying of Food*, D. Boskou and I. Elmadfa (eds.), pp. 205–223, Lancaster, PA: Technomic Publishing Co.

Bognar, A., 1998. Comparative study of frying to other cooking techniques influence on the nutritive value, *Grasas Aceites*, 49(3,4), 250–260.

Boskou, D., 1999. Non-nutrient antioxidants and stability of frying oils, in *Frying of Food*, D. Boskou and I. Elmadfa (eds.), pp. 183–204, Lancaster, PA: Technomic Publishing Co.

Boskou, G., Salta, F.N., Chiou, A., Troullidou, E., and Andrikopoulos, N.K., 2006. Content of *trans,trans*-2,4-decadienal in deep-fried and pan-fried potatoes, *Eur. J. Lipid Sci. Technol.*, 108, 109–115.

Brenes, M., Garcia, A., and Dobarganes, M.C., 2002. Influence of thermal treatments simulating cooking processes on the polyphenol content of virgin olive oil, *J. Agric. Food Chem.*, 50, 5962–5967.

Buchgraber, M. and Ulberth, F., 2001. The determination of *trans* octadecenoic acids by silver-ion chromatography-gas chromatography: An intercomparison of methods, *J. Assoc. Offic. Anal. Chem.*, 84, 1490–1498.

Carlos Espin, J., Soler-Rivas, C., and Wichers, H.J., 2000. Characterization of the total free radical scavenger capacity of vegetable oils and oil fractions using 2,2-diphenyl-1-picrylhydrazyl radical, *J. Agric. Food. Chem.*, 48(3), 648–656.

Carrasco-Pancorbo, A., Cerretani, L., Bendini, A., Segura-carretero, A., Lercker, G., and Gutierrez, F., 2007. Evaluation of the influence of thermal oxidation on the phenolic composition and on the antioxidant activity of extra virgin olive oils, *J. Agric. Food Chem.*, 55, 4771–4780.

Chardigny, J.M., Bretillon, L., and Sebedio, J.L., 2001. New insights in health. Effects of *trans* a-linolenic acid isomers in humans, *Eur. J. Lipid Sci. Technol.*, 103(7), 478–482.

Che Man, Y.B. and Setiowaty, G., 1999. Determination of anisidine value in thermally oxidized palm olein by Fourier Transform Infrared Spectroscopy, *J. Am. Oil Chem. Soc.*, 76(2), 243–247.

Chiou, A., Karpathiou, V., Ghioxari, A., and Andrikopoulos, N.K., 2008. Oleuropein in French fries during the successive deep-frying in sunflower oil enriched with olive leaf extract, *6th Eurofed Lipid Congress*, Athens, Greece, *Book of Abstracts*, FRY-010.

Choe, E. and Min, D.B., 2007. Chemistry of deep-fat frying oils, *J. Food Sci.*, 72, R77–R86.

Christie, W.W. and Dobson, G., 2000. Formation of cyclic fatty acids during the frying process, *Eur. J. Lipid Sci. Technol.*, 102, 515–520.

Clark, W.L. and Serbia, G.W., 1991. Safety aspects of frying fats and oils, *Food Technol.*, 2, 84–94.

Deyawatee, G., Jhaumeer-Laullo, S.B., and Ravish, M., 2001. Evaluation of soybean oil quality during conventional frying by FTIR and some chemical indices, *Int. J. Food Sci. Nutr.*, 52(1), 31–42.

Dobarganes, M.C. and Marquez-Ruiz, G., 1996. Dimeric and higher oligomeric triglycerides, in: *Deep Frying*, E.G. Perkins and M.D. Erickson (eds.), pp. 89–111, Champaign, IL: AOCS Press.

Dobarganes, M.C. and Marquez-Ruiz, G., 1998. Regulation of used frying fats and validity of quick tests for discarding the fats, *Grasas Aceites*, 49, 331–336.

Dobarganes, M.C., Marquez-Ruiz, G., and Perez-Camino, M.C., 1993. Thermal stability and frying performance of genetically modified sunflower seed oil, *J. Agric. Food. Chem.*, 41, 678–681.

Dolezal, M., Dvorakova, L., Zelinkova, Z., and Velisek, J., 2008. Analysis of potato product lipids for 3-MCPD esters, *6th Eurofed Lipid Congress*, Athens, Greece, *Book of Abstracts*, p. 325.

Dutta, P.C., Przybylski, L.A., Appelquist, L.A., and Eskin, N.A.M., 1996. Formation and analysis of oxidized sterols in frying fat, in *Deep Frying*, E.G. Perkins and M.D. Erickson (eds.), pp. 112–150, Champaign, IL: AOCS Press.

EFSA, 2008. http://www.efsa.europa.eu/EFSA/Statement 3-MCPD.

El Anany Ayman Mohamed, 2007. Influence of pomegranate (*punica granatum*) peel extract on the stability of sunflower oil during deep-frying process, *Elect. J. Food Plants Chem.*, 2, 13–19.

Farag, R.S., Mahmoud, E.A., and Basuny, A.M., 2006. Use crude olive leaf extract as a natural antioxidant for the stability of sunflower oil during heating, *Int. J. Food Sci. Technol.*, 42, 107–115.

Fiselier, K., Bazzocco, D., Gama-Baumgartner, F., and Grob, K., 2006. Influence of the frying temperature on acrylamide formation in French fries, *Eur. Food Res. Technol.*, 222, 414–419.

Fujisaki, M., Mohri, S., Endo, Y., and Fujimoto, K., 2000. The effect of oxygen concentration on oxidative deterioration in heated high-oleic safflower oil, *J. Am. Oil. Chem. Soc.*, 77 (3), 231–234.

Gasparoli, A., 1998. The formation of new compounds, *Grasas Aceites*, 49(3,4), 303–309.

Gertz, C., 2000. Chemical and physical parameters as quality indicators of used frying fats, *Eur. J. Lipid Sci. Technol.*, 102, 566–572.

Gertz, C., Klosterman, S., and Kochhar, S.P., 2000. Testing and comparing oxidative stability of vegetable oils and fats at frying temperature, *Eur. J. Lipid Sci. Technol.*, 102, 543–551.

Gloria, H. and Aguilera, J. M., 1998. Assessment of the quality of heated oils by differential scanning calorimetry, *J. Agric. Food Chem.*, 46, 1363–1368.

Gomez-Alonso, S., Fregapane, G., Salvador, M.D., and Gordon, M.H., 2003. Changes in the phenolic composition and antioxidant activity of virgin olive oil during frying, *J. Agric. Food Chem.*, 51, 667–672.

Grandgirard, A., Sebedio, J.L., and Fleury, J., 1984. Geometrical isomerization of linolenic acid during heat treatment of vegetable oils, *J. Am. Oil Chem. Soc.*, 61, 1563–1568.

Guillen, M.D. and Uriarte, P.S., 2008a. Study by ^{1}H nuclear magnetic resonance of the evolution of virgin olive oil composition under frying conditions, *6th Eurofed Lipid Congress*, Athens, Greece, *Book of Abstracts*, FRY-004.

Guillen, M.D. and Uriarte, P.S., 2008b. Evidence of the very early formation in sunflower oil, under frying conditions, of 4-hydroxy-*trans*-2-nonenal and other toxic substances, *6th Eurofed Lipid Congress*, Athens, Greece, *Book of Abstracts*, FRY-004.

Hedegaard, R.V., Granby, K., Frandsen, H., Thygensen, J., and Skibsed, L.H., 2008. Acrylamide in bread. Effect of prooxidants and antioxidants, *Eur. Food Res. Technol.*, 227, 519–525.

Henon, G., Vigneron, P.Y., Stoclin, B., and Caigniez, J., 2001. Rapeseed oil deodorisation study using the response surface methodology, *Eur. J. Lipid Sci. Technol.*, 103, 467–477.

Henry, C.J.K., 1998. Impact of fried foods on micronutrient intake, with special reference to fat and protein, *Grasas Aceites*, 49, 336–339.

Irwandi, J., Che Man, Y.B., and Kitts, D.D., 2000. Use of natural antioxidants in refined palm olein during repeated deep-fat frying, *Food Res. Int.*, 33, 501–508.

Juarez, M.D. and Samman, N., 2006. Evaluation of chemical changes in diet lipids produced by the thermal processing, Nutritional effects, *4th Eurofed Lipid Congress*, Madrid, Spain, *Book of Abstracts*, FRY-015, p. 409.

IUPAC, 1992a. Standard method 2.507: Determination of polar compounds in frying fats, *Standard Methods for the Analysis of Oils, Fats and Derivatives*, 7th edition, International Union of Pure and Applied Chemistry, Blackwell Scientific Publications, Oxford, U.K.

IUPAC, 1992b. Standard method 2.508: Determination of polar polymerized triglycerides in oils and fats by high performance liquid chromatography, *Standard Methods for the Analysis of Oils, Fats and Derivatives*, 7th edition, International Union of Pure and Applied Chemistry, Blackwell Scientific Publications, Oxford, U.K.

Kalantzakis, G., Blekas, G., Peglidou, K., and Boskou, D., 2006. Stability and radical–scavenging activity of heated olive oil and other vegetable oils, *Eur. J. Lipid Sci. Technol.*, 108, 329–335.

Kmiecik, D., Korczak, J., Rudzinska, M., and Kobus, J., 2008. Changes of phytosterols during heating rapeseed oil, *6th Eurofed Lipid Congress*, Athens, Greece, *Book of Abstracts*, FRY-002.

Kochhar, S.P., 2000a. Stabilisation of frying oils with natural antioxidative components, *Eur. J. Lipid Sci. Technol.*, 102, 552–559.

Kochhar, S.P., 2000b. Stable and healthful frying oil for the 21st century, *INFORM*, 11, 642–647.

Koketsu, M. and Satoh, Y., 1997. Antioxidant activity of green tea polyphenols in edible oils, *J. Food Lipids*, 4, 1–9.

Konings, E.M., Ashby, P., Hamlet, C.G., and Thompson, G.A.K., 2007. Acrylamide in cereal and cereal products: A review on progress in level reduction, *Food Addit. Contam.*, 24 suppl., 47–59.

Le Quere, J.L. and Sebedio, J.L., 1996. Cyclic monomers of fatty acids, in *Deep Frying*, E.G. Perkins and M.D. Erickson (eds.), pp. 49–88, Champaign, IL: AOCS Press.

Lea, L.J., Hepburn, P.A., Wolfreys, A.M., and Baldrick, P., 2004. Safety evaluation of phytosterol esters. Part. 8. Lack of genotoxicity and subchronic toxicity with phytosteroloxides, *Food Chem. Toxicol.*, 42, 771–783.

Mahungu, S.M., Artz, W.E., and Perkins, E.G., 1999. Oxidation products and metabolic processes, in: *Frying of Food*, D. Boskou and I. Elmadfa (eds.), pp. 25–46, Lancaster, PA: Technomic Publishing Co.

Marmesat, S., Velasco, J., Marquez-Ruiz, G., and Dobarganes, M.C., 2007, A rapid method for determination of polar compounds in used frying fats and oils, *Grasas Aceites*, 58, 179–184.

Marmesat, S., Velasco, L., Ruiz-Mendez, M.V., Fernandez-Martinez, J.M., and Dobarganes, M., 2008. Thermostability of genetically modified sunflower oils differing in fatty acid and tocopherol compositions, *Eur. J. Lipid Sci. Technol.*, 110, 776–782.

Marquez-Ruiz, G. and Dobarganes, M.C., 1996. Nutritional and physiological effects of used frying oils, in *Deep Frying*, E.G. Perkins and M.D. Erickson (eds.), pp. 160–182, Champaign, IL: AOCS Press.

Mehta, V. and Swinburn, B., 2001. A review of factors affecting fat absorption in Hot Chips, *Crit. Rev. Food Sci. Nutr.*, 44, 133–154.

Mestdagh, F.J., Meuleaner, De, Van Poucke, C., Detavernier, C., Cromhout, C., and Van Petersen, C., 2005. Influence of oil type on the amounts of acrylamide generated in a model system and in French fries, *J. Agric. Food Chem.*, 53, 6170–6174.

Morton, I. D., 1998. Geography and history of the frying process, *Grasas Aceites*, 49, 247–249.

Napolitano, A., Morales, F., Sacchi, R., and Fogliano, V., 2008. Relationship between olive oil phenolic compounds and acrylamide formation in fried crisps, *J. Agric. Food Chem.*, 56, 2034–2040.

Nasirullah, 2001. Development of deep frying edible vegetable oils, *J. Food Lipids*, 8, 295–304.

Nasirullah and Rangaswamy, B.L., 2005. Oxidative stability of healthful frying medium and uptake of inherent nutraceuticals during deep frying, *J. Am. Oil Chem. Soc.*, 82, 753–757.

Nawar, W.W., 1998. Volatile components of the frying process, *Grasas Aceites*, 49(3,4), 271–274.

Nogala-Kalucka, M., Korczak, J., Elmadfa, I., and Wagner, K.-H., 2005. Effect of alpha- and delta-tocopherol on the oxidative stability of a mixed hydrogenated fat under frying conditions, *Eur. Food Res. Technol.*, 221, 291–297.

Normand, L., Eskin, N.A.M., and Przybylski, R., 2001. Effect of tocopherols on the frying stability of regular and modified canola oils, *J. Am. Oil Chem. Soc.*, 78(4), 369–373.

Pantzaris, T.P., 1999. Palm oil frying, in *Frying of Food*, D. Boskou and I. Elmadfa (eds.), pp. 253–268, Lancaster, PA: Technomic Publishing Co.

Paul, S. and Mittal, G.S., 1997. Regulating the use of degraded oil/fat in deep-fat/oil food frying, *Crit. Rev. Food Sci. Nutr.*, 37, 635–662.

Pellegrini, N., Visioli, F., Buratti, S., and Brighenti, F., 2001. Direct analysis of total antioxidant activity of olive oil and studies on the influence of heating, *J. Agric. Food Chem.*, 49, 2532–2538.

Perkins, E.G., 1996. Volatile odor and flavor components formed in deep frying, in *Deep Frying*, E.G. Perkins and M.D. Erickson (eds.), pp. 43–49, Champaign, IL: AOCS Press.

Persson, E., 2003. Influence of antioxidants in virgin olive oil on the formation of heterocyclic amines in fried beefburgers, *Food Chem. Toxic.*, 1, 1587–1589.

Petukhov, I., Malcomson, L.J., Przybylski, R., and Armstrong, L., 1999. Frying performance of genetically modified canola oils, *J. Am. Oil Chem. Soc.*, 76, 627–632.

Pokorny, J., 1999. Changes of nutrients at frying temperatures, in *Frying of Food*, D. Boskou and I. Elmadfa (eds.), pp. 69–102, Lancaster, PA: Technomic Publishing Co.

Pokorny, J., Panek, J., and Trojakova, L., 2001. Effect of food component changes during frying on the nutrition value of fried food, *International Congress of Nutrition*, Vienna, Austria, *Annals of Nutrition and Metabolism*, 388, I. Elmadfa and J. Konig (eds.), Karger, Basel, Switzerland.

Quiles, J.L., Ramirez-Tortosa, C., Alfonso Gomez, J., Huertas, J.R., Mataix, J. et al., 2002. Role of vitamin E and phenolic compounds in the antioxidant capacity, measured by ESR, of virgin olive oil and sunflower oil after frying, *Food Chem.*, 461–469.

Ringseis, R. and Eder, K., 2008. Effects of dietary oxidized fats on gene expression in mammals, *INFORM*, 19, 657–659.

Romero, A., Cuesta, C., and Sanchez-Muniz, F.-J., 2000. Cyclic fatty acid monomers and thermoxidative alteration compounds formed during frying of frozen foods in extra virgin olive oil, *J. Am. Oil Chem. Soc.*, 77, 1169–1175.

Rudzinska, M., Uchman, W., and Wasowicz, E., 2005. Plant sterols in food technology, *Acta Sci. Pol., Technol. Aliment.*, 4, 147–156.

Sanchez-Muniz, F.J. and Bastida, S., 2001. Polar content vs. oligomer content determination in the discarding of edible oils used for frying, *Annals of Nutrition and Metabolism*, 45 (suppl. 1), 387 [I. Elmadfa and J. König (eds.), Basel, Karger].

Sanchez-Muniz, F.J., Bastida, S.A., and Gonzalez-Munoz, 2006. In *vitro* and in vivo enzyme studies for the assessment of thermal alteration in edible fats and oils, *4th Euro Fed Lipid Congress, Frying Oils and Frying Products*, Madrid, Spain, *Book of Abstracts*, p. 12.

Sánchez-Gimeno, A.C., Negueruela, A.I., Benito, M., and Vercet, A., 2008. Some physical changes in Bajo Aragón extra virgin olive oil during the frying process, *Food Chem.*, 110, 654–658.

Sebedio, J.L. and Chardigny, J.M., 1996. Physiological effects of trans and cyclic fatty acids, in: *Deep Frying*, E.G. Perkins and M.D. Erickson (eds.), pp. 183–209, Champaign, IL: AOCS Press.

Sebedio, J.L., Astorg, P.O., Septier, Ch., and Grandgirard, A., 1987. Quantitative analysis of polar components in frying oils by the Iatroscan thin-layer chromatography–flame ionization technique, *J. Chromatogr.*, 405, 371–376.

Simonne, A.H. and Eitenmiller, R.R., 1998. Retention of vitamin E and added retinyl palmitate in selected vegetable oils during deep-fat frying and in fried breaded products, *J. Agric. Food Chem.*, 46, 5273–5277.

Steel, C.J., Dobarganes, M.C., and Barrera-Arrelano, D., 2006. Formation of polymerization compounds during thermal oxidation of cotton seed oil, partially hydrogenated cotton seed oil and their blends, *Grasas Aceites*, 57, 284–291.

Svendsen, K., Jensen, N.H., Silvertsen, I., and Sjaastad, A.K., 2002. Exposure to cooking fumes in restaurant kitchens in Norway, *Ann. Occup. Hyg.*, 46, 395–400.

Ulmansoy, T., Voeker, T., Moczka, A., Roberts, P., Schreckengost, B., Wagner, N. et al., 2008. Improving soybean oil for food and industrial uses, *6th Euro Fed Lipid Congress*, Athens, Greece, *Book of Abstracts*, FRY-013.

Van Loon, A.A.M., Linssen, J.P.H., Leader, A., and Voragen, A.G.J., 2006. Anti-radical power gives insight into lipid oxidation during frying, *J. Sci. Food Agric.*, 86, 1446–1451.

Vaquero, M.P., 1998. Minerals, *Grasas Aceites*, 49(3,4), 352–358.

Varela, G. and Ruiz-Roso, B., 1998. Frying process in the relation fat/ degenerative disease, *Grasas Aceites*, 49(3,4), 359–365.

Warner, K. and Mounts, T.L., 1993. Frying stability of soybean and canola oils with modified fatty acid compositions, *J. Am. Oil Chem. Soc.*, 70, 983–988.

Warner, K. and Dunlap, C., 2006. Effects of expeller-pressed physically refined soybean oil on frying oil stability and flavor of French-fried potatoes, *J. Am. Oil Chem. Soc.*, 83, 435–441.

Weisshaar, R., 2008. 3-MCPD-esters in edible fats and oils-a new and world wide problem, *Eur. J. Lipid Sci. Technol.*, 110, 671–672.

Xu, X.-Q., 2000. A new spectrophotometric method for the rapid assessment of deep frying oil quality, *J. Am. Oil Chem. Soc.*, 77(10), 1083–1086.

Yan, P.S. and White, J.P., 2007. Cholesterol oxidation in heated lard enriched with two levels of cholesterol, *J. Am. Oil Chem. Soc.*, 67, 927–931.

Yen, P.-L. and Lu, Y.-F., 2008. Effect of frying on oxidative stress and blood pressure in spontaneously hypertensive rats, *Euro Fed Lipid Congress*, Athens, Greece, *Book of Abstracts*, FRY-001.

22 Lipid–Protein and Lipid–Saccharide Interactions

*Jan Pokorný, Anna Kołakowska,
and Grzegorz Bienkiewicz*

CONTENTS

22.1 INTRODUCTION

Foods and food raw materials primarily consist of hydrophilic substances compatible with the aqueous medium of plant and animal tissues. Lipids, lipophilic vitamins, terpenes, and other nonpolar water-insoluble substances should thus be emulsified or converted into hydrophilic substances by the formation of complexes with other food constituents. The most important complex-forming food components are proteins and saccharides.

22.2 LIPID–PROTEIN COMPLEXES IN RAW MATERIALS

22.2.1 DISPERSIBLE LIPOPROTEINS

Lipoproteins are the most common form of lipid binding in both animal and plant products (LeMeste and Davidou, 1995). Even storage lipids in plant seeds or in animal tissues are, eventually, lipoproteins. They form emulsions, in which lipids and other nonpolar substances are located in the center of such a particle, held together by hydrophobic and other physical forces. Hydrogen bridges can occur, too, but no covalent bonds between the lipid and the protein moiety exist.

The lipid fraction of lipoproteins is surrounded by a layer of natural emulsifiers, such as phospholipids, sterols, or free fatty acids (FFA). Their hydrophobic chains are oriented inside the oil droplet (Figure 22.1), while their polar groups are oriented toward the outer, aqueous phase. Due to relatively low hydrophilic–lipophilic balance (HLB) value, such emulsions are not very stable. They are stabilized by binding to a layer of hydrated proteins, mostly by multiple hydrogen bonds. In wheat dough, protein–lipid bonds cement the gluten network and contribute to the structure of gas-retaining complex, which makes dough leavened. Other polar groups of such a lipoprotein molecule are loosely bound to surrounding water molecules, again through hydrogen bonds.

Because of their polar character, lipoproteins are easily dispersed in water. Lipoproteins in plant storage organs, such as plant seeds and pericarp, or in animal adipose tissue may be tightly packed in the respective cells. Protein–lipid complexes

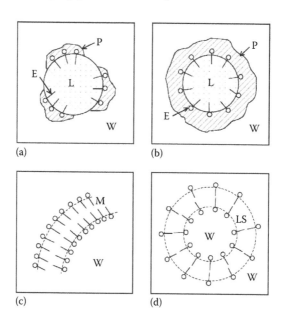

FIGURE 22.1 Types of lipoproteins: (a) low density, (b) high density, (c) lipid membrane, (d) liposome. E, emulsifier (polar lipid); L, lipid phase; LS, liposome; M, lipid membrane; P, protein; W, water phase.

are also known to occur in peanut products, particularly during the storage of peanut seeds and peanut flour.

Other lipoproteins are present in the blood of animals. They are able to carry fatty acids (FA) absorbed from food. Special lipoproteins carry FA from the storage cells to other cells, where they are metabolized. Ratios of the lipid and protein moieties may be very different, as is evident from the comparison of Figure 22.1a and b.

Lipids are lighter than proteins; therefore, lipids of different lipid/protein ratios can be separated by centrifugation (Perkins, 1993). The best known are those of human blood plasma, because of their importance for the study of cardiovascular troubles and other several heavy diseases. They can be divided into very low density (VLDL), low density (LDL), medium density (MDL), and high density (HDL) lipoproteins. The higher the density, the higher polarity the particular lipoprotein possesses. Therefore, HDL have the highest affinity for the aqueous phase. Similar types of lipoproteins are also present in animal tissues. In plants, lipids are not transported; thus, lipoproteins are present only in seeds, pericarp, or germs.

Lipids of the avian yolk contain mostly phospholipids, but also neutral lipids, so that plasma lipoproteins are formed. They are different from blood lipoproteins, both in the FA composition and protein structure (Azagtat and Alli, 2002).

22.2.2 LIPOPROTEIN MEMBRANES

Another form of lipid binding to protein is lipoprotein membranes. Their function is to separate two different aqueous media from one another. Therefore, they consist of a double layer of polar lipids, primarily phospholipids, sterols, and FFA. Their hydrocarbon chains are oriented toward the nonpolar phase, and the polar groups toward a layer of hydrated protein, which in its turn, is oriented toward the aqueous phase (Figure 22.1c). The hydrated protein layer maintains the structure of the membrane, similarly as in case of dispersed lipoproteins. The nonpolar part of the membrane may be covered with a narrow neutral lipid layer. Another polar lipid layer from the other side of the membrane is again covered with hydrated protein. Some emulsions are mixed, e.g., in water droplets of the water phase, small lipid globules are present. The membranes are thus hydrophilic from both sides. In case of liposomes, only a single lipid layer exists (Figure 22.1d).

The same type of membranes can separate dispersed water particles in a continuous lipid phase, e.g., in butter or margarine. In such a case, the layer of emulsifying agents is oriented outside with their nonpolar chains and inside with their polar groups. The polar layer is again stabilized with a layer of hydrated protein.

In animal tissues, intercellular protein membranes separate two neighboring cells of the same tissue. Other membranes are intracellular as they separate tissue particles of the same cell against the cytoplasm. In plant tissues, intercellular walls separating neighboring cells consist of cellulose and other water-insoluble saccharides. Within the cells, intracellular particles are separated in the same way as in animal tissues, i.e., by lipoprotein membranes. Every membrane contains orifices, which may be opened to enable the contact between both phases. The transfer through orifices is necessary for the metabolism.

22.3 LIPID–PROTEIN INTERACTIONS DURING PROCESSING AND STORAGE OF FOOD

22.3.1 Lipid–Protein Interactions during Storage and Food Preparation at Ambient Temperature

During food storage either at ambient temperature or under refrigeration, lipoproteins do not change appreciably. If the material is dehydrated by water evaporation, small channels remain in the food particles. Dehydration naturally destroys hydrogen bonds, and lipoproteins become unstable. Disturbing the structure of proteins present in lipoproteins changes their stability.

During prolonged storage, lipoprotein lipids as well as free lipids are oxidized, especially if the lipid fraction is polyunsaturated, e.g., in fish lipids or vegetable seed oils. The reaction proceeds even on frozen storage, because water in food turns into ice, leaving free channels. Air can enter these channels, its contact area is thus increased, and air oxygen can stimulate the oxidation.

Another phenomenon involves the input of mechanical energy. Cream separated from milk contains lipids dispersed in the aqueous medium. During churning, the emulsion is converted to butter, which is a water-in-lipid system. The lipoprotein membrane has to be reverted, as reported above. Whole milk powder is an important food product, very useful for the infant nutrition. It is often stored for a long time. Free radicals are formed during storage, which affect the lipoprotein bonds and even destroy them. The reconstitution of milk is then impaired (Thomsen et al., 2005).

At higher temperatures, the protein moiety of lipoproteins undergoes denaturation (Anton et al., 2000), which affects the structure and surface properties. Lipids and phospholipids are liberated from their complexes with proteins so that they become more vulnerable for further reactions, such as oxidation and hydrolysis.

In technological and culinary operations, fats, oils, or their emulsions are often added to the material, they are dispersed, and hydrogen bonds between lipid and protein molecules are formed or rearranged. Weak physical forces also play a role. The final effect depends also on the particular technology employed and the composition of food.

The time of storage and type of changes during cooking depend on the form of lipid binding in the tissue. During storage, lipids were released from natural lipid–protein complexes, and new lipoprotein complexes with different strength of bonds were formed. In trouts caught in January, weak bonds of free lipids were observed, while strong bonds appeared only on storage of partially rancid fish. More extensive lipid binding occurred during roasting than during boiling, but only partially, covalent bonding took place, as it required thermal hydrolysis of tissue (Kołakowska et al., 2006).

22.3.2 Changes of Food Lipoproteins during Frying

During frying, water is released from food, being replaced by frying oil (Chapter 21). Average amounts of fat increased, compared with the original food (Table 22.1). Various foods of vegetable and animal origin were compared after 10, 30, and 70 s of

TABLE 22.1

Average Fat Content in Deep-Fried Products (%, Expressed on the Dry Weight Bases)

Type of Fried Food	Fat Content (%)
Potatoes, French fries, chips	15–37
Cereal products, doughnuts	18–32
Bread slices	33–50
Vegetable, breaded or not	34–74
Mushrooms, breaded	65–82
Beef, pork, meat balls, patties	10–24
Chicken, battered and breaded	11–30
Fish, breaded	20–44
Sausages	38–70

frying at 175°C (Makinson et al., 1987). The fat content increased with the increasing time. Foods of plant origin absorbed more fat than foods of animal origin, because plant foods contain more water than animal foods. The highest fat intake was observed in fried vegetables, such as tomato and onion, or mushrooms. Low fat changes were observed in frying beef, pork, lamb, and sausages. Some foods even lost fat, as in frying skinned beef sausage. The fat intake may be reduced by application of batter and breading on fried food.

Frying oil used for deep-fat frying is repeatedly used for many hours or even days. Its degradation may be intensive (up to 30% polar lipids and 10% polymers is the upper limit applied in most countries). Oxidation occurs not only in frying oil, but also in lipids present in fried food, especially in foods containing high-polyunsaturated FA (seeds, vegetable oils, or fish lipids).

Lipoproteins are the most important ones among the causes of browning on storage of heated foods (Hutaopea et al., 2004). Phospholipids, mostly bound in lipoproteins as emulsifying agents, contribute to the browning. The browning is mainly due to the primary amine group in phosphatidylethanolamine, as it was observed in model systems with ribose (Zamora et al., 2005). In real food, the participation of bound phospholipids was observed by studying a system of porcine *Longissimus dorsi* muscle with several amino acids (Ventanas et al., 2007). Red coloration appears during storage of amines and lipids, which then turn brown. The red coloration is due to primary interaction products, which then polymerize into brown macromolecular compounds (Nakamura et al., 1998). A combination of oxidative browning of lipoproteins and nonenzymic browning was observed in other laboratories (Zamora and Hidalgo, 2005).

22.3.3 EFFECT OF OXIDIZED LIPOPROTEINS ON OTHER FOOD COMPONENTS

Food lipids, either free or bound in lipoproteins (Pokorný, 1998, 1999), are oxidized on storage or by heating, primarily following the free radical mechanism (Chapter 9). The free radicals formed in these processes can react both with lipids and proteins, or with lipoproteins (Saeed et al., 1999). Both reactions start at the same time, but lipoproteins are the most widely distributed reactants in foods (Soyer and Hultin, 2000). Dimers or even higher polymers are produced by addition to oxidized monomers.

Lipid hydroperoxides, with an adjacent double bond, are formed as the first step of oxidation reactions, where the reactive groups are located near the center of the alkyl chain. Hydroperoxides are very unstable, forming several different products, such as aldehydes or aldehydo acids. They can react with various active groups of proteins (Yuan et al., 2007). The reactions may proceed even at ambient temperature, especially in presence of water (Kouřimská et al., 1993). The reactivity of hydroperoxides may be higher than that of the secondary oxidation products, particularly by easy cleavage with the formation of free radicals. It was observed in the case of meat (Janitz et al., 1990). The number of potential combinations is thus quite high (Table 22.2). Additional multiple hydrogen bonds are formed between the lipid and protein moieties, which decrease the solubility of the product (Eymard et al., 2009; Kawai et al., 2007; Kołakowska and Szczygielski, 1994).

Covalent bonds between oxidized lipids and protein are also formed (Figure 22.2). For example, in frozen bream, about 35% total lipids were bound to protein through covalent bonds (Kołakowska et al., 1995). The imine bonds between proteins and lipids are the most evident example. Lysine is the most active amino acid as it possesses a free amine group, and aldehydes are the most active groups of oxidized lipids so that they easily form an imine bond:

Protein—CH_2—N=CH—CH_2—Lipid

Macromolecular pigments were formed in model experiments consisting of lysine and (*E*)-4,5-epoxy-(*E*)-2-heptenal (Hidalgo and Zamora, 1993). Imines and similar

---

TABLE 22.2

Functional Groups of Oxidized Lipids and Proteins Likely to React in Lipid–Protein Interactions

Functional Groups of Oxidized Lipids	Functional Groups of Proteins
Hydroperoxide, peroxide	Amine, amide
Keto, ketol	Thiol, sulfide, disulfide
Aldehyde, aldehydo acid	Peptide
Epoxy, dihydrofuran	Hydroxyl
Hydroxyl	Phenol
Carboxyl	Indole

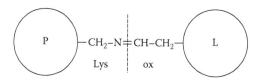

FIGURE 22.2 Example of protein–lipid covalent bonds. L, lipid phase; Lys, residue of bound lysine; ox, residue of a carbonyl chain of oxidized lipid; P, protein.

compounds containing unsaturated bonds are polymerized, and the reaction results in the formation of insoluble substances called melanoidins (Pokorný, 1981).

Another sensitive amino acid is tryptophan, because of the indole group present. Its losses were higher in deep-fried turkey meat than in roasted or microwave cooked meat (Borowski et al., 1986). Losses ranged between 65.6% in pork and 46.2% in white chicken muscle, respectively.

In the surface layer of fried food, where the temperature is high, bound or free serine, cysteine, and cystine are converted into the respective dehydroprotein intermediate. It reacts with the 6-amine group of lysine to form lysinoalanine, which is unavailable for the human organism.

The content of complexes of oxidized lipids and proteins may be quite high, e.g., 1 g bovine serum albumin bound 3.4–3.6 g oxidized methyl oleate (Seguchi, 1986). The complexes of oxidized lipids show moderate antioxidant activity, such as those with carnosine or histidine (Decker et al., 2001). For this reason, small amounts of oxidized lipids may do no harm. On the contrary, they may have beneficial effect on the oxidative stability of food. In most fresh lipids, the content of oxidation products may amount to 1.0% in some cases. Phospholipids contribute to interaction with protein because of their polar groups.

The participation of these reactions in nonenzymic browning should be mentioned (Pokorný, 1981). The browning is often recognized by the formation of fluorescence substances, which is probably produced by N=C bonds. Pyrroles formed in reactions of oxidized lipids with proteins are brown and fluorescing compounds (Zamora et al., 2000). The color intensity rapidly increases with increasing temperature and with increasing degree of unsaturation. The character of products and their active groups are similar to those of melanoidins. The melanoidins are the last step of the Maillard reaction (Hidalgo et al., 1999).

Because fish lipids are rather unsaturated (Chapter 15), the lipid–protein interaction products are deep brown. Melanoidins are oxidized with oxidized lipids and partially loose nitrogen by retro-aldolization. The remaining pigments become increasingly soluble in the lipidic phase (Pokorný et al., 1990). They form stains on the surface of white fish muscle, which are objectionable from the standpoint of appearance of those products stored under refrigeration or at frozen storage. Similar changes sometimes occur in stored white poultry muscle. On storage of apples, scald-like stains are produced by reaction of membrane lipids with amino acids (Burmeister and Dilley, 1995).

22.4 EFFECT OF LIPID–PROTEIN INTERACTIONS ON FUNCTIONAL PROPERTIES AND QUALITY OF FOOD

22.4.1 Effect on the Nutritional Value of Food

Some biologically active substances, such as enzymes, are inactivated under heating conditions, which increases the stability of foods under following storage. Antinutritional proteins, such as avidin in egg white or trypsin inhibitor in legumes, are inactivated, too, at least partially. Trypsin inhibitors in tempeh, a fermented product from soybeans, or in other soybean products are completely destroyed during deep fat frying.

The nutritional value and digestibility changes depend on the food material, e.g., they are more pronounced in sorghum than in maize products (Vivas-Rodríguez et al., 1990). The nutritional value of protein is not significantly affected by deep frying. Battered catfish was deep-fried at 159°C, and the protein efficiency ratio (PER) was about the same in catfish roasted in a rotating hot air oven (Ibrahim and Unklesbay, 1986). Fish of various species were cooked, smoked, or fried. The protein digestibility and net protein utilization (NPU) were very satisfactory. French frying did not considerably affect the protein (including lipoprotein) fraction (Steiner-Asiedou et al., 1991).

Native lipoproteins are easily digestible, because they are enzymatically hydrolyzed nearly completely before the absorption in the small intestine. Their nutritional value is equal to that of the respective lipids and proteins bound in lipoproteins, when they are consumed separately.

On the contrary, the nutritional value of oxidized lipids–protein interaction products is substantially lower than that of original lipoproteins. The main reason is their lower digestibility. Most covalent bonds formed in the interaction are not hydrolyzed by proteases under the conditions of digestion in human subjects. The 6-amine group of bound lysine is particularly sensitive to interaction with carbonylic oxidation products, such as aldehydes or ketols (Janitz et al., 1990), and the resulting imine bonds substantially reduce the lysine availability. Other amino acids, such as tyrosine, tryptophan, and methionine, are also partially converted into unavailable products.

When allergenic proteins have reacted with oxidized lipids, some interaction products may retain their allergenicity (Doke et al. 1989).

The impaired solubility is also important, as it contributes to decrease of digestibility. In stored frozen fish, the decrease of extractivity and solubility of proteins correlated with the amount of oxidized lipids bound to protein (Sikorski and Kołakowska, 1996). Similar results were obtained in the study of other fish, such as sardine and megrim (Careche and Tejada, 1990). In another laboratory (Xiong and Decker, 1995), the functionality of muscle food was found damaged by reactions with oxidized lipids.

The heavy feeling in one's stomach after consumption of fried food may be partially caused by the presence of undigestible oxidized lipid-protein complexes than by oxidized frying oil only.

22.4.2 Effect on the Sensory Value of Fried Food

The subject was discussed in the example of wheat rolls (Römer et al., 2008). Increase of lipid peroxidation decreases the sensory value. Lipid peroxides reacted with amino acid residues in proteins and changed the amino acid profile during baking. The profile was influenced mainly by the baking time.

Fats and oils, either present in food or added to food raw materials during culinary operations or applied after cooking, improve the flavor and texture of the meal. Other factors, such as technological possibilities, have also some effect. The surface properties are important in the process of food preparation, too. The lipid films and droplets assist in retaining gases during dough fermentation, thus improving the volume, softness, appearance, and acceptability of many bakery products. During chewing of high-fat foods, fats are emulsified into a relatively thick fluid. Lipoproteins behave similarly. Consumers perceive the high viscosity as very acceptable and appreciate the texture of such a product.

On the contrary, the interaction products of oxidized lipids with food proteins are less soluble, less easily moistened, and emulsified. The texture of the morsel after food ingestion is then deteriorated. The functional properties of beef proteins are deteriorated by oxidized lipids (Farouk and Swan, 1998). The feeling of the chewed morsel is rated lower because of lower viscosity.

The content of salt-soluble proteins decreased proportionally to increasing lipid peroxidation in frozen fish (Sarma et al., 1998), as lipid oxidation proceeds in frozen food more rapidly than in a refrigerated product. Fortunately, lipid oxidation–protein complexes in such a high amount are only exceptions to become an important factor in human nutrition.

Problems might arise in the case of dry food products, because they contain channels left after removal of water. Oxygen then easily penetrates to lipids and lipoproteins located near the surface of the channels. Lipid oxidation products do react with proteins. The resulting condensation products affect the dispersibility after the dry food is reconstituted.

Oxidized lipoproteins have an unfavorable effect on the appearance because of browning reactions. The problem was discussed earlier. After roasting, baking, or frying, the surface turns deeply brown. The browning is a negative factor in some products, such as dried milk, white fish, and poultry muscles. Stains on their surfaces are surely considered as unfavorable. On the contrary, reddish brown color is appreciated in bakery products, fried meat, and fish.

The flavor of oxidized lipid–protein interactions is sometimes a defect, especially in foods served cold. It may become a preference in heated foods, as they produce flavor profiles typical for the particular food, and expected by the consumers.

22.5   LIPID–SACCHARIDE INTERACTIONS

22.5.1   Reactions of Lipids and Oxidized Lipids with Saccharides

In foods, lipids usually co-occur with proteins and saccharides in interacting biological structures. In nature, they occur in plant seeds and tubers, as well as in biological membranes of higher organisms. Biological transformations and technological factors may induce numerous changes in lipids and in their interactions with non-lipid components.

In addition to starch obtained from tubers of plants such as potatoes, other major starch sources include cereal grains. Starch, the main grain polysaccharide component, is accompanied by proteins, lipids, water, enzymes, and numerous other compounds. They may interact with one another, form various combinations, and catalyze different reactions, both during storage and during various technological processes.

Reactions of native lipids during processing of cereals or potatoes may be divided into those catalyzed by enzymes present in the medium and those not involving enzymes.

Most enzymatic reactions involved are associated with hydrolytic or oxidative pathways, or with isomerization of double bonds of the lipid carbon chains. Nonenzymatic transformations are basically restricted to oxidation and isomerization. The rate of nonenzymatic lipid hydrolysis is extremely low and depends on pH, temperature, and appropriate amount of water. This reaction gains importance during technological processes that disrupt the natural structure of the medium, when water becomes more available; then, the probability of interactions between the reagents increases.

22.5.2 AMYLOSE AND LIPID

Associations between amylose (AM) and surfactants have been frequently described (Bulpin et al., 1982, 1987; Godet et al., 1995; Tufvesson et al., 2003b). Best known is the inclusion AM complex with FA. FA or monoacylglycerol, compounds combining hydrophobic and hydrophilic properties, may penetrate inside the amylose helix to form a stable complex (Figure 22.3). Such complexes are most often formed when AM conformation changes as a result of different technological procedures. AM availability for substances that form such a complex varies and depends on many factors, e.g., the type of starch, AM to amylopectin ratio, degree of AM depolymerization, or the type and structure of substances forming the complex.

Kaur and Singh (2000) studied the formation of AM-FA complexes during cooking of rice flour with FFA: palmitic, muriatic, and stearic. The authors focused on effects of AM-lipid complexes on starch solubility and gelatinization. They observed that, as the time of cooking (95°C) the rice flour with FFA was elongated, the amount of newly formed AM complexes increased. They also demonstrated that the type of FA was important for the formation of complexes; myristic acid was more available than stearic acid.

Effects of FA addition to starch on the formation of complexes were explored also by Karkalas and Raphalides (1986). They demonstrated effects of medium pH on the complex formation; at pH 12, C_8 and C_{10} acids failed to complex with AM, but pH reduction to 4.6 resulted in the complexes being formed.

Important is also the length of the acyl chain itself. As demonstrated by Godet et al. (1995), lauric acid (C_{12}) and palmitic acid (C_{16}) formed stable V-type complexes with AM, whereas myristic acid (C_8) either formed no complex whatsoever or, if a complex was formed, it was far from stable and was soluble in solvents used in the study. Such V-type (inclusive) complexes of AM with FA and/or monoacylglycerols formed with lipids—native or added to the system—were described by,

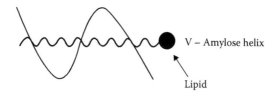

FIGURE 22.3 Amylose–lipid inclusion complex.

i.a., Godet et al. (1993), Snape et al. (1998), Desurmax et al. (1999), Tufvesson and Eliasson (2000), Fanta et al. (2002), and Hyun-Jung et al. (2003).

The formation of an inclusive AM complex depends on the AM chain length, i.e., degree of depolymerization. As demonstrated by Gelders et al. (2004), who studied variously depolymerized (DP 20 to DP 950) complexes of AM with glycerol mono-stearate (GMS) and docosanoic acid (C_{22}), the amount of newly formed AM-lipid complexes increased with AM chain length. In their study involving C_8, C_{12}, and C_{16} FA, Godet et al. (1995), too, observed the formation of such complexes with vari-ously depolymerized AM. The FA recovery from such complexes ranged from 0.7% for DP 30 and C_8 FA to 45.5% for DP 900 and C_{16}.

Thermal properties of AM complexes with amphiphilous substances are most often explored with calorimetric measurements (Biliaderis and Seneviratne, 1990; Karkalas et al., 1995; Tufvesson et al., 2003a).

Thermal processing of an AM-lipid system significantly affects the complex for-mation. Details of the effect were studied, with DSC at 15°C–114°C, by Tufvesson et al. (2003a,b). They addressed the formation and stability of AM complexes with monoacylglycerols (MAG) and FFA, emerging both as amorphous (I) and crystalline (II) form. Comparison of the FA (C_3 to C_{22}) complex analyzed with MAG (C_{10} to C_{18}) showed FA to be generally more amenable than MAG to form complexes (particu-larly their form I) during thermal processing. On the other hand, MAG, particularly its short lipid chain species, more easily produced form II (crystalline) complexes.

As shown by all those studies, the formation and thermal stability as well as melt-ing and crystallization enthalpies of the complexes depend on the chemical nature (chain length and saturation, polarity) of the complexing ligands, AM polymeriza-tion, and complexation conditions (temperature, time, and solvents used).

22.5.3 Amylopectin and Lipids

The majority of relevant papers deal with surfactant interactions with AM. However, amylopectin (AMP), too, can form complexes with various surfactants and lipids, both native and added. This opinion is firmly supported by studies on surfactant binding by AMP in solution (Lundqvist et al., 2002). An indirect support is provided by calorimetric studies on various wax starches accompanied by polar and hydro-phobic compounds.

Like amylose, amylopectine, too, can form complexes via its external. It is capa-ble of binding amounts of FFA similar to those complexed by AM, but the actual binding depends on the AMP molecular structure size (Lundqvist et al., 2002).

22.5.4 Effects of Lipid Freshness on Lipid–Saccharide Interactions

Freshness of the lipids present in a system is extremely important for the reaction of lipids with saccharides. Important, too, is the presence of lipolytic enzymes and lipo-oxygenase. Acylglycerol lipolysis leads to the emergence of MAG or FFA structures, which can easily react with hydrophobic groups of starch helices. FFA, particularly those with unsaturated bonds (e.g., linoleic acid), become susceptible to external factors and lipooxygenase. As demonstrated by Graveland (1970) and Lehtinen et al.

(2000), the amount of linoleic acid was reduced during dough kneading in the presence of lipooxygenase due to the acid being bonded by, i.e., starch rather than due to chemical changes in the acid itself. Bienkiewicz and Kołakowska (2004) showed the degree of lipid oxidation to affect lipid–starch interactions. In fish lipids-AM-water model systems, lipids—oxidized by mixing the system—were more resistant to complexing with AM than fresh lipids. However, heating and freezing of the systems resulted in the lipids being more resistant to extraction from the system than fresh lipids. A different pathway of lipid–starch complex formation was observed when fish lipids interacted with AMP in identical conditions (Figure 22.4). Lipids, oxidized by mixing the system, were more easily and stronger bonded by AMP than fresh lipids. However, after a month-long frozen storage (–18°C) of such a system, oxidized lipids were more amenable to extraction from the system, compared with fresh lipids (Bienkiewicz and Kołakowska, 2003). Differences in lipid complexation with AM and AMP were earlier reported also by Huang and White (1993) and Villwock et al. (1999).

Electrostatic forces prevent the polar carboxyl groups of FA and MAG from getting inside the starch helix (Snape et al., 1998). Although FFA and/or MAG cannot penetrate the helix, penetration is possible on both ends of the AM chain (Godet et al., 1996).

Bonds of that type enhance polymerization of saccharide chains to form cross-linking bonds (Seidel et al., 2001).

In foods, lipids are capable of reacting, although not directly, with reducing sugars and are involved in the complex Maillard reactions. Similarly to other carbonyl compounds, reducing sugars can, in the presence of lipid oxidation products, react with amine groups to form macromolecular pigments having properties similar to those of melanoids. The products may show antioxidant properties and may be formed during, e.g., baking. In their study of model systems, Mastracola et al. (2000) showed that the presence of soybean oil and glucose in the system resulted in the formation of more brown Maillard reaction products than in systems containing no reducing sugar.

Sucrose, a saccharide without reducing properties, has eight hydroxyl groups that can act as potential esterification sites for FA. Esterification may result in low-calorie lipid substitutes, e.g., Olestra. However, complexes of that type do not occur naturally in foods. For such complex to be produced, it is necessary to use high concentration of the substrates and to apply temperature of the order of 150°C. This is also a way to form sucrose esters, less saturated by FA, which may serve as emulsifiers (Stauffer, 2001).

22.6 FINAL REMARKS

Interactions between lipids and proteins, lipids and saccharides, and between all three components affect the biological properties and technological utility as well as sensory and nutritive qualities of foods. The strength of bonds formed between food components during processing and storage of raw materials and products depends on the composition, degree of hydrolysis (proteolysis), and oxidation of lipids, proteins, and saccharides. Disregarding effects of the interactions on extractability of food

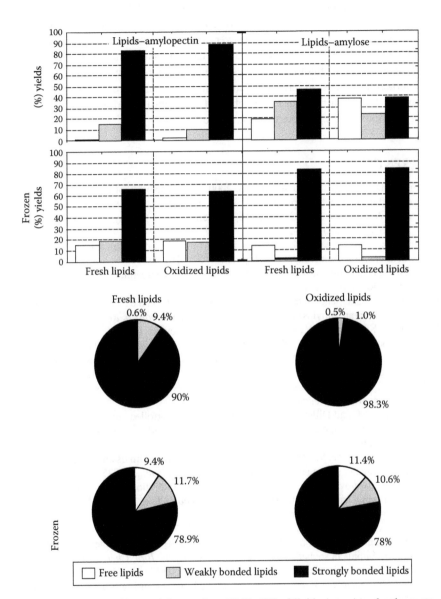

FIGURE 22.4 Fish lipid–starch interaction. Yields [%] of lipids (upper) and polyunsaturated FA (PUFA) (bottom) from lipid–starch system, by selective extraction. (Adapted from Bienkiewicz, G. and Kołakowska, A., Fish lipids–amylopectin starch interactions, *21st Nordic Lipid Symposium Lipidforum,* Bergen, Norway, June 5–8, 2001; Bienkiewicz, G. and Kołakowska, A., Fish lipids–amylose interactions, *24th World Congress International Society for Fat Research (ISF),* Berlin, Germany, September 16–20, 2001, p. 53.)

components may lead to substantial errors in the assessment of quality and quantity of some essential food components, e.g., long chain n-3 PUFA in fish products.

The role of the interactions in question in food quality is being discussed and considered to be mainly negative. However, the interactions are widely used in food technologies such as baking, emulsied sausage, and sauce manufacture, including the most advanced technological processes such as microencapsulation. In addition, the interactions discussed have application in the manufacture of low-calorie foods because of their contribution to reduction of digestibility and availability of high-energy components, e.g., increase in starch resistance.

REFERENCES

Anton, M., Denmat, M., and Gandemer, G. 2000. Thermostability of hen egg yolk granules: Contribution of native structure of granules, *J. Food Sci.*, 65: 581.

Azagtat, A. A. and Alli, I. 2002. Protein-lipid interactions in food systems: A review, *Int. J. Food Sci. Nutr.*, 53: 249–260.

Bienkiewicz, G. and Kołakowska, A. 2003. Effects of lipid oxidation on fish lipids–amylopectin interaction, *Eur. J. Lipid Sci. Technol.*, 105,8: 410–418.

Bienkiewicz, G. and Kołakowska, A. 2004. Effects of thermal treatment on fish lipids–amylase interaction, *Eur. J. Lipid Sci. Technol.*, 106: 376–381.

Biliaderis, C. G. and Seneviratne, H. D. 1990. Solute effects on the thermal stability of glycerylmonostearate-amylose complex superstructures, *Carbohydr. Res.*, 208: 199–213.

Borowski, I., Kozikowski, W., Rotkiewicz, W., and Amarowicz, R. 1986. Influence of cooking methods on the nutritive value of turkey meat, *Nahrung*, 30: 987–993.

Bulpin, P. V., Welsh, E. J., and Morris, E. R. 1982. Physical characterization of Amylose-fatty acid complexes in starch granules and in solution, *Starch/Staerke*, 34,10: 335–339.

Bulpin, P. V., Cutler, A. N., and Lips, A. 1987. Cooperative binding of sodium myristate to amylose, *Macromolecules*, 20,1: 44–49.

Burmeister, D. M. and Dilley, D. R. 1995. A scald-like controlled atmosphere storage disorder of Empire apples—Chilling injury induced by carbon dioxide, *Postharvest Biol. Technol.*, 6: 1.

Careche, M. and Tejada, M. 1990. Effect of neutral and oxidized lipids on protein functionality in megrim and sardine during frozen storage, *Food Chem.*, 37: 275–287.

Decker, E. A., Ivanov, V., Ben, Z. Z., and Frei, B. 2001. Inhibition of low-density lipoprotein oxidation by carnosine and histidine, *J. Agric. Food Chem.*, 49: 511–516.

Desurmax, A., Bouvier, J. M., and Burri, J. 1999. Effect of free fatty acids addition on corn grits extrusion cooking, *Cereal Chem.*, 71,5: 699–704.

Doke, S., Nakamura, M., and Torii, S. 1989. Allergenicity of food proteins interacted with oxidized lipids in soybean-sensitive individuals, *Agric. Biol. Chem.*, 53: 1231–1235.

Eymard, S., Baron, C. P., and Jacobsen, C. 2009. Oxidation of lipid and protein in horse mackerel (*Trachurus trachurus*) mince and washed minces during processing and storage, *Food Chem.*, 114: 57–65.

Fanta, G. F., Felker, F. C., and Shorgen, R. L. 2002. Formation of crystalline aggregates in slowly-cooled starch solution prepared by steam jet cooking, *Carbohydr. Polym.*, 48,2: 161–170.

Farouk, M. M. and Swan, J. E. 1998. Effect of muscle condition before freezing and simulated chemical changes during frozen storage on protein functionality in beef, *Meat Sci.*, 50: 245–256.

Fukuda, M., Kumisada, Y., and Toyosewa, I. 1989. Some properties and *in vivo* digestibility of fried and roasted soybean proteins, *Nihon Eiyo Shokuryo Gakkaishi*, 42: 305–311.

Gelders, G. G., Vandrestukken, T. C., Goesaert, H., and Delcour, J. A. 2004. Amylose–lipid complexation: A new fractionation method, *Carbohydr. Polym.*, 56(4): 447–458.

Godet, M. C., Delaget, M. M., and Buleon, A. 1993. Molecular modelling of the specific interactions involved in the amylose complexation by fatty acids, *Int. J. Biol. Macromol.*, 15(1): 11–16.

Godet, M. C., Tran, V., Colonna, P., Buleon, A., and Pezolet, M. 1995. Inclusion/exclusion of fatty acids in amylose complexes as a function of the fatty acid chain length, *Int. J. Biol. Macromol.*, 17(6): 405–408.

Godet, M. C., Bouchet, B., Colonna, P., Gallant, D. J., and Buleon, A. 1996. Crystalline amylose fatty acid complex: Morphology and crystal thickness, *J. Food Sci.*, 61(6): 1196–1201.

Graveland, A. 1970. Enzymatic oxidation of linoleic acid and glycerol 1-monolynoleate in doughs and flour-water suspension, *J. Am. Oil Chem. Soc.*, 47(9): 352–361.

Huang, J. J. and White, P. J. 1993. Waxy corn starch: Monoglyceride interaction in model system, *Cereal Chem.*, 70: 42–47.

Hutaopea, E. B., Parkányiová, L., Parkányiová, J., Miyahara, M., Sakurai, H., and Pokorný, J. 2004. Browning reactions between oxidized vegetable oils and amino acids, *Czech. J. Food Sci.*, 22: 99–107.

Hyun-Jung, Ch., Hyo-Young, J., and Seung-Taik, L. 2003. Effects of acid hydrolysis and deffating on crystallinity and pasting properties of freeze-thaved high amylose corn starch, *Carbohydr. Polym.*, 54(4): 449–455.

Ibrahim, N. and Unklesbay, N. F. 1986. Comparison of heat processing by rotating hot air and deep fat frying: Effect on selected nutrient content and protein quality of catfish, *J. Can. Diet. Assoc.*, 47: 26–31.

Janitz, W., Pyrcz, J., Grzeszkowiak, B., and Berghofer, E. 1990. Effect of technological treatment on reactions oxidized fats with muscle proteins, *Fleischwirtschaft*, 70: 206.

Karkalas, J. and Raphalides, S. 1986. Quantitative aspects of amylose–lipid interaction, *Carbohydr. Res.*, 157: 215–234.

Karkalas, J., Ma, S., Morrison, W. R., and Pethrick, R. A. 1995. Some factors determining the thermal properties of amylose inclusion complexes with fatty acids, *Carbohydr. Res.*, 268(2): 233–247.

Kaur, K. and Singh, N. 2000. Amylose–lipid complex formation during cooking of rice flour, *Food Chem.*, 71(4): 511–517.

Kawai, Y., Takeda, S., and Terao, J. 2007. Lipidomic analysis for lipid peroxidation-derived aldehydes using gas-chromatography–mass spectrophotometry, *Chem. Res. Toxicol.*, 33: 201–209.

Kołakowska, A. and Szczygielski, M. 1994. Stabilization of lipids in minced fish by freeze texturization, *J. Food Sci.*, 59: 88–90.

Kołakowska, A., Kołakowski, E., and Szczygielski, M. 1995. Effect of undirectional freezing on lipid changes during storage of minced bream, *Proceedings of the 19th International Congress of Refrigeration*, the Hague, the Netherlands, 2: 196.

Kołakowska, A., Ziobrowski, P., and Wróbel, P. 2006. Application of FTIR spectrophotometry to the study of meat tissues of trout during storage and thermal treatment, *XXXVII Scientific Session of the Committee: Nutritional Series*, Gdynia, Poland, Polish Academy of Sciences, Abstract p. 288.

Kouřimská, L., Pokorný, J., Jaschke, A., and Réblová, Z. 1993. Reactions of linoleic acid hydroperoxide with muscle proteins under storage conditions. In *Food Proteins: Structure and Functionality*, Schwenke, K. D. and Mothes, R. (Eds.), Verlag Chemie, Weinheim, Germany, pp. 236–240.

LeMeste, M. and Davidou, S. 1995. Lipid-protein interactions in foods. In *Ingredient Interactions*, Gaokar, A. G. (Ed.), Marcel Dekker, New York, pp. 235–367.

Lundqvist, H., Eliasson, A.-C., and Olofsson, G. 2002. Binding of hexadecyltrimethylam-monium bromide to starch polysaccharides. Part I. Surface tension measurements, *Carbohydr. Polym.*, 49(1): 43–55.

Makinson, J. H., Greenfield, H., Wong, M. L., and Wills, R. B. H. 1987. Fat uptake during deep-fat frying of coated and uncoated foods, *J. Food Comp. Anal.*, 1: 93–101.

Mastracola, D., Munari, M., Cioroi, M., and Lewici, C. R. 2000. Interaction between Miliard reaction products and lipid oxidation in starch based model system, *J. Sci. Food Agric.*, 80: 684–690.

Nakamura, T., Hama, Y., Tanaka, R., Taira, K., and Hatate, H. 1998. A new red coloration induced by the reaction of oxidized lipids with amino acids, *J. Agric. Food Chem.*, 46: 1316–1320.

Perkins, E. 1993. *Lipoproteins*. AOCS Press, Champaign, IL.

Pokorný, J. 1981. Browning from lipid-protein interaction, *Proc. Food Nutr. Sci.*, 5: 421–428.

Pokorný, J. 1998. Substrate influence on the frying process, *Grasas Aceites*, 49: 265–270.

Pokorný, J. 1999. Changes of nutrients at frying temperatures. In *Frying of Food*, Boskou, D. and Elmadfa, I. (Eds.), Technomic Publishing, Lancaster, PA, pp. 69–103.

Römer, J., Majchrzak, D., and Elmadfa, I. 2008. Interaction between amino acid degradation, sensory attributes and lipid peroxidation in wheat rolls, *Eur. J. Lipid Sci. Technol.*, 110: 554–562.

Saeed, S., Fawthrop, S. A., and Howell, N. K. 1998. Electron spin resonance study of free radi-cal transfer in fish lipid-protein interaction, *J. Sci. Food Agric.*, 79: 1809–1816.

Sarma, J., Srikar, L. N., and Reddy, G. V. 1998. Comparative effects of frozen storage on bio-chemical changes in pink perch and oil sardine, *J. Food Sci. Technol. India*, 35: 255–258.

Seguchi, M. 1986. Lipid binding by protein in films heated on glass beads and prime wheat starch, *Cereal Chem.*, 63: 311–315.

Seidel, Ch., Kulicke, W., Hess, Ch., Hartman, B., Lechner, D., and Lazik, W. 2001. Influence of the cross-linking agent on the gel structure of starch derivatives, *Starch*, 53,7, abstracts.

Sikorski, Z. E. and Kołakowska, A. 1996. Changes in proteins in frozen stored fish. In *Seafood Proteins*, Sikorski, Z. E., Sun Pan, B., and Shahidi, F. (Eds.), Chapman and Hall, New York, pp. 99–112.

Snape, C. E., Morrison, W. R., Maroto-Valer, M. M., Karklas, J., and Pethrick, R. A. 1998. Solid state ^{13}C NMR investigation of lipids ligands in V amylose inclusion complexes, *Carbohydr. Polym.*, 36: 225–237.

Soyer, A. and Hultin, H. O. 2000. Kinetics of oxidation of the lipids and proteins of cod sarco-plasmic reticulum, *J. Agric. Food Chem.*, 48: 2127–2134.

Steiner-Asiedou, M., Asiedu, D., and Njan, L. R. 1991. Effect of local processing methods on three fish species from Ghana, II, *Food Chem.*, 41: 227–236.

Thomsen, M. K., Lauridsen, L., Skibsted, L. H., and Risbo, J. 2005. Two types of radicals in whole milk powder. Effect of lactose crystallization, lipid oxidation, and browning reac-tion, *J. Agric. Food Chem.*, 53: 1805–1811.

Tufvesson, F. and Eliasson, A. C. 2000. Formation and crystallization of amylose–monoglyc-eride complex in a starch matrix, *Carbohydr. Polym.*, 43(4): 359–365.

Tufvesson, F., Wahlgren, M., and Eliasson, A. C. 2003a. Formation of amylose–lipid com-plexes and effects of temperature treatment. Part 1. Monoglycerides, *Starch*, 55(2): 61–71.

Tufvesson, F., Wahlgren, M., and Eliasson, A. C. 2003b. Formation of amylose–lipid complexes and effects of temperature treatment. Part 2. Fatty Acids, *Starch* 55(3–4): 138–149.

Ventanas, S., Esteves, M., Dalgado, C. L., and Ruiz, J. 2007. Phospholipid oxidation, non-enzymatic browning development, and volatile compounds generation in model systems containing liposomes from porcine *Longissimus dorsi* and selected amino acids, *Eur. Food Res. Technol.*, 225: 665–675.

Villwock, V., Eliasson, A.-Ch., Silverio, J., and BeMiller, J. N. 1999. Starch–lipid interaction in common waxy *ae du*, and *ae su2* maize starches examined by differential scanning calorimetry, *Cereal Chem.*, 76(2): 292–298.

Vivas-Rodríguez, N. E., Saleliwan, S. O., Waniska, R.O., and Rooney, L. N. 1990. Effect of tortilla chip preparation on the protein fraction of quality protein maize, regular maize and sorghum, *J. Cereal Sci.*, 12: 289–296.

Xiong, Y. L. and Decker, E. A. 1993. Alteration of muscle protein functionality by oxidative and antioxidative processes, *J. Muscle Foods*, 6: 139–160.

Yuan, Q., Zhu, X., and Sayre, L. M. 2007. Chemical nature of stochastic generation of protein-base carbonyls: Metal catalyzed oxidation versus modification by products of lipid oxidation, *Chem. Res. Toxicol.*, 20: 129–139.

Zamora, R. and Hidalgo, F. J. 2005. Coordinate contribution of lipid oxidation and Maillard reaction to the nonenzymic browning, *Crit. Rev. Food Sci. Nutr.*, 45: 49–59.

Zamora, R., Nogales, R., and Hidalgo, F. J. 2005. Phospholipid oxidation and nonenzymatic browning development in phosphatidylethanolamine/ribose model systems, *Eur. Food Res. Technol.*, 220: 459–465.

23 Contaminants of Oils: Analytical Aspects

Fahri Yemiscioglu, Aytaç Saygin Gumuskesen, Semih Otles, and Yildiz Karaibrahimoglu

CONTENTS

23.1 INTRODUCTION

Oils and fats are produced commercially in large quantities. They are also processed and modified for use solely or as ingredients in various foods. Although a large proportion of oils are composed of triacylglycerols, they are multicomponent and complex mixtures of various constituents, some of them harmful. In this chapter, possible contamination sources of oils and fats are discussed. Studies on analysis for their detection are also summarized.

23.2 TRACE METALS

Trace metals are found in oils in various concentrations. The content of heavy metals in fats depends on several factors. Metals are naturally present in seed oils, because they may be absorbed from the soil and fertilizers or environmental pollutants. Lead and copper are potentially present in oils because of environmental contamination. A possibility of metal entry into crude vegetable oils other than the technological one is the environmental exposure to a large variety of elements. They arrive in the plant via deposition as well as bioaccumulation from the soil due to natural metal sources and environmental pollution. The use of calcium phosphate like fertilizers also brings about the risk of trace metal contamination to oils. Especially, the trace element content of olive oils varies according to the origin. This can be used as a basis for geographical characterization. Since natural olive oil is extracted by means of a series of physical operations and is not subjected to any high-temperature-refining process, trace metal content of these oils are thought to be of growth and cultivation origin, which also link these data for the characterization of natural olive oils by using multivariate statistical techniques. Production processes, materials of packaging, and storage of food fats may also be sources of heavy metals contamination.

The oil impurities are removed by means of physical or chemical refining. Most of the trace metals are found in crude oils in the form of salts with phosphatides. Degumming with water and phosphoric acid leads to efficient removal of gums. This results in a significant reduction in the trace metal content of oils. On the other hand, crude oils are stored previously in oil refineries for a certain period and may contain considerable amounts of free fatty acids. The acidity of crude oils is dependent on the storage of seeds and production technique. Especially high acidic oils create problems by interacting with the inner surface of storage tanks. These are reported to be the reasons of metal migration. During alkali neutralization, bleaching, or deodorization, the crude oils are pumped in bulk through a series of equipments. The processing instruments are another source of metal contamination. Several studies emphasize that the type of bleaching clay and bleaching conditions are also determinative on trace metal concentration (Ansari et al. 2009).

Refined oils are hardened by several techniques. Hydrogenation is one of the most widely used techniques for this purpose. It is carried out by using hydrogen in gaseous form, at temperatures ranging between 160°C and 220°C with the aid of catalysts. The most popular hydrogenation catalyst is nickel; also the use of copper, palladium, aluminum, or various alloys are reported. The reaction is terminated at a specified iodine value and melting point. This is followed by filtration to remove the catalyst. The efficiency of filtration and the impurities in the hydrogen gas also affect Ni contamination. The Joint FAO/WHO Expert Committee on Food Additives (JECFA) reduced the tolerable lead consumption per week from 0.05 mg/kg body weight to 0.025 mg/kg in 1993. The nonspecific target level for metal contamination is 100 ng/g. Recent reports regarding the determination of these metals in edible oils at sub-μg/g levels have been based on voltammeter, atomic absorption, and atomic emission spectroscopy (Szłyk and Czerniak 2004, Reyes and Campos 2006).

Trace metal composition is an important criterion for the assessment of quality of vegetable oils. Therefore, the determination of heavy metals in various crude or processed/hardened oils is of crucial importance. The determination of trace elements in edible oils is also important because of both the metabolic role of metals and possibilities for adulteration detection and oil characterization.

The determination involves a sample preparation step to release toxic elements from the fat matrix. The extraction procedure involves the preconcentration of metallic ions followed by matrix removal. This classical approach involves various advantages, such as the lack of calcinations in sample as well as the sample is analyzed in the undissolved form. Moreover, any contamination to the sample that might lead to error in the results is eliminated. Extraction of heavy metals is carried out with hydrochloric acid, Pb-piperazinedithiocarbamate, and potassium cyanide solution. Wet digestion, dry ashing, microwave digestion, and alcoholic solubilization are alternative sample preparation techniques. Atomic absorption spectrometry (AAS) and ion chromatography (IC) are the most used techniques in the analysis of metal ions content of various fat samples. AAS has precisions between 3% and 10% range of standard deviation (RSD). Accuracy values are reported to be 89%–112%, expressed as recoveries that are expressed for various oils and fats. The detection limits are 0.4 mg/kg for Cd(II), 30 mg/kg for Cu(II), and 30 mg/kg for Pb(II). AAS techniques involve sample pretreatment since vegetable oils have high organic content (Lacoste et al. 1993, Allen et al. 1998, Reyes and Campos 2006).

Graphite furnace AAS (GFAAS) is a new feature in the analysis of trace metals in oils and fats. Most atomic spectroscopic techniques require low viscosity and density fluids. In contrast, GFAAS is well fitted even for solids. Proficiency tests have shown that Cu determination in edible oils is still somewhat difficult, with individual recoveries ranging from 10% to 250%, while for Ni, these results range from 50% to 144%. There is also an alternative procedure for the direct GFAAS determination of Cu and Ni in vegetable oils making use of a solid sampling strategy. For this purpose, samples are directly weighed on the graphite platform boat and inserted in the graphite tube. An adequate temperature program permits the calibration by external aqueous analytical curves. Limits of detection for Cu are 0.001 mg/g and 0.002 mg/g for Ni (Hendrikse et al. 1991, Allen et al. 1998, Kowalewska et al. 2005, Ansari et al. 2009).

A simple and fast method for the trace level determination of nickel in ghee is reported. In this work, different methods were applied for the extraction of residual nickel from ghee samples. Using toluene, benzene, and carbon tetrachloride as organic solvents, an acid mixture was used for the extraction of nickel. Extracted nickel was quantified with atomic absorption and colorimetric methods. Among the organic solvents, toluene proved to be the best solvent, mediating a 95% extraction of nickel from ghee samples. Nickel was extracted and determined in 10 different brands of ghee, and in all samples, its amount was well above the permissible limit of WHO (0.2 µg/g). Other metals like Pb, Zn, Cu, and Cd were also determined and their concentrations were found to be much below the WHO permissible limits (Khan et al. 2007).

Inductively Coupled Plasma Atomic Emission (ICP-AES) and AAS are the most common techniques. ICP-mass spectrometry (MS) allows an easy multielement analysis for each single sampling of a complex matrix, such as edible oils. The results showed that the discrimination between olive oils coming from different regions can be made by ICP-MS that involves a rapid determination of 18 elements (Lacoste et al. 1993, Pehlivan et al. 2008). Sample treatments are required in order to eliminate the organic matrix, since the direct analysis faces many drawbacks and usually causes the extinction of the plasma. Another negative effect is represented by the injector blocking due to carbon deposition from incomplete oxidation of organic matrix. These drawbacks are eliminated by the addition of oxygen to the plasma and by a decrease of temperature inside the nebulization chamber. On-line emulsification is an alternative technique for introduction of oil samples directly into ICP, which requires adjustments and modifications in order to proper emulsify the samples into the instrument. Electrothermal vaporization inductively coupled plasma MS is another future for the determination of Zn, Cd, and Pb.

Electrochemical stripping analysis is generally recognized as a technique for measuring the concentrations of trace metals in water, biological fluids, and foods. There are a few reports of its application to oils and fats. The precision is reported to be 1.6%–3.8% RSD. Direct inverse voltammetric method is proposed for the determination of Pb, Cu, and Cd in linseed, soybean, sunflower, and olive oils. Some studies reported the derivative potentiometric stripping analysis used for the determinations of heavy metals in olive oils. The Cd content in all oil samples was below the detection limits. Moreover, a study on the applications of differential pulse anodic/cathodic stripping voltammetry for the determination of Pb and other transition metals in crude oil samples; Cu, Cd, and Pb in vegetable oils; and margarines were described. They were found to be in good agreement with AAS and IC methods. Disadvantage of voltammetric methods is that mercury is used in stripping analysis. The electroanalytical techniques allow the determination of trace metals of nutritional and toxicological interest in a wide range of concentrations. The preconcentration of dissolved metal ions onto the mercury film of the working electrode allowed the achievement of low limits of detection. In addition to the precision and accuracy of electrochemical procedures are also comparable with AAS method (Allen et al. 1998, Kowalewska et al. 2005, Reyes and Campos 2006, Ansari et al. 2009, Cindric et al. 2009).

Trace metal contamination to the crude or processed oils plays an important role from various aspects such as quality characteristics of oils as well as food safety. The crucial point is to assess reliable and accurate methods for analyzing heavy metal concentrations in oils. There are several efforts on this issue. Limiting the trace metal contamination is dependent on less pollutant, proper cultivation, avoiding the contact with poor equipment materials, and a routine measurement of heavy metal concentrations with reliable and accurate analytical techniques in industrial laboratories.

23.3 PESTICIDES

Oil seeds and fruits are contaminated during their growth with numerous pesticides. These pesticides show various physical and chemical properties. Their detection is especially important for raw seeds and fruits, together with naturally consumed oils such as olive oil. Much effort has been applied on the rapid and accurate determination of pesticides in oil bearing materials and especially in olive oil. Other seed oils are refined to remove certain impurities. Especially during physical refining, pesticides are removed to a considerable extent by means of high temperature and vacuum application.

Oilseeds as sources of oils for edible and multiple industrial uses hold an important position in the economy of developed and developing countries. Studies on storage and insect pest control in oil seeds are relatively less when compared with cereal grains. In industrialized countries, aeration techniques appropriate to the storage situation are adopted, whereas in developing countries, the bag-stack storage system is used. These poor packaging techniques lead to insect infestation, deterioration, and mold growth. Fumigation and various pesticides are used to protect these commodities.

Almost 800 compounds having various physical and chemical properties such as polarity and diffusivity have been involved to control undesirable moulds, insects, or various mal formations during the growth of the raw materials. Numerous regulations have been established for the pesticide residues in agricultural products. For this reason, accurate determination of pesticide residues is an important task.

In pesticide testing, the development of multi-residue methods, which allows proper control of a large number of pesticides in a unique analysis, is basically the main applied strategy. However, the different classes and physicochemical properties make the development of methodologies that cover all the analyses under study difficult.

Pesticide testing involves two main stages: the isolation of the pesticides from the matrix (sample treatment) and the analytical method for the determination. Sample treatment, i.e., the extraction of the pesticides and the purification of sample extract, still remains to be the bottleneck of the entire procedure, despite much progress on automation has been accomplished. Sample preparation prior to analysis involves mostly solvent extraction. Various solvents are used for fatty raw materials as in the principle of solid–liquid extraction. Many efforts are reported especially for olive oil in the literature with numerous solvents by liquid–liquid partitioning, like hexane, acetonitrile, acetone, petroleum ether-dichloromethane system. The instruments mostly used after isolation of pesticide residues from liquid oils are LC-MSi, HPLC, and GC-FID. The recovery rates ranged between 70% and 120%; and RSD range between 1% and 11% for olive oil (Sanchez et al. 2003, 2005, Bajpal et al. 2007, Gilbert-López et al. 2009).

QuEChERS method, which is defined as quick, easy, cheap, effective, rugged, and safe, widely extended for fruit and vegetables, has been also applied to fatty vegetable matrices such as avocado and olives. The method is also tested on olive oil by diluting the matrix with water. The procedure is based on a liquid partitioning with acetonitrile followed by a dispersive solid phase extraction (SPE) clean-up procedure with primary secondary amine (PSA). High recovery rates are reported as a result of GC-MS or LC-MS determinations.

Sorbent based methodologies are also applied for sample preparation. Solid-phase extraction is known to be one of the most popular sample preparation techniques. The main principle of this technique is to dissolve the oil in a proper liquid solvent, followed by the removal of interfering substances in a solid column or cartridge. N-alumina, diol, extrelut, and PSA are wire ported for proper use as sorbents. The retention mechanism of N-alumina is adsorption. Diol performs a normal phase retention mechanism. Extrelut absorbs lipophylic compounds in aqueous solutions, whereas PSA works in normal phase separation. Numerous GC-ECD, GC-NPD, and GCMS/MS instrumentation follows the sample preparation with high accuracy (Jongenotter and Janssen 2002, Sanchez et al. 2005, Bajpal et al. 2007, Gonzalez et al. 2007, Gilbert-López et al. 2009). Supercritical carbon dioxide extraction is also a promising tool for pesticide analysis for sample preparation due to being a non-flammable, nontoxic, and cheap fluid.

Most of the working time is spent in preparation of the sample. A fully automated online reversed phase LC-GC system uses a prototype of the automated through oven transfer adsorption desorption interface. It is demonstrated by presenting a new rapid method for the determination of pesticide residue in olive oil, which is injected directly with no sample pretreatment step other than filtration. Methanol-water system is used as the eluent in LC pre-separation step, while the LC fraction containing the pesticides is automatically transferred to GC. Detection limits for the pesticides are reported to vary between 0.18 and 0.44 mg/dm³ with flame ionization detector.

23.4 PROCESSING CONTAMINANTS

Alkenes are processing contaminants related to oils and fats. To determine the contaminations, a study was conducted. A range of edible oils were purchased from a variety of retail outlets and analyzed for their n-alkane ($\Sigma 33/15$) content. The concentrations ranged from 7 to 166 mg/kg, the lowest being walnut oil and the highest sunflower oil. n-Alkanes with chain lengths of C_{27}, C_{29}, and C_{31} predominated in all of the plant oils except olive oil, where C_{23}, C_{25}, and C_{27} were the most significant. The results of this pilot survey suggest that the n-alkanes are indigenous and that their composition may be useful in characterizing a specific plant seed oil. Two of the 42 samples analyzed showed markedly different alkane patterns, in which profiles similar to petroleum-based sources dominated the contribution from odd carbon n-alkanes typical of seed oils (Mc Gill et al. 1993).

23.5 POLYCYCLIC AROMATIC HYDROCARBONS

The occurrence of polycyclic aromatic hydrocarbons (PAHs) in edible oils is attributed mainly to environmental contamination of the vegetable raw material and to contamination coming from some operations carried out during their processing, such as seed drying or solvent extraction. Some authors indicate that the endogenous origin of some PAHs cannot be totally discarded.

The occurrence of PAHs in five samples of olive pomace oil has been studied to determine the contamination degree of this type of oil and to evaluate if specific purification steps must be introduced during its manufacture. The PAHs present have been determined by GC-MS. A high number of PAHs, with a wide range of molecular weights and in very high concentrations, have been found in four of the samples studied. A very high number of alkyl derivatives and, in many cases, in higher concentrations than their respective parent PAHs, have also been identified. One of the samples, however, presented a reduced number of PAHs and in significantly lower concentrations than the others. These findings reveal that it is necessary to introduce adequate cleanup steps in the manufacturing process of olive pomace oil, which can give rise to oils with a relatively low content of PAHs. Some carcinogenic PAHs have also been identified, both unalkylated and alkylated (Moreda et al. 2001, Guillén et al. 2004).

To obtain an idea about the presence of impurities in the solvents employed, samples of two different brands of cyclohexane (CH1 and CH2) were analyzed in scan mode. Total ion chromatograms of samples CH1-A and CH2-A showed that, although the volume of sample CH2-A is approximately half that of CH1-A, the amount of compounds in the former is far higher than in sample CH1-A. Moreover, in cyclohexane CH2, many of these compounds are linear hydrocarbons, some of which interfere with the determination of some PAHs of interest, and they can be difficult to remove when their concentrations are high. It is clear from the results that not only the purity of the solvents must be checked but also their PAH content, because the presence of variable concentrations of these compounds can lead to mistaken data, especially in the case of low PAH content samples. Therefore, it is recommended that commercial solvents be analyzed before use to verify both their purity and the presence of PAHs, either to remove them or to be taken into account in quantification.

Apart from the compounds that can come from the solvents, there are others present in the samples in this study, such as fatty acids, which not only interfere in the determination of some PAHs by GC-MS but are also detrimental to the equipment. For these reasons, when acids are present in the oil extracts, an alkaline treatment is necessary to remove them (Guillén et al. 2004). Although several carcinogenic PAHs have been identified in the samples of this study, it must also be noted that the carcinogenicity mechanism of PAHs is very complex and there are many factors that can determine their final effect. Among the factors that can have an influence on the metabolic activation of PAHs in the organism, some dietary components such as antioxidants can be cited. Some natural antioxidants, although not in so high concentrations as in virgin olive oils, are present in olive pomace oil in higher levels than in other edible vegetable oils. Therefore, the absorption of these compounds together with PAHs may inhibit to a certain extent the oxidation reactions necessary for the metabolic activation of these toxicants.

In another study, factors affecting PAH concentrations in oils and fats, cereals, and related foodstuffs have been investigated. Levels of PAHs were low in retail fish and animal-derived oils and fats, such as butter, where the mean benzo(a)pyrene concentration was $0.06\,\mu g/kg$. Higher and more variable amounts were present in retail vegetable oils for which the mean level of benzo(a)pyrene was $1.29\,\mu g/kg$. Margarine was the major dietary source of PAHs in the oils and fats total diet group, accounting for 70% of the benzo(a)pyrene intake from these commodities. The levels of benzo(a)pyrene were less than $0.1\,\mu g/kg$ in white flour, and similar amounts were found in bread, showing that PAHs are not formed to any significant extent during baking of bread. Higher concentrations of up to $2.2\,\mu g/kg$ benzo(a)pyrene were detected in cereal-derived products containing higher levels of edible oils such as pudding-based desserts, biscuits, and cakes. The presence of vegetable oils as an ingredient also appeared to increase PAH levels in infant formulae as the mean benzo(a)pyrene content of $0.49\,\mu g/kg$ was four times higher than that found in skimmed milk. The mean value in the feed, after reconstituting the formulae with water, would, however, have been less than $0.1\,\mu g/dm^3$ (Dennis et al. 1991).

Investigations of rape seed drying showed no increase in any PAHs when cold or when electrically-heated air was used. Combustion gas drying had no effect for the larger PAHs, such as benzo(a)pyrene, but caused mean increases of between 41% and 126% for fluoranthene, pyrene, and chrysene. These increases did not correlate with reductions in moisture content of the rape seed, implying that the combustion conditions were more important to PAH contamination than the degree of exposure to combustion gases. Concentrations of these three PAHs and also benz(a)anthracene were all significantly reduced by up to a factor of five when crude oils were refined, suggesting that carefully controlled direct drying need not contribute PAHs to refined oils and fats (Dennis et al. 1991).

Further studies on PAH toxicity are necessary, but, meanwhile, caution must prevail, and attempts must be made to minimize the exposure of humans to PAHs. It is necessary to focus the attention on those PAHs suspected to be more carcinogenic, and, of course, alkylated PAHs must be considered in any study on PAHs. Consequently, any regulation on the level of PAHs in edible oils should include all of the carcinogenic PAHs, unalkylated or alkylated. Moreover, laws to limit the presence of PAHs in any type of foodstuff able to be contaminated should be implemented.

23.6 FUNGAL TOXINS

Fungal toxins are detrimental to oil quality due to their toxic effect upon consumption. Storage stands out to be critical, especially for the growth of *Aspergillus* species. These species grow on the seed due to moist conditions in storage houses. The result is aflatoxin contaminated seeds.

Commercial processing of cottonseed requires hexane to extract and recover edible oil. Gossypol and aflatoxin are not removed from extracted meals. A study has been carried out by using aqueous 95% (v/v) aqueous ethanol (EtOH) as solvent. Thus gossypol, aflatoxin, and oil have been extracted with a much less volatile solvent. In this process, cottonseed is pretreated and extracted with ambient 95% EtOH to remove gossypol and then extracted with hot 95% EtOH to extract oil and aflatoxin. Membranes and adsorption columns are used to purify the various extract streams, so that they can be recycled directly. A representative extracted meal contained a total gossypol content of 0.47% (a 70% reduction) and 3 ng/g aflatoxin (a 95% reduction). Residual oil content was approximately 2%. Although the process is technically feasible, it is presently not economical unless a mill has a continual, serious aflatoxin contamination problem. However, if a plant cannot meet the hexane emission standards under the Clean Air Act of 1990, this process could provide a safer solvent that may expand the use and increase the value of cottonseed meal as a feed for nonruminants (Hron et al. 1994).

Cash crops such as groundnut, cashew, cotton, sesame, sugarcane, tobacco, turmeric, and banana play vital role in the growth of economic and nutritional status in India. Of these, groundnut (*Arachis hyposaea* L.) is the most important major crop, which is energy rich, and micro nutrient contents are relatively high compared with other corps. Out of total production, 12% is used as seeds, 8% for edible purpose, 70% extraction of oils, and 10% for export. Hence, it is considered to be an economically important crop. This crop is found to be attacked by a number of fungal diseases. Besides these fungal diseases, groundnut is found to be contaminated with aflatoxins. They are probably the most serious contamination of fungal origin of groundnut on a worldwide scale. Aflatoxins are a group of highly oxygenated heterocyclic compounds with closely related structure of B-1, B-2, G(1) and G(2). These aflatoxins are a closely related group of secondary fungal metabolites that have been epidemiologically implicated carcinogens in humans. Aflatoxin producing fungi are widely distributed in tropical and subtropical areas. Groundnut is highly susceptible to aflatoxin contamination. An attempt has been made to determine the possible variations in the amount of aflatoxin-B present in the groundnuts grown in two different fields treated with chemical fertilizers and organic manure alone independently (Sharkey et al. 1994, Samuel et al. 2002).

23.7 PLASTICIZERS

Foods with at least a few percent of free oil packed in glass jars with metal closures were analyzed for migration of additives, primarily plasticizers, from the gasket of the lid. In a study, 158 samples were collected. In a first step, the composition of the additives in the gaskets was determined. Then the compounds found in the lid were measured in the jar content. Sixty-four percent of the gaskets contained epoxidized soybean oil (ESBO) as principal plasticizer, 22% a phthalate, and 6% substantial amounts of di(2-ethylhexyl) adipate (DEHA). Concentrations in the food reached 1170 mg/kg for ESBO, 270 mg/kg for diisononyl phthalate (DINP), 740 mg/kg for diisodecyl phthalate (DIDP), 825 mg/kg for di(2-ethylhexyl) phthalate (DEHP), and 180 mg/kg for DEHA. Further, elevated concentrations of plasticizers not authorized by the EU were found: diisononyl-cyclohexane-1,2-dicarboxylate (DINCH), 2-ethylhexyl palmitate and stearate, as well as epoxidized linseed oil (ELO). The few samples complying with the European rules contained little or well emulsified oil; some others were probably of very recent production (beginning of shelf life). It is concluded that there was still no lid reliably complying with the European rules (EU and national legislation) (Fankhauser-Noti et al. 2006).

Injector-internal thermal desorption from edible oil or fat is a convenient sample preparation technique for the analysis of solutes in lipids or extracts from fatty foods. The injector temperature is selected to vaporize the solutes of interest while minimizing evaporation of the bulk material of the oil. This technique has been in routine use for pesticides for some time. Now its potential is explored for migrants from food contact materials, such as packaging, into simulant D (olive oil) or fatty/oily food, which means extending the range of application towards less volatile compounds. The performance for high boiling components was investigated for DIDP and diundecyl phthalate (DUP). Since the injector temperature needs to be as high as 260°C, some bulk material of the oil enters the column and must be removed after every analysis. This is achieved by a coated precolumn back flushed towards the end of each analysis. Desorption of the solutes is particularly efficient in the initial phase, when a thin sample film is spread on the liner wall, and is largely determined by the diffusion speed in the oil after the latter has contracted to droplets. An increased carrier gas flow rate during the splitless period supports the transfer into the column. It is concluded that the technique is attractive for migrant analysis, with DUP being at the upper limit of the boiling point. ESBO is widely used as a plasticizer and stabilizer in such polymers [poly(vinyl chloride) (PVC) in particular] commonly adopted for manufacturing of gaskets of the lids for glass jars and plastic films for food packaging. ESBO was determined in foods packed in glass jars closed by lids with a PVC gasket. The methyl ester of a diepoxy linoleic acid isomer was measured, using transesterification directly in the homogenized food and on-line HPLC-GC-FID analysis. Infant foods from the Swiss market consisting of vegetables, potato and rice, or muesli with fruits and berries contained less than 7 mg/kg ESBO, but meat (its fat?) strongly increased ESBO migration up to 86 mg/kg. Some 12% of the products exceeded 15 mg/kg. Austrian and Norwegian samples gave similar results. Edible oil strongly extracts the ESBO from the gasket in food contact within a few weeks. Since this part of the gasket on average contained 91 mg ESBO, the legal limit is likely to be far exceeded whenever the food contains free oil contacting the gasket, such as oily sauces or vegetables and fish in oil. In fact, the mean ESBO concentration in 86 samples was 166 mg/kg, with a maximum of 580 mg/kg (Fankhauser-Noti et al. 2005, Fiselier et al. 2005).

Human exposure to ESBO and its derivatives is likely to occur over a lifetime with a significant variation according to life stage. A reversed phase liquid chromatography interfaced with electrospray ion trap tandem MS method for the determination of ESBO in foods was developed. A simple sample treatment procedure entailing the use of an extraction step with dichloromethane without any further cleanup was proved. Chromatographic separation was performed using two C18 columns with an aqueous acetic acid-acetone-acetonitrile mixture as the mobile phase under gradient conditions. The method was validated in terms of detection limits (4 mg/kg, quantization limits), linearity (established over 2 orders of magnitude), recovery (good mean recoveries, higher than 90% for all of the signals detected), precision (RSD% <8), and trueness. The applicability of the method to the determination of ESBO in different food matrices (in particular those rich in edible oil) was demonstrated, and the performances were compared to those reachable by the commonly well-known GSMS procedure.

In another work, the background of usually high migration of plasticizers from PVC gaskets into oily foods packed in glass jars was investigated. On average, 17 mg gasket material containing 25%–45% plasticizer was in direct food contact per centimeter circumference of the jar rim, ranging from 7 to 33 mg/cm (90 samples). Photographs taken through a microscope suggest that the amount of gasket material in food contact could often be reduced. In products from the market, migration of ESBO sometimes approached the amount present in the gasket in direct food contact, whereas that of phthalates, DEHA, di-(2-ethylhexyl) sebacate (DEHS), and acetyl tributyl citrate (ATBC, Citroflex A) often even included material from underneath and outside the seal to the jar rim. Migration testing with oil at standard conditions (pasteurization/sterilization followed by 10 days at 40°C) yielded migration far below that observed in reality; after 20 days at 60°C, migration was above average in reality, but still did not reach the worst case required by legislation (Fankhauser-Noti et al. 2006).

REFERENCES

Allen, L.B., Siitonen, P.H., and Thompson, H.C. Jr., 1998. Determination of copper, lead, and nickel in edible oils by plasma and furnace atomic spectroscopies. *Journal of American Oil Chemists' Society*, 75/4, 477–481.

Ansari, R., Gul K.T., Jamali, M.K., Arain, M.B., Wagan, M.D., Jalbani, N., Afridi, H.I., and Shah, A.Q., 2009. Variation in accumulation of heavy metals in different varieties of sunflower seed oil with the aid of multivariate technique. *Food Chemistry*, 115, 318–323.

Bajpal, A., Shukla, P., Dixit, B.S., and Banerji, R., 2007. Concentrations of organochlorine insecticides in edible oils from different regions of India. *Chemosphere*, 67/7, 1403–1407.

Cindric, I.J., Zeiner, M., and Steffan, I., 2009. Trace elemental characterization of edible oils by ICP–AES and GFAAS. *Journal of Hazardous Materials*, 165/1–3, 724–728.

Dennis, M.J., Massey, R.C., Cripps, G., Venn, I., Howarth, N., and Lee, G., 1991. Factors affecting the polycyclic aromatic hydrocarbon content of cereals, fats and other food-products. *Food Additives and Contaminants*, 8/4, 517–530.

Fankhauser-Noti, A. and Grob, K., 2006. Migration of plasticizers from PVC gaskets of lids for glass jars into oily foods: Amount of gasket material in food contact, proportion of plasticizer migrating into food and compliance testing by simulation. *Trends in Food Science and Technology*, 17/3, 105–112.

Fankhauser-Noti, A., Fiselier, K., Biedermann, S., Biedermann, M., Grob, K., Armellini, F., Rieger, K., and Skjevrak, I., 2005. Epoxidized soy bean oil (ESBO) migrating from the gaskets of lids into food packed in glass jars. *European Food Research and Technology*, 221/3–4, 416–422.

Fankhauser-Noti, A., Biedermann-Brem, S., and Grob, K., 2006. PVC plasticizers/additives migrating from the gaskets of metal closures into oily food: Swiss market survey June 2005. *European Food Research and Technology*, 223/4, 447–453.

Fiselier, K., Biedermann, M., and Grob, K., 2005. Injector-internal thermal desorption from edible oils. Part 2: Chromatographic optimization for the analysis of migrants from food packaging material. *Journal of Separation Science*, 28/16, 2144–2152.

Gilbert-López, B., García-Reyes, J.F., and Molina-Díaz, A., 2009. Sample treatment and determination of pesticide residues in fatty vegetable matrices: A review. *Talanta*, 79/2, 109–128.

Gonzalez, M.P., Corser, P.I., Cagnasso, M.A., and Urdaneta, A.G., 2007. Organochlorine pesticide residues in 4 types of vegetable oils. *Archivos Latinoamericanos De Nutricion*, 57/4, 397–401.

Guillén, M.D., Sopelana, P., and Palencia, G., 2004. Polycyclic aromatic hydrocarbons and olive pomace oil. *Journal of Agricultural and Food Chemistry*, 52/7, 2123–2132.

Hendrikse, P.W., Slikkerveer, F.J., Folkersma, A., and Dieffenbacher, A., 1991. Determination of lead in oils and fats by direct graphite furnace atomic absorption spectrometry. *Pure Applied Chemistry*, 63, 1183–1190.

Hron, R.J., Kuk, M.S., Abraham, G., and Wan, P.J., 1994. Ethanol extraction of oil, gossypol and aflatoxin from cottonseed. *Journal of the American Oil Chemists' Society*, 71/4, 417–421.

Jongenotter, B. and Janssen, H.G., 2002. On-line GPC-GC analysis of organophosphorus pesticides in edible oils. *LC GC Europe*, 15/6, 338.

Khan, H., Fida, M., and Mohammadzai, I.U., 2007. Estimation of residual nickel and some heavy metals in vanaspati ghee. *Journal of the Chinese Chemical Society*, 54/3, 737–741.

Kowalewska, Z., Izgi, B., Saracoglu, S., and Gucer, S., 2005. Application of liquid-liquid extraction and adsorption on activated carbon of the determination of different forms of metals present in edible oils. *Chemia Analityczna*, 50/6, 1007–1019.

Lacoste, F., Castera, A., and Lespagne, J., 1993. Determination of toxic metals in fats and oils - cadmium, lead, tin, arsenic and chromium (methods and applications). *Revue Francaise Des Corps Gras*, 40/1–2, 19–31.

McGill, A.S., Moffat, C.F., Mackie, P.R., and Cruickshank, P., 1993. The composition and concentration of n-alkanes in retail samples of edible oils. *Journal of the Science of Food and Agriculture*, 61, 357–362.

Moreda, W., Pérez-Camino, M.C., and Cert, A., 2001. Gas and liquid chromatography of hydrocarbons in edible vegetable oils. *Journal of Chromatography A*, 936/1–2, 159–171.

Pehlivan, E., Arslan, G., Gode, F., Altun, T., and Ozcan, M.M., 2008. Determination of some inorganic metals in edible vegetable oils by inductively coupled plasma atomic emission spectroscopy (ICP-AES). *Grasas Y Aceites*, 59/3, 239–244.

Reyes, M.N.M. and Campos, R.C., 2006. Determination of copper and nickel in vegetable oils by direct sampling graphite furnace atomic absorption spectrometry. *Talanta*, 70, 929–932.

Samuel, E.J.J., Duraira, K.S.P., and Mohan, S., 2002. TLC for the detection of aflatoxin in groundnut (*Arachis hyposaea* L.) kernels. *Asian Journal of Chemistry*, 14/2, 874–878.

Sanchez, R., Vázquez, A., Riquelme, D., and Villén, J., 2003. Direct analysis of pesticide residues in olive oil by on-line reversed phase liquid chromatography–gas chromatography using an automated through oven transfer adsorption desorption (TOTAD) interface. *Journal of Agricultural and Food Chemistry*, 5/21, 6098–6102.

Sanchez, R., Cortes, J.M., Villen, J., and Vazquez, A., 2005. Determination of organophosphorus and triazine pesticides in olive oil by on-line coupling reversed-phase liquid chromatography/gas chromatography with nitrogen-phosphorus detection and an automated through-oven transfer adsorption-desorption interface. *Journal of AOAC International*, 88/4, 1255–1260.

Sharkey, A.J., Rooch, O.G., and Coker, D., 1994. A case study on the development of a sampling and testing protocol for aflatoxin levels in edible nuts and oil-seeds. *The Statistician*, 43/2, 267–275.

Szłyk, E. and Czerniak, A.S., 2004. Determination of cadmium, lead, and copper in margarines and butters by galvanostatic stripping chronopotentiometry. *Journal of Agricultural and Food Chemistry*, 52/13, 4064–4071.

Index

Milton Keynes UK
Ingram Content Group UK Ltd.
UKHW021913071024
449327UK00022B/1662